GATE PREP SERIES

GATE

Graduate Aptitude Test in Engineering

2022

Aerospace Engineering

12 Mock Tests

GATE PREP SERIES

GATE

Graduate Aptitude Test in Engineering

2022

Aerospace Engineering

12 Mock Tests

Includes
2 MSQ Practice Papers &
Solved Papers 2020-2021

G. K. PUBLICATIONS (P) LTD.

Title : GATE 2022 - Aerospace Engineering - 12 Mock Tests

Language : English

Editor's Name : Vinit Garg

Typeset & Published by :

Career Launcher Infrastructure (P) Ltd.

A-45, Mohan Cooperative Industrial Area, Near Mohan Estate Metro Station, New Delhi - 110044

Marketed by :

G.K. Publications (P) Ltd.

Plot No. 9A, Sector-27A, Mathura Road, Faridabad, Haryana-121003

ISBN : **978-93-91061-33-3**

Printer's Details : Made in India, New Delhi.

For product information :

Visit ***www.gkpublications.com*** or email to ***gkp@gkpublications.com***

About the Authors

BIPLAB SADHUKHAN

Biplab Sadhukhan received his B.Tech degree in Aerospace Engineering in 2013. He completed his M.Tech from Indian Institute of Technology Kharagpur, in 2017 with an excellence of his class. He is currently working toward the Ph.D. degree at Indian Institute of Technology Kharagpur and working on several projects for Government. of India.

IQBAL SINGH

Iqbal Singh comes from a humble background, born and brought up in the village Lahal of Chamba district of Himachal Pradesh. He did his high and senior secondary school at his native place. He completed B.Tech in Aeronautical engineering from IK Gujral Punjab technical university, Punjab in 2015. After working for 1 year at DRDO Chandigarh, he decided to pursue masters in Aerospace Engineering from Indian Institute of Science, Bangalore. In 2018, he graduated from IISc Bangalore with masters and joined Tata Advanced Systems, Bangalore. Apart from aerospace engineering, he has great passion for music, dance and writing.

He dedicates this book to his parents (Mr. Purushotam Kumar and Sumna Devi) for their love, support and motivation. He would like to thank Sohan Mal, Sc. Vikrant Satya and Dr. S.B.Kandagal for sowing the first seeds of his research career.

PRABHAKAR KUMAR

Prabhakar Kumar has completed B.Tech in Aerospace Engineering with first class degree(distinction) in 2013. After completing bachelor degree, he worked at Hindustan Aeronautics Limited, Aeronautical Development Agency, Bangalore, and several project of Ministry of Defense Government of India. He has completed M.E(Aerodynamics) from Birla Institute of Technology, Mesra, Ranchi in 2018.He would like to thank the Professors of Space Engineering &Rocketry department Dr. Sudip Das, Dr. Mohan Verma, Dr. Partho Mondal, Dr. Rajiv Kumar, Dr. Priyank Kumar, Dr. Shelly for teaching during M.E & Dr. Salil Halder ,Dr. Asim Bose of IIEST Sibpur for teaching during B.Tech.

RANJAY KR SINGH

Ranjay Kr Singh has completed B.E. from aeronautical society of India, in aero-mechanical stream and M.E. from B.I.T. Mesra, Ranchi in space engineering & Rocketry department with specialisation in aerodynamics. At present he is employed as a assistant professor in school of aeronautics, Neemrana, Rajasthan.

He dedicates this book to his family members for their continuous support. He would like to thank prof Rakesh Pandey, his M.E. guide Dr. Sudip Das and all faculty members of the department of space engineering & Rocketry, BIT for their support and motivation.

GATE

Aerospace Engineering

Contents

Preface

GK Publications has been the "Publisher of Choice" to students preparing for GATE, ESE and other technical test prep examinations across the country. GKP's GATE Aerospace Engineering Mock tests 2021 book is to simplify the entire preparation.

Graduate Aptitude Test in Engineering (GATE) is an examination which reflects the relative performance level of a candidate. The GATE score is used for admissions to various post-graduate education programs with financial assistance provided by MHRD and other government agencies. GATE examination is different from college exams which can be studied overnight to pass. It is an exam of deep learning and understanding of concepts and also the application of those concepts.

This book is designed in perception to develop concepts and problem-solving skills, to create a compendium of the concepts of Aerospace Engineering for the students. The books follow the prescribed syllabus of GATE Aerospace Engineering (AE) paper. This book contains the 12 Mock Tests & 2 MSQ Practice Papers based on GATE AE syllabus with detailed solutions to precisely build the skill needed to conquer the fear of branch.

GKP has also launched an android app to provide you an update on all upcoming vacancies in the technical segment. Moreover, the app also has a lot of added content to aid your preparation in the form of vocab builder, GK quiz and so on.

We hope this effort of ours will be helpful in achieving your dreams and securing a government job. If you have any suggestions on improvement of this book, you can write to us at gkp@gkpublications.com.

All the Best!

Team GKP

GATE

Graduate Aptitude Test in Engineering

IIT Institutes

GATE 2021 conducted by
Indian Institute of Technology, Mumbai

GATE 2022 will be conducted by
Indian Institute of Technology, Kharagpur

Indian Institute of Technology, Delhi

Indian Institute of Technology, Chennai

Indian Institute of Technology, Guwahati

Indian Institute of Technology, Roorkee

Indian Institute of Technology, Kanpur

About GATE

The Graduate Aptitude Test in Engineering (GATE) conducted by IISc and IITs has emerged as one of the bench mark tests for engineering and science aptitude in facilitating admissions for higher education (M.Tech./Ph.D.) in IITs, IISc and various other Institutes/Universities/Laboratories in India. With the standard and high quality of the GATE examination in 27 disciplines of engineering and science, Humanities and Social Sciences subjects, it identifies the candidate's understanding of a subject and aptitude and eligibility for higher studies. During the last few years, GATE score is also being used as one of the criteria for recruitment in Government Organizations such as Cabinet Secretariat, and National/State Public Sector Undertakings in India. Because of the importance of the GATE examination, the number of candidates taking up GATE exams has increased tremendously. GATE exams are conducted by the IITs and IISc as a computer based test having multiple choice questions and numerical answer type questions. The questions are mostly fundamental, concept based and thought provoking. From 2017 onwards GATE Exam is being held in Bangladesh, Ethiopia, Nepal, Singapore, Sri Lanka and United Arab Emirates. An Institute with various nationalities in its campus widens the horizons of an academic environment. A foreign student brings with him/her a great diversity, culture and wisdom to share. Many GATE qualified candidates are paid scholarships/assistantship, especially funded by Ministry of Human Resources Development, Government of India and by other Ministries. IIT, Kharagpur is the Organizing Institute for GATE 2022.

Why GATE?

Admission to Post Graduate and Doctoral Programmes

Admission to postgraduate programmes with MHRD and some other government scholarships/ assistantships in engineering colleges/institutes is open to those who qualify through GATE. GATE qualified candidates with Bachelor's degree in Engineering/Technology/Architecture or Master's degree in any branch of Science/Mathematics/Statistics/Computer Applications are eligible for admission to Master/Doctoral programmes in Engineering/Technology/Architecture as well as for Doctoral programmes in relevant branches of Science with MHRD or other government scholarships/ assistantships. Candidates with Master's degree in Engineering/Technology/Architecture may seek admission to relevant Ph.D programmes with scholarship/assistantship without appearing in the GATE examination.

Financial Assistance

A valid GATE score is essential for obtaining financial assistance during Master's programs and direct Doctoral programs in Engineering/Technology/Architecture, and Doctoral programs in relevant branches of Science in Institutes supported by the MHRD or other Government agencies. As per the directives of the MHRD, the following procedure is to be adopted for admission to the post-graduate programs (Master's and Doctoral) with MHRD scholarship/assistantship. Depending upon the norms adopted by a specific institute or department of the Institute, a candidate may be admitted directly into a course based on his/her performance in GATE only **or** based on his/her performance in GATE **and** an admission test/interview conducted by the department to which he/she has applied **and/or** the candidate's academic record. If the candidate is to be selected through test/interview for post-graduate programs, a minimum of 70% weightage will be given to the performance in GATE and the remaining 30% weightage will be given to the candidate's performance in test/interview and/or academic record, as per MHRD guidelines. The admitting institutes could however prescribe a minimum passing percentage of marks in the test/interview. Some colleges/institutes specify GATE qualification as the mandatory requirement even for admission without MHRD scholarship/assistantship.

To avail of the financial assistance (scholarship), the candidate must first secure admission to a program in these Institutes, by a procedure that could vary from institute to institute. Qualification in GATE is also a minimum requirement to apply for various fellowships awarded by many Government organizations. Candidates are advised to seek complete details of admission procedures and availability of MHRD scholarship/assistantship from the concerned admitting institution. The criteria for postgraduate admission with scholarship/assistantship could be different for different institutions. The management of the post-graduate scholarship/assistantship is also the responsibility of the admitting institution. Similarly, reservation of seats under different categories is as per the policies and norms prevailing at the admitting institution and Government of India rules. *GATE offices will usually not entertain any enquiry about admission, reservation of seats and / or award of scholarship / assistantship.*

PSU Recruitments

As many as 37 PSUs are using GATE score for recruitment. It is likely that more number of PSUs may start doing so by next year. Below is the list of PSUs:

MDL, BPCL, GAIL, NLC LTD, CEL, Indian Oil, HPCL, NBPC, NECC, BHEL, WBSEDCL, NTPC, ONGC, Oil India, Power Grid, Cabinet Secretariat, Govt. of India, BAARC, NFL, IPR, PSPCL, PSTCL, DRDO, OPGC Ltd., THDC India Ltd., BBNL, RITES, IRCON, GHECL, NHAI, KRIBHCO, Mumbai Railway Vikas Corporation Ltd. (MRVC Ltd.), National Textile Corporation, Coal India Ltd., BNPM, AAI, NALCO, EdCIL India.

Important :

1. Admissions in IITs/IISc or other Institutes for M.Tech./Ph.D. through GATE scores shall be advertised separately by the Institutes and GATE does not take the responsibility of admissions.
2. Cabinet Secretariat has decided to recruit officers for the post of Senior Field Officer (Tele) (From GATE papers of EC, CS, PH), Senior Research Officer (Crypto) (From GATE papers of EC, CS, MA), Senior Research Officer (S&T) (From GATE papers EC, CS, CY, PH, AE, BT) in the Telecommunication Cadre, Cryptographic Cadre and Science & Technology Unit respectively of Cabinet Secretariat. The details of the scheme of recruitment shall be published in National Newspaper/Employment News by the concerned authority.
3. Some PSUs in India have expressed their interest to utilize GATE scores for their recruitment purpose. The Organizations who intend to utilize GATE scores shall make separate advertisement for this purpose in Newspapers etc.

Who Can Appear for GATE?

Eligibility for GATE

Before starting the application process, the candidate must ensure that he/she meets the eligibility criteria of GATE given in Table.

Eligibility Criteria for GATE 2022

Degree/Program	Qualifying Degree/Examination	Description of Eligible Candidates	Expected Year of Completion
B.E. / B.Tech. / B.Pharm.	Bachelor's degree in Engineering / Technology (4 years after 10+2 or 3 years after B.Sc. / Diploma in Engineering / Technology)	Currently in the 3rd year or higher grade or already completed	2022
B. Arch.	Bachelor's degree of Architecture (5- year course) / Naval Architecture (4- year course) / Planning (4- year course)	Currently in the 3rd year or higher grade or already completed	2023 (for 5-year program), 2022 (for 4-year program)
B.Sc. (Research) / B.S.	Bachelor's degree in Science (Post-Diploma/4 years after 10+2)	Currently in the 3rd year or higher grade or already completed	2022
Pharm. D. (after 10+2)	6 years degree program, consisting of internship or residency training, during third year onwards	Currently in the 3rd/4th/5th/6th year or already completed	2024
M.B.B.S.	Degree holders of M.B.B.S. and those who are in the 5th/6th/7th semester or higher semester of such programme.	5th, 6th, 7th or higher semester or already completed	2022
M. Sc. / M.A. / MCA or equivalent	Master's degree in any branch of Arts/Science/Mathematics/Statistics/ Computer Applications or equivalent	Currently in the first year or higher or already Completed	2022
Int. M.E./ M.Tech. (Post-B.Sc.)	Post-B.Sc Integrated Master's degree programs in Engineering/ Technology (4-year program)	Currently in the 1st / 2nd/3rd/4th year or already completed	2024
Int. M.E./ M.Tech. or Dual Degree (after Diploma or 10+2)	Integrated Master's degree program or Dual Degree program in Engineering/Technology (5-year program)	Currently in the 3rd /4th/5th year or alreadycompleted	2023
B.Sc. / B.A. / B.Com.	Bachelor degree in any branch of Science / Arts / Commerce (3 years program)	Currently in the 3rd year or already completed	2021
Int. M.Sc. / Int. B.S. / M.S.	Integrated M.Sc. or 5-year integrated B.S.-M.S. program	Currently in the 3rd year or higher or already completed	2022
Professional Society Examinations (equivalent to B.E. / B.Tech. / B.Arch.)	B.E./B.Tech./B.Arch. equivalent examinations of Professional Societies, recognized by MoE/UPSC/AICTE (e.g. AMIE by Institution of Engineers-India, AMICE by the Institute of Civil Engineers-India and so on)	Completed Section A or equivalent of such professional courses	NA

In case a candidate has passed one of the qualifying examinations as mentioned above in 2020 or earlier, the candidate has to submit the degree certificate / provisional certificate / course completion certificate / professional certificate / membership certificate issued by the society or institute. In case, the candidate is expected to complete one of the qualifying criteria in 2022 or later as mentioned above, he/she has to submit a certificate from Principal or a copy of marks card for section A of AMIE.

Certificate From Principal

Candidates who have to submit a certificate from their college Principal have to obtain one from his/her institution beforehand and upload the same during the online submission of the application form.

Candidates With Backlogs

Candidates, who have appeared in the final semester/year exam in 2022, but with a backlog (arrears/failed subjects) in any of the papers in their qualifying degree should upload a copy of any one of the mark sheets of the final year,

OR

obtain a declaration from their Principal along with the signature and seal beforehand and upload the same during the online submission of the application form.

GATE Structure

Structure of GATE

GATE 2022 will be conducted on 27 subjects (papers). Table below shows the list of papers and paper codes for GATE 2022. A candidate is allowed to appear in ANY ONE or UP TO TWO papers of the GATE examination. However, note that the combination of TWO papers in which a candidate can appear MUST be from the pre-defined list as given in Table. Also note that for a paper running in multiple sessions, a candidate will be mapped to appear for the examination in one of the sessions ONLY.

List of GATE Papers and Corresponding Codes

GATE Paper	Code	GATE Paper	Code
Aerospace Engineering	AE	Instrumentation Engineering	IN
Agricultural Engineering	AG	Mathematics	MA
Architecture and Planning	AR	Mechanical Engineering	ME
Bio-medical Engineering	BM	Mining Engineering	MN
Biotechnology	BT	Metallurgical Engineering	MT
Civil Engineering	CE	Petroleum Engineering	PE
Chemical Engineering	CH	Physics	PH
Computer Science and Information Technology	CS	Production and Industrial Engineering	PI
Chemistry	CY	Statistics	ST
Electronics and Communication Engineering	EC	Textile Engineering and Fibre Science	TF
Electrical Engineering	EE	Engineering Sciences	XE*
Environmental Science & Engineering	ES	Humanities & Social Sciences	XH**
Ecology and Evolution	EY	Life Sciences	XL***
Geology and Geophysics	GG		

Note: *Environmental Science and Engineering (ES) and Humanities and Social Sciences (XH) are two new papers introduced in GATE 2022.*

*XE Paper Sections	Code	**XH Paper Sections	Code	***XL Paper Sections	Code
Engineering Mathematics (Compulsory) (15 marks)	A	Reasoning and Comprehension (Compulsory) (25 marks)	B1	Chemistry (Compulsory) (25 marks)	P
Any TWO optional Sections (2x35 = 70 marks)		Any ONE optional Section (60 marks)		Any TWO optional Sections (2x30 = 60 marks)	
Fluid Mechanics	B	Economics	C1	Biochemistry	Q
Materials Science	C	English	C2	Botany	R
Solid Mechanics	D	Linguistics	C3	Microbiology	S
Thermodynamics	E	Philosophy	C4	Zoology	T
Polymer Science and Engineering	F	Psychology	C5	Food Technology	U
Food Technology	G	Sociology	C6		
Atmospheric and Oceanic Sciences	H				

*XE (Engineering Sciences), **XH (Humanities & Social Sciences), ***XL (Life Sciences), papers are of general nature and will be comprised of Sections listed in the above table

Note: *Each subject/paper is of total 100 marks. General Aptitude (GA) section of 15 marks is common for all papers. Hence remaining 85 marks are for the respective subject/paper code.*

Combination of Two Papers Allowed to Appear in GATE 2022 (subject to availability of infrastructure and schedule)

Code of The First (Primary) Paper	Codes of Papers Allowed as The Second Paper
AE	XE
AG	ES
AR	CE
BM	BT / XL
BT	BM / XL
CE	AR / ES
CH	CY / PE / XE
CS	MA
CY	CH / XL
EC	IN / PH
EE	IN
ES	AG / CE
EY	XL
GG	MN / PE / PH
IN	EC / EE / PH
MA	CS / PH / ST
ME	XE
MN	GG / XE
MT	PH / XE
PE	CH / GG / XE
PH	EC / GG / IN / MA / MT / ST
PI	XE
ST	MA / PH
TF	XE
XE	AE / CH / ME / MN / MT / PE / PI / TF
XH	----
XL	BM / BT / CY / EY

General Aptitude Questions

All the papers will have a few questions that test the General Aptitude (Language and Analytical Skills), apart from the core subject of the paper.

Duration and Examination Type

All the papers of the GATE 2022 examination will be for 3 hours duration and they consist of 65 questions for a total of 100 marks. Since the examination is an ONLINE computer based test (CBT), at the end of the stipulated time (3-hours), the computer screen will automatically close the examination inhibiting any further action.

Candidates will be permitted to occupy their allotted seats 40 minutes before the scheduled start of the examination. Candidates can login and start reading the instructions 20 minutes before the start of examination. The late login time (if any) recorded by the computer system MUST NOT be beyond 30 minutes from the actual starting time of the examination. Under NO circumstances, will a candidate be permitted to login after 30 minutes from the actual examination starting time. Candidates will NOT be permitted to leave the examination hall before the end of the examination.

Pattern of Question Papers

GATE 2022 may contain questions of THREE different types in all the papers:

(i) **Multiple Choice Questions (MCQ)** carrying 1 or 2 marks each, in all the papers and sections. These questions are objective in nature, and each will have choice of four answers, out of which ONLY ONE choice is correct.

Negative Marking for Wrong Answers: For a wrong answer chosen in a MCQ, there will be negative marking. For 1-mark MCQ, 1/3 mark will be deducted for a wrong answer. Likewise, for 2-mark MCQ, 2/3 mark will be deducted for a wrong answer.

(ii) **Multiple Select Questions (MSQ)** carrying 1 or 2 marks each in all the papers and sections. These questions are objective in nature, and each will have choice of four answers, out of which ONE or MORE than ONE choice(s) is / are correct.

Note: There is NO negative marking for a wrong answer in MSQ questions. However, there is NO partial credit for choosing partially correct combinations of choices or any single wrong choice.

(iii) **Numerical Answer Type (NAT)** Questions carrying 1 or 2 marks each in most of the papers and sections. For these questions, the answer is a signed real number, which needs to be entered by the candidate using the virtual numeric keypad on the monitor (keyboard of the computer will be disabled). No choices will be shown for these types of questions. The answer can be a number such as 10 or –10 (an integer only). The answer may be in decimals as well, for example, 10.1 (one decimal) or 10.01 (two decimals) or –10.001 (three decimals). These questions will be mentioned with, up to which decimal places, the candidates need to present the answer. Also, for some NAT type problems an appropriate range will be considered while evaluating these questions so that the candidate is not unduly penalized due to the usual round-off errors. Candidates are advised to do the rounding off at the end of the calculation (not in between steps). Wherever required and possible, it is better to give NAT answer up to a maximum of three decimal places.

Note: *There is NO negative marking for a wrong answer in NAT questions.*
Also, there is NO partial credit in NAT questions.

Marking Scheme – Distribution of Marks and Questions

General Aptitude (GA) Questions

In all papers, GA questions carry a total of 15 marks. The GA section includes 5 questions carrying 1-mark each (sub-total 5 marks) and 5 questions carrying 2-marks each (sub-total 10 marks).

Question Papers other than GG, XE, XH and XL

These papers would contain 25 questions carrying 1-mark each (sub-total 25 marks) and 30 questions carrying 2-marks each (sub-total 60 marks) consisting of some MCQ type questions, while the remaining may be MSQ and / or NAT questions.

GG (Geology and Geophysics) Paper

Apart from the General Aptitude (GA) section, the GG question paper consists of two parts: Part A and Part B. Part A is compulsory for all the candidates. Part B contains two sections: Section 1 (Geology) and Section 2 (Geophysics). Candidates will have to attempt questions in Part A and questions in either Section 1 or Section 2 of Part B. The choice of Section 1 OR Section 2 of Part B has to be made at the time of online registration in GOAPS. At the examination hall, candidate cannot request for change of section.

Part A consists of 25 questions carrying 1-mark each (sub-total 25 marks and some of these will be MSQ and/or numerical answer type questions while remaining questions will be MCQ type). Either section of Part B (Section 1 and Section 2) consists of 30 questions carrying 2-marks each (sub-total 60 marks and some of these will be MSQ and/or numerical answer type questions while remaining questions will be MCQ type).

XE Paper (Engineering Sciences)

A candidate appearing in the XE paper has to answer the following:

- GA – General Aptitude carrying a total of 15 marks.
- Section A – Engineering Mathematics (Compulsory): This section contains 11 questions carrying a total of 15 marks: 7 questions carrying 1-mark each (sub-total 7 marks), and 4 questions carrying 2-marks each (sub-total 8 marks). Some questions will be of numerical answer type while remaining questions will be MCQ type.
- Any two of XE Sections B to H: The choice of two sections from B to H can be made during the examination after viewing the questions. Only TWO optional sections can be answered at a time. A candidate wishing to change midway of the examination to another optional section must first choose to deselect one of the previously chosen optional sections (B to H). Each of the optional sections of the XE paper (Sections B through H) contains 22 questions carrying a total of 35 marks: 9 questions carrying 1-mark each (sub-total 9 marks) and 13 questions carrying 2-marks each (sub-total 26 marks). Some questions will be of MSQ and/or numerical answer type while remaining questions will be MCQ type.

XH Paper (Humanities and Social Sciences)

A candidate appearing in the XH paper has to answer the following:

- **GA – General Aptitude** carrying a total of 15 marks.
- **Section B1 – Reasoning and Comprehension (Compulsory):** This section contains 15 questions carrying a total of 25 marks: 5 questions carrying 1-mark each (sub-total 5 marks) and 10 questions carrying 2-marks each (sub-total 20 marks). Some questions will be of MSQ and/or numerical answer type while remaining questions will be MCQ type.
- **Any ONE of XH Sections C1 to C6:** The ONE choice of section from C1 to C6 has to be made at the time of online registration in GOAPS. At the examination hall, candidate cannot request for change of section. Each of the optional sections of the XH paper (Sections C1 through C6) contains 40 questions carrying a total of 60 marks: 20 questions carrying 1-mark each (sub-total 20 marks) and 20 questions carrying 2-marks each (sub-total 40 marks). Some questions will be of MSQ and/or numerical answer type while remaining questions will be MCQ type.

XL Paper (Life Sciences)

A candidate appearing in the XL paper has to answer the following:

- **GA – General Aptitude** carrying a total of 15 marks.
- **Section P–Chemistry (Compulsory):** This section contains 15 questions carrying a total of 25 marks: 5 questions carrying 1-mark each (sub-total 5 marks) and 10 questions carrying 2-marks each (sub-total 20 marks). Some questions will be of MSQ and/or numerical answer type while remaining questions will be MCQ type.
- **Any two of XL Sections Q to U:** The choice of two sections from Q to U can be made during the examination after viewing the questions. Only TWO optional sections can be answered at a time. A candidate wishing to change midway of the examination to another optional section must first choose to deselect one of the previously chosen optional sections (Q to U). Each of the optional sections of the XL paper (Sections Q through U) contains 20 questions carrying a total of 30 marks: 10 questions carrying 1-mark each (sub-total 10 marks) and 10 questions carrying 2-marks each (sub-total 20 marks). Some questions will be of MSQ and/or numerical answer type while remaining questions will be MCQ type.

GATE Score

After the evaluation of the answers, the actual (raw) marks obtained by a candidate will be considered for computing the GATE Score. For multi-session papers (subjects), raw marks obtained by the candidates in different sessions will be converted to Normalized marks for that particular subject. Thus, raw marks (for single session papers) or normalized marks (for multi-session papers) will be used for computing the GATE Score, based on the qualifying marks.

Calculation of Normalized Marks for Multi-Session Papers

In GATE 2022 examination, some papers may be conducted in multi-sessions. Hence, for these papers, a suitable normalization is applied to take into account any variation in the difficulty levels of the question papers across different sessions. The normalization is done based on the fundamental assumption that "in all multi-session GATE papers, the distribution of abilities of candidates is the same across all the sessions". This assumption is justified since the number of candidates appearing in multi-session papers in GATE 2022 is large and the procedure for allocation of session to candidates is random. Further, it is also ensured that for the same multi-session paper, the number of candidates allotted in each session is of the same order of magnitude.

Based on the above, and considering various normalization methods, the committee arrived at the following formula for calculating the normalized marks for the multi-session papers.

Nsormalization mark of jth candidate in the ith session $\widehat{M_{ij}}$ is given by

$$\widehat{M_{ij}} = \frac{\bar{M}_t^g - M_q^g}{\bar{M}_{ti} - M_{iq}}(M_{ij} - M_{iq}) + M_q^g$$

where

M_{ij} : is the actual marks obtained by the jth candidate in i^{th} session

$\bar{M}_t^g$: is the average marks of the top 0.1% of the candidates considering all sessions

M_q^g : is the sum of mean and standard deviation marks of the candidates in the paper considering all sessions

$\bar{M}_{ti}$: is the average marks of the top 0.1% of the candidates in the ith session

M_{iq} : is the sum of the mean marks and standard deviation of the ith session

Calculation of GATE Score for All Papers

For all papers for which there is only one session, actual marks obtained by the candidates will be used for calculating the GATE 2022 Score. For papers in multi-sessions, normalized marks will be calculated corresponding to the raw marks obtained by a candidate and the GATE 2022 Score will be calculated based on the normalized marks.

The GATE 2022 score will be computed using the formula given below.

$$\text{GATE Score} = S_q + (S_t - S_q)\frac{(M - M_q)}{(\bar{M}_t - M_q)}$$

where

- M : marks obtained by the candidate (actual marks for single session papers and normalized marks for multi-session papers)
- M_q : is the qualifying marks for general category candidate in the paper
- $\bar{M}_t$: is the mean of marks of top 0.1% or top 10 (whichever is larger) of the candidates who appeared in the paper (in case of multi-session papers including all sessions)
- S_q : 350, is the score assigned to M_q
- S_t : 900, is the score assigned to $\bar{M}_t$

In the GATE 2022 the qualifying marks (M_q) for general category student in each subject will be 25 marks (out of 100) or $\mu + \sigma$, whichever is larger. Here μ is the mean and σ is the standard deviation of marks of all the candidates who appeared in the paper.

After the declaration of results, GATE Scorecards can be downloaded by the GATE qualified candidates ONLY.

The GATE 2022 Committee has the authority to decide the qualifying mark/score for each GATE paper. In case of any claim or dispute with respect to GATE 2022 examination or score, the Courts and Tribunals in Mumbai alone will have the exclusive jurisdiction to entertain and settle them.

GATE Syllabus

GENERAL APTITUDE

Verbal Aptitude

Basic English Grammar: tenses, articles, adjectives, prepositions, conjunctions, verb-noun agreement, and other parts of speech.

Basic Vocabulary: words, idioms, and phrases in context, Reading and comprehension, Narrative sequencing.

Quantitative Aptitude

Data Interpretation: data graphs (bar graphs, pie charts, and other graphs representing data), 2-and 3-dimensional plots, maps, and tables.

Numerical Computation and Estimation: ratios, percentages, powers, exponents and logarithms, permutations and combinations, and series, Mensuration and geometry, Elementary statistics and probability.

Analytical Aptitude

Logic: Deduction and induction, Analogy, Numerical relations and reasoning.

Spatial Aptitude

Transformation of shapes: translation, rotation, scaling, mirroring, assembling, and grouping Paper folding, cutting, and patterns in 2 and 3 dimensions.

Important Note for Candidates: In each of the following subjects the topics have been divided into two categories – Core Topics and Special Topics. The corresponding sections of the question paper will contain 90% of their questions on Core Topics and the remaining 10% on Special Topics.

Section1: Engineering Mathematics

Core Topics:

Linear Algebra: Vector algebra, Matrix algebra, systems of linear equations, rank of a matrix, eigenvalues and eigenvectors.

Calculus: Functions of single variable, limits, continuity and differentiability, mean value theorem, chain rule, partial derivatives, maxima and minima, gradient, divergence and curl, directional derivatives. Integration, Line, surface and volume integrals. Theorems of Stokes, Gauss and Green.

Differential Equations: First order linear and nonlinear differential equations, higher order linear ODEs with constant coefficients. Partial differential equations and separation of variables methods.

Special Topics:

Fourier Series, Laplace Transforms, Numerical methods for linear and nonlinear algebraic equations, Numerical integration and differentiation. Complex analysis. Probability and statistics.

Section 2: Flight Mechanics

Core Topics:

Basics: Atmosphere: Properties, standard atmosphere. Classification of aircraft. Airplane (fixed wing aircraft) configuration and various parts. Pressure altitude; equivalent, calibrated, indicated air speeds; Primary flight instruments: Altimeter, ASI, VSI, Turn-bank indicator. Angle of attack, sideslip; Roll, pitch & yaw controls. Aerodynamic forces and moments.

Airplane performance: Drag polar; takeoff and landing; steady climb & descent; absolute and service ceiling; range and endurance, load factor, turning flight, V-n diagram. Winds: head, tail & cross winds.

Static stability: Stability & control derivatives; longitudinal stick fixed & free stability; horizontal tail position and size; directional stability, vertical tail position and size; lateral stability. Wing dihedral, sweep & position; hinge moments, stick forces.

Special Topics:

Dynamic stability: Euler angles; Equations of motion; Decoupling of longitudinal and lateral-directional dynamics; longitudinal modes; lateral-directional modes.

Section 3: Space Dynamics

Core Topics:

Central force motion, determination of trajectory and orbital period in simple cases. Kepler's laws; escape velocity.

No Special Topics:

Section 4: Aerodynamics

Core Topics:

Basic Fluid Mechanics: Conservation laws: Mass, momentum and energy (Integral and differential form); Dimensional analysis and dynamic similarity; Potential flow theory: sources, sinks, doublets, line vortex and their superposition. Elementary ideas of viscous flows including boundary layers.

Airfoils and wings: Airfoil nomenclature; Aerodynamic coefficients: lift, drag and moment; Kutta-Joukoswki theorem; Thin airfoil theory, Kutta condition, starting vortex; Finite wing theory: Induced drag, Prandtl lifting line theory; Critical and drag divergence Mach number.

Compressible Flows: Basic concepts of compressibility, One-dimensional compressible flows, Isentropic flows, Fanno flow, Rayleigh flow; Normal and oblique shocks, Prandtl-Meyer flow; Flow through nozzles and diffusers.

Special Topics:

Wind Tunnel Testing: Measurement and visualization techniques. Shock - boundary layer interaction.

Section 5: Structures

Core Topics:

Strength of Materials: Stress and strain: Three-dimensional transformations, Mohr's circle, principal stresses, Three-dimensional Hooke's law, Plane stress and strain. Failure theories: Maximum stress, Tresca and von Mises. Strain energy. Castigliano's principles. Statically determinate and indeterminate trusses and beams. Elastic flexural buckling of columns.

Flight vehicle structures: Characteristics of aircraft structures and materials. Torsion, bending and shear of thin-walled sections. Loads on aircraft.

Structural Dynamics: Free and forced vibrations of undamped and damped SDOF systems. Free vibrations of undamped 2-DOF systems.

Special Topics:

Vibration of beams. Theory of elasticity: Equilibrium and compatibility equations, Airy's stress function.

Section 6: Propulsion

Core Topics:

Basics: Thermodynamics, boundary layers, heat transfer, combustion and thermochemistry.

Aerothermodynamics of aircraft engines: Thrust, efficiency, range. Brayton cycle. Engine performance: ramjet, turbojet, turbofan, turboprop and turboshaft engines. Afterburners.

Turbomachinery: Axial compressors: Angular momentum, work and compression, characteristic performance of a single axial compressor stage, efficiency of the compressor and degree of reaction, multi-staging. Centrifugal compressor: Stage dynamics, inducer, impeller and diffuser. Axial turbines: Stage performance.

Rockets: Thrust equation and specific impulse, rocket performance. Multi-staging. Chemical rockets. Performance of solid and liquid propellant rockets.

Special Topics:

Aerothermodynamics of non-rotating propulsion components such as intakes, combustor and nozzle. Turbine blade cooling. Compressor-turbine matching, Surge and stall.

MOCK TEST – 1

AEROSPACE ENGINEERING

Time Allowed: 3:00 Hours | *Maximum Marks: 100*

Read the following instructions carefully before you begin to answer the questions. This Booklet contains questions in English.

INSTRUCTIONS

1. This Booklet contains **65 questions** in total comprising of the following two parts:

   Part -1: Technical Section (85 Marks) *(1-55 Questions)*

   Part -2: General Aptitude Section (15 Marks) *(56-65 Questions)*

2. All questions are compulsory. Q.1-Q.25 & Q.56-Q.60 carry one mark each. Q.26-Q.55 & Q.61-Q.65 carry two marks each.

3. **Types of Questions:**

   (*a*) Multiple Choice Questions(MCQ) carrying 1 or 2 marks each in all papers and sections. These questions are objective in nature, and each will have a choice of four options, out of which the candidate has to mark the correct answer(s).

   (*b*) Numerical Answer Questions of 1 or 2 marks each in all sections. For these questions the answer is a real number, to be entered by the candidate using the virtual keypad. No choices will be shown for these type of questions.

4. For 1-mark multiple-choice questions, 1/3 marks will be deducted for a wrong answer. Likewise, for 2-marks multiple-choice questions, 2/3 marks will be deducted for a wrong answer. There is no negative marking for numerical answer type questions.

MOCK TEST – 1

1. A laminar flow wing has maximum thickness at:-

(a) Leading edge

(b) Trailing edge

(c) Middle camber line

(d) Quarter chord point

2. In a turbofan engine, the mass flow through the bypass duct is 100 kg/s with a bypass ratio of 5. What is the total mass flow rate?

(a) 20 kg/s (b) 100 kg/s

(c) 120 kg/s (d) 600 kg/s

3. The function $f(x) = \frac{1}{1+|x|}$ is

(a) continuous and differentiable

(b) continuous but not differentiable.

(c) not continuous but differentiable.

(d) not continuous and not differentiable.

4. The Laplace transform of the function e^{-2x} is ____

(a) $\frac{1}{2s}$ (b) $\frac{2}{s}$

(c) $\frac{1}{s+2}$ (d) e^{-2s}

5. Given the escape velocity from earth is 11.2 km per sec. If a body is to be projected in a direction making an angle 45° to the vertical, then the escape velocity is ____________.

6. Find the change in volume (ΔV)of the unit cube under the displacement field given.

$$u = \alpha x,\ v = By,\ w = o$$

(a) $dV = \alpha + \beta + 1$ (b) $dV = \alpha^2 + \beta^2 + 1$

(c) $dV = \alpha\beta + \alpha + \beta$ (d) $dV = \alpha\beta + 1 + \beta$

7. Lift on an aircraft climbing vertically up is

(a) Equal to its weight

(b) Zero

(c) Equal to the drag

(d) Equal to the thrust

8. A circular thin uniform ring is rolling down an inclined plane of inclination 30° without slipping. Its linear acceleration along the inclined plane will be

(a) $\frac{g}{2}$ (b) $\frac{g}{3}$

(c) $\frac{g}{4}$ (d) $\frac{2g}{3}$

9. In the matrix equation $Px = q$, which of the following is a necessary condition for the existence of at least one solution for the unknown vector x

(a) Augmented matrix [Pq] must have the same rank as matrix P

(b) Vector q must have only non-zero elements

(c) Matrix P must be singular

(d) Matrix P must be square

10. Match the following

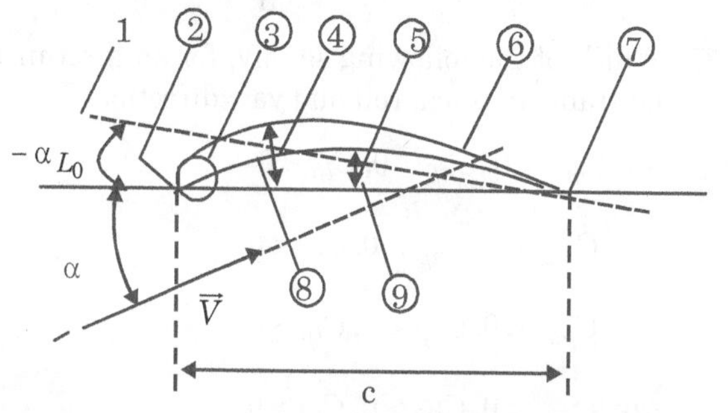

1.	a. Nose circle
2.	b. Chamber mean line
3.	c. Zero lift line
4.	d. Trailing edge

(a) 1 - c, 7 - d, 3 - a, 8 - b

(b) 1 - c, 7 - a, 3 - d, 8 - b

(c) 1 - c, 7 - b, 3 - d, 8 - c

(d) 1 - b, 7 - c, 3 - a, 8 - d

11. In couette flow, if two infinite horizontal plates are 1m apart with upper plate moving with 2 m/s and the fluid's dynamic viscosity is $\mu = 7 \times 10^{-3}\ Ns/m^2$, then find the shear stress between the two fluid layer ?

12. For maximum coefficient of lift $C_{L_{max}} = 1.5$, Aspect Ratio, AR = 5 and $\tau = 0.5$ the change in elevator deflection $\Delta\delta_e$ is

(a) –9.9° (b) –10.9°

(c) –11.9° (d) –12.9°

13. Which of the method is valid for thrust augmentation through liquid injection process?

i. By injecting water or alcohol, we increase the total mass rate through the nozzle, which increase the change in momentum and hence thrust

ii. By spraying water at the inlet of compressor there is reduction in compressor inlet temperature, which causes in increase of compressor pressure ratio and hence increase in thrust.

(*a*) Only i (*b*) Only ii
(*c*) Both i and ii (*d*) Neither i nor ii

14. In a rectangular cross-section the ratio of maximum shear stress to average shear stress is given as: -

(*a*) $\frac{3}{4}$ (*b*) $\frac{4}{3}$

(*c*) $\frac{3}{2}$ (*d*) $\frac{2}{3}$

15. Which of the following satisfy, for an aircraft to be stable in pitch, roll and yaw directions?

(*a*) $C_{m\alpha} < 0, C_{n\beta} < 0, C_{l\beta} < 0$

(*b*) $C_{m\alpha} < 0, C_{n\beta} > 0, C_{l\beta} < 0$

(*c*) $C_{m\alpha} < 0, C_{n\beta} < 0, C_{l\beta} > 0$

(*d*) $C_{m\alpha} < 0, C_{n\beta} > 0, C_{l\beta} > 0$

16. At a given point on the surface of the aerofoil, the pressure coefficient is – 0.3 at very low speeds. If the freestream Mach Number is 0.8, Calculatc C_p at this point.

(*a*) –0.2 (*b*) 0.2
(*c*) 0.5 (*d*) –0.5

17. For a centrifugal pump, the efficiency η_p is given by the formula

(*a*) $\eta_p = \frac{\omega T}{\rho g H_p Q}$ (*b*) $\eta_p = \frac{\rho g H_p Q}{\omega T}$

(*c*) $\eta_p = \frac{2\left(\frac{U}{U_e}\right)}{1+\left(\frac{U}{U_e}\right)}$ (*d*) $\eta_p = \frac{H_p Q}{\omega T}$

18. The Mach angle for a flow at Mach 1.41 is:

(*a*) 30° (*b*) 45°
(*c*) 60° (*d*) 90°

19. The state of plane stress of point in an isotropic material is given by

$$\sigma = \begin{bmatrix} 16 & 5 \\ 5 & -4 \end{bmatrix} MPa$$

The inclination of plane of zero shear stress with respect to specified plane is

(*a*) 31.7° (*b*) 26.6°
(*c*) 13.3° (*d*) 63.4°

20. The angle between two unit-magnitude coplanar vectors *P*(0.866, 0.500, 0) and *Q*(0.259, 0.966, 0) will be

(*a*) 0° (*b*) 30°
(*c*) 45° (*d*) 60°

21. If $M_1 = 2$ and $M_2 = 10$. What is the $\frac{K.E}{I.E}$ ratio ?

(*a*) 5 (*b*) 20
(*c*) 25 (*d*) 50

22. In a gas turbine engine the thermodynamic cycle, heat injection takes place at 12.6 atm and heat rejection takes place at atmospheric pressure, then what is the efficiency of the cycle , given that the heat capacity ratio of the working substance is 1.3.

(*a*) 42.7% (*b*) 61.3%
(*c*) 55.5% (*d*) 34.4%

23. If Airy's stress function is ϕ, which of the following satisfy Airy's stress equation $\nabla^4\phi = 0$?

(*a*) $\phi = x^4y^4$

(*b*) $\phi = y^6 - x^6$

(*c*) $\phi = x^3 + y^3$

(*d*) $\phi = \left(x^2 + y^2\right)^2$

24. In an elastic beam problem, boundary conditions at free end will be

(*a*) $M = 0, V = 0$

(*b*) $u = 0, \frac{\partial u}{\partial x} = 0$

(*c*) $M = 0, u = 0$

(*d*) $V \neq 0, \frac{\partial u}{\partial x} \neq 0$

25. For a matrix $[M]=\begin{bmatrix} \frac{3}{5} & \frac{4}{5} \\ x & \frac{3}{5} \end{bmatrix}$, the transpose of the matrix is equal to the inverse of the matrix, $[M]^T=[M]^{-1}$. The value x is given by

(a) $\frac{-4}{5}$ (b) $\frac{-3}{5}$

(c) $\frac{3}{5}$ (d) $\frac{4}{5}$

26. Consider an aerofoil given is the figure below, the equation of circulation and lift over an aerofoil is given as

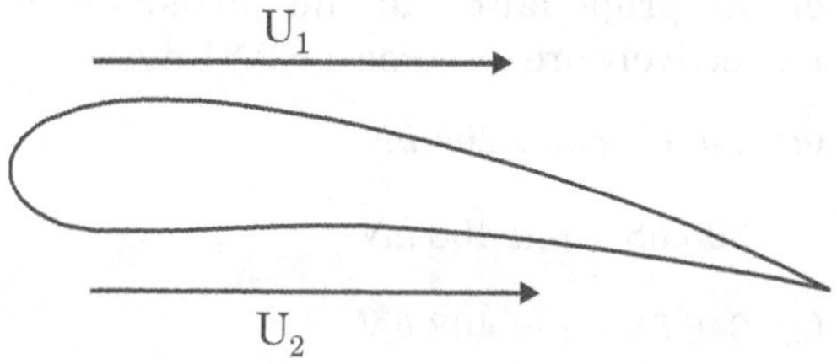

(a) $\Gamma=(u_1-u_2)c$ and $L=\rho.U.(u_1-u_2)c$

(b) $\Gamma=(u_1-u_2)$ and $L=\rho.U.(u_1-u_2)$

(c) $\Gamma=(u_2-u_1)c$ and $L=\rho.U.(u_2-u_1)c$

(d) $\Gamma=(u_2-u_1)$ and $L=\rho.U.(u_2-u_1)$

27. An ideal axial compressor is driven by an ideal turbojet across which the total temperature ratio is 0.333. If the total temperature at the turbine inlet is $T_0=1000$ K and the specific heat of gas $C_p=1$ kJ/kg/K, the power drawn by compressor per unit mass flow of air is approximately-

(a) 330 kW/kg/s

(b) 482 kW/kg/s

(c) 667 kW/kg/s

(d) 786 kW/kg/s

28. For $\sigma=\begin{bmatrix} 2 & 0 & 2 \\ 1 & 2 & 0 \\ 2 & 2 & 1 \end{bmatrix}$ MPa. Find the principle stress $\sigma_1,\sigma_2,\sigma_3$.

(a) 0, 3, 2 (b) 0, 4, 1

(c) 0, 5, 3 (d) 1, 2, 3

29. Mars orbiter sent by ISRO was parked in its first earth orbit of size 264 km × 23903 km. The orbit was then raised to 264 km × 28825 km. Velocity increment was required at perigeeis.

30. Consider the matrix, $M=\begin{pmatrix} 5 & 3 \\ 3 & 5 \end{pmatrix}$ The normalized eigen-vector corresponding to the smallest eigen-value of the matrix M is

(a) $\begin{pmatrix} \frac{\sqrt{3}}{2} \\ \frac{1}{2} \end{pmatrix}$ (b) $\begin{pmatrix} \frac{\sqrt{3}}{2} \\ -\frac{1}{2} \end{pmatrix}$

(c) $\begin{pmatrix} \frac{1}{\sqrt{2}} \\ -\frac{1}{\sqrt{2}} \end{pmatrix}$ (d) $\begin{pmatrix} \frac{1}{\sqrt{2}} \\ \frac{1}{\sqrt{2}} \end{pmatrix}$

31. At a stagnation temperature of 15° C and stagnation pressure 100 kPa enters an axial compressor with an axial flow velocity of 114 m/s and the stage efficiency is 0.80. Following conditions exist at the mean radius of the rotor blade:

Blade Speed = 200 m/s

Inlet blade angle $\beta_1=55^o$

Outlet blade angle $\beta_2=28^o$

$C_p=1005$ J/kg-K and $\gamma=1.4$

What is the stagnation pressure ratio (p_{rs}) for this compressor?

(a) 1.51 (b) 1.41

(c) 1.21 (d) 1.18

32. The minimum required power for cruise flight satisfies

(a) The power required by an airplane will be maximum, when $\frac{C_L^{\frac{3}{2}}}{C_D}$ will be maximum as they both are inversely related.

(b) The power required by an airplane will be minimum, when $\frac{C_L^{\frac{3}{2}}}{C_D}$ will be maximum as they both are inversely related.

(c) The power required by an airplane will be maximum, when $\frac{C_L^{\frac{3}{2}}}{C_D}$ will be minimum as they both are inversely related.

(d) None of the above

33. A thin steel plate 4 mm thick and 1000 mm long is bent into a circular arc which substrate 45° at the center of arc by a bending moment M_0. The maximum bending stress in the plate is .Take $E = 2 \times 10^5\ N/mm^2$

(a) $214\ N/mm^2$ (b) $314\ N/mm^2$

(c) $414\ N/mm^2$ (d) $514\ N/mm^2$

34. For the differential equation

$x^2\frac{d^2y}{dx^2} - 2x\frac{dy}{dx} + 2y = 0$ the general solution is

(a) $y = C_1x + C_2e^x$

(b) $y = C_1 \sin x + C_2 \cos x$

(c) $y = C_1e^x + C_2e^{-x}$

(d) $y = C_1x^2 + C_2x$

35. Calculate the distance 'r' where $U = 1.2\ U_\infty$ at $\theta = 90°$, in terms of radius of cylinder 'a'?

(a) $r = 5a$ (b) $r = \sqrt{5}a$

(c) $r = \frac{a}{5}$ (d) $r = a^5$

36. The principal stress that a material is subjected to is given as 60 MPa, 30 MPa and –30 MPa. Find the Shear Strain energy. If E = 200 GPa, v = 0.33

(a) 14.985 KN–m/m^3 (b) 1.02 KN–m/m^3

(c) 13.998 KN–m/m^3 (d) 56.25 N–m/m^3

37. A rocket of mass m is launched vertically with initial speed u. Assuming negligible air resistance. Then the maximum height attained by the rocket will be

(a) $h = \frac{R^2}{\frac{2gR}{u^2} - 1}$ (b) $h = \frac{R^2}{\frac{2gR}{u^2} + 1}$

(c) $h = \frac{R}{\frac{2gR}{u^2} - 1}$ (d) $h = \frac{R}{\frac{2gR}{u^2} + 1}$

38. The divergence of the velocity field $V = (x^2 + y)\hat{i} + (z - 2xy)\hat{j} + (xy)\hat{k}$ at (1, 1, 1) is ________.

39. In a boundary Layer, for a linear profile, the displacement thickness δ^* is given as

(a) $\delta^* = \delta$ (b) $\delta^* = \frac{\delta}{2}$

(c) $\delta^* = \frac{\delta}{4}$ (d) $\delta^* = \frac{\delta}{6}$

40. Under the ideal condition of fully expanded jet, a rocket engine is tested on a test bed. A nozzle of area 2.5 m^2, has an exhaust velocity of 3.4 km/s. the mass flow rate is 120 kg/s. The specific impulse of the propellant and the thrust developed respectively are (assume g = 9.81 m/s^2)

(a) 145.67 *s and* 204 *kN*

(b) 203.58 *s and* 408 *kN*

(c) 346.58 *s and* 408 *kN*

(d) 456.67 *s and* 204 *kN*

41. A rocket flying at air speed of 1000 m/s. The angle of attack is 300, the analytical sideslip angle β_a is 20° and the experimental sideslip angle, β_e is also 200. The axial velocity with respect to β_a and β_e is

(a) 330.63 m/s and 342.02 m/s

(b) 476.87 m/s and 469.85 m/s

(c) 825.96 m/s and 813.80 m/s

(d) 926.56 m/s and 914.74 m/s

42. In what range should Re (s) remain so that the Laplace transform of the function $e^{(a+2)t+5}$ exists.

(a) Re (s) > a + 2 (b) Re (s) > a + 7

(c) Re (s) < 2 (d) Re (s) > a + 5

43. Aircraft is flying at an altitude where, ambient flow conditions are $P_\infty = 25kN/m^2, T_\infty = 220K$

The temperature is ________ at a point on aircraft where pressure, P = 25 kN/m^2. For air, γ = 1.4

44. At an altitude where the ambient temperature is 15° C, a turbojet engine is flying with a velocity 240 m/s. the stagnation pressure ratio across its diffuser is 0.85. The diffuser efficiency is given as________

45. A cylindrical shell made of mild steel plate having diameter 100 cm is to be subjected to an internal pressure of 2 MPa. If the yield strength of the material is 200 MPa, Calculate the thickness of the plate based on the Maximum Shear Stress Theory, assuming factor of safety to be 3.

(a) 0.015 m (b) 0.0075 m
(c) 0.01299 m (d) 926.56

46. As per property of ellipse, the distance of point P from foci on ellipse is e times distance from latus rectum. Find r in terms of e.

(a) $r = \frac{ek}{1+e\cos\theta}$ (b) $r = \frac{ek}{1-e\cos\theta}$
(c) $r = \frac{e}{1+e\cos\theta}$ (d) $r = \frac{e}{1-e\cos\theta}$

47. The total strain energy per unit volume is given as 14.985 MPa. It the material yields at 150 MPa, Calculate the factor of safety on the total strain energy criterion, if E = 200 GPa, υ = 0.33

(a) 1.02 (b) 2.25
(c) 3.04 (d) 3.75

Linked questions

Consider a stream function as

$$\psi = Ur\sin\theta\left(1-\frac{a^2}{r^2}\right)+\frac{\Gamma}{2\pi}\ln\frac{r}{a}$$

where, $U = 25\ m/s, a = 1\ m, \Gamma = 100\pi\ m^2/s$

48. The stagnation points are at -

(a) 0° or 180° (b) 180° or 30°
(c) 90° or 270° (d) 210° or 330°

Linked questions

49. For what value of $\theta, P = P_\infty\ if\ \Gamma = 0\ m^2/s$

(a) 0° or 180° (b) 90° or 270°
(c) 194° and 346° (d) 210° and 330°

Linked questions 50 and 51.

50. A bipropellant liquid rocket with a nozzle throat diameter of 40 mm operates at a chamber pressure of 50 bar. The characteristic velocity is 1850 m/s. If the fuel to oxidizer ratio of propellant is 2 and the fuel density is 600 kg/m^3. What is the rate of total mass flow and rate of mass flow of fuel?

(a) 5.097 kg/s and 3.276 kg/s
(b) 5.097 kg/s and 2.264 kg/s
(c) 3.396 kg/s and 2.264 kg/s
(d) 3.396 kg/s and 3.276 kg/s

51. What should be the minimum fuel tank volume for a burn time of 7 minutes?

(a) 1.58 m^3 (b) 1.68 m^3
(c) 1.78 m^3 (d) 1.88 m^3

Linked question 52 and 53.

52. Given $a > 0$, we wish to calculate the reciprocal value $\frac{1}{a}$ by Newton-Raphson method for $f(x) = 0$.

The Newton-Raphson algorithm for the function will be

(a) $x_{k+1} = \frac{1}{2}\left(x_k + \frac{a}{x_k}\right)$

(b) $x_{k+1} = \left(x_k + \frac{a}{2}x_k^2\right)$

(c) $x_{k+1} = 2x_k - ax_k^2$

(d) $x_{k+1} = x_k - \frac{a}{2}x_k^2$

53. For a = 7 and starting with $x_0 = 0.2$, the first two iterations will be

(a) 0.11, 0.1299 (b) 0.12, 0.1392
(c) 0.12, 0.1416 (d) 0.13, 0.1428

Linked questions 54 and 55

54. An aeroplane model is mounted in a wind tunnel with angle of attack 12 degrees and sideslip angle, $\beta_e = 0.8^0$. By varying β_e slightly to both sides from the orientation, the change in the yawing moment coefficient with respect to β_e is estimated to be 0.250. Find the angle β_a.

(a) $\beta_a = 37.2^0$ (b) $\beta_a = 20.1^0$
(c) $\beta_a = 8.2^0$ (d) $\beta_a = 4.5^0$

55. Estimate the change in the yawing moment coefficient with respect to β_a.

(a) 0.225 (b) 0.245
(c) 0.265 (d) 0.285

56. "In spite of being warned repeatedly, he failed to correct his ________ behavior."

The word that best fills the blank in the above sentence is

(a) rational (b) reasonable
(c) errant (d) good

57. Choose the word or phrase that best completes the sentence below.

_____________ in the frozen wastes of Arctic takes special equipment.

(*a*) To survive (*b*) Surviving

(*c*) Survival (*d*) That survival

58. How much time does a clock lose in a day, if its hands coincide every 64 minutes?

(*a*) 55 minutes (*b*) 67 minutes

(*c*) 69 minutes (*d*) 96 minutes

59. When a point inside of a tetrahedron (a solid with four triangular surfaces) is connected by straight lines to its corners, how many (new) internal planes are created with these lines?

60. a, b, c, d and e are 5 consecutive even numbers. If the sum of a and d is 162, what is the sum of all the numbers?

(*a*) 310 (*b*) 410

(*c*) 510 (*d*) 610

61. Find the infinite sum of the series

$$1+\frac{4}{7}+\frac{9}{7^2}+\frac{16}{7^3}+\frac{25}{7^4}+............\infty$$

62. If 15% of x is the same as 20% of y, then is

(*a*) 1 : 2 (*b*) 2 : 1

(*c*) 4 : 3 (*d*) 3 : 4

63. Two students appeared at an explanation. One of them secured 9 marks more than the other and his marks was 56% of the sum of their marks. The marks obtained by them are

(*a*) 63 and 12 (*b*) 49 and 26

(*c*) 42 and 33 (*d*) 40 and 35

64. Out of the following 4 sentences, select the most suitable sentence with respect to grammar and usage:

(*a*) Since the report lacked needed information, it was of no use to them.

(*b*) The report was useless to them because there were no needed information in it.

(*c*) Since the report did not contain the needed information, it was not real useful to them.

(*d*) Since the report lacked needed information, it would not had been useful to them.

65. Most experts feel that in spite of possessing all the technical skills required to be a batsman of the highest order, he is unlikely to be so due to lack of requisite temperament. He was guilty of throwing away his wicket several time after working hard to lay a strong foundation. His critics pointed out that until he addressed his problem, success at the highest level will continue to elude him.

Which of the statement(s) below is/are logically valid and can be inferred from the above passage?

i. He was already a successful batsman at the highest level.

ii. He was to improve his temperament in order to become a great batsman.

iii. He failed to make many of his good starts count.

iv. Improving his technical skills will guarantee success.

(*a*) iii and iv (*b*) ii and iii

(*c*) i, ii and iii (*d*) ii only

OMR SHEET

1. Ⓐ Ⓑ Ⓒ Ⓓ	18. Ⓐ Ⓑ Ⓒ Ⓓ	35. Ⓐ Ⓑ Ⓒ Ⓓ	52. Ⓐ Ⓑ Ⓒ Ⓓ
2. Ⓐ Ⓑ Ⓒ Ⓓ	19. Ⓐ Ⓑ Ⓒ Ⓓ	36. Ⓐ Ⓑ Ⓒ Ⓓ	53. Ⓐ Ⓑ Ⓒ Ⓓ
3. Ⓐ Ⓑ Ⓒ Ⓓ	20. Ⓐ Ⓑ Ⓒ Ⓓ	37. Ⓐ Ⓑ Ⓒ Ⓓ	54. Ⓐ Ⓑ Ⓒ Ⓓ
4. Ⓐ Ⓑ Ⓒ Ⓓ	21. Ⓐ Ⓑ Ⓒ Ⓓ	38. Ⓐ Ⓑ Ⓒ Ⓓ	55. Ⓐ Ⓑ Ⓒ Ⓓ
5.	22. Ⓐ Ⓑ Ⓒ Ⓓ	39. Ⓐ Ⓑ Ⓒ Ⓓ	56. Ⓐ Ⓑ Ⓒ Ⓓ
6. Ⓐ Ⓑ Ⓒ Ⓓ	23. Ⓐ Ⓑ Ⓒ Ⓓ	40. Ⓐ Ⓑ Ⓒ Ⓓ	57. Ⓐ Ⓑ Ⓒ Ⓓ
7. Ⓐ Ⓑ Ⓒ Ⓓ	24. Ⓐ Ⓑ Ⓒ Ⓓ	41. Ⓐ Ⓑ Ⓒ Ⓓ	58. Ⓐ Ⓑ Ⓒ Ⓓ
8. Ⓐ Ⓑ Ⓒ Ⓓ	25. Ⓐ Ⓑ Ⓒ Ⓓ	42.	59. Ⓐ Ⓑ Ⓒ Ⓓ
9. Ⓐ Ⓑ Ⓒ Ⓓ	26. Ⓐ Ⓑ Ⓒ Ⓓ	43.	60. Ⓐ Ⓑ Ⓒ Ⓓ
10. Ⓐ Ⓑ Ⓒ Ⓓ	27. Ⓐ Ⓑ Ⓒ Ⓓ	44. Ⓐ Ⓑ Ⓒ Ⓓ	61. Ⓐ Ⓑ Ⓒ Ⓓ
11.	28. Ⓐ Ⓑ Ⓒ Ⓓ	45. Ⓐ Ⓑ Ⓒ Ⓓ	62. Ⓐ Ⓑ Ⓒ Ⓓ
12. Ⓐ Ⓑ Ⓒ Ⓓ	29.	46. Ⓐ Ⓑ Ⓒ Ⓓ	63. Ⓐ Ⓑ Ⓒ Ⓓ
13. Ⓐ Ⓑ Ⓒ Ⓓ	30. Ⓐ Ⓑ Ⓒ Ⓓ	47. Ⓐ Ⓑ Ⓒ Ⓓ	64. Ⓐ Ⓑ Ⓒ Ⓓ
14. Ⓐ Ⓑ Ⓒ Ⓓ	31. Ⓐ Ⓑ Ⓒ Ⓓ	48. Ⓐ Ⓑ Ⓒ Ⓓ	65. Ⓐ Ⓑ Ⓒ Ⓓ
15. Ⓐ Ⓑ Ⓒ Ⓓ	32. Ⓐ Ⓑ Ⓒ Ⓓ	49. Ⓐ Ⓑ Ⓒ Ⓓ	
16. Ⓐ Ⓑ Ⓒ Ⓓ	33. Ⓐ Ⓑ Ⓒ Ⓓ	50. Ⓐ Ⓑ Ⓒ Ⓓ	
17. Ⓐ Ⓑ Ⓒ Ⓓ	34. Ⓐ Ⓑ Ⓒ Ⓓ	51. Ⓐ Ⓑ Ⓒ Ⓓ	

ANSWERS

1. (c)	2. (c)	3. (b)	4. (c)	5. (11.2)	6. (c)	7. (b)	8. (c)	9. (a)	10. (a)
11. (14×10^{-3})	12. (b)	13. (c)	14. (c)	15. (b)	16. (d)	17. (b)	18. (b)	19. (c)	20. (c)
21. (c)	22. (d)	23. (c)	24. (a)	25. (a)	26. (a)	27. (c)	28. (b)	29. (127)	30. (c)
31. (c)	32. (b)	33. (c)	34. (d)	35. (b)	36. (c)	37. (c)	38. (0)	39. (b)	40. (c)
41. (c)	42. (a)	43. (178.36)	44. (0.5)	45. (b)	46. (a)	47. (d)	48. (c)	49. (d)	50. (c)
51. (a)	52. (c)	53. (b)	54. (c)	55. (b)	56. (c)	57. (a)	58. (d)	59. (6)	60. (b)
61. (c)	62. (c)	63. (a)	64. (a)	65. (b)					

EXPLANATIONS

1.

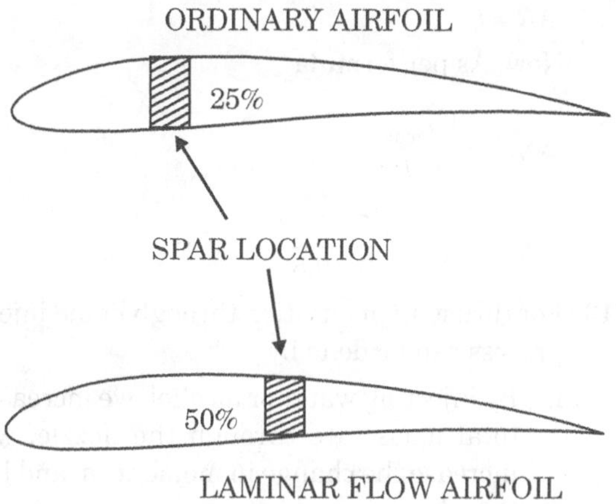

A laminar flow wing has maximum thickness at middle camber line.

2. $\alpha = \frac{\dot{m}_F}{\dot{m}_c} = 5$

$\therefore \dot{m}_c = \frac{\dot{m}_F}{5}$ and $\dot{m}_a = \dot{m}_c + \dot{m}_F$

$$\Rightarrow \dot{m}_a = \frac{\dot{m}_F}{5} + \dot{m}_F$$

$$\Rightarrow \dot{m}_a = \dot{m}_F\left(\frac{1}{5}+1\right)$$

$$\Rightarrow \dot{m}_a = \dot{m}_F \times \frac{6}{5}$$

$$\Rightarrow \dot{m}_a = 100 \times \frac{6}{5} = 120\ kg/s$$

3. $f(x) = \frac{1}{1+|x|}$

$$\text{L.H.S} = \lim_{x\to 0^-} f(x) = \frac{1}{1+|0|} = 1$$

$$\text{R.H.S} = \lim_{x\to 0^+} f(x) = \frac{1}{1+|0|} = 1$$

$\therefore$ L.H.S = R.H.S = $f(x)$

So, at $x = 0$, $f(x)$ is continuous

Derivative

$$f'(x) = \frac{d}{dx}\left((1+|x|)^{-1}\right)$$

Now applying chain rule

let $f = u^{-1}$

$\therefore u = (1+|x|)$

$$\Rightarrow \frac{d}{du}(u^{-1})\frac{d}{dx}(1+|x|)$$

$$\Rightarrow \left(\frac{-1}{u^2}\right)\left(\frac{x}{|x|}\right)$$

$$\Rightarrow \left(\frac{-1}{(1+|x|)^2}\right)\left(\frac{x}{|x|}\right)$$

$$\Rightarrow -\frac{x}{|x|(1+|x|)^2}$$

At $x = 0$,

Left derivative ≠ Right derivative

So, $f(x)$ is not differentiable

4. As per formula of Laplace transformation

For $f(t) = e^{at}$

$$L(f) = \frac{1}{s-a}$$

As per given question, for $f(t) = e^{-2t}$

$$L(f) = \frac{1}{s+2}$$

5. The escape velocity is given by,

$$\sqrt{2gR} = \sqrt{2 \times 9.81 \times 6378 \times 10^3} = 11.2\ km/s$$

6. Given,

$u = \alpha x, v = \beta y, w = 0$

$\frac{\partial u}{\partial x} = \alpha$

$\frac{\partial v}{\partial y} = \beta$

$\frac{\partial w}{\partial z} = 0$

Volume $= 1 \times 1 \times 1 = 1m^3$

Now New volume $= (1+\alpha)(1+\beta)(1) = 1+\alpha\beta+\alpha+\beta$

Net change in volume

$dV = 1+\alpha\beta+\alpha+\beta-1$

$dV = \alpha\beta+\alpha+\beta$

7. Lift on an aircraft climbing vertically up is zero.

8. Acceleration on an inclined plane

$$a = \frac{g\sin\theta}{\left(1+\frac{I}{MR^2}\right)}$$

For circular rings: $I = MR^2$

so by putting the value in above equation we get,

$$\therefore\ a = \frac{g\sin\theta}{2} = \frac{g\sin 30^\circ}{2} = \frac{g}{4}$$

9. Rank $[P_q]$ = Rank $[P]$ is necessary for existence of at least one solution to $P_x = q$.

10.

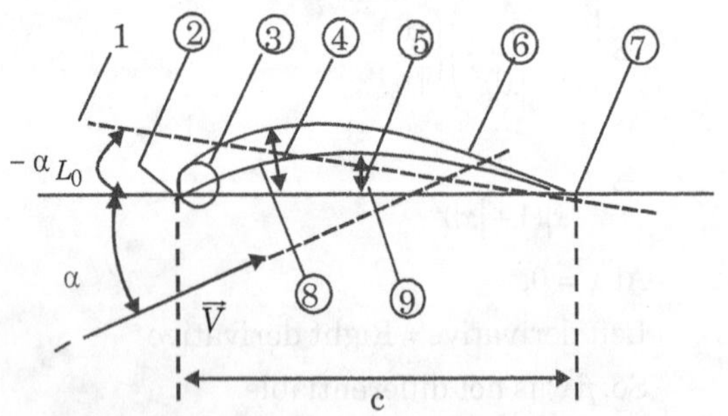

Profile geometry

1 - Zero lift line;

2 - Leading edge;

3 - Nose circle;

4 - Max. thickness;

5 - Camber;

6 - Upper surface;

7 - Trailing edge;

8 - Camber mean-line;

9 - Lower surface;

11. $u = 0\ at\ y = 0$

$u = 2\ m/s\ at\ y = 1\ m$

As per equation

$$\tau = \mu\frac{\partial u}{\partial y} = \mu\frac{V}{h} = 7\times10^{-3}\times2 = 14\times10^{-3}\ N/m$$

$\therefore \tau = 14\times10^{-3}\ N/m$

12. Given,

$\tau = 0.5$

$C_{L_m} = 1.5$

$AR = 5$

Now, As per formula

$$\Delta\delta_e = -\frac{C_{L_{max}}}{\pi AR\tau}$$

$$\Delta\delta_e = -\frac{1.5}{3.14\times5\times0.5} = -10.9^\circ$$

13. For thrust augmentation through liquid injection process can be done by

i. By injecting water or alcohol, we increase the total mass rate through the nozzle, which increase the change in momentum and hence thrust

ii. By spraying water at the inlet of compressor there is reduction in compressor inlet temperature, which causes in increase of compressor pressure ratio and hence increase in thrust.

14. For rectangular cross-section

$$\tau_{max} = \frac{\frac{P}{2}}{\frac{bh^3}{12}}\times\frac{h^2}{4}$$

$$\Rightarrow \tau_{max} = \frac{P}{2}\times\frac{12}{4b\times h}$$

$$\Rightarrow \tau_{max} = \frac{3}{2}\times\frac{P}{bh} \text{ and}$$

$$\tau_{avg} = \frac{total\ load}{area} = \frac{P}{bh}$$

$$\therefore \frac{\tau_{max}}{\tau_{avg}} = \frac{3}{2}\times\frac{P}{bh}\times\frac{bh}{P}$$

$$\Rightarrow \frac{\tau_{max}}{\tau_{avg}} = \frac{3}{2}$$

15. For longitudinal stability $C_{m\alpha}$ must be negative i.e., pitch stability

or lateral stability $C_{n\beta}$ must be positive i.e., roll stability

For directional stability $C_{l\beta}$ must be negative i.e., yaw stability

16. Given,

$$C_{p0} = -0.3$$

$$M_{\infty} = 0.8$$

$$C_p = \frac{C_{p0}}{\sqrt{1-M_{\infty}^2}} = \frac{-0.3}{\sqrt{1-(0.8)^2}} = -0.5$$

17. Actual Head, $H_p = H_t - h_L$

Actual Fluid Power $\dot{W}_f = \rho g H_p Q$

Brake (impeller) Power $\dot{W}_p = \omega T$

Efficiency $\eta_p = \frac{\dot{W}_f}{\dot{W}_p} = \frac{\rho g H_p Q}{\omega T}$

18. The Mach Angle is defined as $\mu = \sin^{-1}\left(\frac{1}{M}\right)$

For M = 1.41

$$\mu = \sin^{-1}\left(\frac{1}{1.41}\right)$$

$$\mu = \sin^{-1}(0.707) = 45°$$

19. Given,

$$\sigma = \begin{bmatrix} 16 & 5 \\ 5 & -4 \end{bmatrix} MPa$$

At plane of zero shear stress,

$$\tan 2\theta = \frac{2\tau_{XY}}{\sigma_{XX} - \sigma_{YY}}$$

$$= \frac{2 \times 5}{16-(-4)} = \frac{10}{20} = \frac{1}{2}$$

$$\Rightarrow 2\theta = \tan^{-1}(0.5) = 26.56°$$

$$\Rightarrow \theta = \frac{26.56}{2} = 13.28°$$

20. $\vec{P} = 0.866\hat{i} + 0.500\hat{j} + 0\hat{k}$

$$\vec{Q} = 0.259\hat{i} + 0.966\hat{j} + 0\hat{k}$$

$$\therefore \vec{P}.\vec{Q} = |\vec{P}||\vec{Q}|\cos\theta$$

$$\Rightarrow |P| = |Q| = 1$$

So, $(0.866\hat{i} + 0.500\hat{j} + 0\hat{k}).(0.259\hat{i} + 0.966\hat{j} + 0\hat{k})$

$$= \sqrt{(0.866)^2 + (0.5)^2} \times \sqrt{(0.259)^2 + (0.966)^2}.\cos\theta$$

$$\therefore \cos\theta = \frac{0.866 \times 0.259 + 0.5 \times 0.966}{\sqrt{1} \times \sqrt{1}} = 0.707$$

$$\theta = \cos^{-1}(0.707) = 45°$$

21. $\frac{\left(\frac{K.E}{I.E}\right)_{M_2}}{\left(\frac{K.E}{I.E}\right)_{M_1}} = \frac{M_2^2}{M_1^2}$

$$\Rightarrow \frac{\left(\frac{K.E}{I.E}\right)_{M_2}}{\left(\frac{K.E}{I.E}\right)_{M_1}} = \frac{100}{4} = 25$$

22. Given,

$$\gamma = 1.2 \text{ and } \frac{p_1}{p_2} = \frac{1}{12.6}$$

The Brayton cycle is the thermodynamic cycle employed in gas turbine engines. The cycle operates between two pressure limits and consists of isobaric and isentropic processes.

Efficiency of an Bryton cycle is given as

$$\eta = 1 - \frac{T_1}{T_2} = 1 - \left(\frac{p_1}{p_2}\right)^{\frac{\gamma-1}{\gamma}}$$

$$\Rightarrow \eta = 1 - \left(\frac{1}{12.6}\right)^{\frac{1.2-1}{1.2}} = 0.344 = 34.4\ \%$$

23. For

$$\phi = x^4 y^4$$

$$\frac{\partial^4 \phi}{\partial x^4} = 24y^4$$

$$\frac{\partial^4 \phi}{\partial y^4} = 24x^4$$

$$2\frac{\partial^2}{\partial x^2}\frac{\partial^2 \phi}{\partial y^2} = 288x^2y^2$$

$$\nabla^4\phi = \frac{\partial^4 \phi}{\partial x^4} + \frac{\partial^4 \phi}{\partial y^4} + 2\frac{\partial^2}{\partial x^2}\frac{\partial^2 \phi}{\partial y^2} \neq 0$$

Therefore, Airy's stress function is not valid.

For

$$\phi = y^6 - x^6$$

$$\frac{\partial^4 \phi}{\partial x^4} = -360x^2$$

$$\frac{\partial^4 \phi}{\partial y^4} = 360y^2$$

$$2\frac{\partial^2}{\partial x^2}\frac{\partial^2 \phi}{\partial y^2} = 0$$

$$\nabla^4\phi = \frac{\partial^4 \phi}{\partial x^4} + \frac{\partial^4 \phi}{\partial y^4} + 2\frac{\partial^2}{\partial x^2}\frac{\partial^2 \phi}{\partial y^2} \neq 0$$

Therefore, Airy's stress function is not valid.

For,

$$\phi = x^3 + y^3$$

$$\frac{\partial^4 \phi}{\partial x^4} = 0$$

$$\frac{\partial^4 \phi}{\partial y^4} = 0$$

$$2\frac{\partial^2}{\partial x^2}\frac{\partial^2 \phi}{\partial y^2} = 0$$

$$\nabla^4\phi = \frac{\partial^4 \phi}{\partial x^4} + \frac{\partial^4 \phi}{\partial y^4} + 2\frac{\partial^2}{\partial x^2}\frac{\partial^2 \phi}{\partial y^2} = 0$$

Hence, Airy's stress function is valid.

24. At free end, displacement u will be maximum, so

$u \neq 0, \frac{\partial u}{\partial x} = 0,$ also free end; there will not be any moment or shear force present.

25. Given,

$$[M]^T = [M]^{-1}$$

$$\Rightarrow M^T M = I$$

$$\Rightarrow \begin{bmatrix} \frac{3}{5} & x \\ \frac{4}{5} & \frac{3}{5} \end{bmatrix}\begin{bmatrix} \frac{3}{5} & \frac{4}{5} \\ x & \frac{3}{5} \end{bmatrix} = \begin{bmatrix} 1 & 0 \\ 0 & 1 \end{bmatrix}$$

$$\Rightarrow \begin{bmatrix} \left(\frac{3}{5}\right)^2 + x^2 & \left(\frac{3}{5}.\frac{4}{5}\right) + \frac{3}{5}x \\ \left(\frac{4}{5}.\frac{3}{5}\right) + \frac{3}{5}x & \left(\frac{4}{5}\right)^2 + \left(\frac{3}{5}\right)^2 \end{bmatrix} = \begin{bmatrix} 1 & 0 \\ 0 & 1 \end{bmatrix}$$

$\Rightarrow$ Compare both sides a_{12}

$$a_{12}\left(\frac{3}{5}\right)\left(\frac{4}{5}\right) + \left(\frac{3}{5}\right)x = 0$$

$$\Rightarrow \frac{3}{5}x = -\frac{3}{5}.\frac{4}{5}$$

$$\Rightarrow x = -\frac{4}{5}$$

26. Considering the aerofoil as given is question

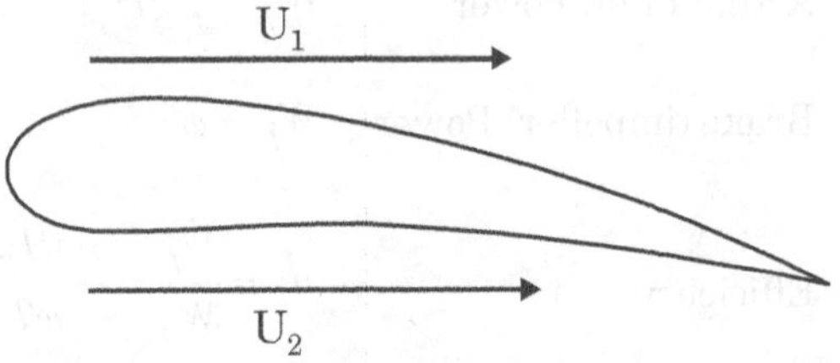

As the thin aerofoil has thickness tending to zero, hence the circulation over the aerofoilcan be approximately written as,

$$\Gamma = (u_1 - u_2)c \ as \ n \to 0$$

Here, c is the chord of the aerofoil.

The lift can be calculated per unit span as,

$$L = \rho.U.\Gamma$$

$$L = \rho.U.(u_1 - u_2)c$$

27. Power drawn by compressor = Power output of turbine

Power output of turbine $= \dot{m}C_p\left(T_{exit} - T_{inlet}\right)$

$$= \dot{m}C_pT_{inlet}(0.333 - 1)$$

$$= -\dot{m}C_pT_{inlet}(0.667)$$

As work is done by the turbine, since power output is negative

Therefore,

$$\frac{P}{\dot{m}} = C_P \times T_{inlet} \times 0.667$$

$$\Rightarrow \frac{P}{\dot{m}} = 1 \times 1000 \times 0.667 = 667\ kW/kg/s$$

28. The eigen value problem is,

$$[[\sigma]-\sigma[I]]\times\vec{n}=0$$

$$[[\sigma]-\sigma[I]]=0$$

$$\begin{vmatrix}2-\sigma & 0 & 2\\ 1 & 2-\sigma & 0\\ 2 & 2 & 1-\sigma\end{vmatrix}=0$$

$$\Rightarrow(2-\sigma)(2-\sigma)(1-\sigma)+2(2-2(2-\sigma))=0$$

$$\Rightarrow(4+\sigma^2-4\sigma)(1-\sigma)+4-8+4\sigma=0$$

$$\Rightarrow 4-4\sigma+\sigma^2-\sigma^3-4\sigma+4\sigma^2-4+4\sigma=0$$

$$\Rightarrow -\sigma^3+5\sigma^2-4\sigma=0$$

The principle stresses are obtained as:

$$\Rightarrow \sigma_1=0, \sigma_2=4, \sigma_3=1$$

29. Given,

$R_p = 264\ km$

$R_{a_1} = 23903\ km$

$R_{a_2} = 28825\ km$

$R_e = 6378\ km$

$r_p = 6378 + 264 = 6642\ km$

$r_{a_1} = 6378 + 23903 = 30281\ km$

$r_{a_2} = 6378 + 28825 = 35203\ km$

Increment in velocity is given as,

$$\Delta V = V_2 - V_1$$

$$\Delta V = \sqrt{\frac{2GMr_{a_2}}{r_p(r_p+r_{a_2})}} - \sqrt{\frac{2GMr_{a_1}}{r_p(r_p+r_{a_1})}}$$

$$=\sqrt{\frac{2\times 6.67\times 10^{-11}\times 5.98\times 10^{24}\times 35203\times 10^3}{6642\times 10^3\times(6642+35203)\times 10^3}}-$$

$$\sqrt{\frac{2\times 6.67\times 10^{-11}\times 5.98\times 10^{24}\times 30281\times 10^3}{6642\times 10^3\times(6642+30281)\times 10^3}}$$

$$= 10015.69 - 9888.94$$

$$= 126.75\ m/s \approx 127\ m/s$$

30. Given,

$$M=\begin{pmatrix}5 & 3\\ 3 & 5\end{pmatrix}$$

Eigen values are given as,

$$[A-\lambda]I = 0$$

Here,

$$\begin{pmatrix}5-\lambda & 3\\ 3 & 5-\lambda\end{pmatrix}=0$$

$$\Rightarrow (5-\lambda)^2-9=0$$

$$\Rightarrow \lambda^2-10\lambda+16=0$$

$$\Rightarrow (\lambda-2)(\lambda-8)=0$$

$$\Rightarrow \lambda=2,8$$

Eigenvector is given as
Taking the smallest eigenvalue i.e.$\lambda = 2$

$$\begin{pmatrix}5-\lambda & 3\\ 3 & 5-\lambda\end{pmatrix}\begin{pmatrix}x\\ y\end{pmatrix}=0$$

$$\Rightarrow\begin{pmatrix}3 & 3\\ 3 & 3\end{pmatrix}\begin{pmatrix}x\\ y\end{pmatrix}=0$$

Dividing the lower column by upper, we get,

$$\Rightarrow\begin{pmatrix}3 & 3\\ 0 & 0\end{pmatrix}\begin{pmatrix}x\\ y\end{pmatrix}=0$$

$$\Rightarrow 3x+3y=0$$

$$\Rightarrow x=-y$$

$$\therefore \text{Eigenvectors}, v=\begin{pmatrix}-1\\ 1\end{pmatrix}$$

Normalization of Eigen vector is given by,

$$\chi=\frac{x}{\|x\|}$$

$$\chi=\frac{1}{|\sqrt{1+1}|}=\frac{1}{\sqrt{2}}\text{and}$$

$$=\frac{1}{|\sqrt{1+1}|}=\frac{-1}{\sqrt{2}}$$

31. Given,

$$T_{01} = 273 + 15 = 288\ K$$

$$U = 200\ m/s$$

$$C_a = 114\ m/s$$

$$\beta_1 = 55°$$

$$\beta_2 = 28°$$

$$\gamma = 1.4$$

$$C_p = 1005\ J/kg - K$$

$$T_{03} - T_{01} = \frac{UC_a(\tan\beta_1 - \tan\beta_2)}{C_p}$$

$$\Rightarrow T_{03} - T_{01} = \frac{200 \times 114 \times \left(\tan 55^o - \tan 28^o\right)}{1005}$$

$$\Rightarrow T_{03} - T_{01} = 20.33$$

As,

$$\frac{P_{03}}{P_{01}} = \left[1 + \eta_e \frac{T_{03} - T_{01}}{T_{01}}\right]^{\frac{\gamma}{\gamma-1}}$$

$$\Rightarrow \frac{P_{03}}{P_{01}} = \left[1 + .8 \times \frac{20.33}{288}\right]^{\frac{1.4}{0.4}}$$

$$\therefore \frac{P_{03}}{P_{01}} = 1.21$$

32. Power required minimum for cruise flights

$$P_{req} = T_{req}V = DV = \frac{WV}{C_L/C_D} = \sqrt{\frac{2W^3C_D^2}{S\rho C_L^3}}$$

$$\Rightarrow P_{req} \propto \frac{1}{\frac{C_L^{\frac{3}{2}}}{C_D}}$$

It states that, the power required by an airplane will be minimum, when $\frac{C_L^{\frac{3}{2}}}{C_D}$ will be maximum as they both are inversely related.

33. For a thin plate,

$$I = \frac{bt^3}{12} = \frac{b \times 4^3}{12}$$

Also we know that,

$$\frac{M}{I} = \frac{E}{R} = \frac{\sigma}{y}$$

and $\Delta l = y\theta$

$$R\theta = 1000$$

$$\Rightarrow R \times \frac{\pi}{4} = 1000$$

$$\Rightarrow R = 1273.33\ mm$$

Again,

$$\frac{E}{R} = \frac{\sigma}{y}$$

$$Here, y = \frac{4}{2} = 2\ mm$$

$$\Rightarrow \frac{\sigma}{2} = \frac{2 \times 10^5}{1273.23}$$

$$\Rightarrow \sigma = 314\ N/mm^2$$

34. $x^2\frac{d^2y}{dx^2} - 2x\frac{dy}{dx} + 2y = 0$

Let us write it as,

$$x^2y'' - 2xy' + 2y = 0$$

Now let,

$$y = x^m$$

$$\Rightarrow x^2m(m-1)x^{(m-2)} - 2xmx^{(m-1)} + 2x^m = 0$$

$$\Rightarrow x^2m(m-1)x^{(m-2)} - 2xmx^{(m-1)} + 2x^m = 0$$

$$\Rightarrow m^2 - m - 2m + 2 = 0$$

$$\Rightarrow m^2 - 3m + 2 = 0$$

$$\Rightarrow (m-1)(m-2) = 0$$

$$\Rightarrow m = 1, 2$$

Therefore, the solution is

$$y = C_1x + C_2x^2$$

35. As $\theta = 90°$, thus $u_r = 0$

Hence,

$$u_\theta = -U_\infty\left(1+\frac{a^2}{r^2}\right)$$

As θ and u_θ is positive anticlockwise, also given,

$u_\theta = -1.2U_\infty$

Therefore,

$$\Rightarrow -1.2U_\infty = -U_\infty\left(1+\frac{a^2}{r^2}\right)$$

$$\Rightarrow 1.2 = \left(1+\frac{a^2}{r^2}\right)$$

$$\Rightarrow 0.2 = \frac{a^2}{r^2}$$

$$\Rightarrow r^2 = \frac{a^2}{0.2}$$

$$\Rightarrow r = \sqrt{5}a$$

36. Given,

$\sigma_1 = 60\ MPa$

$\sigma_2 = 30\ MPa$

$\sigma_3 = -30\ MPa$

Shear strain energy per unit volume is

$$= \frac{1}{12G}\left[(\sigma_1-\sigma_2)^2+(\sigma_2-\sigma_3)^2+(\sigma_3-\sigma_1)^2\right]$$

Also,

$$G = \frac{E}{2(1+\nu)} = \frac{200\times10^9}{2\times1.33} = 75.01\ GPa$$

Therefore, Shear strain energy per unit volume is

$$= \frac{10^{12}}{12\times76.923\times10^9}$$

$$\left[(60-30)^2+(30+30)^2+(-30-60)^2\right]$$

$$= 13.998\ KN-m/m^3$$

37. $$\frac{-GMm}{(R+h)} = \frac{-GMm}{R}+\frac{1}{2}mu^2$$

$$\Rightarrow \frac{-GMm}{(R+h)}\times\frac{R^2}{R^2} = \frac{-GMm}{R}+\frac{1}{2}mu^2$$

$$\Rightarrow \frac{-GM}{(R+h)}\times\frac{R^2}{R^2} = \frac{-GM}{R}+\frac{1}{2}u^2$$

$$\Rightarrow -g\left[\frac{R^2-R^2-Rh}{(R+h)}\right] = \frac{1}{2}u^2$$

$$\Rightarrow \frac{h}{(R+h)} = \frac{1}{2gR}u^2$$

$$\Rightarrow \frac{(R+h)}{h} = \frac{2gR}{u^2}$$

$$\Rightarrow \frac{R}{h}+1 = \frac{2gR}{u^2}$$

$$\Rightarrow h = \frac{R}{\frac{2gR}{u^2}-1}$$

38. The divergence of the velocity field

$V = (x^2+y)\hat{i}+(z-2xy)\hat{j}+(xy)\hat{k}\ at\ (1,1,1)$

is given as

$$\text{Divergence, } \nabla V = \frac{\partial V}{\partial x}+\frac{\partial V}{\partial y}+\frac{\partial V}{\partial z}$$

$$\frac{\partial V}{\partial x} = 2x$$

$$\frac{\partial V}{\partial y} = -2x$$

$$\frac{\partial V}{\partial z} = 0$$

$$\therefore\ \nabla V_{(1,1,1)} = 2-2 = 0$$

39. For linear profile

$$u = \frac{U_\infty y}{\delta}$$

$$\Rightarrow \frac{u}{U_\infty} = \frac{y}{\delta}$$

Displacement Thickness, δ^*

$$\delta^* = \int_0^\delta \left(1 - \frac{y}{\delta}\right) dy = \left[y - \frac{y^2}{2\delta}\right]_0^\delta$$

$$\Rightarrow \delta^* = \delta - \frac{\delta}{2}$$

$$\Rightarrow \delta^* = \frac{\delta}{2}$$

40. Given,

$A_e = 2.5\ m^2$

$U_e = 3.4\ km/s$

$\dot{m} = 120\ kg/s$

$g = 9.81\ m/s^2$

Thrust of rocket engine

$$F = \dot{m}U_e + (p_e - p_\infty)A_e$$

Given that it is fully expanded $\Rightarrow p_e = p_\infty$

Now,

$$F = \dot{m}U_e$$

$$\Rightarrow F = 120 \times 3.4$$

$$\Rightarrow F = 408\ kN$$

Specific Impulse, $I_s = \dfrac{F}{\dot{m}g}$

$$= \frac{408 \times 10^3}{120 \times 9.81} = 346.58\ \text{s}$$

41. Given,

$\beta_a = 20°$

$\beta_e = 20°$

$\alpha = 30°$

$V = 1000\ m/s$

For analytical sideslip angle $\beta_a = 20°$

Axial velocity is given as

$$V_{axial} = \frac{V}{\sqrt{1 + \tan^2\alpha + \tan^2\beta_a}}$$

$$\Rightarrow V_{axial} = \frac{1000}{\sqrt{1 + (0.577350)^2 + (0.363970)^2}}$$

$$\Rightarrow V_{axial} = 825.96\ m/s$$

For experimental sideslip angle $\beta_e = 20°$

$$V_{axial} = V\cos\alpha\cos\beta_e$$

$$\Rightarrow V_{axial} = 1000 \times \cos(30) \times \cos(20)$$

$$\Rightarrow V_{axial} = 1000 \times 0.866025 \times 0.93969$$

$$\Rightarrow V_{axial} = 813.80\ m/s$$

42. $f(t) = e^{(a+2)t+5} = e^5 e^{(a+2)}$

$$F(s) = \left[\frac{1}{s - (a+2)}\right]e^5$$

$\therefore$ For Laplace Transform to exist, $\text{Re}(s) > a + 2$

43. Assuming isentropic flow over the aircraft
Using isentropic flow relations-

$$\left(\frac{T_2}{T_1}\right)^{\gamma/\gamma-1} = \frac{p_2}{p_1}$$

$$\Rightarrow \left(\frac{T}{220}\right)^{3.5} = \frac{12}{25}$$

$$\Rightarrow T = 178.36\,K$$

44. Given,

$T = 15°\,C = (15 + 273) = 288\ K$

$r_d = 0.85$

$U = 240\ m/s$

$\gamma = 1.4$

Now, Mach number is given as

$$M = \frac{U}{\sqrt{\gamma RT}} = \frac{240}{\sqrt{1.4 \times 287 \times 288}} = 0.705$$

The diffuser efficiency is given by

$$\eta_d = \frac{\left(1 + \frac{\gamma - 1}{2}M^2\right) r_d^{\frac{\gamma-1}{\gamma}} - 1}{\frac{\gamma - 1}{2}M^2}$$

$$\Rightarrow \eta_d = \frac{\left(1 + \frac{1.4 - 1}{2} \times 0.705^2\right) \times (0.85)^{\frac{1.4-1}{1.4}} - 1}{\frac{1.4 - 1}{2} \times 0.705^2}$$

$$\Rightarrow \eta_d = 0.498 \approx 0.5$$

45. Hoop Stress, $\sigma_\theta = \frac{pd}{2h} = \frac{2\times1}{2h} = \frac{1}{h} MPa$

Longitudinal Stress,

$$\sigma_l = \frac{pd}{4h} = \frac{2\times1}{4h} = \frac{0.5}{h} MPa$$

According to Maximum Shear Stress Theory

$$\frac{\sigma_\theta - \sigma_l}{2} = \frac{\sigma}{2}$$

$$\Rightarrow \frac{1-0.5}{2h} = \frac{200}{6}$$

$$\Rightarrow \frac{0.5}{2h} = \frac{200}{6}$$

$$\Rightarrow h = 0.0075$$

46.

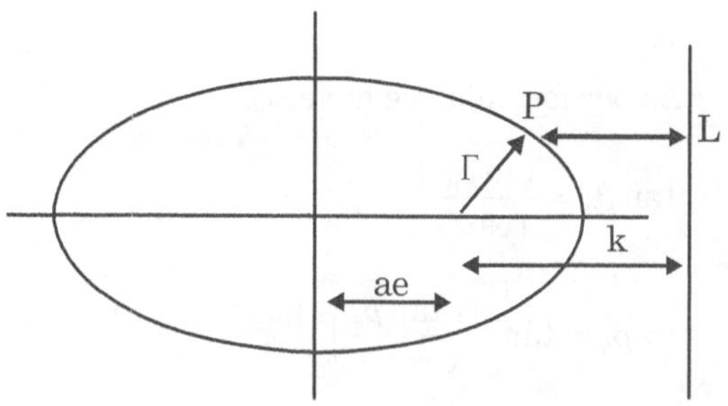

The distance of point P from focii on ellipse is e times the distance from latus rectum.

$$|FP| = e|PL|$$

$$r = e(k - r\cos\theta)$$

$$r + re\cos\theta = ek$$

$$r = \frac{ek}{1+e\cos\theta}$$

47. Given,

Total strain energy per unit volume = 14.985 KN-m/m^3

Now,

Strain energy per unit volume under uni axial loading is

$$= \frac{\sigma^2}{2E} = \frac{150^2\times10^{12}}{2\times200\times10^9} = 56.25\ KN-m/m^3$$

Factor of safety

$$= \frac{56.25}{14.985} = 3.75$$

48. At stagnation point, $u_r = 0, u_\theta = 0$

$$u_\theta = -U\left(1+\frac{a^2}{r^2}\right)\sin\theta - \frac{\Gamma}{2\pi r}$$

$$\Rightarrow 0 = -U\left(1+\frac{a^2}{r^2}\right)\sin\theta - \frac{\Gamma}{2\pi r}$$

$$\Rightarrow 0 = -25(1+1)\sin\theta - \frac{100\pi}{2\pi\times1}$$

$$\Rightarrow 0 = -50\sin\theta - 50$$

$$\Rightarrow \sin\theta = -1$$

$$\Rightarrow \theta = \sin^{-1}(-1)$$

$$\Rightarrow \theta = 270°$$

49. $C_p = \frac{P-P_\infty}{\frac{1}{2}\rho U_\infty^2} = 0$

$$C_p = 1 - 4\sin^2\theta$$

$$\Rightarrow 0 = 1 - 4\sin^2\theta$$

$$\Rightarrow \sin^2\theta = \frac{1}{4}$$

$$\Rightarrow \theta = \sin^{-1}\left(\pm\frac{1}{2}\right)$$

$$\Rightarrow \theta = 30°,\ 150°,\ 210°,\ 330°$$

50. Given,

$$p = 50\ bar$$

$$A_t = 40\ mm = 40\times10^{-3} m$$

$$v^* = 1850\ m/s$$

$$\frac{\dot m_f}{\dot m_{ox}} = 2$$

$$\rho_f = 600\ kg/m^3$$

As per formula of characteristic velocity,

$$\dot m = \frac{pA_t}{v*} = \frac{50\times10^5\times\left(\frac{\pi}{4}\times\left(40\times10^{-3}\right)^2\right)}{1850}$$

$$\Rightarrow \dot m = 3.396\ kg/s \text{ and}$$

$$\dot m_f = \frac{\frac{\dot m_f}{\dot m_{ox}}}{\frac{\dot m_f}{\dot m_{ox}}+1}\times \dot m$$

$$= \frac{2}{2+1}\times3.396 = 2.264\ kg/s$$

51. Given,

$p = 50\ bar$

$A_t = 40\ mm = 40 \times 10^{-3} m$

$v^* = 1850\ m/s$

$\frac{\dot{m}_f}{\dot{m}_{ox}} = 2$

$\rho_f = 600\ kg/m^3$

If the required fuel volume is V, then for a burn time of $t = 7$ min $= 7 \times 60 = 420\ s$

$$V = \frac{\dot{m}_f \times t}{\rho_f} = \frac{2.264 \times 420}{600} = 1.58\ m^3$$

52. To calculate $\frac{1}{a}$ using N-R method,

Let us consider,

$x = \frac{1}{a}$

$\Rightarrow a = \frac{1}{x}$

$\Rightarrow \frac{1}{x} - a = 0$

$\therefore f(x) = \frac{1}{x} - a = 0$

Now,

$f'(x) = -\frac{1}{x^2}$

$f(x_k) = \frac{1}{x_k} - a$

$f'(x_k) = -\frac{1}{x_k^2}$

From $N-R$ method

$$x_{k+1} = x_k - \frac{f(x_k)}{f'(x_k)}$$

$$\Rightarrow x_{k+1} = x_k - \frac{\frac{1}{x_k} - a}{-\frac{1}{x_k^2}}$$

Simplyfying we get,

$$x_{k+1} = 2x_k - ax_k^2$$

53. For $a = 7$ iteration

The equation becomes,

$x_{k+1} = 2x_k - ax_k^2$

$\Rightarrow x_{k+1} = 2x_k - 7x_k^2$

$x_0 = 0.2$

$x_1 = 2x_0 - 7x_0^2 = 2 \times 0.2 - 7(0.2)^2 = 0.12$

$x_2 = 2x_1 - 7x_1^2 = 2 \times 0.12 - 7(0.12)^2 = 0.1392$

54. Given,

$\alpha = 12°$

$\beta_e = 8°$

As per formula, we have,

$$\tan \beta_a = \frac{\tan \beta_e}{\cos \alpha}$$

$$\Rightarrow \beta_a = \tan^{-1}\left(\frac{\tan \beta_e}{\cos \alpha}\right)$$

$$\Rightarrow \beta_a = \tan^{-1}\left(\frac{0.140541}{0.978148}\right) = 0.1424 = 8.2°$$

55. From previous question we have

$\alpha = 12°$

$\beta_e = 8°$

and $\beta_a = 8.2°$

Now,

As per formula,

$$\frac{\partial C}{\partial \beta_a} = \frac{\cos \alpha}{\cos^2 \beta_a + \cos^2 \alpha \sin^2 \beta_a} \frac{\partial C}{\partial \beta_e}$$

$$\frac{\partial C}{\partial \beta_a} = \frac{0.978148}{(0.98935)^2 + (0.978148)^2 (0.142220)^2} \times$$

$0.250 = 0.245$

56. In spite of being warned repeatedly, he failed to correct his errant behavior.

57. To survive in the frozen wastes of Arctic takes special equipment.

58. As it says it meets at 64 mins,but a hour is of 60 mins so it is 4 mins slow

Therefore, 24 × 4 = 96

59. Tetrahedron has 4 corner points. Therefore, it forms 4 planes. Now, we add an internal point making a total of 5 points. Any three combination of points leads to a plane and thus we can get $^5C_3 = 10$ planes. So, newly created planes = 10 – 4 = 6.

60. Suppose 5 numbers are

$a = x - 4,$

$b = x - 2,$

$c = x,$

$d = x + 2,$

$e = x + 4$

Now,

As per given question, sum of a and d gives

$a + d = x - 4 + x + 2 = 162$

$or\ 2x = 164$

$\Rightarrow x = 82$

Threfore,

Sum of all numbers are

78 + 80 + 82 + 84 + 86 = 410

61. Let $x = 1 + \frac{4}{7} + \frac{9}{7^2} + \frac{16}{7^3} + \frac{25}{7^4}$ ...(1)

Multipling x by $\frac{1}{7}$

$$\frac{x}{7} = \frac{1}{7} + \frac{4}{7^2} + \frac{9}{7^3} + \frac{16}{7^4} \quad ...(2)$$

Substracting (1) from (2) we get,

$$\frac{6x}{7} = 1 + \frac{3}{7} + \frac{5}{7^2} + \frac{7}{7^3} + \frac{9}{7^4} \quad ...(3)$$

Multipliying (3) by $\frac{1}{7}$ we get,

$$\frac{6x}{49} = \frac{1}{7} + \frac{3}{7^2} + \frac{5}{7^3} + \frac{7}{7^4} \quad ...(4)$$

Substractomg (4) from (3) we get,

$$\frac{36x}{49} = 1 + \frac{2}{7} + \frac{2}{7^2} + \frac{2}{7^3} + \frac{2}{7^4} ...\text{infinity}$$

$$\frac{36x}{49} = 1 + \frac{2}{7}\left(\frac{1}{1 - \frac{1}{7}}\right) = 1 + \frac{1}{3} = \frac{4}{3}$$

$$x = \frac{4}{3} \times \frac{49}{36} = \frac{49}{27}$$

62. $15\%\ of\ X = \frac{15X}{100}$

$20\%\ of\ Y = \frac{20Y}{100}$

Therefore,

$\frac{15X}{100} = \frac{20Y}{100}$

$\Rightarrow 15X = 20Y$

$\Rightarrow \frac{X}{Y} = \frac{20}{15} = \frac{4}{3}$

63. Let,

Their marks be $x + 9$ x

$x + 9 = \frac{56}{100}(x + 9 + x)$

$\Rightarrow 25 \times (x + 9) = 14 \times (2x + 9)$

$\Rightarrow 3x = 99$

$\Rightarrow x = 33$

Therefore,

$x + 9 = 42$

64. Correct versions of other 3:

B. The report was useless to them because there was no needed information in it.

C. Since the report did not contain the needed information, it was not really useful to them.

D. Since the report lacked needed information, it would not have been useful to them.

65. possessing all the technical skills" - iv is false

"throwing away his wicket several time after working hard to lay a strong foundation" - iii is true

"he is unlikely to be so due to lack of requisite temperament" - ii is true

"success at the highest level will continue to elude him" - i is false

So, (b) is the correct answer.

MOCK TEST – 2

AEROSPACE ENGINEERING

Time Allowed: 3:00 Hours *Maximum Marks: 100*

Read the following instructions carefully before you begin to answer the questions. This Booklet contains questions in English.

INSTRUCTIONS

1. This Booklet contains **65 questions** in total comprising of the following two parts:

   Part -1: Technical Section (85 Marks) *(1-55 Questions)*

   Part -2: General Aptitude Section (15 Marks) *(56-65 Questions)*

2. All questions are compulsory. Q.1-Q.25 & Q.56-Q.60 carry one mark each. Q.26-Q.55 & Q.61-Q.65 carry two marks each.

3. **Types of Questions:**

   (*a*) Multiple Choice Questions(MCQ) carrying 1 or 2 marks each in all papers and sections. These questions are objective in nature, and each will have a choice of four options, out of which the candidate has to mark the correct answer(s).

   (*b*) Numerical Answer Questions of 1 or 2 marks each in all sections. For these questions the answer is a real number, to be entered by the candidate using the virtual keypad. No choices will be shown for these type of questions.

4. For 1-mark multiple-choice questions, 1/3 marks will be deducted for a wrong answer. Likewise, for 2-marks multiple-choice questions, 2/3 marks will be deducted for a wrong answer. There is no negative marking for numerical answer type questions.

MOCK TEST – 2

1. The shear modulus G for given Young modulus, E = 200 GPa and v = 0.33 is
   (a) 85 GPa (b) 80 GPa
   (c) 75 GPa (d) 70 Pa

2. A body travelling in a hyperbolic orbit around a planet will have
   (a) Zero speed at infinite distance from planet
   (b) Finite speed at infinite distance from planet
   (c) Infinite speed at infinite distance from planet
   (d) Will always be at finite distance from planet

3. The value of the definite integral $\int_1^e \ln x dx$ is equal to _____.

4. NACA 6412 aerofoil indicates that
   (a) It has maximum camber of 4 % located 60 % (0.6 chords) from the leading edge.
   (b) It has maximum camber of 6 % located 40 % (0.4 chords) from the leading edge.
   (c) It has maximum camber of 1 % located 20 % (0.2 chords) from the leading edge.
   (d) It has maximum camber of 2 % located 10 % (0.1 chords) from the leading edge.

5. For a rocket engine with inlet velocity U and exhaust velocity Ue the thrust in Newton is given by the formula
   (a) $F = \dot{m}(U - U_e)$ (b) $F = \dot{m}(U_e - U)$
   (c) $F = \dot{m}U$ (d) $F = \dot{m}U_e$

6. For an isothermal region of the atmosphere. The isothermal pressure is given as
   (a) $\frac{p}{p_1} = e^{[g_0/(RT)](h-h_1)}$
   (b) $\frac{p}{p_1} = e^{-[g_0/(RT)](h-h_1)}$
   (c) $\frac{p}{p_1} = e^{-[g_0/(RT)](h_1-h)}$
   (d) $\frac{p_1}{p} = e^{-[g_0/(RT)](h-h_1)}$

7. Mack tuck is an aerodynamic effect, in which aircraft tends to
   (a) Pitch upwards as the airflow around the wing reaches subsonic speeds
   (b) Pitch downwards as the airflow around the wing reaches supersonic speeds
   (c) Pitch downwards as the airflow around the wing reaches subsonic speeds
   (d) Pitch upwards as the airflow around the wing reaches supersonic speeds

8. Which of the following is true for centrifugal compressor?
   I. It makes use of the centripetal force developed on the fluid particles to obtain the required pressure difference for the gas.
   II. The centripetal force experienced by the gas changes, as the gas flows through the compressor.
   (a) Only I
   (b) Only II
   (c) Both I and II
   (d) Neither I nor II

9. When the Newton-Raphson method is applied to solve the equation $f(x) = x^3 + 2x - 1 = 0$, the solution at the end of the first iteration with the initial guess value as $x_0 = 1.2$ is
   (a) – 0.82 (b) 0.49
   (c) 0.705 (d) 1.69

10. In a I-shaped cross-section with D as the total height, the shear stress for a flange is given as: -
    (a) $\tau_{flange} = \frac{P}{2I}\left(\frac{D^2}{4} - y^2\right)$
    (b) $\tau_{flange} = \frac{P}{2I}\left(\frac{D^2}{4} + y^2\right)$
    (c) $\tau_{flange} = \frac{P}{I}\left[1 + \frac{1}{2}\left(\frac{D^2}{4} - y^2\right)\right]$
    (d) $\tau_{flange} = \frac{P}{I}\left[1 + \frac{1}{2}\left(\frac{D^2}{4} + y^2\right)\right]$

11. A planet has twice the radius but the mean density is 1/4th as compared to earth. What is the radio of the escape velocity from the earth to that from the planet?

(*a*) 3:1 (*b*) 1:2

(*c*) 1:1 (*d*) 2:1

12. The annual precipitation data of a city is normally distributed with mean and standard deviation as 1000 mm and 200 mm, respectively. The probability that the annual precipitation will be more than 1200 mm is

(*a*) < 50 % (*b*) 50 %

(*c*) 75 % (*d*) 100 %

13. For a damped harmonic motion, the frequency of damped vibration is given by

(*a*) $\omega_d = \left(\sqrt{1-\zeta}\right)\omega_n$ (*b*) $\omega_d = \left(\sqrt{1-\zeta^2}\right)\omega_n$

(*c*) $\omega_d = \left(\sqrt{1+\zeta^2}\right)\omega_n$ (*d*) $\omega_d = \left(1-\zeta^2\right)\omega_n$

14. The potential function for the corner flow around two surfaces inclined with an angle of 90° and 180°, where Az = 1 is ____________.

15. The ideal static pressure coefficient of a diffuser with an area ratio of 1.5 is:

(*a*) 0.25 (*b*) 0.35

(*c*) 0.45 (*d*) 0.55

16. A solution of the following differential equation is given by $\frac{d^2y}{dx^2} - 5\frac{dy}{dx} + 6y = 0$

(*a*) $y = e^{2x} + e^{-3x}$ (*b*) $y = e^{2x} + e^{3x}$

(*c*) $y = e^{-2x} + e^{3x}$ (*d*) $y = e^{-2x} + e^{-3x}$

17. What is the speed of sound for isothermal flow?

(*a*) $a = \sqrt{RT^2}$ (*b*) $a = \sqrt{RT}$

(*c*) $a = RT$ (*d*) $a = \sqrt{\gamma RT}$

18. The normal velocity of a rocket flying at air speed of 1000 m/s at an angle of attack 300 with the experimental sideslip angle, β_e of 200 is given as

(*a*) 300.63 m/s (*b*) 342.02 m/s

(*c*) 469.85 m/s (*d*) 813.80 m/s

19. For the Euler's Theory which of the following assumption is NOT valid.

(*a*) The cross-section of the column is uniform and axial strain should be assumed negligible.

(*b*) The weight of column is negligible and ends are frictionless

(*c*) The material of the column is homogenous, linear and isotropic.

(*d*) The column and load may not be exactly axial.

20. For a given Mass Ratio, $MR = \frac{m_2}{m_1}$, the propellant Mass Ratio of a rocket engine is given as (where m_1 = total mass of the rocket at takeoff and m_2 = final mass of the rocket after burnout)

(*a*) $\zeta = 1 - MR$

(*b*) $\zeta = 1 + MR$

(*c*) $\zeta = MR$

(*d*) $\zeta = 1$

21. What is the value of M_2 after normal shock if air is flowing with Mach number $M_1 = 7$.

(*a*) 0.333 (*b*) 0.189

(*c*) 0.157 (*d*) 0.1502

22. If the Earth stops rotating (Neglecting the effect of the revolution), then gravity of pole and equation will

(*a*) Increase (*b*) Decrease

(*c*) Remain Same (*d*) Changes irratically

23. Match the following regimes

a. Incompressible	I. $M < 0.3$
b. Compressible (subsonic)	II. $0.7 < M < 1.3$
c. Transonic	III. $M > 4$
d. Supersonic	IV. $0.3 < M < 0.9$
e. Hypersonic	V. $1.3 < M < 4.0$

(*a*) a - 1, b - 2, c - 3, d - 4, e - 5

(*b*) a - 1, b - 4, c - 2, d - 5, e - 3

(*c*) a - 1, b - 4, c - 5, d - 2, e - 3

(*d*) a - 1, b - 3, c - 4, d - 2, e - 5

24. The equation $\begin{bmatrix} 2 & -2 \\ 1 & -1 \end{bmatrix}\begin{bmatrix} x_1 \\ x_2 \end{bmatrix} = \begin{bmatrix} 0 \\ 0 \end{bmatrix}$ has

(*a*) No solution

(*b*) Only one solution $\begin{bmatrix} x_1 \\ x_2 \end{bmatrix} = \begin{bmatrix} 0 \\ 0 \end{bmatrix}$

(*c*) Non-zero unique solution

(*d*) Multiple solutions

25. Which of the following is true for shear center?

(*a*) At shear center, the twisting will not produce bending and bending will not produce twisting.

(*b*) At shear center, the twisting will produce bending and bending will produce twisting.

(*c*) At shear center, the twisting will not produce bending but bending will produce twisting.

(*d*) At shear center, the twisting will produce bending but bending will not produce twisting.

26. For a uniform flow inclined at an angle α the *u* and *v* component are given as

(*a*) $u = U \sin\alpha$ and $v = U \cos\alpha$

(*b*) $u = U \cos\alpha$ and $v = U \sin\alpha$

(*c*) $u = U$ and $v = U \sin\alpha$

(*d*) $u = U \sin\alpha$ and $v = U$

27. There are two perpendicular plane, where stresses are acting as:

$\sigma_{xx} = -90 MPa$ and $\sigma_{yy} = 150\ MPa$ also given $\sigma_1 = 100\ MPa$ and $\sigma_2 = -140\ MPa.$ the principle stresses are The shear stress, τ acting on the planes is__________.

28. A box has ten identical sized balls. Seven of these balls are black in color and the rest three are red. Three balls are picked from the box one after another without replacement. The probability that two of the balls are black and one is red is equal to ______.

29. An aircraft is performing a coordinated turn manoeuvre at a bank angle of 30° and forward speed of 1000 m/s. Assume $g = 9.8$ m/s^2. The load factor and turn radius respectively are -

(*a*) $\frac{2}{\sqrt{3}}$ and 1.76 km (*b*) $\sqrt{3}$ and 17.6 km

(*c*) 2 and 0.18 km (*d*) $\frac{2}{\sqrt{3}}$ and 17.6 km

30. A combat aircraft engine is equipped with an afterburner followed by a variable area convergent nozzle (operating with the nozzle chocked). The exhaust gas temperature is 620 K when afterburner is off and 5589 K, when it is on. When afterburner is turned on, (assuming the total pressure will remain same, and the specific heat ratio (γ) remains constant), approximately by what factor must the nozzle area be changed?

(*a*) 0.5 (*b*) 1

(*c*) 2 (*d*) 3

31. A satellite is revolving around earth is 400 km x 1200 km elliptical orbit. What is velocity of satellite at its apoapsis and periapsis?

(*a*) 7.87 km/s and 6.82 km/s

(*b*) 7.04 km/s and 7.87 km/s

(*c*) 7.04 km/s and 6.82 km/s

(*d*) 7.56 km/s and 7.04 km/s

32. The Laplace transform of a function *f* (*t*) is

$\dfrac{2}{\dfrac{1}{s^2(s+1)}}$. The function *f* (*t*) is

(*a*) $t - 1 + e^{-1}$ (*b*) $t + 1 + e^{-1}$

(*c*) $1 + e^{-1}$ (*d*) $2t + e^t$

33. What is the position of aerodynamic center for a general aerofoil.

(*a*) 0.5 c (*b*) 0.4 c

(*c*) 0.25 c (*d*) 0.125 c

34. A rectangle beam 8 cm × 3 cm is 1m long. It carries a load 2 *KN* at mid-span. Determine the maximum bending stress induced in the beam provided that the beam is simply supported.

(*a*) 41.66 MPa (*b*) 66.42 MPa

(*c*) 33.14 MPa (*d*) 89.56 MPa

35. Two concentric cylinders of infinite length is having the radius of $r_1 = 2m,\ and\ r_2 = 5m.$ The incompressible fluid is filled in the annulus region of these cylinders. Calculate the induced fully developed velocity profile of the fluid when the fluid starts rotating due to the shear forces produced by the cylindrical wall of inner and outer cylinder. The tangential velocity of inner and outer cylinder is respectively?

$u_1 = 0\ m/s, u_2 = 5\ m/s$

(*a*) $u_\theta = -1.19\left(r + \frac{4}{r}\right)$

(*b*) $u_\theta = 1.19\left(r - \frac{4}{r}\right)$

(*c*) $u_\theta = -1.38\left(r - \frac{6.89}{r}\right)$

(*d*) $u_\theta = 1.38\left(r + \frac{6.89}{r}\right)$

36. At 10 km a jet engine is operating at a Mach number of 0.7. The efficiency of the air intake is 0.8 and that of the compressor is 0.87. The total temperature (in *K*) at the exits of the air intake and the compressor respectively are

(Ambient pressure = 26.5 kPa; Ambient temperature =250 K; Gas Constant, $\gamma = 14$; $p_{rc} = 7$)

(*a*) 274.5 and 509.13

(*b*) 234.9 and 486.8

(*c*) 274.5 and 486.8

(*d*) 234.9 and 509.13

37. A cylindrical shell made of mild steel plate having diameter 100 cm is to be subjected to an internal pressure of 2 MPa. If the yield strength of the material is 200 MPa, Calculate the thickness of the plate on the basis of the Maximum Principal Stress Theory, assuming factor of safety to be 3.

(*a*) 0.015 m

(*b*) 0.0075 m

(*c*) 0.01299 m

(*d*) 0.1169 m

38. If the velocity profile given in boundary layer over the flat plate as

$$u = U \sin\left(\frac{ay}{\delta}\right)$$

Calculate possible values of a

(*a*) π (*b*) $\frac{\pi}{2}$

(*c*) $\frac{3\pi}{2}$ (*d*) 2π

39. The sideslip velocity of a rocket flying at air speed of 1000 m/s at an angle of attack 300 with the analytical sideslip angle β_a and the experimental sideslip angle, β_e of 20° is given as

(*a*) 330.63 m/s and 342.02 m/s

(*b*) 476.87 m/s and 469.85 m/s

(*c*) 825.96 m/s and 813.80 m/s

(*d*) 926.56 m/s and 914.74 m/s

40. Let X be a random variable following normal distribution with mean +1 and variance 4. Let Y be another normal variable with mean -1 and variance unknown. If $P(X \le -1) = P(Y \le 2)$ the standard deviation of *Y* is

(*a*) 3 (*b*) 2

(*c*) $\sqrt{2}$ (*d*) 1

41. Calculate equivalent value of spring constant (k) for the following figure:

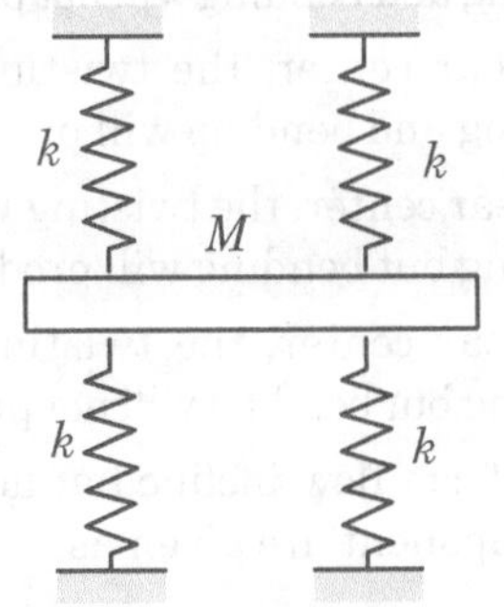

(*a*) k (*b*) $2k$

(*c*) $4k$ (*d*) $\frac{k}{2}$

42. The ratio of two planet's *A* and *B* radii and the ratio of two planet's gravity are $k_1 = \frac{R_A}{R_B}$ and $k_2 = \frac{g_A}{g_B}$. Then the ratio of their escape velocity from the surface of their respective planets will be -

(*a*) $k_1 \times k_2$ (*b*) $\sqrt{\frac{k_1}{k_2}}$

(*c*) $\sqrt{k_1 \times k_2}$ (*d*) $\sqrt{\frac{k_2}{k_1}}$

43. In an external laminar boundary layer flow over large flat plate

$\delta = 5\ mm, x_{LE} = 0.2\ m$. Find δ at $x_{LE} = 3.2\ m$.

44. A certain airport receives on an average of 4 aircrafts per hour. What is the probability that no aircraft lands in a particular 2 hr.period?

(*a*) 8 (*b*) e^{-8}

(*c*) e^{8} (*d*) e^{-7}

45. A total temperature found to have $K = 0.9$ over a wide range of operating conditions. A pitot tube inserted into a supersonic flow indicates a pressure of 250 *KPa*, a static probe indicates 50 *KPa*. If the indicated temperature of the total temperature probe is 650 *K* when inserted into the same supersonic flow. What is the free stream temperature (in *K*)?

(*a*) 430 *K* (*b*) 399 *K*

(*c*) 230 *K* (*d*) 688 *K*

46. The stagnation pressure and stagnation temperature inside the combustion chamber of a liquid rocket engine are 1.5 MPa and 2500 K respectively. The burned gases have $\gamma = 1.2$ and $R = 692.83\ J/kg-K$. The flow at the exit of the nozzle is isentropic. The mass flow rate of the gases out of the nozzle is 18 kg/s. The throat area of the converging diverging nozzle is ________.

47. Consider the following system of equations in three real variables x_1, x_2 and x_3

$$2x_1 - x_2 + 3x_3 = 1$$

$$3x_1 - 2x_2 + 5x_3 = 2$$

$$-x_1 - 4x_2 + x_3 = 3$$

This system of equations has

(*a*) No solution

(*b*) A unique solution

(*c*) More than one but a finite number of solutions

(*d*) An infinite number of solutions

48. For a harmonic motion $X_1 = 10\cos\omega t$ and $X_2 = 15\cos(\omega t + 2)$. The value of magnitude of resultant vector A is given as

(*a*) 11.14 (*b*) 12.14

(*c*) 13.14 (*d*) 14.14

49. For a harmonic motion $X_1 = 10\cos\omega t$ and $X_2 = 15\cos(\omega t + 2)$.The value of angle is given as

(*a*) 74.58° (*b*) 64.44°

(*c*) 58.34° (*d*) 53.78°

50. An aircraft with the wing aspect ratio $AR_w = 15$ has a zero lift drag coefficient $C_{D_0} = 0.0322$. The Oswald's efficiency factor is $e = 0.5$. What is the lift coefficient of aircraft when the thrust required for steady flight is minimum?

(*a*) 0.66 (*b*) 0.72

(*c*) 0.77 (*d*) 0.87

51. An aircraft with the wing aspect ratio $AR_w = 15$ has a zero lift drag coefficient $C_{D_0} = 0.0322$. The Oswald's efficiency factor is $e = 0.5$. What is the glide angle that results in maximum range in a power-off glide?

(*a*) 2.578° (*b*) 3.646°

(*c*) 4.233° (*d*) 5.646°

52. The particle coming out at $t = 0$ from a point at (1,1) What is the provided streamline equation at the point?

(*a*) $y = x^t$

(*b*) $y = x$

(*c*) $2\ln y = (\ln x)^2$

(*d*) $y = e^x$

53. The particle coming out at $t = 0$ from a point at (1,1)

What is the pathline equation at the given point?

(*a*) $y = x^t$ (*b*) $y = x$

(*c*) $2\ln y = (\ln x)^2$ (*d*) $y = e^x$

54. A satellite is launched into Earth orbit where its launch vehicle burns out at an altitude of 250 km. At burnout, the satellite's velocity is 7950 m/s with zenith angle equal to 89 degrees. Calculate the satellite's altitude at perigee and apogee.

(*a*) 239 km and 971 km

(*b*) 12995 km and 13728 km

(*c*) 250 km and 1008 km

(*d*) 115 km and 654 km

55. Calculate the eccentricity of the orbit for the satellite is launched into Earth orbit where its launch vehicle burns out at an altitude of 250 km. At burnout, the satellite's velocity is 7950 m/s with zenith angle equal to 89 degrees.

(*a*) 0.03

(*b*) 0.04

(*c*) 0.05

(*d*) 0.06

56. The ways in which this game can be played__________ potentially infinite.

(*a*) is

(*b*) is being

(*c*) are

(*d*) are being

57. A person suffering from Alzheimer's disease short-term memory loss.

(*a*) experienced

(*b*) has experienced

(*c*) is experiencing

(*d*) experiences

58. If $xy = 9$, $yz = 16$ and $xz = 25$, then what is the value of $(x + y + z)$?

(*a*) 12 (*b*) 12.4

(*c*) 12.8 (*d*) 13

59. A deck of 52 playing cards is shuffled, and the cards are turned up one at a time until the first ace appears. Is the next card-that is, the card following the first ace-more likely to be the ace of spades or the two of clubs?

60. Manoj takes twice as much time as Ajay and thrice as much as Vijay to finish a piece of work. Together they finish the work in 1 day. What is the time taken by Manoj to finish the work?

(*a*) 6 days (*b*) 3 days

(*c*) 2 days (*d*) none

61. A firm is selling its product at Rs. 60 per unit. The total cost of production is Rs. 100 and firm is earning a total profit of Rs. 500. Later, the total cost increased by 30%. By what percentage should the price be increased to maintain the same profit level?

(*a*) 5 (*b*) 10

(*c*) 15 (*d*) 30

62. A container contain spirit to water ratio of 2:3 and other container contain 4:7 spirit to water ratio. If both mixtures are combined together in a large container what will be the new spirit to water ratio.

(*a*) $\frac{21}{49}$ (*b*) $\frac{21}{34}$

(*c*) $\frac{5}{6}$ (*d*) $\frac{60}{61}$

63. Average age of class of 39 students is 15 years. If the age of teacher is included the average is increased by 3 months. Age of the teachers is _____.

64. A smart city integrates all modes of transport, uses clean energy and promotes sustainable use of resources. It also uses technology to ensure safety and security of the city, something which critics argue, will lead to a surveillance state.

Which of the following can be logically inferred from the above paragraph?

i. All smart cities encourage the formation of surveillance states.

ii. Surveillance is an integral part of a smart city.

iii. Sustainability and surveillance go hand in hand in a smart city.

iv. There is a perception that smart cities promote surveillance.

(*a*) (i) and (iv) only (*b*) (ii) and (iii) only

(*c*) (iv) only (*d*) (i) only

65. Let T be a set of integers {3, 11, 19, 27, ... 451, 459, 467} and *S* be a subset of T such that the sum of no two elements of *S* is 470. The maximum number of elements in *S* is?

(*a*) 32 (*b*) 28

(*c*) 29 (*d*) 30

OMR SHEET

1. Ⓐ Ⓑ Ⓒ Ⓓ
2. Ⓐ Ⓑ Ⓒ Ⓓ
3. ________
4. Ⓐ Ⓑ Ⓒ Ⓓ
5. Ⓐ Ⓑ Ⓒ Ⓓ
6. Ⓐ Ⓑ Ⓒ Ⓓ
7. Ⓐ Ⓑ Ⓒ Ⓓ
8. Ⓐ Ⓑ Ⓒ Ⓓ
9. Ⓐ Ⓑ Ⓒ Ⓓ
10. Ⓐ Ⓑ Ⓒ Ⓓ
11. Ⓐ Ⓑ Ⓒ Ⓓ
12. Ⓐ Ⓑ Ⓒ Ⓓ
13. Ⓐ Ⓑ Ⓒ Ⓓ
14. ________
15. Ⓐ Ⓑ Ⓒ Ⓓ
16. Ⓐ Ⓑ Ⓒ Ⓓ
17. Ⓐ Ⓑ Ⓒ Ⓓ
18. Ⓐ Ⓑ Ⓒ Ⓓ
19. Ⓐ Ⓑ Ⓒ Ⓓ
20. Ⓐ Ⓑ Ⓒ Ⓓ
21. Ⓐ Ⓑ Ⓒ Ⓓ
22. Ⓐ Ⓑ Ⓒ Ⓓ
23. Ⓐ Ⓑ Ⓒ Ⓓ
24. Ⓐ Ⓑ Ⓒ Ⓓ
25. Ⓐ Ⓑ Ⓒ Ⓓ
26. Ⓐ Ⓑ Ⓒ Ⓓ
27. ________
28. ________
29. Ⓐ Ⓑ Ⓒ Ⓓ
30. Ⓐ Ⓑ Ⓒ Ⓓ
31. Ⓐ Ⓑ Ⓒ Ⓓ
32. Ⓐ Ⓑ Ⓒ Ⓓ
33. Ⓐ Ⓑ Ⓒ Ⓓ
34. Ⓐ Ⓑ Ⓒ Ⓓ
35. Ⓐ Ⓑ Ⓒ Ⓓ
36. Ⓐ Ⓑ Ⓒ Ⓓ
37. Ⓐ Ⓑ Ⓒ Ⓓ
38. Ⓐ Ⓑ Ⓒ Ⓓ
39. Ⓐ Ⓑ Ⓒ Ⓓ
40. Ⓐ Ⓑ Ⓒ Ⓓ
41. Ⓐ Ⓑ Ⓒ Ⓓ
42. Ⓐ Ⓑ Ⓒ Ⓓ
43. ________
44. Ⓐ Ⓑ Ⓒ Ⓓ
45. Ⓐ Ⓑ Ⓒ Ⓓ
46. ________
47. Ⓐ Ⓑ Ⓒ Ⓓ
48. Ⓐ Ⓑ Ⓒ Ⓓ
49. Ⓐ Ⓑ Ⓒ Ⓓ
50. Ⓐ Ⓑ Ⓒ Ⓓ
51. Ⓐ Ⓑ Ⓒ Ⓓ
52. Ⓐ Ⓑ Ⓒ Ⓓ
53. Ⓐ Ⓑ Ⓒ Ⓓ
54. Ⓐ Ⓑ Ⓒ Ⓓ
55. Ⓐ Ⓑ Ⓒ Ⓓ
56. Ⓐ Ⓑ Ⓒ Ⓓ
57. Ⓐ Ⓑ Ⓒ Ⓓ
58. Ⓐ Ⓑ Ⓒ Ⓓ
59. ________
60. Ⓐ Ⓑ Ⓒ Ⓓ
61. Ⓐ Ⓑ Ⓒ Ⓓ
62. Ⓐ Ⓑ Ⓒ Ⓓ
63. ________
64. Ⓐ Ⓑ Ⓒ Ⓓ
65. Ⓐ Ⓑ Ⓒ Ⓓ

ANSWERS

1. (c)	2. (b)	3. (1)	4. (b)	5. (d)	6. (b)	7. (b)	8. (c)	9. (c)	10. (a)
11. (c)	12. (a)	13. (b)	14. (1)	15. (d)	16. (b)	17. (b)	18. (c)	19. (d)	20. (a)
21. (c)	22. (a)	23. (c)	24. (d)	25. (a)	26. (b)	27. (50)	28. (0.525)	29. (a)	30. (d)
31. (b)	32. (a)	33. (c)	34. (a)	35. (b)	36. (a)	37. (a)	38. (b)	39. (a)	40. (a)
41. (c)	42. (c)	43. (20)	44. (b)	45. (b)	46. (0.024)	47. (b)	48. (d)	49. (a)	50. (d)
51. (c)	52. (a)	53. (c)	54. (a)	55. (c)	56. (c)	57. (d)	58. (c)	59. (1/52)	60. (a)
61. (d)	62. (b)	63. (25)	64. (c)	65. (c)					

EXPLANATIONS

1. Given,

$$E = 200\ GPa = 200 \times 10^9\ Pa$$

$$\nu = 0.33$$

$$G = \frac{E}{2(1+\nu)}$$

$$\Rightarrow G = \frac{200 \times 10^9}{2 \times 1.33}$$

$$\Rightarrow G = 75.01\ GPa$$

2. A hyperbolic orbit has $l = 1$, and its total energy is positive. Hence at infinity, when potential energy is zero Kinetic Energy is positive.

$$P.E = -\frac{GMm}{r} = -\frac{GMm}{p}(1 + e\cos\theta)$$

$$K.E = \frac{GMm}{2p}\left(1 + 2e\cos\theta + e^2\right)$$

$$P.E + K.E = \frac{GMm}{p}\left(-1 - e\cos\theta + \frac{1+e^2}{2} + e\cos\theta\right)$$

$$\Rightarrow P.E + K.E = \frac{GMm}{2p}\left(e^2 - 1\right)$$

3. $\int_1^e \ln x dx$

$$\int_1^e \ln x dx = [x\ln x - x]_1^e$$

$$\Rightarrow [e\ln e - e] - [\ln 1 - 1]$$

$$\Rightarrow [e - e] - [0 - 1] = 0 + 1 = 1$$

4. In *NACA* 6412, here 6 signifies the maximum camber, 4 signifies the location of maximum camber in tenths of chord i.e. maximum 6 % camber locates at 40 % from the leading edge.

5. Considering the engine now, using Newton's second law

$$F = \dot{m}(U_e - U) + (P_i - P_e)A$$

In rockets, U and U_e is very high

Also $P_i = 0.1\ atm$

$$\therefore \quad \dot{m}(U_e - U) \gg (P_i - P_e)A$$

thus, $$F = \dot{m}(U_e - U)$$

For rockets,

$$F = \dot{m}U_e$$

6. Let us consider an isothermal region

Now using hydrostatic equation

$$dp = -\rho g_0 dh$$

Divide the above equation by equation of state, we get

$$\frac{dp}{p} = \frac{-\rho g_0 dh}{\rho RT} = \frac{-g_0}{RT}dh$$

Integrating above equation

$$\int_{p_1}^{p} \frac{dp}{p} = \frac{-g_0}{RT}\int_{h_1}^{h} dh$$

$$\Rightarrow \ln\frac{p}{p_1} = \frac{-g_0}{RT}(h - h_1)$$

$$\frac{p}{p_1} = e^{-[g_0/(RT)](h-h_1)}$$

where p_1 is the pressure at altitude of isothermal region.

7. Mach tuck is an aerodynamic effect whereby the nose of an aircraft tends to pitch downward as the airflow around the wing reaches supersonic speeds; the aircraft will first experience this effect at significantly below Mach 1. This speed is known as the critical Mach number of the wing.

8. For centrifugal compressor both are true

I. It makes use of the centripetal force developed on the fluid particles to obtain the required pressure difference for the gas.

II. The centripetal force experienced by the gas changes, as the gas flows through the compressor.

9. $f(x) = x^3 + 2x - 1 = 0$

$$f'(x) = 3x^2 + 2$$

$$f'(x_0) = 3(1.2)^2 + 2 = 6.32$$

$$f(x_0) = (1.2)^3 + 2 \times 1.2 - 1 = 3.128$$

$$f'(x_1) = x_0 - \frac{f(x_0)}{f(x_1)} = 1.2 - \frac{3.728}{6.32} = 0.705$$

10. As per formula

$$\tau = \frac{Pa\bar{y}}{Ib}$$

Here, $y > \frac{D}{2}$

For flange

$$\tau_{flange} = \frac{P\left[\left(\frac{D}{2} - y\right) \times b \times \left(y + \frac{1}{2}\left(\frac{D}{2} - y\right)\right)\right]}{I \times b}$$

$$\Rightarrow \tau_{flange} = \frac{P}{2I} \times \left(\frac{D}{2} - y\right)\left(2y + \left(\frac{D}{2} - y\right)\right)$$

$$\Rightarrow \tau_{flange} = \frac{P}{2I}\left(\frac{D^2}{4} - y^2\right)$$

11. Given,

Radius of the planet = $2R$

Mean density of planet = $\frac{1}{4}\rho$

As per formula,

$$M = \rho \times \frac{4}{3}\pi R^3$$

$$\frac{V_{e_{Planet}}}{V_{e_{earth}}} = \sqrt{\frac{2GM_{planet}}{R_{Planet}}} \times \sqrt{\frac{R_{earth}}{2GM_{earth}}}$$

$$= \sqrt{\frac{2 \times G \times \rho \times 4 \times \pi \times (2R)^3}{4 \times 3 \times 2R}} \times$$

$$\sqrt{\frac{R \times 3}{2 \times G \times \rho \times 4 \times \pi \times R^3}}$$

$$= \sqrt{\frac{8}{8}} \times 1 = 1$$

12. The annual precipitation is normally distributed with $\mu = 1000\ mm\ and\ \sigma = 200\ mm$

$$p(x > 1200) = p\left(z > \frac{1200 - 1000}{200}\right)$$

$$= p(z > 1)$$

where z is the standard normal variate.

In normal distribution

Now, since $p(-1 < z < 1) \approx 0.68$

(= 68 % of data is within one standard deviation of the mean)

$$p(0 < z < 1) = \frac{0.68}{2} = 0.34$$

$$\therefore p(z > 1) = 0.5 - 0.34$$

$$= 0.16 \simeq 16\%$$

Therefore, the choice is clearly < 50 %.

13. For an underdamped system

$$\left(\zeta < 1\ or\ \frac{c}{2m} < \sqrt{\frac{k}{m}}\right)$$

where ζ is the damping ratio and c is the damping constant.

For this condition

$(\zeta^2 - 1)$ is negative and roots are s_1 and s_2 which can be expressed as

$$s_1 = \left(-\zeta + i\sqrt{1 - \zeta^2}\right)\omega_n$$

$$s_2 = \left(-\zeta - i\sqrt{1 - \zeta^2}\right)\omega_n$$

The solution for the equation is given as

$$x(t) = e^{-\zeta\omega_n t}$$

$$x(t) = e^{-\zeta\omega_n t}\begin{Bmatrix}(C_1 + C_2)\cos\left(\sqrt{1-\zeta^2}\right)\omega_n t + i \\ (C_1 - C_2)\sin\left(\sqrt{1-\zeta^2}\right)\omega_n t\end{Bmatrix}$$

The motion described in the above equation is a damped harmonic motion which is having an angular frequency as $\left(\sqrt{1-\zeta^2}\right)\omega_n$ which is the frequency of damped vibration.

14. If $\theta = \frac{\pi}{2}$

as $n = \frac{\pi}{\theta}$

$\Rightarrow \omega = Az^2$

and

If $\theta = \pi$

as $n = \frac{\pi}{\theta}$

$\Rightarrow \omega = Az$

15. The ideal static pressure rise coefficient for a diffuser is given as

For Area Ratio $(AR) = 1.5$

$$c_p = 1 - \frac{1}{AR^2}$$

$$\Rightarrow c_p = 1 - \frac{1}{(1.5)^2}$$

$$\Rightarrow c_p = 0.55$$

16. $A.E. \Rightarrow D^2 - 5D + 6 = 0$

$(D-2)(D-3) = 0$

$D = 2, 3$

$\therefore y = e^{2x} + e^{3x}$

17. For isothermal flow (T = constant)

$p = \rho RT$

where RT = constant

$dp = d\rho RT$

$$\frac{dp}{d\rho} = RT$$

$$\Rightarrow a = \sqrt{RT}$$

18. Given,

$\beta_e = 20°$

$\alpha = 30°$

$V = 1000\ m/s$

As per formula

$V_{sideslip} = V \sin\beta_e$

$\Rightarrow V_{sideslip} = 1000 \times \sin(20)$

$\Rightarrow V_{sideslip} = 1000 \times 0.342020$

$\Rightarrow V_{sideslip} = 342.02\ m/s$

19. Assumptions for Euler's theory

- The material of the column is homogenous, linear and isotropic.
- The weight of column is negligible and ends are frictionless.
- The cross-section of the column is uniform and axial strain should be assumed negligible.
- The initial condition on the column is that, it should be straight and the applied load should be axial.

20. Given,

$$MR = \frac{m_2}{m_1}$$

$$\zeta = \frac{\text{propellant mass at takeoff}}{\text{total mass of the rocket at takeoff}}$$

$$\Rightarrow \zeta = \frac{m_p}{m_1}$$

$$\Rightarrow \zeta = \frac{m_1 - m_2}{m_1} = 1 - \frac{m_2}{m_1}$$

$$\Rightarrow \zeta = 1 - MR$$

21. Using formula,

$$M_2^2 = \frac{2 + (\gamma - 1)M_1^2}{2\gamma M_1^2 - (\gamma - 1)}$$

For, $M_1 = 7$

and $\gamma = 1.4$ for air

$$M_2^2 = \frac{2 + (1.4 - 1) \times (7)^2}{2 \times 1.4 \times (7)^2 - (1.4 - 1)}$$

$$M_2^2 = \frac{2 + 0.4 \times 49}{2 \times 1.4 \times 49 - 0.4} = \frac{21.6}{136.8} = 0.157$$

22. For $v = 0, g_p$ will remain same but ge will increase as

$$g_e = g - R\omega^2$$

$$\Rightarrow mg = \frac{GM}{r^2} - \frac{mv^2}{r}$$

As, $v = 0$ therefore, $mg = \frac{GM}{r^2}$, thus g has increased.

23. Incompressible - $M < 0.3$
Compressible (subsonic) - $0.3 < M < 0.9$
Transonic - $0.7 < M < 1.3$
Supersonic - $1.3 < M < 4.0$
Hypersonic - $M > 4$

24. $$\begin{bmatrix} 2 & -2 \\ 1 & -1 \end{bmatrix}\begin{bmatrix} x_1 \\ x_2 \end{bmatrix} = \begin{bmatrix} 0 \\ 0 \end{bmatrix}$$

$$2x_1 - 2x_2 = 0$$

$$x_1 - x_2 = 0$$

$$\Rightarrow x_1 = x_2$$

25. At shear center, the twisting will not produce bending and bending will not produce twisting.

26. Consider a Uniform flow inclined at an angle α

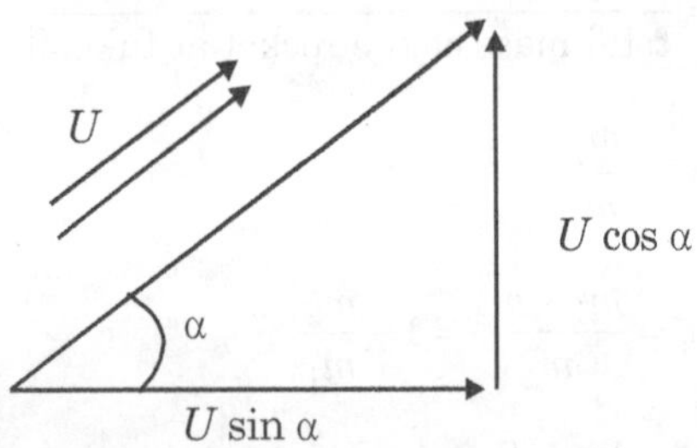

For uniform flow at an angle, complex potential function can be written as

$$\omega = \phi + i\psi = Uze^{-i\alpha}$$

$$\Rightarrow \omega = U(x + iy)(\cos\alpha - i\sin\alpha)$$

On Comparing

$$\phi = Ux\cos\alpha + Uy\sin\alpha$$

$$\psi = Uy\cos\alpha - Ux\sin\alpha$$

Also,

$$\frac{d\omega}{dz} = u - iv = Ue^{-i\alpha}$$

$$\Rightarrow u - iv = U(\cos\alpha - i\sin\alpha)$$

$$\therefore u = U\cos\alpha$$

and $v = U\sin\alpha$

27. Given,

$$\sigma_{xx} = -90MPa \text{ and } \sigma_{yy} = 150\ MPa$$

$$\sigma_1 = 100\ MPa \text{ and } \sigma_2 = -140\ MPa.$$

$$\left[\begin{bmatrix} 150 & \tau \\ \tau & -90 \end{bmatrix} - \begin{bmatrix} 100 & 0 \\ 0 & -140 \end{bmatrix}\right]\begin{bmatrix} n_x \\ n_y \end{bmatrix} = 0$$

$$\Rightarrow \begin{bmatrix} 50 & \tau \\ \tau & 50 \end{bmatrix}\begin{bmatrix} n_x \\ n_y \end{bmatrix} = 0$$

Taking the magnitute

$$\begin{vmatrix} 50 & \tau \\ \tau & 50 \end{vmatrix} = 0$$

$$\Rightarrow 2500 - \tau^2 = 0$$

$$\Rightarrow \tau = 50\ MPa$$

28. Given,

Total No. of balls = 10

No. of black balls = 7

No. of red balls = 3

No. of ways three balls will be taken out of the box, $^{10}C_3 = \frac{10 \times 9 \times 8}{3 \times 2 \times 1} = 120$

No. of ways two black balls are taken out of the box, $^7C_2 = \frac{7 \times 6}{2 \times 1} = 21$

No. of ways one red ball is taken out of the box,

$$= {}^3C_1 = \frac{3}{1} = 3$$

P (two balls are black and one is red)

$$= \frac{^7C_2 \times {}^3C_1}{^{10}C_3} = \frac{21 \times 3}{120} = \frac{21}{40} = 0.525$$

29. In a coordinated, turn manoeuvre,

$$L\cos\phi = w$$

$$\Rightarrow \text{Load factor} = \frac{L}{W}$$

$$n = \frac{1}{\cos\phi} = \frac{1}{\cos 30} = \frac{2}{\sqrt{3}}$$

Component of lift will provide necessary force of turn manoeuvring

$$L\sin\phi = \frac{mv^2}{R}$$

$$\Rightarrow w\tan\phi = \frac{mv^2}{R}$$

$$\Rightarrow R = \frac{mv^2}{w\tan\phi}$$

$$\Rightarrow R = \frac{v^2}{g\tan\phi} = \frac{100^2}{9.8\times\tan 30^0}$$

$$\Rightarrow R = 1.76\ km$$

30. As per formula, we know,

$$\frac{\dot{m}}{A^*} \propto \frac{1}{T_{0\,exit}^{1/2}}$$

Therefore, for given ṁ

$$A^* \propto T_{0\,exit}^{1/2}$$

For nozzle as well,

$$A_N \propto T_{0\,exit}^{1/2}$$

Now

$$\frac{(A_N)_{AB}}{(A_N)_{without\ AB}} = \sqrt{\frac{5580}{620}} = 3$$

31. Given,

$R_a = 1200\ km$

$R_p = 400\ km$

Let, Radius of the Earth, $r = 6400\ km$

$G = 6.67\times10^{-11}$

$M = 5.98\times10^{24}$

Now, $r_a = r + R_a = 6400 + 1200$

$= 7600\ km = 7600\times10^3\ m$

$r_p = r + R_p = 6400 + 400$

$= 6800\ km = 6800\times10^3\ m$

Velocity at apoapsis V_a is given as

$$V_a = \sqrt{\frac{2GM\times r_p}{r_a\times(r_a + r_p)}}$$

$$= \sqrt{\frac{2\times6.67\times10^{-11}\times5.98\times10^{24}\times6800\times10^3}{7600\times10^3\times(6800+7600)\times10^3}}$$

$= 7049.3\ m/s = 7.04\ km/s$

$$V_p = \sqrt{\frac{2GM\times r_a}{r_p\times(r_a + r_p)}}$$

$$V_p = \sqrt{\frac{2\times6.67\times10^{-11}\times5.98\times10^{24}\times7600\times10^3}{6800\times10^3\times(6800+7600)\times10^3}}$$

$= 7868.63\ m/s = 7.86\ km/s$

32. $f(t) = L^{-1}\left[\frac{1}{s^2(s+1)}\right]$

$$\frac{1}{s^2(s+1)} = \frac{A}{s} + \frac{B}{s^2} + \frac{C}{s+1}$$

$$\frac{1}{s^2(s+1)} = \frac{A(s+1) + B(s+1) + C(s^2)}{s^2(s+1)}$$

Matching coefficient of s^2, s and

constant in numerator

we get,

$A + C = 0$

$A + B = 0$

$B = 1$

and $A = -1, B = 1, C = 1$

So, $f(t) = L^{-1}\left[\frac{-1}{s} + \frac{1}{s^2} + \frac{1}{s+1}\right]$

$= -1 + t + e^{-t}$

$= t - 1 + e^{-t}$

33. As per formula

$$C_{mLE} = C_{m_x} - \frac{x}{c}C_N = C_{m_{ac}} - \frac{X_{ac}}{c}C_N$$

$$\Rightarrow C_{m_{ac}} = C_{m_x} + \left(\frac{X_{ac}}{c} - \frac{x}{c}\right)C_N$$

But, we know that aerodynamic center is independent of angle of attack. Therefore,

$$\frac{\partial C_{m_{ac}}}{\partial \alpha} = 0$$

$$\frac{\partial C_{m_{ac}}}{\partial \alpha} = \frac{\partial C_{m_x}}{\partial \alpha} + \left(\frac{X_{ac}}{c} - \frac{x}{c}\right)\frac{\partial C_N}{\partial \alpha} = 0$$

Thus,

$$\frac{X_{ac}}{c} = \frac{x}{c} - \frac{\frac{\partial C_{m_x}}{\partial \alpha}}{\frac{\partial C_N}{\partial \alpha}}$$

From the above expression, it can be inferred that the location of the aerodynamic center can be determined by the derivative of normal force coefficient and moment coefficient about any arbitrary point.

The position of aerodynamic center is nearly constant at quarter chord for most of the aerofoils

i.e., $\frac{X_{ac}}{c} = 0.25$

34. Maximum bending moment,

$$M = \frac{Wl}{4} = \frac{1\times 2}{4} = 0.5\ KN - m$$

$$I = \frac{bh^3}{12} = \frac{8\times 3^3 \times 10^{-8}}{12} = 18\times 10^{-8} m^4$$

Maximum bending stress will occur at $y = 1.5$ cm

$$\frac{M}{I} = \frac{\sigma}{y} \qquad \sigma = \frac{M}{I}y$$

$$\sigma = \frac{0.5\times 10^3 \times 1.5\times 10^{-2}}{18\times 10^{-8}} = 41.66\ MPa$$

35. $\omega_1 = \frac{u_1}{r_1} = \frac{0}{2} = 0\ rad/s$

$$\omega_2 = \frac{u_2}{r_2} = \frac{5}{5} = 1\ rad/s$$

As per formula

$$u_\theta = \frac{\omega_2 - \omega_1 \frac{R_1^2}{R_2^2}}{1 - \frac{R_1^2}{R_2^2}}.r + \frac{R_1^2(\omega_1 - \omega_2)}{1 - \frac{R_1^2}{R_2^2}}.\frac{1}{r}$$

$$u_\theta = 1.19\left(r - \frac{4}{r}\right)$$

36. At the exit of intake,

$$T_{01} = T\left[1 + \frac{\gamma - 1}{2}M^2\right]$$

$$\Rightarrow T_{01} = 250\times\left[1 + \frac{1.4-1}{2}\times(0.7)^2\right]$$

$$\Rightarrow T_{01} = 274.5\ K$$

At the exit of compressor,

$$T_{02} = T_{01}\left[1 + \frac{1}{\eta_c}\left(p_{rc}^{\frac{\gamma-1}{\gamma}} - 1\right)\right]$$

$$\Rightarrow T_{02} = 274.5\times\left[1 + \frac{1}{0.87}\times\left(7^{\frac{1.4-1}{1.4}} - 1\right)\right]$$

$$\Rightarrow T_{02} = 509.13\ K$$

37. Hoop Stress, $\sigma_\theta = \frac{pd}{2h} = \frac{2\times 1}{2h} = \frac{1}{h}MPa$

Longitudinal Stress, $\sigma_l = \frac{pd}{4h} = \frac{2\times 1}{4h} = \frac{0.5}{h}MPa$

Allowable Stress in simple tension,

$$\sigma = \frac{200}{3}MPa$$

Now,

According to Maximum Principal Stress Theory

$$\frac{1}{h} = \frac{200}{3}$$

$$h = 0.015\ m$$

38. As $y = \delta, u = U$

Therefore, using the velocity profile equation we get,

$$u = U\sin\left(\frac{ay}{\delta}\right)$$

$$\sin(a) = 1$$

$$a = \frac{\pi}{2}$$

39. Given,

$$\beta_a = 20°$$

$$\beta_e = 20°$$

$\alpha = 30°$

$V = 1000\ m/s$

For analytical sideslip angle $\beta_a = 20°$

Sideslip velocity is given as

$$V_{sideslip} = \frac{V \tan \beta_a}{\sqrt{1 + \tan^2 \alpha + \tan^2 \beta_a}}$$

$$\Rightarrow V_{sideslip} = \frac{1000 \times 0.363970}{\sqrt{1 + (0.577350)^2 + (0.363970)^2}}$$

$$\Rightarrow V_{sideslip} = 300.63\ m/s$$

For experimental sideslip angle $\beta_e = 20°$

$$V_{sideslip} = V \sin \beta_e$$

$$\Rightarrow V_{sideslip} = 1000 \times \sin(20)$$

$$\Rightarrow V_{sideslip} = 1000 \times 0.342020$$

$$\Rightarrow V_{sideslip} = 342.02\ m/s$$

40. Given,

$\mu_X = 1, \sigma_X^2 = 4$

$\Rightarrow \sigma_X = 2$

Also given,

$\mu_Y = -1$ *and* σ_Y is unknown

and $p(X \le -1) = p(Y \ge 2)$

Converting into standard normal variates

$$p\left(z \le \frac{-1 - \mu_X}{\sigma_X}\right) = p\left(z \ge \frac{2 - \mu_Y}{\sigma_Y}\right)$$

$$p\left(z \le \frac{-1 - 1}{2}\right) = p\left(z \ge \frac{2 - (-1)}{\sigma_Y}\right)$$

$$p(z \le -1) = p\left(z \ge \frac{3}{\sigma_Y}\right)$$

Now, we know that in standard distribution

$$p(z \le -1) = p(z \ge 1)$$

Therefore, we get,

$$\frac{3}{\sigma_Y} = 1$$

$$\Rightarrow \sigma_Y = 3$$

41.

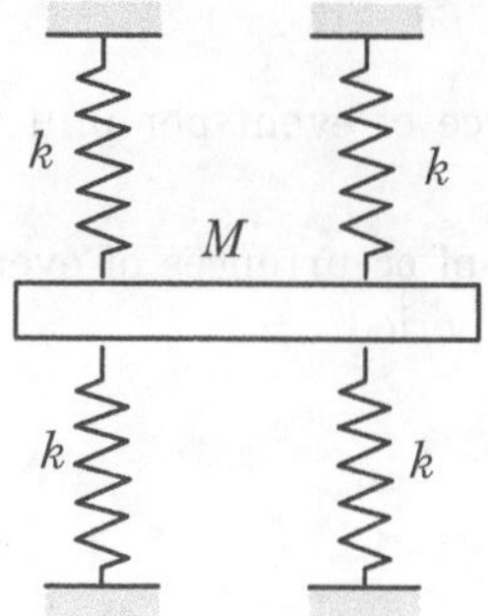

In the picture above two springs are in parallel and below two springs are in parallel. Therefore, they can be combined with both having 2k as spring constant.

Above spring combine into

$$k_1 = k + k = 2k$$

Below spring combine into

$$k_2 = k + k = 2k$$

Now, these two are also in parallel to each other, giving

$$k_{eq} = k_1 + k_2$$

$$\Rightarrow k_{eq} = 2k + 2k$$

$$\Rightarrow k_{eq} = 4k$$

42. $V_e = \sqrt{2gR}$

$$\frac{V_{e1}}{V_{e2}} = \frac{\sqrt{2g_1R_1}}{\sqrt{2g_2R_2}}$$

$$\frac{V_{e1}}{V_{e2}} = \sqrt{k_1 \times k_2}$$

Ratio of escape velocities

43. $\delta = \dfrac{4.62x}{\sqrt{Re_x}} = \dfrac{4.62x}{\sqrt{\dfrac{\rho U x}{\mu}}}$

$$\Rightarrow \delta \propto x^{1/2}$$

$$\Rightarrow \frac{\delta_1}{\delta_2} = \left(\frac{x_1}{x_2}\right)^{1/2}$$

$$\Rightarrow \frac{5}{\delta_2} = \left(\frac{0.2}{3.2}\right)^{1/2}$$

$$\Rightarrow \frac{5}{\delta_2} = \frac{1}{4}$$

$$\Rightarrow \delta_2 = 20\ mm$$

44. Given equation,

α = rate of occurrence of event per unit time = 4/hr

λ = average number of occurrences of event in specified observation period

$= \alpha\Delta t$

In this case

$\alpha = 4/\text{hr}$ and $\Delta t = 2h$

$\lambda = 4 \times 2 = 8$

Now we wish that no aircraft should land for 2 hrs, i.e. $x = 0$

$$P(x=0) = \frac{e^{-\lambda}\lambda^0}{0!} = \frac{e^{-8}8^0}{0!} = e^{-8}$$

45. Given,

$K = 0.9$

$P_{01} = 250\ KPa$

$P_1 = 50\ KPa$

$T_{0\infty} = 650\ K$

As per given in question,

$$\frac{P_1}{P_{01}} = \frac{50}{250} = 0.2$$

By solving

$$\frac{P_{02}}{P_1} = \frac{\gamma+1}{2}M_1^2\left[\frac{(\gamma+1)^2 M_1^2}{4\gamma M_1^2 - 2(\gamma-1)}\right]^{\frac{-\gamma}{\gamma-1}}$$

we find $M_1 = M_\infty = 1.87056$

Also,

$$T_{0\infty} = \left(1+\frac{\gamma-1}{2}M_\infty^2\right)T_\infty$$

$$\therefore \frac{T_{0\infty}}{T_\infty} = 1+\frac{1.4-1}{2}\times 1.87056^2 = 1.6997$$

and $\dfrac{T_\infty}{T_{0\infty}} = 0.58830$

Now,

$$K = \frac{T_{0,indicated} - T_\infty}{T_{0\infty} - T_\infty}$$

$$\Rightarrow K = \frac{T_{0,indicated} - T_\infty}{\dfrac{T_\infty}{0.5883} - T_\infty}$$

$$\Rightarrow 0.9 = \frac{650 - T_\infty}{0.69981T_\infty}$$

$$\Rightarrow 0.629829T_\infty + T_\infty = 650$$

$$\Rightarrow 1.629829T_\infty = 650$$

$$T_\infty = \frac{650}{1.629829} = 398.8 \approx 399\ K$$

46. Given,

$T_0 = 2500\ K$

$P_0 = 1.5\ MPa = 1.5\times 10^6\ Pa$

$\gamma = 1.2$

$R = 692.83\ J/kg-K$

$\dot{m} = 18\ kg/s$

As per formula, we get

$$\frac{\dot{m}_p\sqrt{T_0}}{A^*P_0}\sqrt{\frac{R}{\gamma}} = \left(\frac{2}{\gamma+1}\right)^{\frac{(\gamma+1)}{2(\gamma-1)}}$$

$$A^* = \frac{\dot{m}_p\sqrt{T_0}}{P_0\left(\frac{2}{\gamma+1}\right)^{\frac{(\gamma+1)}{2(\gamma-1)}}}\sqrt{\frac{R}{\gamma}}$$

$$\Rightarrow A^* = \frac{18\times\sqrt{2500}}{1.5\times 10^6\times\left(\frac{2}{1.2+1}\right)^{\frac{(1.2+1)}{2(1.2-1)}}}\sqrt{\frac{692.83}{1.2}}$$

$$\Rightarrow A^* = 0.024\ m^2$$

47. The augmented matrix for the given system is

$$\begin{bmatrix} 2 & -1 & 3 \\ 3 & -2 & 5 \\ -1 & -4 & 1 \end{bmatrix} = \begin{bmatrix} 1 \\ 2 \\ 3 \end{bmatrix}$$

Using gauss-elimination method on above matrix, we get,

$$\begin{bmatrix} 2 & -1 & 3 \\ 3 & -2 & 5 \\ -1 & -4 & 1 \end{bmatrix} = \begin{bmatrix} 1 \\ 2 \\ 3 \end{bmatrix} \xrightarrow[R_3+\frac{1}{2}R_1]{R_2-\frac{3}{2}R_1} \begin{bmatrix} 2 & -1 & 3 \\ 0 & -1/2 & 1/2 \\ 0 & -9/2 & 5/2 \end{bmatrix}$$

$$= \begin{bmatrix} 1 \\ 1/2 \\ 7/2 \end{bmatrix} \xrightarrow{R_3 - 9R_2} \begin{bmatrix} 2 & -1 & 3 \\ 0 & -1/2 & 1/2 \\ 0 & 0 & -2 \end{bmatrix} = \begin{bmatrix} 1 \\ 1/2 \\ -1 \end{bmatrix}$$

$$\text{Rank}(\langle A|B\rangle) = 3$$

$$\text{Rank}([A]) = 3$$

Therefore, $\text{Rank}(\langle A|B\rangle) = \text{Rank}([A])$ = number of variables.

The system has unique solutions.

48. Given,

$$X_1 = 10\cos\omega t$$

$$X_2 = 15\cos(\omega t + 2)$$

Comparing with

$$X_1 = A_1 \cos\omega t$$

and $X_2 = A_2 \cos(\omega t + \phi)$

We get,

$$\phi = 2$$

$$rad = 114.59°$$

$$A_1 = 10, A_2 = 15$$

As per formula

$$A = \sqrt{(A_1 + A_2\cos\phi)^2 + (A_2\sin\phi)^2}$$

$$\Rightarrow A = \sqrt{(10 + 15\cos 114.59)^2 + (15\sin 114.59)^2}$$

$$\Rightarrow A = \sqrt{(3.758)^2 + (13.63)^2}$$

$$\Rightarrow A = \sqrt{200.0653}$$

$$\Rightarrow A = 14.14$$

49. Given,

$$X_1 = 10\cos\omega t$$

$$X_2 = 15\cos(\omega t + 2)$$

Comparing with

We get,

$$\phi = 2$$

$$rad = 114.59^o$$

$$A_1 = 10, A_2 = 15$$

$$X_1 = A_1 \cos\omega t$$

and $X_2 = A_2 \cos(\omega t + \phi)$

As per formula

$$\alpha = \tan^{-1}\left(\frac{A_2 \sin\phi}{A_1 + A_2\cos\phi}\right)$$

$$\Rightarrow \alpha = \tan^{-1}\left(\frac{15\sin 114.59}{10 + 15\cos 114.59}\right)$$

$$\Rightarrow \alpha = \tan^{-1}\left(\frac{13.63}{3.758}\right)$$

$$\Rightarrow \alpha = 74.58°$$

50. Given,

$$AR_w = 15$$

$$C_{D_0} = 0.0322$$

$$e = 0.5$$

As per formula,

$$K = \frac{1}{\pi A R e}$$

$$\Rightarrow K = \frac{1}{\pi \times 15 \times 0.5} = 0.0424$$

For mimimum thrust

$$C_L = \sqrt{\frac{C_{D_0}}{K}} = \sqrt{\frac{0.0322}{0.0424}} = 0.87$$

51. Given,

$$AR_w = 15$$

$$C_{D_0} = 0.0322$$

$$e = 0.5$$

Now, For maximum range glider should be flying at an angle ϕ

Therefore,

$$C_L = \sqrt{\frac{C_{D_0}}{K}}$$

Also,

$$C_D = C_{D_0} + KC_L^2$$

$$\Rightarrow C_D = C_{D_0} + C_{D_0}$$

$\Rightarrow C_D = 2C_{D_0}$

$\Rightarrow C_D = 2\times 0.0322 = 0.0644$

From previous question

$C_L = 0.87$

Now,

$$\tan\phi = \frac{1}{C_L/C_D}$$

$$\Rightarrow \tan\phi = \frac{1}{0.87/0.0644}$$

$$\therefore \phi = \tan^{-1}\left(\frac{0.0644}{0.87}\right) = 4.233°$$

52. Consider a flow where,

$$\frac{v}{dy} = \frac{u}{dx} = \frac{w}{dz}$$

$$\Rightarrow \frac{v}{u} = \frac{dy}{dx}$$

Considering an unsteady 2D velocity field

$\vec{V} = x\hat{i} + y\hat{j}$

Let us say thaat the particle is coming out from (1,1) at t=0, hence

$u = x$

$v = yt$

$$\therefore \frac{v}{u} = \frac{dy}{dx}$$

$$\Rightarrow \frac{dy}{dx} = \frac{yt}{x}$$

$$\Rightarrow \int\frac{dx}{x} = \int\frac{dy}{yt}$$

$$\Rightarrow \ln x = \frac{1}{t}\ln y$$

$$\Rightarrow t\ln x = \ln y + c$$

Using initial condition of particle location (1,1) at t = 0, we get

$$\Rightarrow y = x^t$$

This is streamline equation

At different 't' interval the streamline equation changes.

53. Consider the actual path the flow follows

$$\frac{dx}{dt} = u, \frac{dy}{dt} = v, \frac{dz}{dt} = w$$

Considering an unsteady 2D velocity field

$\vec{V} = x\hat{i} + y\hat{j}$

Let us say thaat the particle is coming out from (1,1) at t=0, hence

$u = x$

$v = yt$

From above equation,

$$\Rightarrow \ln x = t$$

$$\Rightarrow \ln y = \frac{t^2}{2}$$

On substituting the value of $t = ln x$ we get,

$$\Rightarrow 2ln\ y = (ln\ x)^2$$

This represents the pathline equation at $t = 0$ and at point (1,1)

54. $r_1 = (6378.14 + 250)\times 1000 = 6{,}628{,}140\ m$

$v_1 = 7950\ m/s$

$\phi = 89^o$

$$\left(\frac{R_p}{r_1}\right)_{1,2} = \frac{-C \pm \sqrt{C^2 + 4\times(1-C)\sin^2\phi}}{2\times(1-C)}$$

$$\text{where } C = \frac{2\times GM}{r_1 \times v_1^2}$$

$$\therefore C = \frac{2\times 3{,}986005\times 10^{14}}{\left(6{,}628140\times(7950)^2\right)}$$

$$\left(\frac{R_p}{r_1}\right)_{1,2}$$

$$= \frac{-1.903 \pm \sqrt{1.903^2 + 4 \times (1-1.903)\sin^2 89^\circ}}{2 \times (1-1.903)}$$

$$\left(\frac{R_p}{r_1}\right)_{1,2} = 0.9984 \text{ and } 1.1089$$

Perigee Radius, $R_p = R_{p1} = r_1 \times \left(\frac{R_p}{r_1}\right)_1$

$R_p = 6{,}628{,}140 \times 0.996019$

$R_p = 6{,}617{,}534\ m$

Altitude at perigee

$$= \frac{6{,}617{,}534}{1000} - 6378.14 = 239.39\ km$$

Apogee Radius, $R_a = R_{p2} = r_1 \times \left(\frac{R_p}{r_1}\right)_2$

$R_a = 6{,}628{,}140 \times 1.1089$

$R_a = 7{,}349{,}944\ m$

Altitude at apogee

$$= \frac{7{,}349{,}944}{1000} - 6378.14 = 971.80446\ km$$

55. $r_1 = (6378.14 + 250) \times 1000 = 6{,}628{,}140\ m$

$v_1 = 7950\ m/s$

$\phi = 89^o$

$R_p = 6{,}617{,}534\ m$

$R_a = 7{,}349{,}944\ m$

$$e = \frac{R_a - R_p}{R_a + R_p}$$

$$\Rightarrow e = \frac{7{,}349{,}944 - 6{,}617{,}534}{7{,}349{,}944 + 6{,}617{,}534}$$

$\Rightarrow e = 0.0524$

56. The subject here is "the ways" and therefore, "are" has to be used

57. A patient of Alzheimer's disease experiences forgetfulness or loss of short-term memory that interferes with his/her daily life.

58. $x = \frac{9}{y}$ ... (1)

$z = \frac{16}{y}$ ... (2)

$z = \frac{25}{x}$ ... (3)

comparing 2 and 3

$$\frac{25}{9/y} = \frac{16}{y} = y^2 = \frac{16 \times 9}{25} = \frac{12}{5}$$

substituting in other equations we get

$$x = \frac{45}{12} \text{ and } z = \frac{80}{12}$$

Adding we get,

12.82

59. Total ways to arrange a deck of 52 cards = 52!

Total ways when ace of spades appears just after ace = 51!

Total ways when two of clubs appears just after two of clubs = 51!

Probability of both the scenarios $= \frac{51!}{52!} = \frac{1}{52}$

60. Manoj takes twice as much time as Ajay and thrice as much as Vijay Manoj takes x days, Ajay takes $\frac{x}{2}$ days and vijay takes $\frac{x}{3}$ days together they work then finished in 1 day

$$\frac{1}{x} + \frac{2}{x} + \frac{3}{x} = 1$$

$$\Rightarrow \frac{6}{x} = 1$$

$\Rightarrow x = 6$

61. Mean profit percentage $= \frac{\text{net profit}}{\text{cost price}}$,

Also, net profit margin $= \frac{\text{net profit}}{\text{revenue}}$

$$= \frac{\text{net profit}}{\text{net profit} + \text{cost price}}$$

Both will produce the same answer, 30%.

62. Total spirit $= \frac{2}{5} + \frac{4}{11}$

$$= \frac{22+20}{55} = \frac{42}{55}$$

Total water $= 2 - \frac{42}{55}$

$$= \frac{110-42}{55} = \frac{68}{55}$$

So, New spirit to water is

$$= \frac{42}{55} \times \frac{55}{68} = \frac{42}{68} = \frac{21}{34}$$

63. Avg age of 39 students is 15 years.

$$\Rightarrow \frac{x_1 + x_2 + ... + x_{39}}{39} = 15 \quad ...(1)$$

Now,

Include the teacher's age, T

$$\Rightarrow \frac{x_1 + x_2 + ... + x_{39} + T}{40} = 15 + 0.25 \quad ...(2)$$

From equ (1)

$$x_1 + x_2 + ... + x_{39} = 585$$

From equ (2)

$$x_1 + x_2 + ... + x_{39} + T = 620$$

Therefore,

$$585 + T = 620$$

$$T = 25$$

64. "It also uses technology to ensure safety and security of the city, something which critics argue, will lead to a surveillance state."

critics argue because of perception that smart city promotes surveillance that makes statement (iv) correct.

There is nothing mention in passage about encouragement about formation of surveillance states so that makes statement (i) false.

and, "will lead to a surveillance state tis makes" this line makes statement (ii) false.

and there is nothing mention about "both Sustainability and surveillance" in passage so statement (iii) is wrong too.

Hence, only statement (iv) is correct which is option (c)

65. Given,

a = 3

d = 8

and $l = 467$

As per formula

$$a + (n+1)d = l$$

$$\Rightarrow 3 + (n+1) \times 8 = 467$$

$$\Rightarrow (n+1) \times 8 = 467 - 3$$

$$\Rightarrow n = \frac{464}{8} - 1 = \frac{464-8}{8} = 57$$

$n = 57$

Now,

As sum of a term quidistant from ends are 470 except the middle one

$$\therefore \frac{57}{2} = 28.5 \approx 29$$

MOCK TEST – 3

AEROSPACE ENGINEERING

Time Allowed: 3:00 Hours — *Maximum Marks: 100*

Read the following instructions carefully before you begin to answer the questions. This Booklet contains questions in English.

INSTRUCTIONS

1. This Booklet contains **65 questions** in total comprising of the following two parts:

   Part -1: Technical Section (85 Marks) *(1-55 Questions)*

   Part -2: General Aptitude Section (15 Marks) *(56-65 Questions)*

2. All questions are compulsory. Q.1-Q.25 & Q.56-Q.60 carry one mark each. Q.26-Q.55 & Q.61-Q.65 carry two marks each.

3. **Types of Questions:**

   (*a*) Multiple Choice Questions(MCQ) carrying 1 or 2 marks each in all papers and sections. These questions are objective in nature, and each will have a choice of four options, out of which the candidate has to mark the correct answer(s).

   (*b*) Numerical Answer Questions of 1 or 2 marks each in all sections. For these questions the answer is a real number, to be entered by the candidate using the virtual keypad. No choices will be shown for these type of questions.

4. For 1-mark multiple-choice questions, 1/3 marks will be deducted for a wrong answer. Likewise, for 2-marks multiple-choice questions, 2/3 marks will be deducted for a wrong answer. There is no negative marking for numerical answer type questions.

MOCK TEST – 3

1. In an unpowered glide of an aircraft having weight W, lift L and drag D, the equilibrium glide angle is defined as-

(*a*) $\tan^{-1}\left(\frac{W}{L}\right)$ (*b*) $\tan^{-1}\left(\frac{L}{W}\right)$

(*c*) $\tan^{-1}\left(\frac{D}{L}\right)$ (*d*) $\tan^{-1}\left(\frac{L}{D}\right)$

2. The thermal efficiency of a turbojet engine moving with velocity U and having exhaust velocity Ue with respect to engine is given by

(*a*) $\dfrac{U_e^2\left[1-\left(\frac{U}{U_e}\right)^2\right]}{2fQ_R}$ (*b*) $\dfrac{U_e^2\left[1+\left(\frac{U}{U_e}\right)^2\right]}{2fQ_R}$

(*c*) $\dfrac{1-\left(\frac{U}{U_e}\right)}{f\left(\frac{Q_R}{U^2}\right)\left(\frac{U}{U_e}\right)}$ (*d*) $\dfrac{2\left(\frac{U}{U_e}\right)}{1+\left(\frac{U}{U_e}\right)}$

3. NACA 12018 aerofoil indicates its distance of maximum camber from leading edge

(*a*) 12 % of the chord

(*b*) 10 % of the chord

(*c*) 22 % of the chord

(*d*) 18 % of the chord

4. The value $\lim_{x\to 0}\frac{e^x-1}{\sin x}$ of is equal to______.

5. A column of length L and flexure rigidity EI has one end fixed and other end hinged. The critical buckling load for the column is

(*a*) $\frac{\pi^2 EI}{(0.5L)^2}$ (*b*) $\frac{\pi^2 EI}{(0.7L)^2}$

(*c*) $\frac{\pi^2 EI}{(L)^2}$ (*d*) $\frac{\pi^2 EI}{(2L)^2}$

6. The second law of Kepler stating constancy of aerial velocity of a planet is a consequence of the law of Conservation of

(*a*) angular momentum

(*b*) linear momentum

(*c*) energy

(*d*) none of these

7. An aeroplane model is mounted in a wind tunnel with angle of attack 40° and sideslip angle, $\beta_e = 30°$ By varying β_e slightly to both sides from the orientation, the change in the yawing moment coefficient with respect to β_e is estimated to be 0.250. Find the angle β_a

(*a*) $\beta_a = 37°$ (*b*) $\beta_a = 20.1°$

(*c*) $\beta_a = 8.2°$ (*d*) $\beta_a = 4.5°$

8. For a complex number $Z = \frac{1}{2}+\frac{\sqrt{3}}{2}i$, the value of Z^6 is

(*a*) $-\left(\frac{1}{2}+\frac{\sqrt{3}}{2}i\right)$ (*b*) –1

(*c*) $\left(\frac{1}{2}-\frac{\sqrt{3}}{2}i\right)$ (*d*) 1

9. Supercritical aerofoils are defined for

(*a*) 5- series (*b*) 4- series

(*c*) 8- series (*d*) 6- series

10. In a centrifugal compressor, the work done factor ψ for the case of isentropic flow is

(*a*) 0 (*b*) 0.5

(*c*) 1 (*d*) 1.5

11. In a circular cross-section the ratio of maximum shear stress to average shear stress is given as: -

(*a*) $\frac{3}{4}$ (*b*) $\frac{4}{3}$

(*c*) $\frac{3}{2}$ (*d*) $\frac{2}{3}$

12. Roots of the algebraic equation

$x^3 + x^2 + x + 1 = 0$ are

(*a*) (+1, +j, –j)

(*b*) (+1, –1, +1)

(*c*) (0, 0, 0)

(*d*) (–1, +j, –j)

13. What is true for the maximum velocity of the rocket at constant burnout?

I. Constant gravitational force and Specific Impulse

II. Steady flow through the rocket nozzle

III. Drag force is negligible.

(*a*) I and II (*b*) II and III

(*c*) I and III (*d*) I, II and III

14. Which one of the following criteria leads to maximum turn rate and minimum radius in a steady pull down flight?

(*a*) Highest possible load factor and highest possible velocity

(*b*) Lowest possible load factor and lowest possible velocity

(*c*) Highest possible load factor and lowest possible velocity

(*d*) Lowest possible load factor and highest possible velocity

15. In a circular pipe of radius R carrying a laminar flow, the ratio of average velocity to the maximum velocity in the conduct is

(*a*) 0.50 (*b*) 1.00

(*c*) 0.67 (*d*) 0.33

16. Which of the following is the right combination for an aeroplane to have stable trim flight?

(*a*) $C_{m_0} > 0$ and $C_{m_\alpha} > 0$

(*b*) $C_{m_0} > 0$ and $C_{m_\alpha} < 0$

(*c*) $C_{m_0} < 0$ and $C_{m_\alpha} > 0$

(*d*) $C_{m_0} < 0$ and $C_{m_\alpha} < 0$

17. The general solution of $\frac{d^2y}{dx^2} + y = 0$ is

(*a*) $y = P\cos x + Q\sin x$

(*b*) $y = P\cos x$

(*c*) $y = P\sin x$

(*d*) $y = P\sin^2 x$

18. A satellite is an elliptical orbit has time period equal to a satellite in Geo- stationary orbit (GEO). What is the semi- major axis of this orbit?

(*a*) 42226 km (*b*) 37254 km

(*c*) 35856 km (*d*) Data insufficient

19. Due to presence of taper in the structure of aircraft wing, which of the following remains unchanged compared to the case with no taper at same load applied?

(*a*) Axial stress in longitudinal

(*b*) Shear stress due to applied bending moment

(*c*) Shear flow due to torsional moment applied

(*d*) None of the above

20. Which of the following remains almost constant, in a phugoid manoeuvre?

(*a*) Pitch angle (*b*) Aircraft speed

(*c*) Altitude (*d*) Angle of attack

21. For a Convergent- Divergent nozzle, the initial stagnation pressure P_0 and Temperature T_0 is 241.3 KPa and 611 K respectively. For an exit Mach number $M_e = 2.15$, what is the exit pressure P_e? (Given $\gamma = 1.34$)

(*a*) 126.58 KPa (*b*) 68.95 KPa

(*c*) 24.54 KPa (*d*) 13.78 KPa

22. Which of the following tends to impart rolling stability to an aircraft?

(*a*) Anhedral

(*b*) Wings positioned at top of fuselage

(*c*) Swept forward wing

(*d*) None of the above

23. Which of the following holds true, for maximum range of gliding flight?

(*a*) $C_{D_0} = KC_L^2$ (*b*) $C_{D_0} = 3KC_L^2$

(*c*) $C_{D_0} = \frac{KC_L^2}{3}$ (*d*) $C_{D_0} = \frac{KC_L^2}{2}$

24. In a thin walled tube as shown in figure. The wall thickness is t =0.005 m and the average radius is 0.3 m. Calculate the torsional constant for closed and open section?

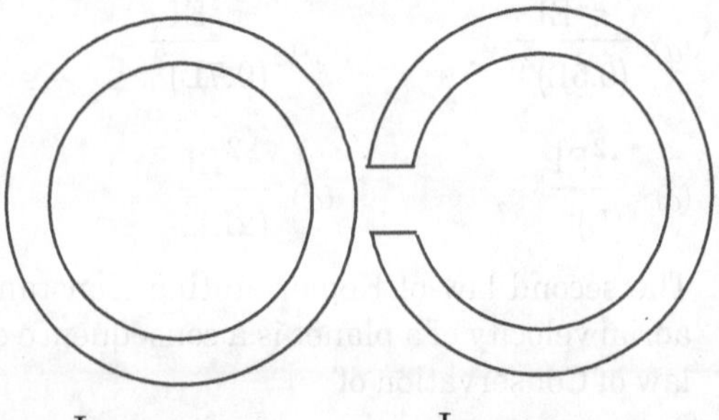

(*a*) $\frac{J_1}{J_2} = 9{,}400$ (*b*) $\frac{J_1}{J_2} = 10{,}800$

(*c*) $\frac{J_1}{J_2} = 11{,}300$ (*d*) $\frac{J_1}{J_2} = 12{,}500$

25. Which of the following satisfy, for an aircraft to be stable in yaw direction?

(*a*) $C_{l\beta} < 0$ (*b*) $C_{l\beta} > 0$

(*c*) $C_{l\beta} = 0$ (*d*) $C_{l\beta} = C_{m\alpha}$

26. Consider a stream function as

$$\psi = Ur\sin\theta\left(1 - \frac{a^2}{r^2}\right) + \frac{\Gamma}{2\pi}\ln\frac{r}{a}$$

where, U = 25 m/s, a = 1 m, $\Gamma = 100\pi\ m^2/s$

Then the value of u_r, u_θ at $\theta = \pi$ is –

(*a*) $u_\theta = 0$ m/s, $u_r = 0$ m/s

(*b*) $u_\theta = 50$ m/s, $u_r = 0$ m/s

(*c*) $u_\theta = -50$ m/s, $u_r = 0$ m/s

(*d*) $u_\theta = 100$ m/s, $u_r = 0$ m/s

27. The change in P.E. to throw a mass from surface of earth to height h where h=nR is, where R is radius of earth :-

(*a*) $mgR\left(\frac{n}{n-1}\right)$ (*b*) mgRn

(*c*) $mgR\left(\frac{n^2}{n^2-1}\right)$ (*d*) $mgR\left(\frac{n}{n+1}\right)$

28. $u = 0.02x + 0.02y - 0.01z$,

$v = 0.01y - 0.02z$,

$w = -0.01x + 0.01z$

For the initial points at A(0, 0, 0) and B(5, 0, 0). Find the distance between two points A and B after applying the above displacement field induced by the stress.

(*a*) 5.11 units (*b*) 6.11 units

(*c*) 7.11 units (*d*) 8.11 nits

29. A box contains 2 washers, 3 nuts and 4 bolts. Items are drawn from the box at random one at a time without replacement. The probability of drawing 2 washers first followed by 3 nuts and subsequently the bolts is

(*a*) $\frac{2}{315}$ (*b*) $\frac{1}{630}$

(*c*) $\frac{1}{1260}$ (*d*) $\frac{1}{2520}$

30. An aircraft in a steady level flight at a forward speed of 50 m/s suddenly rolls by 180° and becomes inverted. If no other changes are made to the configuration or controls of the aircraft, the nature of the subsequent flight path taken by the aircraft and its characteristics parameter (assume $g = 9.81\ m/s^2$) are

(*a*) Straight line path with a speed of 50 m/s

(*b*) Upward circular path with a speed of 50 m/s and radius 127.4 m

(*c*) Downward circular path with speed of 50 m/s and radius 127.4 m.

(*d*) Upward circular path with a speed of 25 m/s and radius 254.8 m

31. Which of the following is true as a working of a Ramjet Engine?

I. Ramjet engine has no rotating component

II. In ramjet engine compressor is not used, the compression is achieved by ram effect.

III. Ram compression does not depend on efficiency of inlets to decelerate the incoming air.

(*a*) I, III

(*b*) I, II

(*c*) II, III

(*d*) I, II, III

32. In a boundary Layer, for a linear profile, the momentum thickness θ is given as

(*a*) $\theta = \delta$ (*b*) $\theta = \frac{\delta}{2}$

(*c*) $\theta = \frac{\delta}{4}$ (*d*) $\theta = \frac{\delta}{6}$

33. For a function f(x), the values of the function in the interval [0, 1] are given in the table below.
x f(x)

x	f(x)
0.0	1.0
0.2	1.24
0.4	1.56
0.6	1.96
0.8	2.44
1.0	3.0

The value of the integral $\int_0^1 f(x)dx$ according to the trapezoidal rule is ________.

34. For an actual density, ρ and density at sea level, ρ_s. What is relationship between calibrated and true speed?

(a) $\frac{V_{true}}{V_{calibrated}} = \sqrt{\frac{\rho_s}{\rho}}$ (b) $\frac{V_{true}}{V_{calibrated}} = \frac{\rho_s}{\rho}$

(c) $\frac{V_{true}}{V_{calibrated}} = \sqrt{\frac{\rho}{\rho_s}}$ (d) $\frac{V_{true}}{V_{calibrated}} = \frac{\rho}{\rho_s}$

35. A cylindrical shell made of mild steel plate having diameter 100 cm is to be subjected to an internal pressure of 2 MPa. If the yield strength of the material is 200 MPa, Calculate the thickness of the plate on the Maximum Shear Strain Energy Theory, assuming factor of safety to be 3.

(a) 0.015 m (b) 0.0075 m

(c) 0.01299 m (d) 0.1169 m

36. The kinetic energy of an earth bounded satellite in circular orbit around earth with increasing altitude

(a) Increases

(b) Decreases

(c) Remains constant

(d) Increases exponentially with altitude

37. If the standard deviation of the spot speed of vehicles in a highway is 8.8 kmph and the mean speed of the vehicles is 33 kmph, the coefficient of variation in speed is

(a) 0.1517 (b) 0.1867

(c) 0.2666 (d) 0.3646

38. The $\frac{C_L}{C_D}$ ratio corresponding to maximum endurance during steady level flight configuration is (correct to first decimal place)

Given, Aspect ratio = 5

Zero lift drag coefficient = 0.019

Span efficiency factor = 0.81

(a) 12.2 (b) 13.4

(c) 11.2 (d) 10.1

39. A multistage axial compressor operating at an overall adiabatic efficiency of 0.8 develops a total pressure ratio of 10. The total temperature at inlet is 288 K and stagnation enthalpy rise across each stage is 31.5 kJ/kg. What is the polytrophic efficiency for each stage? Given $\gamma = 1.4$

(a) 0.85 (b) 0.091

(c) 0.93 (d) 0.97

40. A cantilever beam of length 3 m carrying a point load of 100 kN at a distance of 2 m from fixed end. If $I = 10^8 \text{ mm}^4$ and $E = 2 \times 10^6 \text{ N/mm}^2$. The deflection of the beam at its free end is ________.

41. Air of Mach number M_1 =3.5 flow through an oblique shock which is at an angle $\beta = 40°$. The value of M_2 is ________.

42. A rocket having specific thrust 240 s is flying at a speed of 850 m/s. In a time when its weight becomes half, what will be the velocity? Neglect any drag or gravitational field.

(a) 1600 m/s (b) 1738.6 m/s

(c) 2158.6 m/s (d) 2480.28/s

43. For the given figure with $w_1 = w_2 = 2$ kN/m and length L = 1.5 m with a = b = c = 0.5 m. The bending moment x = 0.75 m at is

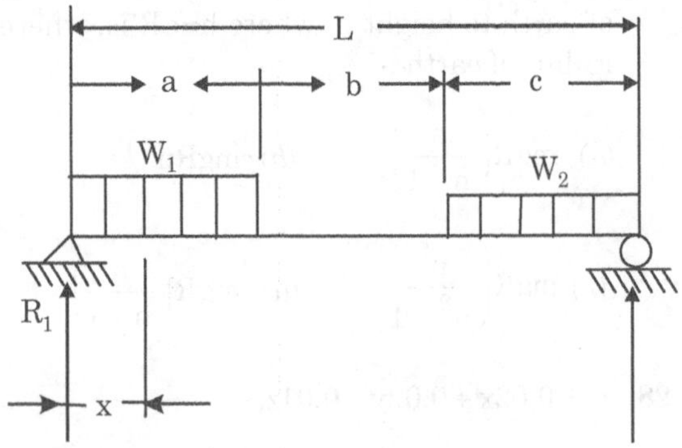

(a) 0.125 kNm (b) 0.25 kNm

(c) 0.5 kNm (d) 0.75 kNm

44. What is the effect of flaps on the $C_L - \alpha$ curve of the aircraft?

(a) $C_{L max}$ increases

(b) Stalling angle increases

(c) The slope of $C_L - \alpha$ curve increases

(d) Zero lift increases

45. Which of the following is a solution of the differential equation $\frac{d^2y}{dx^2} + p\frac{dy}{dx} + (q+1) = 0$.

Given, p = 4, q = 3.

(a) e^{-3x} (b) xe^{-x}

(c) xe^{-2x} (d) x^2e^{-2x}

46. For an aircraft flying at a velocity, at absolute ceiling of aircraft

(*a*) Rate of climb is 100 ft/min

(*b*) Rate of climb is zero

(*c*) Rate of climb is maximum

(*d*) Rate of climb is more than 150 ft/min

47. For the differential equation $\frac{d^2y}{dx^2}+6\frac{dy}{dx}+8x=0$ with initial conditions $x(0)=1$ and $\frac{dy}{dx}_{t=0}=0$, the solution is

(*a*) $x(t)=2e^{-6t}-e^{-2t}$ (*b*) $x(t)=2e^{-2t}-e^{-4t}$

(*c*) $x(t)=-e^{-6t}+2e^{-4t}$ (*d*) $x(t)=e^{-2t}-e^{-4t}$

48. If the degree of reaction at mid-radius of compressor is 0.5 and hub-to- tip ratio is 0.8, then the degree of reaction of the compressor at blade tip is

(*a*) 0.24 (*b*) 0.68

(*c*) 1.2 (*d*) 1.9

49. Consider flow over a finite wing with elliptical circulation distribution: -

$$\Gamma(y)=100\sqrt{1-\left(\frac{2y}{4}\right)^2}\ m^2/s$$

If the free stream velocity is 100 m/s with total wing span 4m, then the induced angle of attack will be?

(*a*) 0.125 rad

(*b*) –0.125 rad

(*c*) $0.125\sqrt{1-\left(\frac{y}{2}\right)^2}$ rad

(*d*) $-0.125\sqrt{1-\left(\frac{y}{2}\right)^2}$ rad

50. For an initial value problem $y'=f(t,y)$ and $y(t_0)=y_0$, the 4th order Runge-kutta method is given by the equations

$$y_{n+1}=y_n+\frac{h}{6}(k_1+2k_2+2k_3+k_4)$$

$$t_{n+1}=t_n+h$$

If $k_1=f(t_n,y_n)$

Then $k_4=?$

(*a*) $k_4=f\left(t_n+\frac{h}{2},y_n+hk_3\right)$

(*b*) $k_4=f\left(t_n+\frac{h}{2},y_n+\frac{hk_3}{2}\right)$

(*c*) $k_4=f(t_n+h,hk_3)$

(*d*) $k_4=f(t_n+h,y_n+hk_3)$

51. The principal stress that a material is subjected to is given as 60 MPa, 30 MPa and -30 MPa. The total strain energy is_____________.
If E = 200 GPa, $\nu=0.33$

52. The principal stress that a material is subjected to is given as 60 MPa, 30 MPa and -30 MPa. The Volumetric strain energy is ____________.
If E = 200 GPa, $\nu=0.33$

53. If the stagnation pressure p_0 is decreased by 0.6 times its value, and if the stagnation Temperature T_0 at compressor inlet is increased 1.44 times, then by what factor should the mass flow rate be decreased in the compressor to achieve the same pressure ratio.

(*a*) 1.63

(*b*) 1.83

(*c*) 2.0

(*d*) 2.11

54. Which of the following satisfy, for an aircraft to be stable in pitch and roll directions?

(*a*) $C_{m\alpha}<0$ and $C_{n\beta}<0$

(*b*) $C_{m\alpha}>0$ and $C_{n\beta}>0$

(*c*) $C_{m\alpha}>0$ and $C_{n\beta}>0$

(*d*) $C_{m\alpha}<0$ and $C_{n\beta}<0$

55. The cross-sectional area $A_{cs}=0.5\ m^2$ with an end burning grain of a solid rocket motor. The propellant density $\rho_p=2000\ kg/s$ and a linear regression rate $\dot{r}=7mm/s$. If the specific impulse of the propulsion system is $I_{sp}=200\ s$, the thrust produced by the motor is approximately.

(*a*) 4.8 kN (*b*) 6.2 kN

(*c*) 9.3 kN (*d*) 13.7 kN

56. The volume of a sphere of diameter 1 unit is ______ than the volume of a cube of side 1 unit.

(*a*) Least (*b*) Less

(*c*) Lesser (*d*) Low

57. Choose the most appropriate phrase from the options given below to complete the following sentence.

The aircraft _________ take off as soon as its flight plan was filed.

(*a*) Is allowed to (*b*) Will be allowed to

(*c*) Was allowed to (*d*) Has been allowed to

58. In any given year, the probability of an earthquake greater than Magnitude 6 occurring in the Garhwal Himalayas is 0.04. The average time between successive occurrences of such earthquakes is ____ years.

59. Which of the following options is the closest in meaning to the sentence below?

"As a woman, I have no country."

(*a*) Women have no country.

(*b*) Women are not citizens of any country.

(*c*) Women's solidarity knows no national boundaries.

(*d*) Women of all countries have equal legal rights.

60. The average of 50 numbers is 30. If two numbers, 35 and 40 are discarded, then the average of the remaining number is nearly?

61. A window is made up of a square portion and an equilateral triangle portion above it. The base of the triangular portion coincides with the upper side of the square. If the perimeter of the window is 66 m, the area of the window in m^2 is ___________.

(*a*) 1.43 (*b*) 2.06

(*c*) 2.68 (*d*) 2.88

62. A team P of 20 engineers can complete work or task in 32 days. Another team Q of 16 engineers can complete same task in 30 days.Then the ratio of working capacity of 1 member of P to the 1 member of Q is

(*a*) 2 : 5 (*b*) 3 : 5

(*c*) 3 : 2 (*d*) 4 : 3

63. We have a 40% acid solution and we want 75 liters of a 15% acid solution. How much water should we put into the 40% solution to do this?

(*a*) 46.875 L (*b*) 47.875 L

(*c*) 48.875 L (*d*) 49.875 L

64. The Tamil version of_________ John Abraham-starrer Madras Cafe _________ cleared by the Censor Board with no cuts last week, but the film's distributor _______ no takers among the exhibitors for a release in Tamilnadu _______ this Friday

(*a*) Mr., was, found, on

(*b*) a, was, found, at

(*c*) the, was, found, on

(*d*) a, being, find at

65. Three men and three rakhsasas arrive together at a ferry crossing to find a boat with an oar, but no boatman. The boat can carry one or at the most two persons, for example, one man and one rakhsasas, and each man or rakhsasas can row. But if at any time, on any bank, (including those who maybe are in the boat as it touches the bank) rakhsasas outnumber men, the former will eat up the latter. If all have to go to the other side without any mishap, what is the minimum number of times that the boat must cross the river?

(*a*) 7 (*b*) 9

(*c*) 11 (*d*) 13

OMR SHEET

1. Ⓐ Ⓑ Ⓒ Ⓓ	18. Ⓐ Ⓑ Ⓒ Ⓓ	35. Ⓐ Ⓑ Ⓒ Ⓓ	52.
2. Ⓐ Ⓑ Ⓒ Ⓓ	19. Ⓐ Ⓑ Ⓒ Ⓓ	36. Ⓐ Ⓑ Ⓒ Ⓓ	53. Ⓐ Ⓑ Ⓒ Ⓓ
3. Ⓐ Ⓑ Ⓒ Ⓓ	20. Ⓐ Ⓑ Ⓒ Ⓓ	37. Ⓐ Ⓑ Ⓒ Ⓓ	54. Ⓐ Ⓑ Ⓒ Ⓓ
4.	21. Ⓐ Ⓑ Ⓒ Ⓓ	38. Ⓐ Ⓑ Ⓒ Ⓓ	55. Ⓐ Ⓑ Ⓒ Ⓓ
5. Ⓐ Ⓑ Ⓒ Ⓓ	22. Ⓐ Ⓑ Ⓒ Ⓓ	39. Ⓐ Ⓑ Ⓒ Ⓓ	56. Ⓐ Ⓑ Ⓒ Ⓓ
6. Ⓐ Ⓑ Ⓒ Ⓓ	23. Ⓐ Ⓑ Ⓒ Ⓓ	40.	57. Ⓐ Ⓑ Ⓒ Ⓓ
7. Ⓐ Ⓑ Ⓒ Ⓓ	24. Ⓐ Ⓑ Ⓒ Ⓓ	41.	58.
8. Ⓐ Ⓑ Ⓒ Ⓓ	25. Ⓐ Ⓑ Ⓒ Ⓓ	42. Ⓐ Ⓑ Ⓒ Ⓓ	59. Ⓐ Ⓑ Ⓒ Ⓓ
9. Ⓐ Ⓑ Ⓒ Ⓓ	26. Ⓐ Ⓑ Ⓒ Ⓓ	43. Ⓐ Ⓑ Ⓒ Ⓓ	60.
10. Ⓐ Ⓑ Ⓒ Ⓓ	27. Ⓐ Ⓑ Ⓒ Ⓓ	44. Ⓐ Ⓑ Ⓒ Ⓓ	61. Ⓐ Ⓑ Ⓒ Ⓓ
11. Ⓐ Ⓑ Ⓒ Ⓓ	28. Ⓐ Ⓑ Ⓒ Ⓓ	45. Ⓐ Ⓑ Ⓒ Ⓓ	62. Ⓐ Ⓑ Ⓒ Ⓓ
12. Ⓐ Ⓑ Ⓒ Ⓓ	29. Ⓐ Ⓑ Ⓒ Ⓓ	46. Ⓐ Ⓑ Ⓒ Ⓓ	63. Ⓐ Ⓑ Ⓒ Ⓓ
13. Ⓐ Ⓑ Ⓒ Ⓓ	30. Ⓐ Ⓑ Ⓒ Ⓓ	47. Ⓐ Ⓑ Ⓒ Ⓓ	64. Ⓐ Ⓑ Ⓒ Ⓓ
14. Ⓐ Ⓑ Ⓒ Ⓓ	31. Ⓐ Ⓑ Ⓒ Ⓓ	48. Ⓐ Ⓑ Ⓒ Ⓓ	65. Ⓐ Ⓑ Ⓒ Ⓓ
15. Ⓐ Ⓑ Ⓒ Ⓓ	32. Ⓐ Ⓑ Ⓒ Ⓓ	49. Ⓐ Ⓑ Ⓒ Ⓓ	
16. Ⓐ Ⓑ Ⓒ Ⓓ	33. Ⓐ Ⓑ Ⓒ Ⓓ	50. Ⓐ Ⓑ Ⓒ Ⓓ	
17. Ⓐ Ⓑ Ⓒ Ⓓ	34. Ⓐ Ⓑ Ⓒ Ⓓ	51.	

ANSWERS

1. (*d*)	2. (*a*)	3. (*b*)	4. 1	5. (*b*)	6. (*a*)	7. (*a*)	8. (*c*)	9. (*c*)	10. (*c*)
11. (*b*)	12. (*d*)	13. (*d*)	14. (*c*)	15. (*a*)	16. (*b*)	17. (*a*)	18. (*a*)	19. (*c*)	20. (*d*)
21. (*c*)	22. (*b*)	23. (*a*)	24. (*b*)	25. (*a*)	26. (*b*)	27. (*d*)	28. (*a*)	29. (*c*)	30. (*c*)
31. (*b*)	32. (*d*)	33. (*c*)	34. (*a*)	35. (*c*)	36. (*b*)	37. (*c*)	38. (*c*)	39. (*a*)	40. (23.33)
41. (2.01)	42. (*d*)	43. (*b*)	44. (*a*)	45. (*c*)	46. (*b*)	47. (*b*)	48. (*b*)	49. (*a*)	50. (*d*)
51. (14.985)	52. (1.02)	53. (*c*)	54. (*d*)	55. (*d*)	56. (*b*)	57. (*c*)	58. (25)		
59. (*c*)	60. (29.6875)	61. (*b*)	62. (*d*)	63. (*a*)	64. (*c*)	65. (*b*)			

EXPLANATIONS

1. For an unpowered glide, there is no acceleration in X and Y directions

So, $L = w\cos\theta$ and $D = w\sin\theta$

$$\Rightarrow \tan\theta = \frac{D}{L}$$

$$\Rightarrow \theta = \tan^{-1}\left(\frac{D}{L}\right)$$

2. $$\eta_T = \frac{\text{Energy released by the engine}}{\text{Energy available from the fuel in the engine}}$$

$$\eta_T = \frac{\frac{1}{2}\dot{m}_e U_e^2 - \frac{1}{2}\dot{m}_i U^2}{\dot{m}_{fuel} Q_R}$$

$$\eta_T = \frac{\frac{1}{2}(\dot{m}_i + \dot{m}_{fuel})U_e^2 - \frac{1}{2}\dot{m}_i U^2}{\dot{m}_{fuel} Q_R}$$

where, Q_R = Calorific value of the fuel

$$\eta_T = \frac{\dot{m}_i\left[(1+f)\frac{U_e^2}{2} - \frac{U}{2}\right]}{\dot{m}_{fuel} Q_R} \approx \frac{U_e^2\left[1-\left(\frac{U}{U_e}\right)^2\right]}{2fQ_R}$$

3. In NACA 12018, Second and third digits i.e. 20, when divided by 2, gives the distance of maximum camber from the leading edge (as percent of chord). Therefore, maximum camber is 10 % of the chord.

4. $$\lim_{x\to 0}\frac{e^x - 1}{\sin x}$$

As indeterminate form of type 0/0, then we apply L'Hospital's rule

$$\lim_{x\to 0}\frac{e^x - 1}{\sin x} = \lim_{x\to 0}\frac{\frac{d}{dx}(e^x - 1)}{\frac{d}{dx}\sin x}$$

$$\Rightarrow \lim_{x\to 0}\frac{e^x - 0}{\cos x} = \lim_{x\to 0}\frac{e^x}{\cos x} = \frac{e^0}{\cos 0} = 1$$

5. For one end fixed and other hinged, we have,

$$P = \frac{\pi^2 EI}{\left(\frac{L}{\sqrt{2}}\right)^2} = \frac{\pi^2 EI}{(0.7L)^2}$$

6. According to Kepler's second law, Area swept out by radial vector from the primary planet to satellite in equal interval of time is constant, i.e

$$\frac{dA}{dt} = \text{constant}$$

$$h = \vec{v} \times \vec{r}$$

$$h = vr\sin\theta$$

$$\therefore \frac{dA}{dt} = \frac{1}{2} \times v \times r\sin\theta = \frac{1}{2}h = \text{constant}$$

where, h is the angular momentum which is conserved.

7. Given, $\alpha = 40°$

$\beta_e = 30°$

As per formula, we have,

$$\tan\beta_a = \frac{\tan\beta_e}{\cos\alpha}$$

$$\Rightarrow \beta_a = \tan^{-1}\left(\frac{\tan\beta_e}{\cos\alpha}\right)$$

$$\Rightarrow \beta_a = \tan^{-1}\left(\frac{0.577350}{0.766044}\right) = 0.645850 = 37°$$

8. $$Z = \frac{1}{2} + \frac{\sqrt{3}}{2}i$$

$$Z^2 = \left(\frac{1}{2} + \frac{\sqrt{3}}{2}i\right)\left(\frac{1}{2} + \frac{\sqrt{3}}{2}i\right) = \frac{-1}{2} + \frac{\sqrt{3}}{2}i$$

$$Z^3 = \left(-\frac{1}{2} + \frac{\sqrt{3}}{2}i\right)\left(\frac{1}{2} + \frac{\sqrt{3}}{2}i\right) = \frac{1}{2} - \frac{\sqrt{3}}{2}i$$

$$Z^4 = \left(\frac{1}{2} - \frac{\sqrt{3}}{2}i\right)\left(\frac{1}{2} + \frac{\sqrt{3}}{2}i\right) = \frac{-1}{2} - \frac{\sqrt{3}}{2}i$$

$$Z^5 = \left(-\frac{1}{2}-\frac{\sqrt{3}}{2}i\right)\left(\frac{1}{2}+\frac{\sqrt{3}}{2}i\right) = \frac{-1}{2}+\frac{\sqrt{3}}{2}i$$

$$Z^6 = \left(-\frac{1}{2}+\frac{\sqrt{3}}{2}i\right)\left(\frac{1}{2}+\frac{\sqrt{3}}{2}i\right) = \frac{1}{2}-\frac{\sqrt{3}}{2}i$$

9. A supercritical airfoil is an airfoil designed, primarily, to delay the onset of wave drag in the transonic speed range. Supercritical airfoils are characterized by their flattened upper surface, highly cambered aft section, and larger leading edge radius compared with NACA 6-series laminar airfoil shapes. Standard wing shapes are designed to create lower pressure over the top of the wing. The camber of the wing determines how much the air accelerates around the wing.

Generally, 8 series are nomenclature for Supercritical Aerofoils.

10. For a centrifugal compressor, Work done factor ψ is defined as the ratio of actual work done to theoretical work done., ψ ranges from 1.035 - 1.04, for isentropic case ψ = 1.

11. For a circular cross section

$$\tau_{max} = \frac{4}{3}\times\frac{P\left(r^2\right)}{\pi r^4}$$

$$\Rightarrow \tau_{max} = \frac{4}{3}\times\frac{P}{\pi r^2}$$

$$\text{and } \tau_{avg} = \frac{\text{total load}}{\text{area}} = \frac{P}{\pi r^2}$$

$$\therefore \frac{\tau_{max}}{\tau_{avg}} = \frac{4}{3}\times\frac{P}{\pi r^2}\times\frac{\pi r^2}{P}$$

$$\Rightarrow \frac{\tau_{max}}{\tau_{avg}} = \frac{4}{3}$$

12. –1 is one of the roots since

$$x^3 + x^2 + x + 1 = 0$$

$$(-1)^3 + (-1)^2 + (-1) + 1 = 0$$

By polynomial division

$$\frac{x^3 + x^2 + x + 1}{\{x-(-1)\}} = x^2 + 1$$

$$\Rightarrow x^3 + x^2 + x + 1 = \left(x^2+1\right)(x+1)$$

so the roots are

$$(-1, +j, -j)$$

13. For the maximum velocity of the rocket at constant burnout the following assumptions are-

I Constant gravitational force and Specific Impulse

II Steady flow through the rocket nozzle

III Drag force is negligible.

14. For a steady pull down flight

The turn rate is defined as $\psi = \frac{g}{V}(n+1)$

Therefore, $\psi \propto n,\ \psi \propto \frac{1}{V}$

$$\text{Radius of turn} = \frac{V}{\psi_{max}} = \frac{V^2}{g(n+1)}$$

15. $v_{avg} = \frac{v_{max}}{2}$

$$\frac{v_{avg}}{v_{max}} = \frac{1}{2} = 0.5$$

where, v_{avg} is average velocity and v_{max} maximum velocity respectively.

16. For longitudinal stability, $\frac{\partial C_m}{\partial \alpha} < 0$, so that the pitching moment for any change α is restoring at the trim condition.

In order for $\alpha_{trim} > 0$, C_{m_0} should be positive.

If α increases α_{trim}, negative pitching moment will decrease angle of attack and similar will follow for decrease in α.

17. $\frac{d^2y}{dx^2} + y = 0$

$$D^2 + 1 = 0$$

$$D = \pm i = 0 \pm 1i$$

$\therefore$ General solution is

$$y = e^{0x}\left[C_1\cos(1\times x) + C_2\sin(1\times x)\right]$$

$$= C_1\cos x + C_2\sin x$$

$$= P\cos x + Q\sin x$$

where P and Q are some constants.

18. Given,

T = 24 hours =86400 s

As per question,

$$T^2 = \frac{4\pi^2 a^3}{GM}$$

$$a = \sqrt[3]{\frac{T^2 \times GM}{4\pi^2}}$$

$$= \sqrt[3]{\frac{(86400)^2 \times 6.67 \times 10^{-11} \times 5.97 \times 10^{24}}{4 \times \pi^2}}$$

= 42226 km

19. Due to presence of taper in wing, axial stress in longitudinal also lead to a component of shear forces and hence also share part of shearing loads. Thus, both the axial loads carried by longitudinal as well as shear is altered due to presence of taper. But, the torsional moment is taken primarily by the skin and not by longitudinal won't develop any significant axial loads, as anyhow the net shear force due to components of axial force in longitudinal in the plane of shear must be zero.

20. Phugoid is a longitudinal long period oscillation where whole aircraft moves in a sinusoidal path. There is exchange of kinetic and potential energy. Angle of attack remains almost the same, while there are large variations in altitude and speed.

21. Given, P_0= 241.3 KPa

$T_0 = 611$ K

$M_e = 2.15$

$\gamma = 1.34$

As per formula, $= \left(1 + \frac{\gamma - 1}{2} M_e^2\right)^{\frac{\gamma}{\gamma - 1}}$

$$\Rightarrow \frac{P_0}{P_e} = \left(1 + \frac{1.34 - 1}{2} \times 2.15^2\right)^{\frac{1.34}{1.34-1}}$$

$$\Rightarrow \frac{P_0}{P_e} = 1.785825^{3.9411} = 9.82972$$

$$\Rightarrow \frac{P_e}{P_0} = \frac{1}{9.82972} = 0.101736$$

$\therefore P_e = 0.101736 \times P_0 = 0.101736 \times 241.3$

= 24.54 KPa

22. Positive dihedral, high mounted wing and swept back wings provide rolling stability to aircraft.

23.

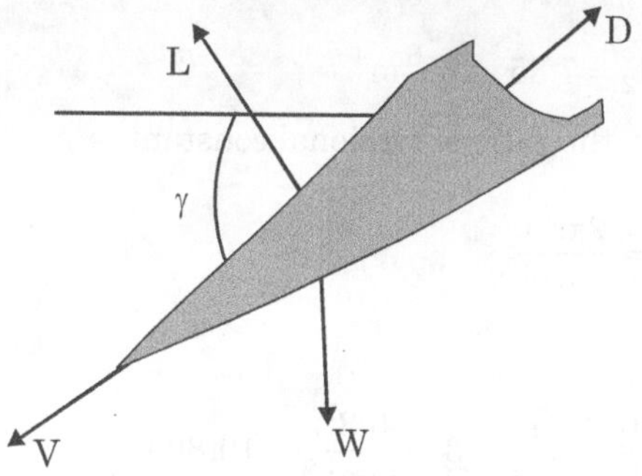

$L = W\cos\gamma$

$D = W\sin\gamma$

$$\frac{D}{L} = \frac{W\sin\gamma}{W\cos\gamma}$$

$$\tan\gamma = \frac{C_D}{C_L} = \frac{C_{D_0} + KC_L^2}{C_L}$$

For maximum range of gliding flight γ and $\tan\gamma$ should be minimum, so we wish to minimize the function.

$$\frac{C_{D_0}}{C_L} + KC_L = 0$$

$$\Rightarrow \frac{d}{dC_L}\left(\frac{C_{D_0}}{C_L} + KC_L\right) = 0$$

$$\Rightarrow -\frac{C_{D_0}}{C_L^2} + K = 0$$

$$\Rightarrow C_{D_0} = KC_L^2$$

24. For the closed section $J_1 = \frac{4\bar{A}^2}{\oint \frac{ds}{t}} = \frac{4 \times (\pi r^2)^2}{\frac{2\pi r}{t}}$

$$\Rightarrow J_1 = 2\pi r^3 t$$

$$\Rightarrow J_1 = 2 \times \pi \times 0.3^3 \times 0.005 = 8.48 \times 10^{-4}\, m^4$$

For the open section

$$J_2 = \frac{bt^3}{3}$$

$$\Rightarrow J_2 = \frac{2\pi r t^3}{3}$$

$$\Rightarrow J_2 = \frac{2 \times \pi \times 0.3 \times (0.005)^3}{3}$$

$$\Rightarrow J_2 = 7.85 \times 10^{-8} m^4$$

Now, the ratio of torsional constant

$$\frac{J_1}{J_2} = \frac{2\pi r^3 t}{\frac{2\pi r t^3}{3}}$$

$$\Rightarrow \frac{J_1}{J_2} = 3\frac{r^2}{t^2} = 3 \times \frac{0.3^2}{0.005^2} = 10{,}800$$

25. For directional stability $C_{l\beta}$ must be negative i.e., yaw stability.

26. $u_r = U\left[1 - \frac{a^2}{r^2}\right]\cos\theta$

when $\theta = \pi$ and $a = 1, r = 1$

$\Rightarrow u_r = 0$ m/s

Now, $u_\theta = -U\left(1 + \frac{a^2}{r^2}\right)\sin\theta - \frac{\Gamma}{2\pi r}$

$$\Rightarrow u_\theta = -25(1+1)\sin\pi - \frac{100\pi}{2\pi}$$

$$\Rightarrow u_\theta = 0 - 50$$

$$u_\theta = 50 \text{ m/s}$$

27. Potential Energy at height h is $\frac{GMm}{R_e + h}$ and

$h = nR$

$$P.E = \frac{-GMm}{(n+1)R}$$

Change in P.E. is

$$\Delta P.E = \frac{-GMm}{(n+1)R} - \left(\frac{-GMm}{R}\right)$$

$$\Delta P.E = \frac{-GMm}{R} \cdot \frac{R}{R}\left[\frac{1}{n+1} - 1\right]$$

$$\Delta P.E = \frac{-GMm}{R} \cdot \frac{R}{R}\left[\frac{1-n-1}{n+1}\right]$$

$$\Delta P.E. = mgR\left(\frac{n}{n+1}\right)$$

28. Given, $u = 0.02x + 0.02y - 0.01z$,

$v = 0.01y - 0.02z$,

$w = -0.01x + 0.01z$

Therefore, Normal strain Displacement

$$\frac{\partial u}{\partial x} = 0.02, \frac{\partial v}{\partial y} = 0.01, \frac{\partial w}{\partial z} = 0.01$$

Displacement at A (0, 0, 0)

$u = 0.02(0) + 0.02(0) - 0.01(0)$

$\therefore u = 0, v = 0, w = 0$

Displacement at B (5, 0, 0)

$u = 0.02(5) + 0.02(0) - 0.01(0)$

$\therefore u = 0.1, v = 0, w = -0.05$

New location of B is (5.1, 0, –0.05)

Distance between new locations is,

$$D = \sqrt{(5.1-0)^2 + 0^2 + (-0.05-0)^2} = 5.11 \text{ units}$$

29. Box contains 2 washers, 3 nuts and 4 bolts

p (2 washers, then 3 nuts, then 4 bolts)

$$= \left(\frac{2}{9} \times \frac{1}{8}\right) \times \left(\frac{3}{7} \times \frac{2}{6} \times \frac{1}{5}\right) \times \left(\frac{4}{4} \times \frac{3}{3} \times \frac{2}{2} \times \frac{1}{1}\right)$$

$$= \frac{1}{1260}$$

30. The aircraft takes roll over by 180° from steady level flight. Therefore, in the initial situation is in steady level flight. There was no change took place. So, after rolling velocity will be same, angle of attack will be same.

Now angle of attack will be negative to previous condition. Therefore, direction of lift will be changed. Both lift and weight force act in downward direction and aircraft has forward velocity. So, this will have circular motion.

Lift will produce necessary centripetal force.

Given as $L + mg = \frac{mv^2}{R}$

$$\Rightarrow 2mg = \frac{mv^2}{R}$$

$$\Rightarrow R = \frac{v^2}{2g}$$

$$\Rightarrow R = 127.4 \text{ m}$$

31. In a Ramjet Engine compressor is not used, the compression is achieved by ram effect. Ramjet engine has no rotating component Ram compression directly depend on efficiency of inlets to decelerate the incoming air without much pressure loss before entering into the combustion zone.

32. For linear profile $u = \frac{U_\infty y}{\delta}$

$$\Rightarrow \frac{u}{U_\infty} = \frac{y}{\delta}$$

Momentum thickness,

$$\theta = \int_0^\delta \frac{y}{\delta}\left(1 - \frac{y}{\delta}\right) dy$$

$$\Rightarrow \theta = \int_0^\delta \left(\frac{y}{\delta} - \frac{y^2}{\delta^2}\right) dy$$

$$\theta = \int_0^\delta \frac{y}{\delta}\left(1 - \frac{y}{\delta}\right) dy$$

$$\Rightarrow \theta = \left[\frac{y^2}{2\delta} - \frac{y^3}{3\delta^2}\right]_0^\delta$$

$$\Rightarrow \theta = \frac{\delta}{2} - \frac{\delta}{3}$$

$$\Rightarrow \theta = \frac{\delta}{6}$$

33. Given, $\int_0^1 f(x)dx$

x	f(x)
0.0	1.0
0.2	1.24
0.4	1.56
0.6	1.96
0.8	2.44
1.0	3.0

According to trapezoidal method,

$$\int_{x_0}^{x_n} y dx = \frac{h}{2}\left[y_0 + 2(y_1 + y_2 + ... + y_n) + y_n\right]$$

$$h = \frac{b-a}{n} = \frac{1-0}{5} = \frac{1}{5}$$

$$I = \frac{1}{10}[1 + 2(1.24 + 1.56 + 1.96 + 2.44) + 3]$$

$$I = 1.84.$$

34. $p_0 = p + \frac{1}{2}\rho v_1^2$

$$\Rightarrow v_1 = \sqrt{\frac{2(p_0 - p)}{\rho}}$$

$$\Rightarrow v_{actual} = \sqrt{\frac{2(p_0 - p)}{\rho}}$$

p is the static pressure measured by altimeter and T is air temperature at that point.

For calibrated speed which is observed in airspeed indicater uses ρ_s instead of ρ.

$$\therefore v_{calibrated} = \sqrt{\frac{2(p_0 - p)}{\rho_s}}$$

Therefore,

$$\frac{v_{actual}}{v_{calibrated}} = \sqrt{\frac{\rho_s}{\rho}}$$

35. Hoop Stress, $\sigma_\theta = \frac{pd}{2h} = \frac{2 \times 1}{2h} = \frac{1}{h}$MPa

Longitudinal Stress, $\sigma_l = \frac{pd}{4h} = \frac{2 \times 1}{4h} = \frac{0.5}{h}$MPa

$$\sigma_1^2 + \sigma_2^2 - \sigma_1\sigma_2 = \sigma^2$$

$$\Rightarrow \frac{1}{h^2}\left[(1)^2 + (0.5)^2 - 1 \times 0.5\right] = \left(\frac{200}{3}\right)^2$$

$$\Rightarrow \frac{1}{h^2}[1 + 0.25 - 0.5] = \frac{40000}{9}$$

$$\Rightarrow h^2 = \frac{0.75 \times 9}{40000} = 1.687 \times 10^{-4}$$

$$\Rightarrow h = 0.01299 \text{ m}$$

36. Kinetic Energy of a satellite in an orbit around earth is $KE = \frac{1}{2}mv_0^2 = \frac{GM_e m}{2(R_e + h)}$

$R = R_e + h$

$\therefore KE \propto \frac{1}{R_e + h}$

Thus, KE decreases as altitude increases.

37. Given,
Standard deviation, $\sigma = 8.8$ km/h
Mean speed, $\mu = 33$ km/h

$CV = \frac{\sigma}{\mu}$

$= \frac{8.8}{33}$

$= 0.2666$

38. Given, $AR = 5$
$e = 0.81$

$C_{D_0} = 0.019$

Maximum endurance case corresponds to minimum power in steady level flight.

$$P = Dv = W\left(\frac{C_D}{C_L}\right)v \qquad [\because W = L]$$

$$\therefore v = \sqrt{\frac{W}{\frac{1}{2}\rho S C_L}}$$

$$\Rightarrow P = \sqrt{\frac{W^3}{\frac{1}{2}\rho S}} \times \left(\frac{C_D}{C_L^{3/2}}\right)$$

Maximum endurance occurs when $\frac{C_L^{3/2}}{C_D}$ is maximum.

$$\Rightarrow \frac{d}{dC_L}\left(\frac{C_{D_0} + \frac{C_L^2}{\pi e AR}}{C_L^{3/2}}\right) = 0$$

At minimum power

$$\Rightarrow C_L = \sqrt{3\pi e AR C_{D_0}}$$

$$\therefore C_D = 4C_{D_0}$$

Now, for min imum power

$$\left(\frac{C_L}{C_D}\right) = \sqrt{\frac{3\pi e AR}{16C_{D_0}}} = \sqrt{\frac{3 \times \pi \times 0.81 \times 5}{16 \times 0.019}} = 11.20$$

39. Given, $\gamma = 1.4$

$\eta_c = 0.8$

$r = 10$

As per formula, we get,

$$\eta_c = \frac{r^{\frac{\gamma-1}{\gamma}} - 1}{r^{\frac{\gamma-1}{\gamma\eta_p}} - 1}$$

$$\Rightarrow r^{\frac{\gamma-1}{\gamma\eta_p}} - 1 = \frac{r^{\frac{\gamma-1}{\gamma}} - 1}{\eta_c}$$

$$\Rightarrow 10^{\frac{1.4-1}{1.4\eta_p}} - 1 = \frac{10^{\frac{1.4-1}{1.4}} - 1}{0.8}$$

$$\Rightarrow 10^{\frac{0.4}{1.4\eta_p}} = 1 + \frac{0.93}{0.8} = 2.1625$$

$$\Rightarrow \frac{0.4}{1.4\eta_p} = \log(2.1625) = 0.3349$$

$$\Rightarrow \eta_p = \frac{0.4}{1.4 \times 0.3349} = 0.8531$$

40. Given, $P = 100 \times 10^3$ N

$a = 2 \text{ m} = 2 \times 10^3$ mm

$L = 3 \text{ m} = 3 \times 10^3$ mm

$I = 10^8 \text{ mm}^4$

$E = 2 \times 10^6 \text{ N/mm}^2$

The relation gives deflection at the free end is

$$y_A = \frac{Pa^3}{3EI} + \frac{Pa^2}{2EI}(L - a)$$

Applying the values, we get

$y_A = 13.33 + 10 = 23.33$ mm

41. Let us first find θ, which is angle of disturbance

$$\tan\theta = \frac{2\cot\beta\left[M^2\sin^2\beta - 1\right]}{M_1^2(\gamma + \cos 2\beta) + 2}$$

$$\theta = \tan^{-1}\left[\frac{2\cot(40)\left[(3.5)^2\sin^2(40) - 1\right]}{(3.5)^2(1.4 + \cos 2(40)) + 2}\right] = 24.4°$$

Now,

$$M_{2,n}^2 = \frac{2 + (\gamma - 1)M_1^2\sin^2\beta}{2\gamma M_1^2\sin^2\beta - (\gamma - 1)}$$

$$\Rightarrow M_{2,n}^2 = \frac{2 + (1.4 - 1)\times 3.5^2\sin^2(40)}{2\times 1.4\times 3.5^2\sin^2(40) - (1.4 - 1)}$$

$$\Rightarrow M_{2,n}^2 = \frac{2 + 0.4\times 12.25\times 0.413}{2\times 1.4\times 12.25\times 0.413 - (0.4)}$$

$$= \frac{4.0237}{13.765} = 0.292$$

$$\Rightarrow M_{2,n} = 0.5406$$

$$M_2 = \frac{M_{2,n}}{\sin(\beta - \theta)} = \frac{0.5406}{\sin(40 - 24.4)} = 2.01$$

42. Given, $I_{sp} = 240$ s

$g = 9.8\ m/s^2$

$u_i = 850$ m/s

For rocket flight without drag and gravity is given as

$$\Delta u = I_{sp} g \ln\left(\frac{m_0}{m_f}\right)$$

$$\therefore u_f - 850 = 240\times 9.8\times \ln\left(\frac{m_0}{m_0/2}\right)$$

$$\Rightarrow u_f = 850 + 2352\times \ln 2$$

$$\Rightarrow u_f = 2480.28\ m/s$$

43. Given, $w_1 = w_2 = 2$ kN/m

L = 1.5 m

a = b = c = 0.5 m

x = 0.75 m

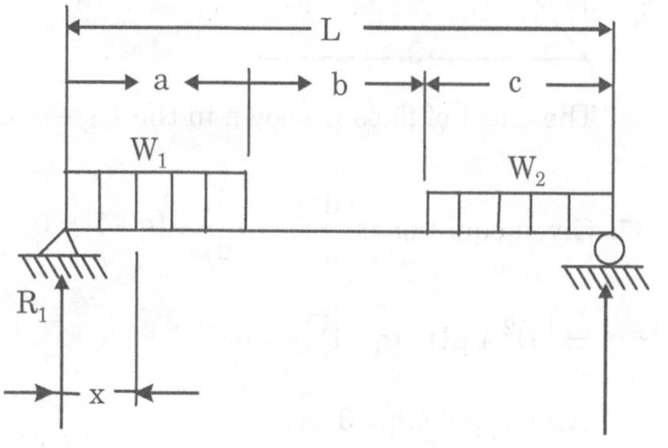

Taking moment along 1 i.e, $\sum M_1$

$$\therefore R_1 = \frac{w_1 a(2L - a) + w_2 c^2}{2L}$$

$$\Rightarrow R_1 = \frac{2\times 0.5\times(2\times 1.5 - 0.5) + 2\times 0.5^2}{2\times 1.5}$$

$$\Rightarrow R_1 = 1\ kN$$

Taking moment along 2 i.e. $\sum M_2 = 0$

$$\therefore R_2 = \frac{w_1 c(2L - c) + w_2 a^2}{2L}$$

$$\Rightarrow R_2 = \frac{2\times 0.5\times(2\times 1.5 - 0.5) + 2\times 0.5^2}{2\times 1.5}$$

$$\Rightarrow R_2 = 1\ kN$$

Now, Taking moment at $x = \frac{R_1}{w_1}$

where $R_1 < w_1 a$

i.e. $M_{max} = \frac{R_1^2}{2w_1} = \frac{1}{2\times 2} = 0.25$ kNm

Therefore, Bending moment at x

$$M_x = R_1 x - \frac{w_1 a}{2}(2x - a)$$

$$\Rightarrow M_x = 1\times 0.75 - \frac{2\times 0.5}{2}(2\times 0.75 - 0.5)$$

$$\Rightarrow M_x = 0.25\ kNm$$

44.

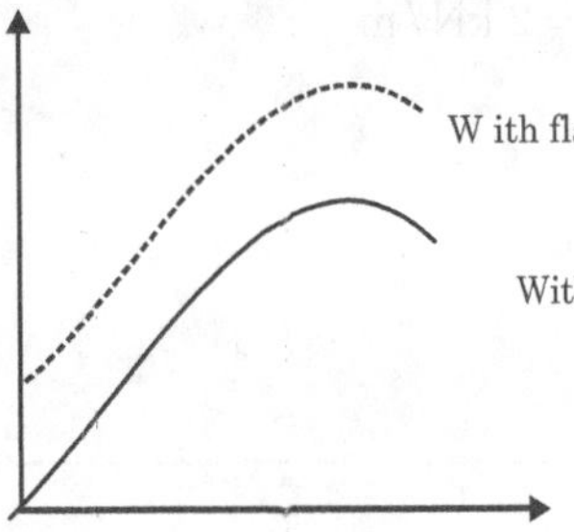

The effect of flaps is shown in the $C_L - \alpha$ curve

45. Given equation is $\frac{d^2y}{dx^2} + p\frac{dy}{dx} + (q+1) = 0$

$$\Rightarrow \left[D^2 + pD + (q+1)\right]y = 0$$

Assign $p = 4, q = 3$

$$\therefore \left(D^2 + 4D + 4\right)y = 0$$

$$D^2 + 4D + 4 = 0$$

$$(D+2)^2 = 0$$

$$\therefore D = -2, -2$$

$$\Rightarrow y = (c_1x + c_2)e^{-2x}$$

Out of choices given, $y = xe^{-2x}$

here, $c_1 = 1$ and $c_2 = 0$

46. Absolute ceiling of an aircraft is defined as an altitude at which power available is equal to power required to fly at that altitude. Rate of climb is zero and aircraft cannot go any higher for same flight velocity, simply because it doesn't have sufficient power available to increase the altitude further.

Rate of climb

$$= \frac{\text{Power available} - \text{Power required}}{\text{Weight of the aircraft}}$$

47. $\frac{d^2y}{dx^2} + 6\frac{dy}{dx} + 8x = 0$

$x(0) = 1$ and $\frac{dy}{dx}_{t=0} = 0$

$$D^2 + 6D + 8 = 0$$

$$(D+4)(D+2) = 0$$

$$D = -2 \text{ and } D = -4$$

$\therefore$ Solution is $x = C_1e^{-2t} + C_2e^{-4t}$

$\because x(0) = 1$

We have $C_1 + C_2 = 1$

$$\frac{dx}{dt} = -2C_1e^{-2t} - 4C_2e^{-4t}$$

$$\left[\frac{dx}{dt}\right]_{t=0} = 0$$

$$-2C_1 - 4C_2 = 0$$

Solving (i) and (ii)

we have, $C_1 = 2$ and $C_2 = -1$

So the solution is $x(t) = 2e^{-2t} - e^{-4t}$

48. Given, $\frac{r}{r_{tip}} = 0.8$

$R_{mid} = 0.5$

The degree of reaction varies with the radius as

$$R = 1 - \frac{k}{r^2} = 1 - \frac{k_1}{\left(r/r_{tip}\right)^2}$$

Putting $\frac{r}{r_{tip}} = 0.8$, to obtain

$k_1 = 0.64 \times 0.5 = 0.32$ $\quad [\because R_{mid} = 0.5]$

now, Putting $\frac{r}{r_{tip}} = 1$, to obtain

$$R_{tip} = 1 - k_1 = 1 - 0.32 = 0.68$$

49. $\Gamma(y) = 100\sqrt{1 - \left(\frac{2y}{4}\right)^2}\ m^2/s$

Here, $\Gamma_0 = 100, b = 4$

$\alpha_i = \frac{\Gamma_0}{2bv_\infty}$ (as the distribution is elliptical)

Induced angle of attack,

$$\alpha_i = \frac{100}{2 \times 4 \times 100} = 0.125 \text{ rad}$$

50. For a fourth order Runge-Kutta Method

$$y_{n+1} = y_n + \frac{h}{6}(k_1 + 2k_2 + 2k_3 + k_4)$$

$$t_{n+1} = t_n + h$$

where, $k_1 = f(t_n, y_n)$

$$k_2 = f\left(t_n + \frac{h}{2}, y_n + \frac{hk_1}{2}\right)$$

$$k_3 = f\left(t_n + \frac{h}{2}, y_n + \frac{hk_2}{2}\right)$$

$$k_4 = f(t_n + h, y_n + hk_3)$$

51. Given, $\sigma_1 = 60$ MPa

$\sigma_2 = 30$ MPa

Total strain energy per unit volume is

$$= \frac{1}{2E}\left[\sigma_1^2 + \sigma_2^2 + \sigma_3^2 - 2\nu(\sigma_1\sigma_2 + \sigma_2\sigma_3 + \sigma_3\sigma_1)\right]$$

$$= \frac{(10^6)^2}{2 \times 200 \times 10^9} \times$$

$$\left[\begin{matrix}60^2 + 30^2 + (-30)^2 - 2 \times 0.33 \\ ((60 \times 30) + (30 \times -30) + (-30 \times 60))\end{matrix}\right]$$

$$= 2.5 \times [5400 + 594]$$

$$= 14.985 \text{ KN} - \text{m/m}^3$$

52. Given, $\sigma_1 = 60$ MPa

$\sigma_2 = 30$ MPa

$\sigma_3 = -30$ MPa

Volumetric strain energy per unit volume is

$$= \frac{1}{3}(\sigma_1 + \sigma_2 + \sigma_3)^2\left(\frac{1-2\nu}{2E}\right)$$

$$= \frac{1}{3}(60 + 30 - 30)^2 \times 10^{12} \times \left(\frac{1 - 0.66}{2 \times 200 \times 10^9}\right)$$

$$= 1.02 \text{ KN} - \text{m/m}^3$$

53. The compressor characteristics curve is a measure of performance in terms of pressure ratio across it as a function of the non-dimensional mass flow rate. $\frac{\dot{m}\sqrt{T_0}}{p_0}$ is the main term in the non-dimensionalized mass flow rate. Therefore, for same pressure ratio we equate the quantity

$$\Rightarrow \frac{\dot{m}' \times \sqrt{1.44T_0}}{p_0} = \frac{\dot{m}\sqrt{T_0}}{p_0}$$

$$\Rightarrow \dot{m}' = \frac{\dot{m}}{2}$$

Therefore, the mass flow rate should be decreased by the factor of 2.0 to keep a constant pressure ratio.

54. For longitudinal stability $C_{m\alpha}$ must be negative i.e., pitch stability

For lateral stability $C_{n\beta}$ must be negative i.e., roll stability

55. Given, $A_{cs} = 0.5 \text{ m}^2$

$\rho_p = 2000$ kg/s

$\dot{r} = 7$ mm/s

$I_{sp} = 200$ s

$$\text{Thrust} = \dot{m}U_e = \dot{m}gI_{sp}$$

Now, $\dot{m} = A_{cs} \times \rho_p \times \dot{r}$

$$\Rightarrow \dot{m} = 0.5 \times 2000 \times 0.007$$

$$\Rightarrow \dot{m} = 7 \text{ kg/s}$$

Therefore, Thrust $= 7 \times 200 \times 9.81$

Thrust = 13.73 kN

56. The volume of a sphere of diameter 11 unit is less than the volume of a cube of side 11 unit.

57. The aircraft was allowed to take off as soon as its flight plan was filed.

58. Probability of an earthquake greater than magnitude 6 is 0.04.

This means 4 out of 100 years will face an earthquake greater than magnitude 6. So average time between successive earthquakes will be 100/4 = 25 years.

59. Option A - Wrong as this statement means that no country is welcome to women as against the meaning conveyed by the given sentence.

Option B - Wrong - the given statement says nothing about citizenship.

Option C - Correct - expresses the same meaning as given sentence.

Option D - Wrong - this may or may not be universally true, but in no way expresses the same meaning as given statement.

60. As per given question,

$$\frac{\text{Sum of total numbers}}{50} = 30$$

$\therefore$ Sum of total numbers = 1500

After 35 and 40 are discarded, we get,

1500 – 35 – 40 = 1425

$$\therefore \text{New Average} = \frac{1425}{48} = 29.6875$$

61. Perimeter = 6 m

5a = 6 m

a = 65 m

a = 1.2 m

Area =Area (triangle) + Area (square)

$$\text{Area} = \frac{\sqrt{3}}{4}.a^2 + a^2$$

$$\text{Area} = \frac{\sqrt{3}}{4}.1.2^2 + 1.2^2$$

Area = 2.06 m .

62. Total work done (P team) = $20 \times 32 = 640$ engineer-days.

Total work done (Q team) = $16 \times 30 = 480$ engineer-days.

$$\text{Ratio } \frac{\text{1 member of P}}{\text{1 member of Q}} = \frac{640}{480} = \frac{4}{3} = 4:3$$

63. Let, after adding 'x' litre of water, we get 75 litres of a 15% acid solution.

Before adding these 'x' litre of water we had (75 – x) litre of 40% acid solution.

$$\therefore 40\% \text{ of } (75 - x) = 15\% \text{ of } 75$$

$$\Rightarrow (75 - x) = \frac{11.25}{0.40}$$

$$\Rightarrow x = 75 - \frac{11.25}{0.40}$$

$$\Rightarrow x = \frac{30 - 11.25}{0.40} = 46.875 \text{ L}^2$$

64. The Tamil version of the John Abraham- starrer Madras Cafe was cleared by the Censor Board with no cuts last week, but the film's distributor found no takers among the exhibitors for a release in Tamilnadu on this Friday.

Article "the" is used to define a particular object here (The Movie).

65. Let us consider M for man and R for rakhsas.

Step 1: Two rakhsasas row the boat from one side to another. So, one rakshas will reach other side and one rakhsas will have to row back.

So, first side now (3M + 2R) and second side (1R). Therefore, boat crosses 2 times.

Step 2: 2R row to other side and leaves one R there and comes back to first side. So, now first side has (3M + 1R) and second side has (2R). Therefore, boat crosses 2 times.

Step 3: 2 M row to other side and gets down at other side. So, first side has (1R + 1M) and second side has (2M + 2R). Therefore, boat crosses 1 time.

Step 4: Now, 1M and 1R will have to row back from second side, and they will take 1M + 1M to other side. So, in the first side, we now have 2R. Therefore, boat crosses 2 times.

Step 5: Last step, 2R will row to the other side. Therefore, boat crosses 1 time.

Total time boat crosses = 8 + 1 = 9 times.

MOCK TEST – 4

AEROSPACE ENGINEERING

Time Allowed: 3:00 Hours | *Maximum Marks: 100*

Read the following instructions carefully before you begin to answer the questions. This Booklet contains questions in English.

INSTRUCTIONS

1. This Booklet contains **65 questions** in total comprising of the following two parts:

   Part -1: Technical Section (85 Marks) *(1-55 Questions)*

   Part -2: General Aptitude Section (15 Marks) *(56-65 Questions)*

2. All questions are compulsory. Q.1-Q.25 & Q.56-Q.60 carry one mark each. Q.26-Q.55 & Q.61-Q.65 carry two marks each.

3. **Types of Questions:**

   (*a*) Multiple Choice Questions(MCQ) carrying 1 or 2 marks each in all papers and sections. These questions are objective in nature, and each will have a choice of four options, out of which the candidate has to mark the correct answer(s).

   (*b*) Numerical Answer Questions of 1 or 2 marks each in all sections. For these questions the answer is a real number, to be entered by the candidate using the virtual keypad. No choices will be shown for these type of questions.

4. For 1-mark multiple-choice questions, 1/3 marks will be deducted for a wrong answer. Likewise, for 2-marks multiple-choice questions, 2/3 marks will be deducted for a wrong answer. There is no negative marking for numerical answer type questions.

MOCK TEST – 4

1. Matrix A = $\begin{bmatrix} 2 & 1 \\ 2 & 1 \end{bmatrix}$ and B = $\begin{bmatrix} 3 & 2 \\ 2 & 1 \end{bmatrix}$. What is the determinant of matrix A × B?

(*a*) ∞ (*b*) –1

(*c*) 1 (*d*) 0

2. Minimum of $G(x) = 4x^2 - 8x + 10$ occurs at

(*a*) x = 0 (*b*) x = 9

(*c*) x = ∞ (*d*) x = 1

3. $(\sin\theta + \cos\theta)^2 - \sin^2\theta - \cos^2\theta - 2\sin\theta\cos\theta$ is equal to

(*a*) ∞ (*b*) 1

(*c*) –1 (*d*) 0

4. x + y + z = 1, 2x + 3y + 4z = 10

above system of equation has

(*a*) Unique solution (*b*) Infinite solution

(*c*) 2 solutions (*d*) No solution

5. Check whether given statement is true or false:

Range of a Glider when lift to weight ratio is maximum ________

6. For roll stability of high wing airplane which of following configuration is favourable

(*a*) wing with dihedral angle

(*b*) wing with anhedral angle

(*c*) straight wing

(*d*) tapered wing

7. Condition for steady level flight is :

(*a*) L = W, T = L (*b*) T = W, L = D

(*c*) L = W, T = D (*d*) None of these

8. Which control surface is generally used for rolling:

(*a*) flap (*b*) flaperons

(*c*) ailerons (*d*) throttle

9. For aircraft longitudinal static stability

(*a*) $\frac{dC_{CM}}{dC_M} < 0$ (*b*) $\frac{dC_M}{d\alpha} < -1$

(*c*) $\frac{dC_M}{d\alpha} < 0$ (*d*) $\frac{dC_M}{d\alpha} > 0$

10. Stall velocity of aircraft is given by.

(*a*) $V_{stall} = \sqrt{\frac{2W}{C_L \times \rho S^2}}$ (*b*) None of these

(*c*) $V_{stall} = \sqrt{\frac{2C_L}{W\rho S}}$ (*d*) $V_{stall} = \sqrt{\frac{2W}{\rho S C_{L_{max}}}}$

11. Velocity which is required for a statelike to remain in orbit is called.

(*a*) Escape velocity (*b*) Orbital velocity

(*c*) Critical velocity (*d*) Velocity of orbit

12. Escape velocity of satellite is given by :

(*a*) $V_{escape} = \sqrt{\frac{GR}{M}}$ (*b*) $V_{escape} = \sqrt{\frac{2GM}{R^2}}$

(*c*) $V_{escape} = \sqrt{\frac{2GM}{R}}$ (*d*) None of these

13. Which statement is true

Across a normal shock

(*a*) Total temperature remains constant

(*b*) Total pressure remains constant

(*c*) Static temperature remains constant

(*d*) Static pressure remains constant

14. Critical Mach Number of an airfoil is always

(*a*) less than drag divergence Mach Number

(*b*) Greater than drag - divergence Mach Number

(*c*) Equal to drag - divergence Mach Number

(*d*) None of these

15. Total pressure of a flow is given by :

(*a*) $P_t = P_{st}\left[1 + \frac{\gamma - 1}{2}M^2\right]^{\frac{\gamma}{\gamma-1}}$

(*b*) $P_t = \left[1 + \left(\frac{\gamma - 1}{2}\right)M^2\right]$

(*c*) $P_t = P_{st}\left[\frac{\gamma - 1}{2}M^2\right]$

(*d*) None of these

16. Reynolds number of flow over a plate of length l at a distance x is given by:

(*a*) $\frac{\rho V x}{\mu}$ (*b*) $\frac{\rho V d}{\mu}$

(*c*) $\frac{\mu}{\rho x}$ (*d*) $\frac{\rho V \mu}{x}$

17. Which of the following airfoils are used to increase drag - divergence mach number?

(*a*) symmetric airfoil (*b*) high camber airfoil

(*c*) laminar airfoil (*d*) super critical airfoils

18. Damping ratio of a spring mass system is greater than 1. This is a

(*a*) underdamped system

(*b*) overdamped system

(*c*) critically damped system

(*d*) undamped system

19. For plain stress problem :

(*a*) $\sigma_{xx}, \sigma_{yy}, \sigma_{zz} = 0$

(*b*) only $\sigma_{zz} = 0$

(*c*) only σ_{xx} and $\sigma_{yy} = 0$

(*d*) $\tau_{xz}, \tau_{yz}, \sigma_{zz} = 0$

20. For an isotropic material young's modulus is 4 times greater than shear modulus. What is the Poisson's ratio?

(*a*) 0.4 (*b*) 0.2

(*c*) 1 (*d*) 0.13

21. Maximum deflection for simply supported beam under point load at center is given by :

(*a*) $\frac{PL^3}{3EI}$ (*b*) $\frac{PL^3}{4EI}$

(*c*) $\frac{PL^3}{30EI}$ (*d*) $\frac{PL^3}{48EI}$

22. A beam made up of two material with young's modulus and moment of Inertia $E_A = 100 \times 10^9$ Pa, $E_B = 50 \times 10^9$ Pa and $I_A = 6 \times 10^{-9}$ m^4, $I_B = 8 \times 10^{-9}$ m^4 is subjected to pure bending moment 20kN-m. Radius of curvature of the beam deflection is _______ .

23. Turbojet engine of aircraft operates on

(*a*) Otto cycle (*b*) Carnot cycle

(*c*) Brayton cycle (*d*) Rankine cycle

24. Which of the following engines has highest overall effeciency in mach number range of 0.6 – 0.86?

(*a*) Turbojet (*b*) Tuboshaft

(*c*) Turboprop (*d*) Turbofan

25. A constant Area-duct combustor of Aircraft Gas tuxine engine has inlet stagnation temperature of 540 K and outlet stagnation temperature of 1200K. Heat addition is causing flow inside combustor. Assuming flow is 1–D, incompressible, and frictionless; Pressure loss factor of combustor, is ______ .

26. Eigen values of matrix $A = \begin{bmatrix} 2 & 2 \\ 2 & 3 \end{bmatrix}$ are

(*a*) 1, 2 (*b*) 3, 1

(*c*) 4, 2 (*d*) 4.5, 0.4

27. Solving integral $f = \int_0^1 (x^2 + 2)dx$ using trapezoidal rule with $\Delta x = 0.5$ gives f = ________

28. Let $F(s) = \frac{(S+5)}{(S+3)(S+4)}$

Partial fraction of F(S) is given by

(*a*) $\frac{1}{(S+3)} + \frac{1}{(S+4)}$ (*b*) $\frac{3}{(S+3)} + \frac{4}{(S+4)}$

(*c*) $\frac{2}{(S+3)} - \frac{1}{(S+4)}$ (*d*) None of these

29. Laplace inverse of $f(S) = \frac{S+5}{(S+3)(S+4)}$ is

(*a*) $e^{4t} + e^{-3t}$ (*b*) $e^{-3t} + e^{-4t}$

(*c*) $2e^{-3t} + e^{-4t}$ (*d*) None of these

30. Which of the following is a linear differential equation?

(*a*) $\frac{d^2y}{dx^2} + \frac{dy}{dx} + 4y = 0$

(*b*) $\frac{d^2y}{dt^2} + \frac{dy}{dt} - 4y = 0$

(*c*) $m\frac{d^2y}{dt^2} + \frac{cdx}{dt} - kx = 0$

(*d*) All of these

31. An aircraft flying at maximum lift coefficient of $C_{lmax} = 0.5$ has weight of 500 kg. If surface area of wing is S = 5m^2 and is flying at sea-level, what is the stall velocity of aircraft?

(*a*) 55 m/sec.

(*b*) 56.56 m/sec.

(*c*) 225 m/sec.

(*d*) 100 m/sec.

32. The lateral directional characteristic equation for an airplane has given following roots.

$\lambda_1 = -0.67, \lambda_2 = -0.006, \lambda_{3,4} = -0.02 \pm \sqrt{3.2}j$

Which of the root Corresponds to the lightly damped spiral mode.

(*a*) $\lambda_1 = -0.67$ (*b*) $\lambda_2 = -0.006$

(*c*) $\lambda_3 = -0.02 + \sqrt{3.2}j$ (*d*) $-0.02 - \sqrt{3.2}j$

33. For an aircraft longitudinal static stability margin (SM) is between 5% to 10% of mean aerodynamic chord and $\frac{dC_m}{dC_L} = 0.2$, $\overline{X}_{CG} = 0.25$ [from experiment]. What is the most forward location of center of gravity permitted to fulfill the requirement : (Given $\frac{dC_m}{dC_L} = -SM$)

(*a*) $0.34\overline{C}$ (*b*) $0.45\overline{C}$

(*c*) $0.4\overline{C}$ (*d*) None of these

34. From data of Q.33 what is most aft location of C.G.

(*a*) 0.4C (*b*) 0.3C

(*c*) 0.35C (*d*) 0.45C

35. Maximum excess power of and aircraft is given to be 14×10^4 W. If aircraft weighs 500kg. What is maximum rate of climb of this aircraft.

(*a*) 20.57 (*b*) 21.57

(*c*) 27.57 (*d*) 28.57

36. Absolute ceiling of aircraft is defined as the altitude :

(*a*) at which $\left(\frac{R}{C}\right)_{max.}$ is 100 ft/min

(*b*) at which $\left(\frac{R}{C}\right)_{max.}$ is 0.508 m/sec

(*c*) at which $\left(\frac{R}{C}\right)_{max.}$ is zero

(*d*) none of these

37. A planet has radius of 40×10^5m and accelerection due to gravity g = 8m/sec². What is the escape velocity of object, given that mass of planet is 10^{16} kg.

(*a*) 0.56 m/sec (*b*) 0.55 m/sec

(*c*) 0.58 m/sec (*d*) 0.57 m/sec

38. Orbital velocity of a satellite depends upon :

(*a*) Mass of planet

(*b*) distance of object from center of planet

(*c*) None of 1 and 2

(*d*) Both of 1 and 2

39. 10mm thick laminar boundary layer is developed over a flat plate parallel to the flow at a distance 0.5m downward of leading edge. Thickness of boundary layer at 0.8m downstream of flow is :

(*a*) 10mm (*b*) 20mm

(*c*) 30mm (*d*) 12.64mm

40. Maximum thickness as %age of chord for the NACA 25014 airfoil is ______ .

41. Pressure at a point of wing in found to be 8×10^4 Pa when freestream velocity and pressure was 100 m/sec. and 10^5 Pa respectively what is the coefficient of pressure at that point if flow is incompressible and density of freestream is 1.225 kg/m²

(*a*) 3.26 (*b*) 2.36

(*c*) 6.23 (*d*) –3.26

42. Consider a C – D Nozzle with steady, inviscid flow and a normal shock at divergent portion. Mach number at downstream of the shock

(*a*) increases (*b*) decreases

(*c*) remains constant (*d*) none of these

43. Consider a flow with stream function.

$$\varphi = \frac{1}{8}\,\mathrm{In}\left[x^2 + y^2\right]$$

What is tangential velocity?

(*a*) $\frac{1}{2r}$ (*b*) $\frac{1}{3r}$

(*c*) $\frac{1}{4r}$ (*d*) 4r

44. Reynold number for a turbulent boundary layer was found to be 10^4, on a flat plate at distance of 0.8 m from leading edge of plate. What is skin-friction coefficient?

(*a*) 0.0114

(*b*) 0.0113

(*c*) 0.0124

(*d*) 0.031

45. What is the response for a spring mass damper system with mass = 10kg and spring constant $K = 10^4 \frac{N}{m}$, if it is excited by a harmonic force of 100N at 5Hz frequency given that damping ratio of system is 0.2.

(*a*) 0.024m (*b*) 0.025m

(*c*) 0.26 (*d*) 0.46

46. A pinned-pinned column of rectangular cross-section of length 1m, width 0.2m and height 0.1m is made up of steel material with young's modulus of 210 GPa. What is the critical Buckling load for the column?

(*a*) 34560.2 kN (*b*) 34720.4 kN

(*c*) 34543.6 kN (*d*) 3400.0 kN

47. For a thin walled closed tube of thickness 3mm, outer and inner radius of 55mm and 52mm, torsional constant J is ________ mm⁴

48. A 2 – D rectangular plate is under plane stress condition. This is made up of an isotropic material with yield strength 250 MPa. Tensile stresses in two perpendicular directions are given as 200 MPa and 150 MPa. Using Von-mises failure theory, predict the failure of plate.

(*a*) fail (*b*) safe

(*c*) Buckle (*d*) non of these

49. Composite materials are preferred in Aerospace Industries because of

(*a*) Cheaper

(*b*) Strength to weight ratio

(*c*) Easy fabrication

(*d*) Availability

50. A shaft made up of two materials, has inner material with G_A = 50 GPa and radius 10mm and outer material with G_B = 100 GPa and radius 20mm. If length of shaft is 2m, Torsional slifﬁness $\frac{T}{\phi}$ (Where ϕ is angle of twis t)_____kNmm

51. What is equivalence ratio for combustion of octane fuel and air oxidizer with following stoichiometric chemical reaction given that fuel air ratio is 0.03.

$4C_8H_{18} + (250_2 + 94N_2) \rightarrow$
$32CO_2 + 36H_2O + 94N_2$

(*a*) 0.23 (*b*) 0.29
(*c*) 0.227 (*d*) 0.243

52. Inlet Mach number of a diffuser of an aircraft engine is 0.8 which is decelerated to 0.32 at compressor inlet. If diffuser pressure recovery ratio is 0.9 when isentropic efficiency of diffuser is _____ (γ = 1.4)

53. In an rocket engine, exit mass flow rate of hot gas coming out of Nozzle is 800 kg/sec with 2000 m/sec. Exit velocity & cross-sectional area of nozzle is 0.7 m^2. If nozzle operates at optimum condition. Then thrust produced by rocket is _____ kN

Linked question (54-55)

Entering and existing velocity from an compressor of gas turbine engine is 137 m/s and 39 m/s. Stagnation pressure at inlet is 0.8 MPa and at outlet 3.66 MPa. Assuming that specific heat ratio is constant i.e. γ = 1.4

54. Compressor pressure ratio is equal to

(*a*) 4.2 (*b*) 4.1
(*c*) 4.32 (*d*) 4.57

55. If isentropic efficiency of compressor 0.87. Then what is polytropic efficiency?

(*a*) 0.7 (*b*) 0.83
(*c*) 0.89 (*d*) 0.93

56. Fate smiles _____ those who untringly grapple with stark realities of life.

(*a*) With (*b*) Over
(*c*) On (*d*) Round

57. Catching the earlier train will give us the _____ to do some shopping.

(*a*) Chance (*b*) Luck
(*c*) Possibility (*d*) Occasion

58. Which of sentence given below has error :

(*a*) We discussed about the problem so thoroughly
(*b*) On the eve of the examination
(*c*) That I found it very easy to work it out
(*d*) No error

59. In first 10 overs of a cricket game, the run rate was only 3.2. What should be the run rate in the remaining 40 overs to reach the target of 282 runs?

(*a*) 6.25
(*b*) 5.5
(*c*) 7.4
(*d*) 5

60. Which of following statement is not correct?

(*a*) log(2 × 4 × 6) = log2 + log4 + log6
(*b*) $\log_5 1 = 0$
(*c*) log(3 + 4) = log(3 × 4)
(*d*) $\log_5 5 = 1$

61. Rearrange the words below to make meaningful sentence: When he

P. did not know
Q. he was nervous and
R. heard the hue and cry at midnight
S. what to do

(*a*) RQPS (*b*) QSPR
(*c*) SQPR (*d*) PQRS

62. Choose the sentence which best expresses given sentence in active/passive voice.

After driving professor kumar to the museum, she dropped him at his hotel.

(*a*) After being driven to the museum, professor kumar was dropped at his hotel
(*b*) Professor kumar was being driven dropped at his hotel
(*c*) Statement is grammatically not correct.
(*d*) None of these

63. Pick out best one, which complete given sentence.

Despite his best efforts to conceal the anger _____.

(*a*) We could detect that he was very happy
(*b*) He failed to give us an impression of his agony
(*c*) People came to know that he was annoyed
(*d*) None of these

64. The angle of elevation of a ladder leaning against a wall is of 65° and foot of ladder is 13.4m away from wall. The length of ladder is

(*a*) 7m
(*b*) 8.9m
(*c*) 31.7m
(*d*) 10.31m

65. A girl takes 6 hours in walking to a place and coming back on a ride. She would have gained 2 hours by takig ride for both sides. Time she would take if she walks both side?

(*a*) 11 hrs (*b*) 13 hrs
(*c*) 12.2 hrs (*d*) 8 hrs

OMR SHEET

1. Ⓐ Ⓑ Ⓒ Ⓓ	18. Ⓐ Ⓑ Ⓒ Ⓓ	35. Ⓐ Ⓑ Ⓒ Ⓓ	52. ☐
2. Ⓐ Ⓑ Ⓒ Ⓓ	19. Ⓐ Ⓑ Ⓒ Ⓓ	36. Ⓐ Ⓑ Ⓒ Ⓓ	53. ☐
3. Ⓐ Ⓑ Ⓒ Ⓓ	20. Ⓐ Ⓑ Ⓒ Ⓓ	37. Ⓐ Ⓑ Ⓒ Ⓓ	54. Ⓐ Ⓑ Ⓒ Ⓓ
4. Ⓐ Ⓑ Ⓒ Ⓓ	21. Ⓐ Ⓑ Ⓒ Ⓓ	38. Ⓐ Ⓑ Ⓒ Ⓓ	55. Ⓐ Ⓑ Ⓒ Ⓓ
5. ☐	22. ☐	39. Ⓐ Ⓑ Ⓒ Ⓓ	56. Ⓐ Ⓑ Ⓒ Ⓓ
6. Ⓐ Ⓑ Ⓒ Ⓓ	23. Ⓐ Ⓑ Ⓒ Ⓓ	40. ☐	57. Ⓐ Ⓑ Ⓒ Ⓓ
7. Ⓐ Ⓑ Ⓒ Ⓓ	24. Ⓐ Ⓑ Ⓒ Ⓓ	41. Ⓐ Ⓑ Ⓒ Ⓓ	58. Ⓐ Ⓑ Ⓒ Ⓓ
8. Ⓐ Ⓑ Ⓒ Ⓓ	25. ☐	42. Ⓐ Ⓑ Ⓒ Ⓓ	59. Ⓐ Ⓑ Ⓒ Ⓓ
9. Ⓐ Ⓑ Ⓒ Ⓓ	26. Ⓐ Ⓑ Ⓒ Ⓓ	43. Ⓐ Ⓑ Ⓒ Ⓓ	60. Ⓐ Ⓑ Ⓒ Ⓓ
10. Ⓐ Ⓑ Ⓒ Ⓓ	27. ☐	44. Ⓐ Ⓑ Ⓒ Ⓓ	61. Ⓐ Ⓑ Ⓒ Ⓓ
11. Ⓐ Ⓑ Ⓒ Ⓓ	28. Ⓐ Ⓑ Ⓒ Ⓓ	45. Ⓐ Ⓑ Ⓒ Ⓓ	62. Ⓐ Ⓑ Ⓒ Ⓓ
12. Ⓐ Ⓑ Ⓒ Ⓓ	29. Ⓐ Ⓑ Ⓒ Ⓓ	46. Ⓐ Ⓑ Ⓒ Ⓓ	63. Ⓐ Ⓑ Ⓒ Ⓓ
13. Ⓐ Ⓑ Ⓒ Ⓓ	30. Ⓐ Ⓑ Ⓒ Ⓓ	47. ☐	64. Ⓐ Ⓑ Ⓒ Ⓓ
14. Ⓐ Ⓑ Ⓒ Ⓓ	31. Ⓐ Ⓑ Ⓒ Ⓓ	48. Ⓐ Ⓑ Ⓒ Ⓓ	65. Ⓐ Ⓑ Ⓒ Ⓓ
15. Ⓐ Ⓑ Ⓒ Ⓓ	32. Ⓐ Ⓑ Ⓒ Ⓓ	49. Ⓐ Ⓑ Ⓒ Ⓓ	
16. Ⓐ Ⓑ Ⓒ Ⓓ	33. Ⓐ Ⓑ Ⓒ Ⓓ	50. ☐	
17. Ⓐ Ⓑ Ⓒ Ⓓ	34. Ⓐ Ⓑ Ⓒ Ⓓ	51. Ⓐ Ⓑ Ⓒ Ⓓ	

ANSWERS

1. (*d*)	2. (*d*)	3. (*d*)	4. (*d*)	5. (False)	6. (*a*)	7. (*c*)	8. (*c*)
9. (*c*)	10. (*d*)	11. (*b*)	12. (*c*)	13. (*a*)	14. (*a*)	15. (*a*)	16. (*a*)
17. (*d*)	18. (*b*)	19. (*d*)	20. (*c*)	21. (*d*)	22. (ρ = 0.05m)		23. (*c*)
24. (*d*)	25. (PLF = 0)		26. (*d*)	27. (f = 2.375)		28. (*c*)	29. (*c*)
30. (*d*)	31. (*b*)	32. (*b*)	33. (*c*)	34. (*d*)	35. (*d*)	36. (*c*)	37. (*d*)
38. (*d*)	39. (*d*)	40. (.14)	41. (*d*)	42. (*a*)	43. (*c*)	44. (*a*)	45. (*b*)
46. (*c*)	47. (2886439.567)		48. (*a*)	49. (*b*)	50. (12959.06)		51. (*c*)
52. (ηd= 0.73)		53. (1600)	54. (*d*)	55. (*c*)	56. (*c*)	57. (*a*)	58. (*a*)
59. (*a*)	60. (*c*)	61. (*a*)	62. (*a*)	63. (*c*)	64. (*c*)	65. (*d*)	

EXPLANATIONS

1. $A \times B = \begin{bmatrix} 2 & 1 \\ 2 & 1 \end{bmatrix} \begin{bmatrix} 3 & 2 \\ 2 & 1 \end{bmatrix}$

$= \begin{bmatrix} 6+2 & 4+1 \\ 6+2 & 4+1 \end{bmatrix} = \begin{bmatrix} 8 & 5 \\ 8 & 5 \end{bmatrix}$

$\therefore$ determinant $|A \times B| = 40 - 40 = 0$

2. $G(x) = 4x^2 - 8x + 10$

$G'(x) = 8x - 8$

$G''(x) = 8$ = Positive [$\therefore$ minimum of G(x)]

Minimum of G(x) :

$G'(x) = 8x - 8 = 0$

$\boxed{x = \frac{8}{8} = 1}$

3. $(\sin\theta + \cos\theta)^2 - \sin^2\theta - \cos^2\theta - 2\sin\theta\cos\theta$

Using property $\boxed{\sin^2\theta + \cos^2\theta = 1}$,

$(a + b)^2 = a^2 + b^2 + 2ab$

$= \sin^2\theta + \cos^2\theta + 2\sin\theta\cos\theta - \sin^2\theta - \cos^2\theta$

$- 2\sin\theta\cos\theta$

$= 1 - 1 = 0$

4. There are 3 variable or unknowns in above equation, we need at least 3 equations to solve them but here we have only 2 equations available, thus this system has no solution.

5. False

Because to maximize the range of a glider lift to drag ratio should be maximum in place of left to weight.

$R \propto \frac{L}{D}$

6. Wing with dihedral angle

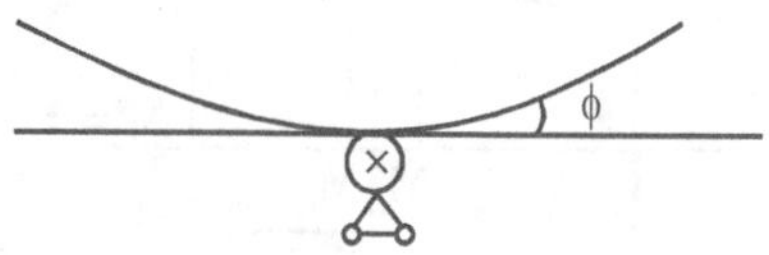

ϕ = dihedral angle

Dihedral angle helps aircraft to come to original position whenever there is a roll disturbance.

7.

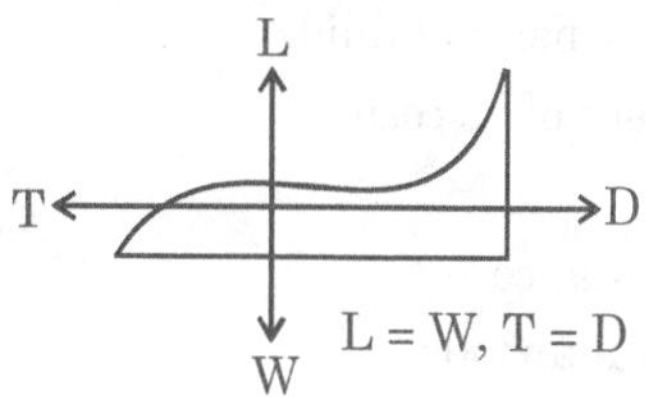

9.

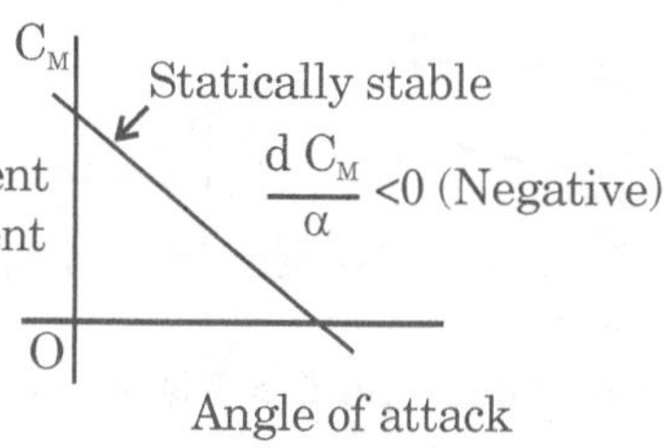

10. Stall speed occurs at maximum lift coefficient

$$V_{Stall} = \sqrt{\frac{2W}{\rho S C_{L_{max}}}}$$

ρ = density of freestream air

S = surface area

$C_{L_{max}}$ = maximum lift coefficient

W = Weight of aircraft

12. Escape velocity $V_{escape} = \sqrt{\frac{2GM}{R}}$ m/s.

Where, G = = 6.673 × 10^{-11} Nm^2/Kg^2, Universal Gravitational constant

M = Mass of planet (kg)

R = Radius of planet (m)

13. Across a normal shock only total enthalpy thus total temperature remains constant. All other quantities varies.

14. $M_{critical}$ is always less than $M_{drag\text{-}diverg}$ for an airfoil.

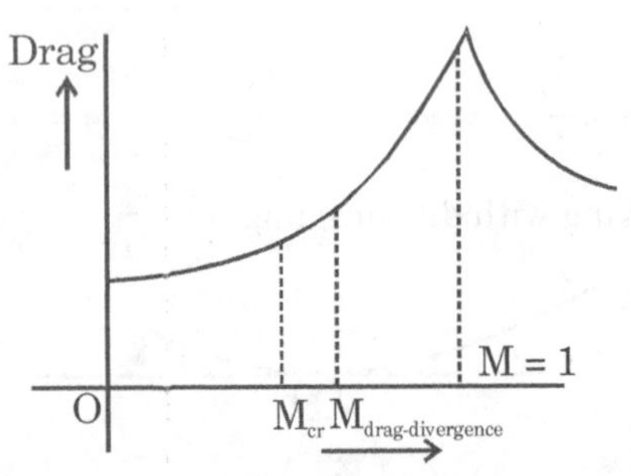

Mach Number

16. $Re = \frac{\rho V x}{\mu}$

Where, ρ = density of fluid,

μ = coefficient of viscosity

V = Velocity

x = plate distance

17. super critical airfoils

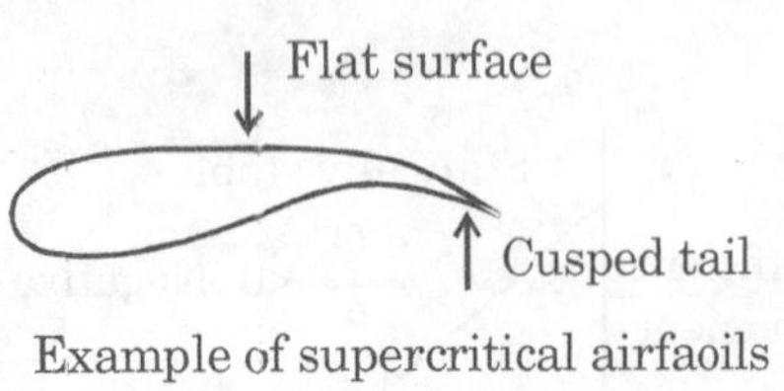

Example of supercritical airfaoils

18. Damping ratio $\zeta > 1$ is overdamped

$\zeta < 1$ underdemped

$\zeta = 1$ critically damped

$\zeta = 0$ undamped

19. $\sigma_{zz} = \tau_{xz} = \tau_{yz} = 0$ for plane stress

$\sigma_{xx}, \sigma_{yy}, \tau_{xy}$, are non zero

20. We know that

$$G = \frac{E}{2(1+\nu)}$$

$$\nu = \left[\frac{E}{2G} - 1\right]$$

$$= \left[\frac{4G}{2G} - 1\right] = 1$$

$$\boxed{\nu = 1}$$

E = Young's Modulus

G = Shear Modulus

ν = Poission's ratio

E = 4G

21. $$\boxed{\delta_{max} = \frac{pl^3}{48EI}}$$

P = Point Load

L = Length of Beam

E = Young's Modulus

I = 2nd moment of Area

22. Radius of curvature of such a beam is given by

$$\rho = \frac{E_A I_A + E_B I_B}{M}$$

Bending moment 20×10^3 Nm

$$\rho = \frac{100 \times 10^9 \times 6 \times 10^{-9} + 50 \times 10^9 \times 8 \times 10^{-9}}{20{,}000}$$

$$\boxed{\rho = 0.05m}$$

24. Turbofan has highest overall efficiency in this Mach number range. Turbofan Produes thrust in combination of Nozzle thrust and fan thrust. Fan thrust is mainly produced using large mass flow rate which is much eficient way of producing thrust as compare to burning fuel and increasing velocity.

25. Pressure loss factor PLF = $\frac{P_{02} - P_{01}}{P_{01}}$

As, flow is 1–D, incompressible and frictionless thus there are no losses inside combustor.

$\therefore$ $P_{02} = P_{01}$ (Total Pressure remains constant)

$\therefore$ $\boxed{PLF = 0}$

26. For geting eigen values of A, we need to solve characteristic equation

$|A - \lambda I| = 0$

$\Rightarrow \begin{vmatrix} 2-\lambda & 2 \\ 2 & 3-\lambda \end{vmatrix} = 0$

$\Rightarrow (2-\lambda)(3-\lambda) - 4 = 0$

$\Rightarrow \lambda^2 - 5\lambda + 6 - 4 = 0$

$\Rightarrow \lambda^2 - 5\lambda + 2 = 0$

$\therefore\ \lambda_1 = \dfrac{-(-5) + \sqrt{(-5)^2 - 4\times1\times2}}{2\times1}$,

$\lambda_2 = \dfrac{-(-5) - \sqrt{(-5)^2 - 4\times1\times2}}{2\times1}$

$\lambda_1 = 4.56, \lambda_2 = 0.438$

27. Solving : $\Delta n = \Delta x = 0.5$

$y = x^2 + 2$

x	0.0	0.5	1
y	2	2.85	3

using Trapezoidal rule

$f = \int_0^1 (x^2+2)dx = \int_0^1 ydx$

$= \dfrac{\Delta x}{2}\left[y(x=0) + 2y(x=0.5) + y(x=1)\right]$

$= \dfrac{0.5}{2}[2 + 2\times2.25 + 3]$

$\boxed{f = 2.375}$

28. Let $\dfrac{S+5}{(S+3)(S+4)} = \dfrac{A}{(S+3)} + \dfrac{B}{(S+4)}$

$\Rightarrow (S + 5) = (S + 4)A + (S + 3)B$

$\Rightarrow S + 5 = As + 4A + Bs + 3B$

comparing coeffcients both sides,we get

$A + B = a$ (A)

$4A + 3B = 5$ (B)

Solving (A) and (B)

$A = 1 - B$, Putting in (B)

$4(1 - B) + 3B = 5$

$4 - 4B + 3B = 5$

$-B = +1$

$\boxed{B = -1}$

$\Rightarrow A = 1 - (-1) = 2$

$\boxed{\dfrac{S+5}{(S+3)(S+4)} = \dfrac{2}{(S+3)} - \dfrac{1}{(S+4)}}$

29. Partial fraction of $f(S) = \dfrac{S+5}{(S+3)(S+4)}$

$= \dfrac{2}{(S+3)} - \dfrac{1}{(S+4)}$

Laplace Inverse

$\boxed{L^{-1}[f(S)] = F(t) = 2e^{-3t} - e^{-4t}}$

30. All of above equation are linear. Only variable are changing in each equation. For equation to be linear : none of the terms of unknown variable should be multiplied by themselves or each other.

31. Stall velocity :

$V_{stall} = \sqrt{\dfrac{2W}{\rho S C_{L_{max}}}}$

$W = 500\text{ kg} = 500 \times 9.8 = 4900N$

ρ at sea level = 1.225 kg/m^3

$S = 5m^2$

$C_{L_{max}} = 0.5$

$V_{stall} = \sqrt{\dfrac{2\times4900}{1.225\times5\times0.5}}$

$V_{stall} = 56.56$ m/sec.

32. $\lambda_2 = -0.006$ corresponds to lightly damped spiral mode

$\lambda_1 = -0.67$ heavily damped roll mode

$\lambda_{3,4} = -0.02 \pm \sqrt{3.2}j$ Oscillatory motion dutchroll mode

33. $\overline{X}_{CG} = 0.25$

$\dfrac{dC_m}{dC_L} = -0.2 = -SM$

Maximum SM = 10%C = 0.1C

We know,

$\Rightarrow \dfrac{dC_m}{dC_L} = X_n - X_{CG}$

$\boxed{X_{CG} = X_n - 0.05 = 0.5 - 0.05 = 0.45C}$

Most forward location of C.G.

$X_n - X_{CG} = 0.1$

$\boxed{X_{CG} = (X_n - 0.1) = 0.5 - 0.1 = 0.4C}$

34. Most aft location of C.G.

$X_n - X_{Cg} = 0.05$

$\boxed{X_n - 0.05 = 0.5 - 0.05}$

35. Max. Rate of climb $\left(\frac{R}{C}\right)_{max.}$

$$= \frac{\text{maximum excess power}}{\text{weight}} = \frac{14 \times 10^4}{500 \times 9.8}$$

= 28.57 m/sec.

36. Absolute ceiling of aircraft is defined as altitude at which maximum rats of climb is zero.

37. Escape velocity $V_{escape} = \sqrt{\frac{2GM}{r}}$

$r = 40 \times 10^5 m$

$G = 6.67 \times 10^{-11} \frac{Nm^2}{kg^2}$

$M = 10^{16}$ kg.

$$\boxed{V_{escape} = \sqrt{\frac{2 \times 6.67 \times 10^{-11} \times 10^{16}}{40 \times 10^5}} = 0.57 m/sec}$$

38. Orbital velocity $V = \sqrt{\frac{GM}{r}}$

M = mass of planet

r = distance of object to planet center.

39. Laminar boundary layer thickness :

$$\delta = \frac{5x}{\sqrt{Re_x}}$$

Re_x = Reynold number at x distance

$$\delta = \frac{5x}{\sqrt{\frac{\rho Vx}{\mu}}} = 5\sqrt{\frac{x\mu}{\delta V}}$$

$\therefore \delta \propto \sqrt{x}$

$$\Rightarrow \frac{\delta_1}{\delta_2} = \sqrt{\frac{X_1}{X_2}}$$

$$\Rightarrow \delta_2 = \delta_1\sqrt{\frac{X_2}{X_1}}$$

$$= 10\sqrt{\frac{0.8}{0.5}}$$

$\boxed{\delta_2 = 12.64 mm}$

Flow parameters

V = velocity

f = density

μ = coefficient of viscosity

40. Given airfoil is 5 digit NACA series

∴ last 2 digits gives maximum thickness as percentage of chord

∴ $\boxed{\text{maximum thickness} = 14\%C}$ = 0.14C

41. Coefficient of presure $C_p = \frac{P - P_\infty}{\frac{1}{2}\rho V_\infty^2}$

$$= \frac{8 \times 10^4 - 10^5}{\frac{1}{2} \times 1.225 \times (100)^2} = -3.26$$

42. Decreases

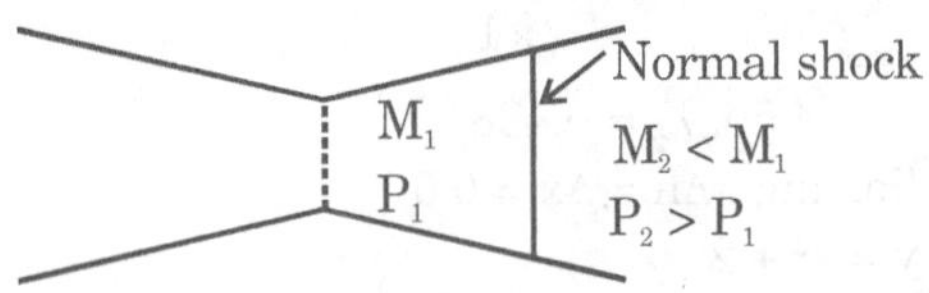

As flow encounter Normal shock waves mach number reduces to subsonic value.

43. $\varphi = \frac{1}{8}\ln\left[x^2 + y^2\right]$

$= \frac{1}{8}\ln r^2$ as $r^2 = x^2 + y^2$

$= \frac{1}{8} \times 2 \times \ln r$

$= \frac{1}{4} \ln r$

∴ Tangential velocity $V_\theta = \frac{\partial \rho}{\partial r}$

$\boxed{V_\theta = \frac{1}{4\rho}}$

44. Skin friction coefficient for turbulent Boundary layer is given by

$$C_f = \frac{0.072}{(Re_L)^{\frac{1}{5}}} = \frac{0.072}{(10^4)^{\frac{1}{5}}}$$

$\boxed{C_f = 0.0114}$

45. Response of SDOF (single degree of freedom)

$$X = \frac{\left|\frac{Fo}{k}\right|}{\sqrt{(1 - \Omega^2)^2 + (2\xi\Omega)^2}}$$

here, mass m = 10 kg spring constant k = 10^4 $\frac{N}{m}$.

$$\Omega = \frac{\omega}{\omega_n} = \frac{2\pi 5}{\sqrt{\frac{10^4}{10}}} = 0.9934$$

ω = forcing frequency = 2π5 rad/sec.

ω_n = Natural frequency = $\sqrt{\frac{K}{m}}$ rad/sec.

F_0 = 100N

ξ = 0.2

$$\therefore \frac{\left(\frac{100}{10^4}\right)}{\sqrt{\left(1-0.99^2\right)^2+\left(2\times0.2\times0.99\right)^2}}$$

$\Rightarrow \boxed{x = 0.025m}$

46. Critical buckling load of this pinned - pinned column is given by :

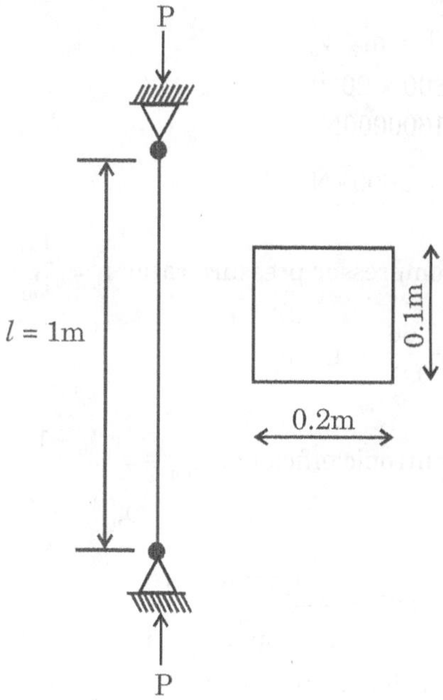

$$P_{Cr.} = \frac{\pi^2 EI}{l^2}$$

$$I = \frac{bd^3}{12} = \frac{0.2\times0.1^3}{12} = 1.66 \times 10^{-5}\ m^4$$

$$P_{Cr.} = \frac{\pi^2\times210\times10^9\times1.66\times10^{-5}}{1^2}$$

= 34543615.4 N = 34543.6 KN

47. For this walled closed tube

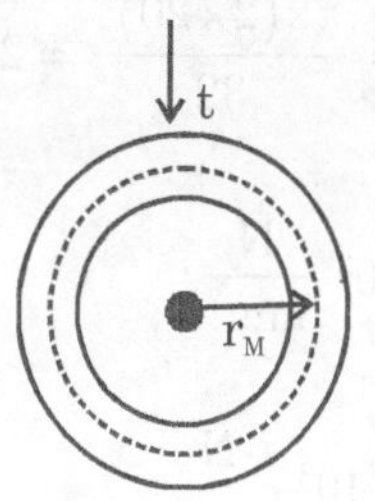

Torsional constant J = $2\pi r_m^3 \times t$

r_m = mean radius

$= \frac{55+52}{2} = 53.5mm$

∴ Torsional constant

J = 2 × π + (53.5)³ × 3

= 2886439.567 mm⁴

48. Von-mises failure theory for 2–D

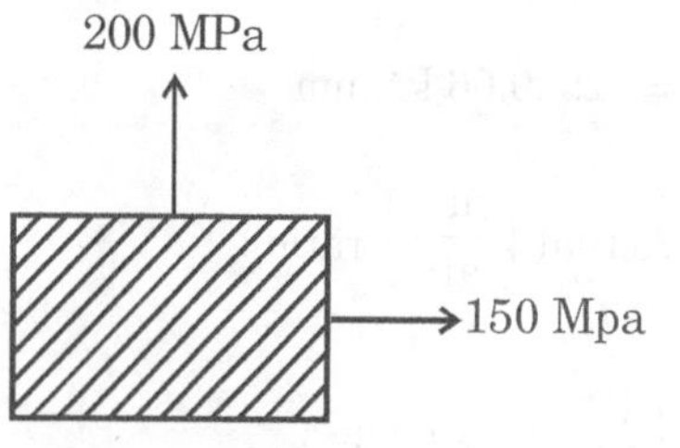

$\sigma_1{}^2 - \sigma_1\sigma_2 + \sigma_2{}^2 \leq \sigma^2$ yield

For plate to be safe

L.H.S. ≤ R.H.S.

L.H.S.

(200)² + (150)² + (200 × 150)

= 92500

R.H.S.

(250)² = 62500

L.H.S. > R.H.S.

Thus plate fails.

49. Composite materials have high strength to weight ratio than any other metals or alloys.

50. Torsional stiffness of such a shaft is given by

$$\frac{T}{\phi} = \frac{G_A J_A}{L} + \frac{G_B J_B}{L}$$

Torsional Constant $J_A = \frac{\pi d_1^4}{32} = \frac{\pi\times(2\times10)^4}{32}$

= 15707.96 mm⁴

$$J_B = \frac{\pi d_2^4}{32} = \frac{\pi \times (2\times 20)^4}{32} = 251327.41 \text{ mm}^4$$

$$G_A = 50 \times 10^3 \frac{N}{mm^2}$$

$$G_B = 100 \times 10^3 \frac{N}{mm^2}$$

L = 2000 mm

$$\therefore \frac{T}{\phi} = \frac{50\times 10^3 \times 15707.96}{2000} + \frac{100\times 10^3 \times 251327.41}{2000}$$

$$\therefore \frac{T}{\phi} = 12959064.5 \text{ Nmm}$$

= 12959.06 kNmm

51. Actual $\left(\frac{\text{fuel}}{\text{air}}\right)$ ratio :

$$\left(\frac{F}{A}\right)_{Actual} = 0.03$$

$$\text{Equivalence Ratio } \phi = \frac{\left(\frac{F}{A}\right) \text{actual}}{\left(\frac{F}{A}\right) \text{stoich}}$$

$\left(\frac{F}{A}\right)$ stoich

$$= \frac{4 \times \text{Atomic wt. of } C_8H_{18}}{25\times \text{Atomic wt. of } O_2 + 94\times \text{Atomic wt of } N_2}$$

$$= \frac{4\times[8\times 12 + 18\times 1]}{25\times 2\times 16 + 94\times 2\times 14} = 0.132$$

$$\boxed{\phi = \frac{0.03}{0.132} = 0.227}$$

52. Isentropic efficiency of diffuser

$$\eta_d = \frac{T_{02}^1 - T_a}{T_{oa} - T_a}$$

$$= \frac{\left(\frac{T_{02}^1}{T_a} - 1\right)}{\left(\frac{T_{0a}}{T_a} - 1\right)} = \frac{\frac{T_{02}^1}{T_{0a}} \times \frac{T_{0a}}{T_a} - 1}{\left(1 + \frac{\gamma - 1}{2} M_a^2\right) - 1}$$

$$= \frac{r_p^{\left(\frac{\gamma-1}{\gamma}\right)} \times \left[1 + \frac{\gamma-1}{2} M_a^2\right] - 1}{\left(\frac{\gamma-1}{2}\right) M_a^2}$$

Given $r_p = 0.9$, $\gamma = 1.4$

$$\eta_d = \frac{(0.9)^{\frac{0.4}{1.4}} \times [1 + 0.2\times 0.8^2] - 1}{0.2\times 0.8^2}$$

$$\boxed{\eta_d = 0.73}$$

53. Rocket Thrust $T = \dot{m}_e v_e + (P_e - P_a) A_e$

At optimum operation $P_e = P_a$

$\therefore T = \dot{m}_e v_e$

= 800 × 2000

= 1600000N

$$\boxed{T = 1600 \text{ kN}}$$

54. Compressor pressure ratio, $r_p = \frac{P_{02}}{P_{01}}$

$$= \frac{3.66}{0.8} = 4.575$$

55. Isentropic efficiency $\eta_{isn} = \dfrac{r_p^{\frac{\gamma-1}{\gamma}} - 1}{r_p^{\frac{\gamma-1}{\gamma\times\eta_{pol}}} - 1}$

$$\Rightarrow 0.87 = \frac{(4.575)^{\frac{0.4}{1.4}} - 1}{(4.575)^{\frac{0.4}{1.4}\times\eta_{pol}} - 1}$$

By solving this,we get

$$\boxed{\eta_{pol} = 0.894}$$

59. Run scored in first 10 overs = 10 × 3.2 = 32

Total runs to chase = 282

Runs left to be scored = 282 – 32 = 250

Over left = 40

Run rate required = $\frac{250}{40}$ = 6.25

60. Left hand side :

log(3 + 4) = log(7)

Right hand side

log(3 × 4) = log3 + log(4)

or

= log(12)

61. When he heard the hue and cry at midnight, he was nervous and did not know what to do.

63. Despite his best efforts to conceal the anger, people came to know that he was annoyed.

64. Considering the figure below

Where BC is wall height

AB is length from foot of ladder to the wall

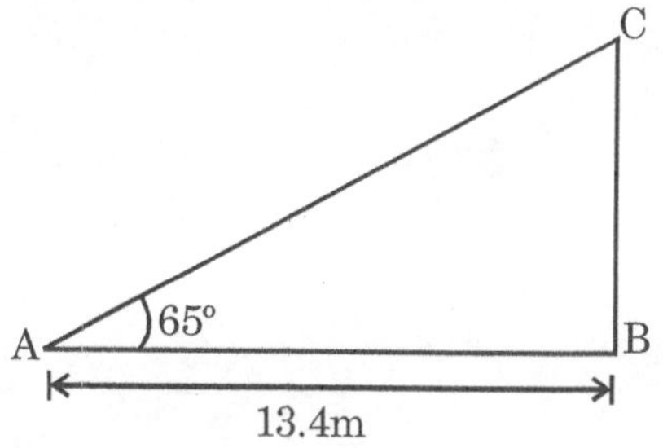

AC is length of ladder

Using Pythagoras theorem

$$\frac{AB}{AC} = \cos(65°)$$

$$AC = \frac{13.4}{\cos(65°)} = 31.707\text{m}$$

65. Time taken for both side in ride is 2 hours less than time needed to walk oneway and riding back.

∴ Time needed for riding one way = time needed for walking one way – 2 hours

Time taken in walking one way and riding back = 6 hours

Hence, she would take to walk both ways

= 6 hours + 2 hours

= 8 hours

MOCK TEST – 5

AEROSPACE ENGINEERING

Time Allowed: 3:00 Hours — *Maximum Marks: 100*

Read the following instructions carefully before you begin to answer the questions. This Booklet contains questions in English.

INSTRUCTIONS

1. This Booklet contains **65 questions** in total comprising of the following two parts:

   Part -1: Technical Section (85 Marks) *(1-55 Questions)*

   Part -2: General Aptitude Section (15 Marks) *(56-65 Questions)*

2. All questions are compulsory. Q.1-Q.25 & Q.56-Q.60 carry one mark each. Q.26-Q.55 & Q.61-Q.65 carry two marks each.

3. **Types of Questions:**

   (*a*) Multiple Choice Questions(MCQ) carrying 1 or 2 marks each in all papers and sections. These questions are objective in nature, and each will have a choice of four options, out of which the candidate has to mark the correct answer(s).

   (*b*) Numerical Answer Questions of 1 or 2 marks each in all sections. For these questions the answer is a real number, to be entered by the candidate using the virtual keypad. No choices will be shown for these type of questions.

4. For 1-mark multiple-choice questions, 1/3 marks will be deducted for a wrong answer. Likewise, for 2-marks multiple-choice questions, 2/3 marks will be deducted for a wrong answer. There is no negative marking for numerical answer type questions.

MOCK TEST – 5

1. The ratio of matrix A determinant and matrix B determinant is __________ if both A and B are 2 × 2 identity matrices.

2. Possible values of x, when $f(x) = x^2 + 2x + 2$ & $f(x) = 0$ are

(a) 2, 2 (b) 3, 3

(c) 4, 3 (d) $-1 \pm i$

3. Solve simultaneous equations

$x + 3y = 4$

$y - 2x = 6$

to get x and y

(a) x = 2, y = –2 (b) x = –2, y = 2

(c) both a and b (d) only a

4. A vector function given by $\vec{v} = 4x\hat{i} + 3xyz\hat{j}$

where $\hat{i}$ and $\hat{j}$ are unit vectors along x and y respectively. What is curl of $\vec{v}$?

(a) $x\hat{i} - 4\hat{k}$ (b) $-3xy\hat{i} + 3yz\hat{k}$

(c) zero (d) None

5. A pitot static tube is used for measurement of

(a) Velocity (b) Pressure

(c) Temperature (d) humidity

6. Glide angle of a glider is given by:

(a) $\tan^{-1}\left(\frac{L}{D}\right)$ (b) $\tan^{-1}\left(\frac{D}{L}\right)$

(c) $\tan^{-1}\left(\frac{L}{W}\right)$ (d) All of these

7. For yawing, which of control surface is necessary?

(a) Aileron (b) rudder

(c) Flaps (d) None of these

8. Which type of airfoil is important to increase the drag divergence Mach number?

(a) laminar airfoil (b) super critical airfoil

(c) both a & b (d) None of these

9. Drag produced, whenever there is shock wave in supersonic flight is called:

(a) wave drag (b) shock drag

(c) skin friction drug (d) All of these

10. Density of air as we move up in atmosphere:-

(a) increases

(b) deceases

(c) Nothing happen to density

(d) constant

11. For a satellite moving in orbit, which of following is constant

(a) velocity

(b) linear momentum

(c) Angular momentum

(d) None of these

12. Relationship between orbital and escape velocity is

(a) $V_{esc} = \sqrt{2}V_{orb}$ (b) $V_{orb} = \sqrt{2}V_{esc}$

(c) both are correct (d) $V_{orb} = 2\,V_{sec}$

13. For a specific fluid specific heat at constant pressure found to be 40% more than specific heat at constant volume. Ratio γ is equal to

(a) 1.4 (b) 1.2

(c) 2.4 (d) 3.2

14. For supersonic velocity which of following nozzle is used:

(a) Convergent Nozzle

(b) Convergent - Divergent Nozzle

(c) a & b both

(d) None of these

15. Total Temperature of a flow is given by

(a) $T_t = T_{st}\left[1 + \frac{\gamma - 1}{2}M^2\right]^{\frac{\gamma}{\gamma-1}}$

(b) $T_t = T_{st}\left[1 + \frac{\gamma - 1}{2}M^2\right]$

(c) $T_t = T_{st} + \frac{\gamma - 1}{2}M^2$

(d) $T_t = T_{st}$

16. Which of following quantity remains constant across a shock wave.

(a) Total pressure (b) Total enthalpy

(c) Static pressure (d) Static temperature

17. Which of the airfoil is used to reduce skin-friction drag?

(*a*) tim airfoil

(*b*) symmetric airfoil

(*c*) laminar flow airfoil

(*d*) super critical airfoil

18. Which of following cannot be used for rudder:

(*a*) NACA 0012 (*b*) NACA 1212

(*c*) NACA 0032 (*d*) NACA 0034

19. Natural frequency of a spring mass system is 20Hz. What is spring stiffness to mass ratio which produces this?

(*a*) 11.4 (*b*) 11.6

(*c*) 11.2 (*d*) 11.9

20. For a plane stress problem, state of stress is given as

$\sigma_x = 200$ MPa $\tau_{xy} = 20$ MPa

$\sigma_y = 300$ MPa $\tau_{yz} = 0$

$\sigma_z = 130$ MPa $\tau_{xz} = 200$ MPa

if material has yield strength of 350 MPa, check whether material fails.

(*a*) fails

(*b*) safe

(*c*) Problem statement is wrong

(*d*) None of these

21. Maximum deflection for a cantilever beam with point load at end is given by:

(*a*) $\dfrac{PL^3}{3EA}$ (*b*) $\dfrac{PL^3}{3E\rho}$

(*c*) $\dfrac{PL^2}{3EI}$ (*d*) $\dfrac{PL^3}{3EI}$

22. For a fixed - free column, critical buckling load is equal to:

(*a*) $\dfrac{Pl^3}{3EI}$ (*b*) $\dfrac{4\pi^2 EI}{l^2}$

(*c*) $\dfrac{\pi^2 EI}{4l^2}$ (*d*) $\dfrac{\pi^2 EI}{2l^2}$

23. Which of following sections having same cross-section area is best for bending

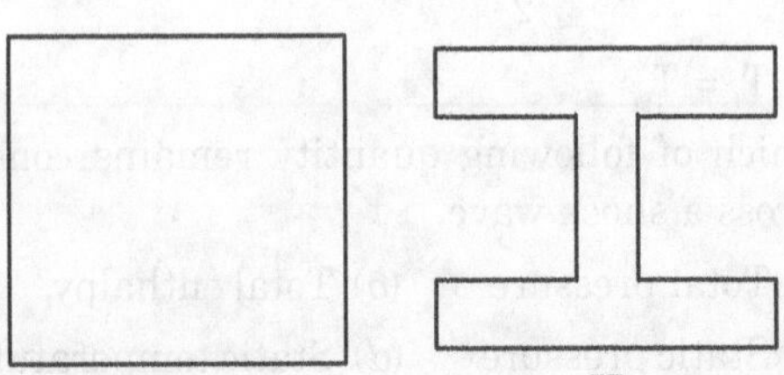

I II

(*a*) I (*b*) II

(*c*) I & II (*d*) None

24. For an impulse turbine static enthalpy change across stage is 20 times more than that of rotor. What is the degree of reaction?

(*a*) 20 (*b*) $\dfrac{1}{2}$

(*c*) zero (*d*) ∞

25. Ratio of flight speed to the exhaust velocity is $\dfrac{1}{2}$ propulsive efficiency of engine is __________%

26. One of the eigen values of $A = \begin{bmatrix} 3 & -4 \\ -2 & 9 \end{bmatrix}$ is $6 + \sqrt{17}$ what is other one?

(*a*) 6 (*b*) $\sqrt{17}$

(*c*) $6 - \sqrt{17}$ (*d*) $\sqrt{17} - 6$

27. Solution of $m\ddot{x} + kx = 0$ is given by x =

(*a*) $C_1 \cos\sqrt{\dfrac{m}{k}}t + C_2 \sin\sqrt{\dfrac{m}{k}}t$

(*b*) $C_1 \cos\sqrt{\dfrac{k}{m}}t$

(*c*) $C_1 \cos\sqrt{\dfrac{k}{m}}t + \sin t$

(*d*) $C_1 \cos\sqrt{\dfrac{k}{m}}t + C_2 \sin\sqrt{\dfrac{k}{m}}t$

28. $\displaystyle\int_2^6 \frac{12x^2}{4x^3 + 3}\,dx$ is equal to

(*a*) 2.3 (*b*) 3.3

(*c*) 3.4 (*d*) 3.2

29. Laplace transform of $e^{3t}\left[e^{4t} - e^{7t}\right] - e^{8t}$ is

(*a*) $\dfrac{1}{(s-7)} - \dfrac{1}{(s-10)} - \dfrac{1}{(s-8)}$

(*b*) $\dfrac{1}{(s-10)} + \dfrac{1}{(s-7)} - \dfrac{1}{(s-8)}$

(*c*) $\dfrac{1}{(s-7)} + \dfrac{1}{(s-10)} + \dfrac{1}{(s-8)}$

(*d*) None of these

30. Evaluate integral $\int_2^4 \left[4x^2 + 4x^2 + 4x + 4\right] dx$ using Simpson's rule with one step?

(*a*) 346.66 (*b*) 346.68

(*c*) 346.67 (*d*) 346.667

31. Time period of phugoid motion was found to be 150 sec, and distance from start of 1 cycle to end of the cycle was 2 km. At what velocity aircraft is doing the motion in m/sec.

(*a*) 13.34 (*b*) 13.334

(*c*) 13.35 (*d*) 13.36

32. Oswald efficiency factor for a wing loading is 0.88 wing is made up of airfoil NACA3217 with chord length of 0.5m. What is the wing span if it has lift coefficient C_L = 1.5 and produces a induce drag coefficient C_{D_i} = 0.2

(*a*) 2.03 (*b*) 3.03

(*c*) 4.03 (*d*) 5.03

33. A Glider glides at 45° angle and has weight of 4000 kg. What is the lift to drag ratio which help it to fly?

(*a*) 1 (*b*) 0.2

(*c*) 3 (*d*) 1.1

34. What is the ratio of range of two turbojet aircraft having lift coefficients C_{L_A} & C_{L_B}. Given that all other parameter for both aircrafts are same &

$$\frac{C_{L_A}}{C_{L_B}} = 2$$

(*a*) $\sqrt{2}$ (*b*) $\frac{1}{\sqrt{2}}$

(*c*) $\frac{\sqrt{2}}{2}$ (*d*) $2\sqrt{2}$

35. For an aircraft zero lift drag coefficient is 1.15. For a maximum $\left(\frac{C_L^{\frac{3}{2}}}{C_D}\right)_{max}$ what should be the induced drag coefficient?

(*a*) 4.35 (*b*) 3.45

(*c*) 6.39 (*d*) 2

36. An aircraft wing is 3 meters above the ground and it has a wing span of 10 metres by what factor induce drag will be reduced due to ground effect?

(*a*) 18.2% (*b*) 18%

(*c*) 17% (*d*) 14%

37. For a pull up maneuver load factor is increased to twice than the original keeping velocity of aircraft same. What is the change in turn rate given n_1 = 3?

(*a*) 0.58 w_1 (*b*) 0.57w_1

(*c*) 0.56 w_1 (*d*) zero

38. A rocket launched with maximum speed of 11 km/sec from a planet having radius of 7000 km and mass of 10^{25} kg. rocket will

(*a*) escape planet (*b*) fall back

(*c*) rotate in orbit (*d*) none of these

39. Using data in 38 question, what will be the orbital velocity of satellite to be injected at 200 km above surface of planet?

(*a*) 20 km/sec (*b*) 5 km/sec

(*c*) 9.6 km/sec (*d*) 15 km/sec

40. Potential function for an incompressible, irrotational flow is given as

$\phi = 34x + 42y$

What is the ratio of velocity components in x and y direction?

(*a*) 0.84 (*b*) 0.809

(*c*) 0.83 (*d*) 0.7

41. Lift coefficient for incompressible flow is 1.67 when freestream Mach number is $M_\infty = 0.65$. What will be compressible flow lift coefficient?

(*a*) 2.19 (*b*) 2.4

(*c*) 2.5 (*d*) 2.6

42. Oblique shock wave with a wave angle of 45° was generated from a wedge angle of 30°. The ratio of Mach number downstream of shock to its normal component is __________

43. Total pressure registered by a pitot static tube is 62732 N/m². If static pressure P_∞ = 40,000 N/m² and free stream density ρ_∞ = 0.63 N/m² and γ = 1.4. Airspeed indicated is __________ m/sec.

44. An aircraft flying with 250 m/sec. at an altitude where freestream temperature is 320K and density is 0.625 kg/m³. Specific heat ratio is 1.4 with and R= 287 J/kgK. Total temperature an aircraft is __________K.

45. For a spring-mass damper system (single degree of freedom). What is the force transmitted to ground if force applied at mass is 200N e^{j20t} and absolute transfer function is 2?

All parameter are m = 200 kg

k = 10^4 N/m

c = 30 N s/m

(*a*) 400 N (*b*) 300 N

(*c*) 200 N (*d*) zero

46. For a fixed-fixed column of length 4 meter and rectangular cross-sectional dimension are 4cm × 4cm. What is critical buckling load? (E = 200 GPa)

(*a*) 105 kN (*b*) 200 kN
(*c*) 300 kN (*d*) 106 kN

47. For a plain strain problem tensile stresses in x and y direction are 150 MPa and 200 MPa. It is evaluated that if stress in z direction increases beyond 310 MPa structure will fail. Stress in z direction is (ν = 0.3)

(*a*) 103 MPa (*b*) 102 MPa
(*c*) 105 MPa (*d*) 106 MPa

48. Airy stress function for particular boundary condition is given as

$\phi = 12x^2 + 13y^2 - 2xy$.

What is the ratio of σ_{xx} and τ_{xy}.

(*a*) 12 (*b*) –12
(*c*) ±12 (*d*) None of these

49. A 2mm thick cylindrical shell with inner diameter of 0.4 m is designed to withstand hoop stress of 150 MPa. If 1000 Pa pressure is applied inside. What is factor of safety?

(*a*) 1400 (*b*) 1500
(*c*) 1501 (*d*) 1502

50. Bending stress for a circular cross-section cantilever beam is ________MPa if maximum bending moment is 1000 Nm and radius of cross-section is 0.2 m.

51. For a turbojet engine flying in air has combustor inlet temperature 500 K and outlet temperature 1200K. Heating value of fuel is 50 MJ/kg. Consider combustion to be 90% efficient, fuel/air ratio of engine is ________.

(C_p = 1004 J Kg K, γ = 1.4)

52. Fuel air-ratio for reaction between octane and air in an combustor is ________,

if reaction is $2C_8H_{18} + 25\,O_2 + 94\,N_2 \rightarrow 16CO_2 + 18\,H_2O + 94N_2$ and equivalence ratio is 0.5

53. For a centrifugal compressor inlet stagnation temperature is 280K and exit stagnation temperature is 380K. If pressure ratio across the compressor is 2.5 and specific heat ratio is 1.4 then what is isentropic efficiency?

(*a*) 80% (*b*) 90%
(*c*) 83.7% (*d*) 92%

54. Mass of a single stage rocket is 4000 kg out of which 300 kg is propellant. If specific impulse of rocket is 400 sec. then what is the maximum velocity that can be achieved by rocket?

(*a*) 305 (*b*) 305.1
(*c*) 305.6 (*d*) 305.4

55. Mass flow rate through exit of a rocket nozzle is 200 kg/sec. with velocity 1900 m/sec. Specific impulse of this rocket is ________ sec if nozzle operates of optimum condition.

56. Which of the following options have error

(*a*) An Indian ship
(*b*) Laden with merchandise
(*c*) Got drowned in the Pacific Ocean.
(*d*) No error.

57. Opposite of ENORMOUS is

(*a*) Soft (*b*) Average
(*c*) Tiny (*d*) Weak

58. Give one word for the sentence given: that which cannot be corrected

(*a*) Unitelligible (*b*) Indelible
(*c*) Illegible (*d*) Incorrigible

59. Average of 20 numbers is zero. How many of them may be greater than zero, at the most?

(*a*) 20 (*b*) 18
(*c*) 19 (*d*) None of these

60. $\text{Log}_5(0)$ = ?

(*a*) 0 (*b*) 1
(*c*) none of these (*d*) 2

61. An error 3% in excess is made while measuring the side of a square. What is percentage of error in the calculated area of square?

(*a*) 5% (*b*) 2%
(*c*) 4% (*d*) 6.09%

62. A person crosses a 680 m long street in 5.5 minutes what is his speed in km/hour.

(*a*) 7.418 km/hour (*b*) 20 km/hour
(*c*) 40 km/hour (*d*) 1 km/hour

63. 10 years ago, P was half of Q's age. If the ratio of their present age is 5 : 6 what will be the total of their present ages?

(*a*) 30 years (*b*) 27.5 years
(*c*) 3 years (*d*) 40 years

64. Which one of following represents the sentence best in direct/indirect speech:

She said that her brother was getting married

(*a*) She said, "Her brother is getting married".
(*b*) She told, "Her brother is getting married".
(*c*) She said, "My brother is getting married".
(*d*) None of these

65. Replace the words with underline to make sentence grammatically correct.

1. The small child does whatever his father <u>was done</u>.

(*a*) has done (*b*) did
(*c*) does (*d*) had done

OMR SHEET

1. [ ]
2. (A) (B) (C) (D)
3. (A) (B) (C) (D)
4. (A) (B) (C) (D)
5. (A) (B) (C) (D)
6. (A) (B) (C) (D)
7. (A) (B) (C) (D)
8. (A) (B) (C) (D)
9. (A) (B) (C) (D)
10. (A) (B) (C) (D)
11. (A) (B) (C) (D)
12. (A) (B) (C) (D)
13. (A) (B) (C) (D)
14. (A) (B) (C) (D)
15. (A) (B) (C) (D)
16. (A) (B) (C) (D)
17. (A) (B) (C) (D)
18. (A) (B) (C) (D)
19. (A) (B) (C) (D)
20. (A) (B) (C) (D)
21. (A) (B) (C) (D)
22. (A) (B) (C) (D)
23. (A) (B) (C) (D)
24. (A) (B) (C) (D)
25. [ ]
26. (A) (B) (C) (D)
27. (A) (B) (C) (D)
28. (A) (B) (C) (D)
29. (A) (B) (C) (D)
30. (A) (B) (C) (D)
31. (A) (B) (C) (D)
32. (A) (B) (C) (D)
33. (A) (B) (C) (D)
34. (A) (B) (C) (D)
35. (A) (B) (C) (D)
36. (A) (B) (C) (D)
37. (A) (B) (C) (D)
38. (A) (B) (C) (D)
39. (A) (B) (C) (D)
40. (A) (B) (C) (D)
41. (A) (B) (C) (D)
42. [ ]
43. [ ]
44. [ ]
45. (A) (B) (C) (D)
46. (A) (B) (C) (D)
47. (A) (B) (C) (D)
48. (A) (B) (C) (D)
49. (A) (B) (C) (D)
50. [ ]
51. [ ]
52. [ ]
53. (A) (B) (C) (D)
54. (A) (B) (C) (D)
55. [ ]
56. (A) (B) (C) (D)
57. (A) (B) (C) (D)
58. (A) (B) (C) (D)
59. (A) (B) (C) (D)
60. (A) (B) (C) (D)
61. (A) (B) (C) (D)
62. (A) (B) (C) (D)
63. (A) (B) (C) (D)
64. (A) (B) (C) (D)
65. (A) (B) (C) (D)

ANSWERS

1. (1)	2. (*d*)	3. (*b*)	4. (*b*)	5. (*a*)	6. (*b*)	7. (*b*)	8. (*b*)	9. (*a*)	10. (*b*)
11. (*c*)	12. (*a*)	13. (*a*)	14. (*b*)	15. (*b*)	16. (*b*)	17. (*c*)	18. (*b*)	19. (*c*)	20. (*c*)
21. (*d*)	22. (*c*)	23. (*b*)	24. (c)	25. (66)	26. (*c*)	27. (*d*)	28. (*d*)	29. (*a*)	30. (*d*)
31. (*b*)	32. (*a*)	33. (*a*)	34. (*a*)	35. (*b*)	36. (*b*)	37. (*a*)	38. (*b*)	39. (*c*)	40. (*b*)
41. (*a*)	42. (3.86)	43. (268.63)	44. (350.47)	45. (*a*)	46. (*a*)	47. (*c*)	48. (*b*)	49. (*b*)	50. (0.16)
51. (0.016)	52. (0.0332)	53. (*c*)	54. (*c*)	55. (193.68)	56. (*c*)	57. (*c*)	58. (*d*)	59. (*d*)	60. (*c*)
61. (*d*)	62. (*a*)	63. (*b*)	64. (*c*)	65. (*c*)					

EXPLANATIONS

1. Both A & B are identity matrices

$$A = B = \begin{bmatrix} 1 & 0 \\ 0 & 1 \end{bmatrix}$$

$|A| = |B| = 1 - 0 = 1$

Ratio of determinants is

$$\frac{|A|}{|B|} = \frac{1}{1} = 1$$

2. $f(x) = x^2 + 2x + 2,\ f(x) = 0$

$\Rightarrow\ x^2 + 2x + 2 = 0$

$$\Rightarrow x = \frac{-2 \pm \sqrt{4 - 4 \times 2}}{2}$$

$$= \frac{-2 \pm \sqrt{-4}}{2} = \frac{-2 \pm 2i}{2}$$

$i = \sqrt{-1}$

$= -1 \pm i$

3. $x + 3y = 4$ ...(A)

$y - 2x = 6$ ....(B)

using (A) $x = 4 - 3y$ Put in B

$y - 2[4 - 3y] = 6$

$y - 8 + 6y = 6$

$7y = 14$

$\boxed{y = 2}$

$\Rightarrow\ x = 4 - 3 \times 2 = -2$

$\boxed{x = -2}$

4. $\vec{v} = 4x\hat{i} + 3xyz\hat{j}$

Curl of $\vec{v} = \vec{\nabla} \times \vec{v}$

$$= \begin{vmatrix} \hat{i} & \hat{j} & \hat{k} \\ \frac{\partial}{\partial x} & \frac{\partial}{\partial y} & \frac{\partial}{\partial z} \\ 4x & 3xyz & 0 \end{vmatrix}$$

$$= \hat{i}\left[\frac{\partial}{\partial y}(0) - \frac{\partial}{\partial z}(3xyz)\right] - \hat{j}\left[\frac{\partial}{\partial x}(0) - \frac{\partial}{\partial z}(4x)\right]$$

$$+\hat{k}\left[\frac{\partial}{\partial x}(3xyz) - \frac{\partial}{\partial y}[4x]\right]$$

$$= -3xy\hat{i} + 3yz\hat{k}$$

5. Pitot static tube is used for measuring velocity.

6. $L = \omega \cos\theta$ ....(i)

$D = \omega \sin\theta$ ....(ii)

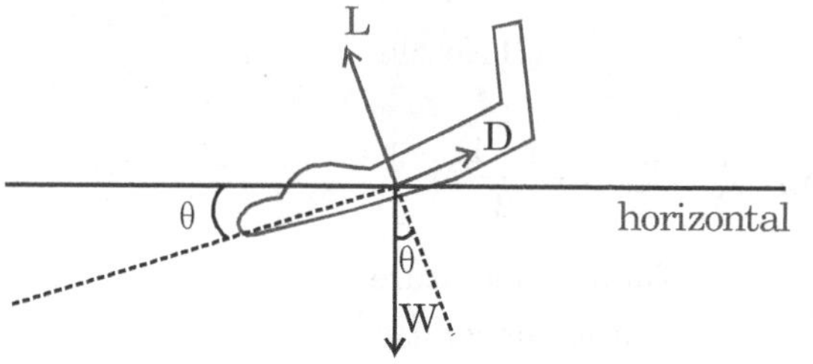

dividing (i) & (ii)

$$\frac{\sin\theta}{\cos\theta} = \frac{D}{L}$$

$$\tan\theta = \frac{D}{L}$$

$$\text{Glide Angle } \theta = \tan^{-1}\left(\frac{D}{L}\right)$$

7. Rudder is main control surface for yawing motion

8. Super critical airfoils are used for increasing the drag divergence Mach number.
9. Drag produced by shock wave is called wave drag.
10. Air density decreases with height.
11. Angular momentum
12. $V_{esc} = \sqrt{\dfrac{2GM}{R^2}}$

$$V_{orb} = \sqrt{\frac{GM}{R^2}}$$

$$\therefore \boxed{V_{esc} = \sqrt{2}.\,V_{orb}.}$$

13. Specific heat at constant pressure = C_p
specific heat at constant volume = C_v.
$C_p = C_v + 40\%\ C_v$.
Specific heat ratio

$$\gamma = \frac{C_p}{C_v} = \frac{C_v + 40\%\ C_v}{C_v}$$

$$= 1 + \frac{40}{100} = 1.4$$

$$\boxed{\gamma = 1.4}$$

14. Convergent - Divergent Nozzle

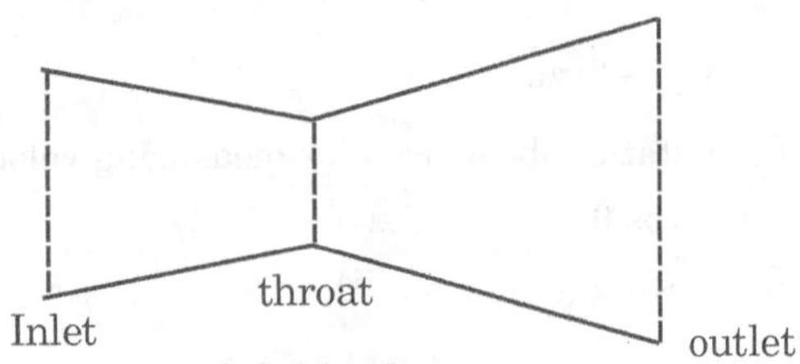

C-D Nozzle

15. $T_t = T_{st}\left[1 + \dfrac{\gamma - 1}{2} M^2\right]$

T_t = Total temperature
Ts_t = static temperature
M = Mach number
γ = specific heat ratio

16. Total enthalpy

M_1	$M_2 < M_1$
T_1	$T_2 > T_1$
T_{01}	$T_{02} = T_{01}$
H_{01}	$H_{02} = H_{01}$
P_1	$P_2 > P_1$

Shock wave

17. Laminar flow airfoil
18. NACA 1212 is cambered airfoil which is reflected by first two digits. Only symmetric airfoils are used for rudder, with zero camber.
19. Natural frequency

$$\omega_n = \sqrt{\frac{k}{m}} = 2 \times \pi \times 20 \text{ rad/sec}$$

$(\omega_n = 2\pi f)$

$$\therefore \quad \frac{k}{m} = \sqrt{2\pi \times 20}$$

$= 11.20$
k = stiffness m = mass
f = frequency in Hz.

20. Problem statement is wrong as for plane stress problems σ_z, τ_{xz}, τ_{yz} should be zero
21. $\dfrac{Pl^3}{3EI}$

where P is point load at end
l = length
E = young's modulus
I = 2nd moment of area

22. $P_{cr} = \dfrac{\pi^2 EI}{4l^2}$
23. II i.e. I section is always a best section for bending as it has more moment of inertia and more mass is placed where we get maximum bending stress.
24. Zero

Degree of reaction is represented by R

$$R = \frac{\text{static enthalpy drop in rotor}}{\text{total enthalpy drop in stage}}$$

For impulse turbine, static enthalpy drop across rotor is always zero.

$$\therefore \boxed{R = 0}$$

25. Propulsive efficiency

$$\eta_p = \frac{2v_i}{v_i + v_e} = \frac{2\left(\dfrac{v_i}{v_e}\right)}{1 + \left(\dfrac{v_i}{v_e}\right)}$$

$$= \frac{2 \times \left(\dfrac{1}{2}\right)}{1 + \left(\dfrac{1}{2}\right)} = \frac{2}{3} = 0.66 = 66\%$$

26. $A = \begin{bmatrix} 3 & -4 \\ -2 & 9 \end{bmatrix}$

For finding eigen value

$|A - \lambda I| = 0$

$\begin{vmatrix} 3-\lambda & -4 \\ -2 & 9-\lambda \end{vmatrix} = 0$

$(3-\lambda)(9-\lambda) - 8 = 0$

$\lambda^2 - 12\lambda + 27 - 8 = 0$

$\lambda^2 - 12\lambda + 19 = 0$

$\lambda_1 = 6 + \sqrt{17}$

$\lambda_2 = 6 - \sqrt{17}$

27. $m\ddot{x} + kx = 0 \Rightarrow \ddot{x} + \frac{k}{m}x = 0$

Characteristic equation

$D^2 + \frac{k}{m} = 0$

$D = \pm\sqrt{\frac{k}{m}}\, i$

$\therefore\ x = C_1 \cos\sqrt{\frac{k}{m}}t + C_2 \sin\sqrt{\frac{k}{m}}t$

28. $\int_2^6 \frac{12x^2}{4x^3+3}$

here denominator $\frac{d}{dx}\left[4x^3+3\right] = 12x^2$

Numerator

$\Rightarrow \ln\left[4x^3+3\right]_2^6$

$= \ln\left[4\times 6^3+3\right] - \ln\left[4\times 2^3+3\right]$

$= \ln[861] - \ln[35]$

$= \ln\left[\frac{861}{35}\right] = \ln[24.6] = 3.2$

29. $L\left[e^{3t}\left(e^{4t} - e^{7t}\right) - e^{8t}\right]$

$= L\left[e^{7t} - e^{10t} - e^{8t}\right]$

$\boxed{L\left[e^{at}\right] = \frac{1}{s-a}}$

$= \frac{1}{s-7} - \frac{1}{s-10} - \frac{1}{s-8}$

30. $I = \int_2^4 \left(4x^3 + 4x^2 + 4x + 4\right) dx$

$\Rightarrow f(x) = \left(4x^3 + 4x^2 + 4x + 4\right)$

One step Simpson's rule

x	2	3	4
f(x)	60	160	340

$\int_{x_0}^{x_n} f(x)\,dx = \frac{h}{3}$

$\left[(y_o + y_n) + 2(y_2 + y_1 + ---- + y_{n-2}) + 4(y_1 + y_3 + ---- y_{n-1})\right]$

here $h = 3 - 2 = 4 - 3 = 1$

$\therefore\ I\int_2^4 f(x)dx = \frac{1}{3}\left[(60+340) + 4\times 160\right] = 346.667$

31. 13.334

Time period T = 150 sec.

wavelength λ = 2 km

$\therefore$ velocity $= \frac{\lambda}{T}$

$= \frac{2\times 10^3}{150} = 13.334$ m/sec.

32. $C_{Di} = \frac{{C_L}^2}{\pi e AR}$

$\Rightarrow AR = \frac{b^2}{s} = \frac{C_L^2}{\pi e C_{Di}}$

$= \frac{1.5^2}{\pi \times 0.88 \times 0.2} = 4.06$

$A.R = \frac{b^2}{s} = \frac{b^2}{b\times c} = \frac{b}{c} = 4.06$

$\therefore\ b = 4.06 \times 0.5$

$\boxed{b = 2.03m}$

33. Glide angle $\theta = \tan^{-1}\left[\frac{1}{\frac{L}{D}}\right]$

$\therefore \frac{L}{D} = \frac{1}{\tan\theta} = \frac{1}{\tan(45^\circ)} = 1$

34. Range $= R = 2\sqrt{\dfrac{2}{\rho_\infty s} \times \dfrac{1}{c_t}\dfrac{C_L^{\frac{1}{2}}}{CD}\left[w_0^{\frac{1}{2}} - w_1^{\frac{1}{2}}\right]}$

$$\therefore \frac{R_A}{R_B} = \frac{C_{L_A}^{\frac{1}{2}}}{C_{L_B}^{\frac{1}{2}}} = \left(\frac{C_{L_A}}{C_{L_B}}\right)^{\frac{1}{2}} = (2)^{\frac{1}{2}} = \sqrt{2}$$

35. For $\left(\dfrac{C_L^{\frac{3}{2}}}{C_D}\right)_{max}$

$$C_{D,0} = \frac{C_{Di}}{3}$$

CD,0 = zero lift coefficient

$C_{D,i}$ = induced drag coefficient

$\therefore C_{D_i} = 3C_{D,0} = 3 \times 1.15 = 3.45$

36. Factor for ground effect

$$\phi = \frac{\dfrac{(16h)^2}{(b)}}{1 + \left(\dfrac{16h}{b}\right)^2}$$

Height of wing h = 3 m

Wing span b = 10 m

$\therefore \phi = 0.82$

Reduction in induced drag

$= (1 - 0.82) \times 100\%$

$= 18\%$

37. Turn rate

$$\omega = \frac{g\sqrt{n-1}}{v_\infty}$$

$$\therefore \frac{\omega_2}{\omega_1} = \frac{\sqrt{n_2 - 1}}{\sqrt{n_1 - 1}} = \sqrt{\frac{6-1}{3-1}}$$

$g_1 = g_2$

$v_{\infty_1} = v_{\infty_2}$

$n_2 = 2n_1 = 6$

$\therefore \boxed{\omega_2 = 1.58\,\omega_1}$

$\therefore$ Change in turn rate

$\Delta\omega = 0.58\, w_1$.

38. Escape velocity of planet

$$v_e = \sqrt{\frac{2GM}{R}}$$

$$= \sqrt{\frac{2 \times 6.67 \times 10^{-11} \times 10^{25}}{7000 \times 10^3}} = 13.804 \text{ km/sec}$$

$\therefore$ Rocket will fall back as it velocity is less than escape velocity

39. Orbital velocity $= \sqrt{\dfrac{GM}{a}}$

her a = R + 200

= 7200 km

$$= \sqrt{\frac{6.67 \times 10^{-11} \times 10^{25}}{7200 \times 10^3}}$$

= 9624.9 m/sec = 9.624 km/sec

40. Velocity in x direction

$$u = \frac{\partial\phi}{\partial x} = 34 \text{ m/sec}$$

velocity in y direction

$$v = \frac{\partial\phi}{\partial y} = 42 \text{ m/sec}$$

$$\therefore \frac{u}{v} = 0.809$$

41. Using compressible rule of Prandtl – Glauert

$$C_l = \frac{C_{lo}}{\sqrt{1 - M_\infty^2}} = \frac{1.67}{\sqrt{1 - 0.65^2}} = 2.19$$

42. $M_{n_2} = M_2 \sin(\beta - \theta)$

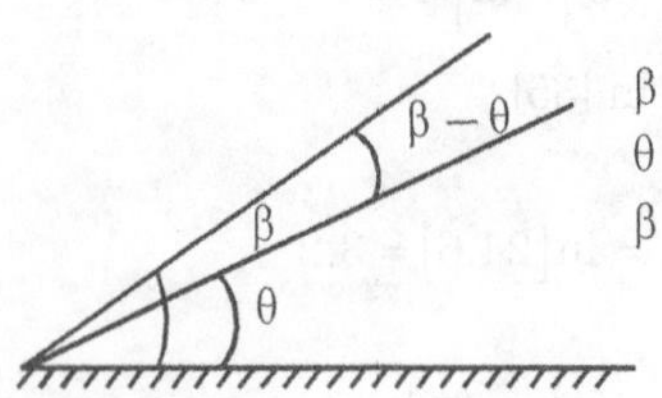

β = 45°
θ = 30°
β – θ = 15°

$$\frac{M_2}{M_{n_2}} = \frac{1}{\sin(\beta - \theta)} = \frac{1}{\sin(15°)} = 3.863$$

43. $V_\infty = \sqrt{\dfrac{2(P_o - P_\infty)}{\rho_\infty}}$

$$= \sqrt{\frac{2 \times [62732 - 40000]}{0.63}}$$

= 268.63 m/sec

44. Mach no. $= \dfrac{v}{a}$

$$= \frac{v}{\sqrt{\gamma RT}} = \frac{2560}{\sqrt{1.4 \times 287 \times 320}} = 0.69$$

∴ Stagnation temperature

$$T_o = T\left(1 + \frac{\gamma - 1}{2} M^2\right)$$

$$= 320\left[1 + \frac{1.4 - 1}{2} 0.69^2\right] = 350.47 \text{ K}$$

45. Absolute transfer function

$$|T| = 2 = \left|\frac{F_T}{F}\right|$$

$F_T = 2 \times 200 = 400$ N

46. For fixed- fixed column

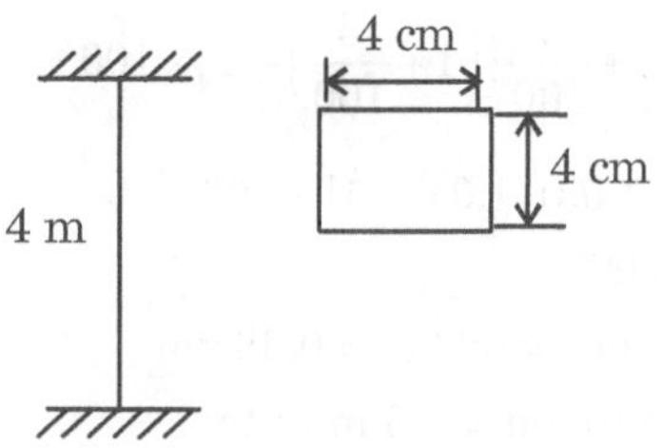

$$P_{c.r.} = \frac{4\pi^2 EI}{l^2}$$

$$= \frac{4 \times \pi^2 \times 200 \times 10^9 \times \left(\frac{4 \times 4^3}{12}\right) \times 10^{-8}}{4^2}$$

$= 105275.75 \text{ N} \approx 105$ KN

47. For plane strain

$\sigma_{zz} = \nu(\sigma_{xx} + \sigma_{yy}) = 0.3\,[150 + 200] = 105$ MPa

∴ Structure will not fail.

48. $\sigma_{xx} = \dfrac{d^2\phi}{dx^2} = 24$

$$\tau_{xy} = \frac{\partial^2\phi}{\partial x\,\partial y} = -2$$

$$\therefore \boxed{\frac{\sigma_{xx}}{\tau_{xy}} = \frac{24}{-2} = -12}$$

49. Hoop stress $= \dfrac{Pd}{2t} = \dfrac{1000 \times 0.4}{0.002 \times 2} = 100000$ Pa

$= 0.1 \times 10^6$ Pa $= 0.1$ MPa

∴ FOS = 1500

50. Bending stress

$$\sigma = \frac{My}{I}$$

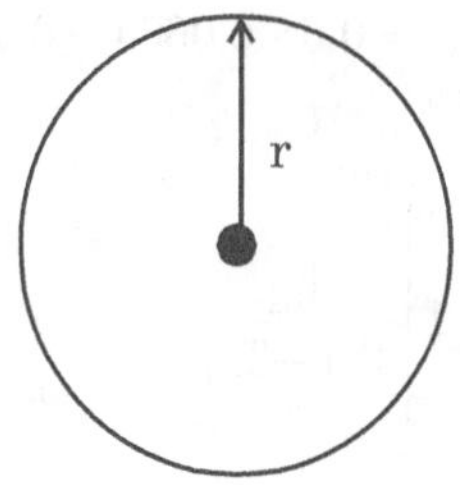

d = 2r

$$= \frac{M \times \frac{d}{2}}{\frac{\pi d^4}{64}}$$

$$= \frac{1000 \times 0.2}{\frac{\pi \times [2 \times 0.2]^4}{64}}$$

$= 159154.94$ Pa $= 0.16$ MPa

51. Combustion efficiency

$$\eta_{cc} = \frac{h_2 - h_1}{\dot{m}_f \times \text{heating value}}$$

$$= \frac{\left(\dot{m}_a + \dot{m}_f\right) C_P T_2 - \dot{m}_a C_P T_1}{\dot{m}_f \times Q}$$

$$\eta_{cc} = \frac{(1+f) C_P T_2 - C_P T_1}{f\,Q}$$

$$fQ \times \eta_{cc} = (1+f) C_P T_2 - C_P T_1$$

$$f\left[Q\,\eta_{cc} - C_p T_2\right] = C_P T_2 - C_P T_1$$

$$f = \frac{C_P T_2 - C_P T_1}{Q\eta_{cc} - C_P T_2}$$

$$= \frac{1004[1200 - 500]}{50 \times 10^6 \times 0.9 - 1004 \times 1200} = 0.016$$

52. $$\left(\frac{f}{a}\right)_{stochiometric} = \frac{2 \times \text{Atomic weight of } C_8H_{18}}{25 \times \text{Atomic weight of } O_2}$$

$+ 94 \times$ Atomic weight of N_2.

$$= \frac{2 \times [8 \times 12 + 18 \times 1]}{25 \times [2 \times 16] + 94 \times 2 \times 14} = 0.0664$$

equivalence ratio

$$\phi = \frac{\left(\frac{f}{a}\right)_{actual}}{\left(\frac{f}{a}\right)_{stoch}} \Rightarrow \left(\frac{f}{a}\right)_{actual} = 0.5 \times 0.0664 = 0.0332$$

53. (c) $$\eta_{isentropic} = \left[r_P^{\frac{\gamma-1}{\gamma}} - 1\right]\frac{T_{01}}{T_{02} - T_{01}}$$

$$= \left[(2.5)^{\frac{1.4-1}{1.4}} - 1\right]\frac{280}{[380 - 280]} = 0.837 = 83.7\%$$

54. Maximum velocity →

$$V_b = gI_{sp} \ln\frac{m_i}{m_i - m_p}$$

if initial velocity is zero

$$= 9.8 \times 400 \times \ln\left[\frac{4000}{4000 - 300}\right] = 305.60 \text{ m/sec}$$

55. Thrust $T = \dot{m}_p V_e + (P_2 - P_a)Ae$

at optimum operation Pe = Pa

∴ T = 200 × 1900 = 380 KN

Specific impulse

$$= \frac{T}{\dot{m}_p \times g} = \frac{380 \times 10^3}{200 \times 9.81} = 193.679 \text{ sec.}$$

56. Got sank in the Pacific Ocean

57. Tiny

58. Incorrigible

59. 19

Avg of 20 number = 0

$$\Rightarrow \frac{\text{Sum of 20 number}}{20} = 0$$

⇒ Sum of 20 number = 0

∴ Maximum 19 numbers can be +ve such that if sum of 19 numbers is x then 20th number can be –x to get total sum zero.

60. None of these

$\log_b(0)$ is undefined

61. Let a be the side of square

error in measuring = 3%

∴ Total side measured = a + 3% a

actual area = a × a

measured area = (a + 3%a) (a + 3%a)

error in area = (a + 3%a)(a + 3%a) – a × a

∴ %age error

$$= \frac{(a + 3\%a)(a + 3\%a) - a \times a}{a \times a} \times 100$$

$$= \frac{a^2[[1 + 3\%](1 + 3\%) - 1]}{a^2} \times 100$$

$$= \left[\left(1 + \frac{3}{100}\right)\left(1 + \frac{3}{100}\right) - 1\right] \times 100$$

= [(1.03) (1.03) – 1] × 100

= 6.09%

62. Distance = 680 m = 0.68 km

Time taken = 5.5 minutes

$$= \frac{5.5}{60} = 0.09166\text{Hr}$$

$$\text{Velocity} = \frac{\text{Distance}}{\text{Time}} = 7.418\text{Km/hr}$$

63. Let present age of P and Q are 5x and 6x respectively

10 years ago

$$[5x - 10] = \frac{1}{2}[6x - 10]$$

5x – 10 = 3x – 5

2x = 5

$$x = \frac{5}{2} = 2.5 \text{ years}$$

⇒ Total of their age

= 5x + 6x = 5 × 2.5 + 6 × 2.5

= 27.5 years

64. She said, “My brother is getting married”.

65. The small child does what ever his father does.

MOCK TEST – 6

AEROSPACE ENGINEERING

Time Allowed: 3:00 Hours | *Maximum Marks: 100*

Read the following instructions carefully before you begin to answer the questions. This Booklet contains questions in English.

INSTRUCTIONS

1. This Booklet contains **65 questions** in total comprising of the following two parts:

   Part -1: Technical Section (85 Marks) *(1-55 Questions)*

   Part -2: General Aptitude Section (15 Marks) *(56-65 Questions)*

2. All questions are compulsory. Q.1-Q.25 & Q.56-Q.60 carry one mark each. Q.26-Q.55 & Q.61-Q.65 carry two marks each.

3. **Types of Questions:**

   (*a*) Multiple Choice Questions(MCQ) carrying 1 or 2 marks each in all papers and sections. These questions are objective in nature, and each will have a choice of four options, out of which the candidate has to mark the correct answer(s).

   (*b*) Numerical Answer Questions of 1 or 2 marks each in all sections. For these questions the answer is a real number, to be entered by the candidate using the virtual keypad. No choices will be shown for these type of questions.

4. For 1-mark multiple-choice questions, 1/3 marks will be deducted for a wrong answer. Likewise, for 2-marks multiple-choice questions, 2/3 marks will be deducted for a wrong answer. There is no negative marking for numerical answer type questions.

MOCK TEST – 6

1. Laplace transform of $y(t) = 3e^{-t} + \frac{4}{5}e^{-3t}$ is

(a) $\frac{3}{s+1} + \frac{4}{5(s+1)}$ (b) $\frac{4}{s+1} + \frac{5}{4(s+1)}$

(c) $\frac{3}{s+1} + \frac{4}{5(s+3)}$ (d) None of the above

2. For Given matrix $A = \begin{bmatrix} 4 & 5 \\ 2 & 3 \end{bmatrix}$, inverse

(a) Exists (b) Does not exist

(c) Not enough data (d) None of these

3. Gradient of function $f(x, y) = 2x^2 y - 3y^2 x$ at (1,1) is

(a) $i - 4\hat{j}$ (b) $\hat{i} + 4\hat{j}$

(c) $i - 8j$ (d) $4\hat{i} - \hat{j}$

4. Which one is a non - linear ordinary differential equation

(a) $\frac{d^2y}{dx^2} + 6\frac{dy}{dx} + y = 0$

(b) $\frac{dy}{dt} + y = f(t)$

(c) $y^2\frac{d^2y}{dx^2} + y\frac{dy}{dx} = 0$

(d) None of these

5. For a given airplane which of these does not vary

(a) Center of gravity (b) Aerodynamic center

(c) Center of pressure (d) None of these

6. For two versions of aircraft lift coefficient was $C_{L_1} = 1.15$ and $C_{L_2} = 1.2$. What is Ratio of lift of the 1st aircraft to 2nd aircraft . [ρ, V, S are same for both]

7. Canard surfaces are mainly used for

(a) Pitching (b) Rolling

(c) Yawing (d) None

8. Indicated Airspeed of an airplane is 60 m/sec. climbing at 35° climb angle. The climb rate for this aircraft is _______ m/sec.

9. Time period of Phugoid oscillation is 20 sec. The velocity at which it is happening is _______ m/sec

10. Turn rate for a pull up maneuver is given by:

(a) $\omega = \frac{g(n+1)}{v}$ (b) $\omega = \frac{g(n-1)}{v}$

(c) $\omega = \frac{(n-1)}{v}$ (d) None of these

11. Across a normal shock wave mach number reduces to:

(a) subsonic (b) remain supersonic

(c) mach is 1 (d) zero

12. Maximum camber of NACA 2412 airfoil is _______ % of chord.

13. Bernoulli's equation is represented by

(a) $P + \frac{1}{2}\rho v^2 = \text{constant}$

(b) $P + \frac{1}{2}\rho v^2 + gz = \text{constant}$

(c) a and b both

(d) None of these

14. Drag - divergence Mach number is always greater than the critical Mach number. Given statement is

(a) True (b) False

(c) Not enough data (d) None

15. Angle of deflection of wedge is always _______ than the shock wave angle

(a) Smaller (b) Larger

(c) Equal (d) All

16. Maximum Amplitude at Resonance for a single degree of freedom system _______ with increase in damping damping ratio

(a) Increases (b) Decreases

(c) Remains constant (d) None

17. For a fixed - pinned column critical buckling load is k time more than pinned - pinned column. k is equal to

(a) $\frac{1}{4}$ (b) 4

(c) 2 (d) 1

18. Natural frequency of a beam with circular crossection ______ with increase in diameter.

(*a*) Increases (*b*) Decrcases

(*c*) Remains same (*d*) None of these

19. For a given state of stress $\sigma_{zz} = 0.45$ MPa and $\epsilon_{yz} = 0.003$ and given that all other stresses are non - zero.

Given problem is a ______ problem.

(*a*) Plane stress

(*b*) Plane strain

(*c*) 3 - Dimensional elasticity

(*d*) None of these

20. Boundary conditions for a cantilever beam are given by

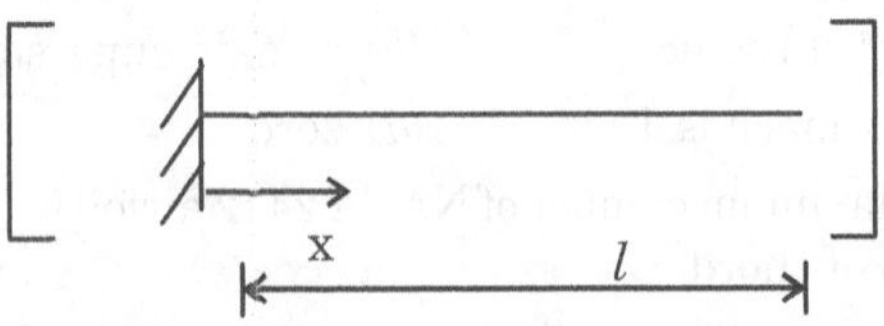

(*a*) $\frac{dy}{dx} = y = 0$ at $x = 0$

(*b*) $\frac{dy}{dx} = 0$ at $x = 0$

(*c*) $\frac{d^3y}{dx^3} = 0$ at $x = l$

(*d*) All of the above

21. Afterburners are mainly use to ______ thrust

(*a*) Increase (*b*) Decrease

(*c*) Keep constants (*d*) None of - these

22. Centrifugal compressors are prone to which of the following loading.

(*a*) Fatigue (*b*) Thermal

(*c*) Vibrations (*d*) a and c

23. Which of following engines cannot be used from take off till the landing of airplane.

(*a*) Turbojet (*b*) Turbooan

(*c*) Ramjet (*d*) Scramjet

24. Which of the following component of aircraft engine has maximum isentropic efficiency comparatively

(*a*) Nozzle (*b*) Compressor

(*c*) Turbine (*d*) Diffuser

25. Rocket Motor generally refers to a rocket with ______ propellant

(*a*) Solid

(*b*) Liquid

(*c*) Composite solid + liquid

(*d*) None of these

26. $Q = \begin{bmatrix} \frac{1}{\sqrt{2}} & \frac{1}{\sqrt{2}} \\ \frac{1}{\sqrt{2}} & -\frac{1}{\sqrt{2}} \end{bmatrix}$

Column of a matrix are

(*a*) orthonormal (*b*) equal

(*c*) orthogonal (*d*) both a and c

27. $A = \begin{bmatrix} 1 & 1 & -2 \\ 1 & -2 & 1 \\ -2 & 1 & 1 \end{bmatrix}$

Eigen values of inverse of A are

(*a*) 1, –1, 0 (*b*) 0, 3, 4

(*c*) 0, 0, 4 (*d*) not enough data

28. If $f = \nabla g = \frac{4x + 3y^2}{\sqrt{6}}\hat{i} + \frac{3xy}{\sqrt{6}}\hat{j}$

Directional derivative of function g in the direction of vector $\vec{v} = \frac{3i - 4j}{\sqrt{6}}$ is

(*a*) $\frac{1}{6}\left[12xy - 9y^2 - 12x\right]$

(*b*) $\frac{1}{6}\left[12x + 9y^2 - 12xy\right]$

(*c*) $\frac{1}{6}\left[9y^2 - 12xy - 2x\right]$

(*d*) $\frac{1}{6}\left[12x - 12xy - 9y^2\right]$

29. $\frac{d^2y}{dt^2} + 9\frac{dy}{dt} + 6y = 0$

Solution of this equation in spacial coordinates is equal to ______ given that y is a function of time only.

(*a*) $y = c_1e^{-3t} + c_2e^{-4t}$

(*b*) $c_1e^{-0.72t} + c_2e^{-2.27t}$

(*c*) cannot be solved

(*d*) None of these

30. $Z_1 = 2 + 3i$, $z_2 = 3 + 4i$, $z_3 = a + bi$ $a, b \in R$

What is the value of $|z_1 + z_2|$?

(*a*) $\sqrt{75}$ (*b*) $\sqrt{73}$

(*c*) $\sqrt{72}$ (*d*) 8.602

31. If Geopotential altitude is given as 100km corresponding geometric altitude in terms of earth's radius is h_G = ________

$$\alpha = \left(\frac{r}{100} - 1\right)$$

(*a*) $\frac{r}{\alpha}$km (*b*) $\frac{\alpha}{r}$km

(*c*) $\frac{r-1}{\alpha}$km (*d*) None

32. For a isothermal layer of atmosphere having temperature 288K, what is the pressure P_2 at height 12 km if pressure at 8 km is $P_1 = 8 \times 10^4$ Pa. (R = 287 J/kgK). P_2 = ________ Pa

33. Which of below given graph represents drag polar:

(*a*)

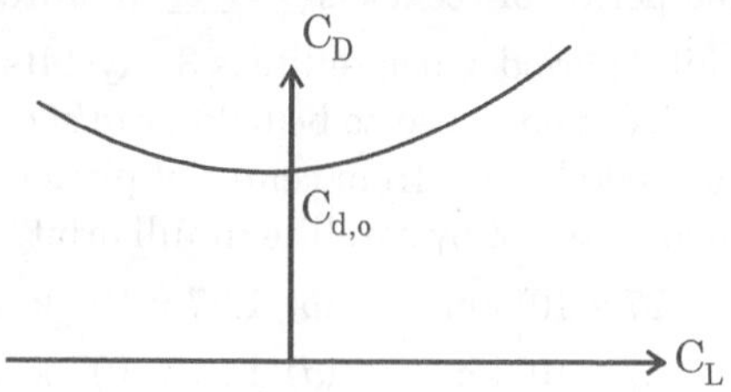

(*b*)

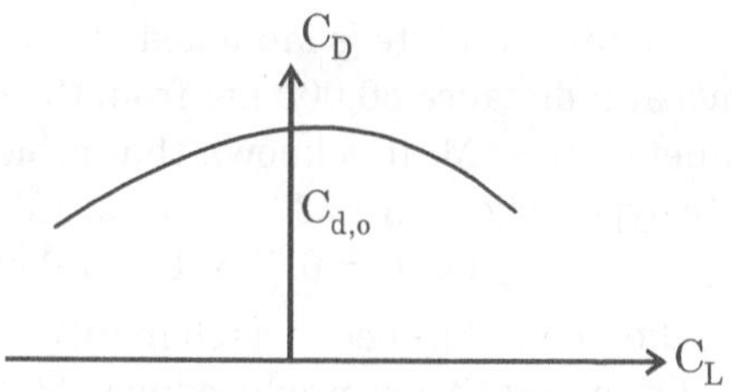

(*c*)

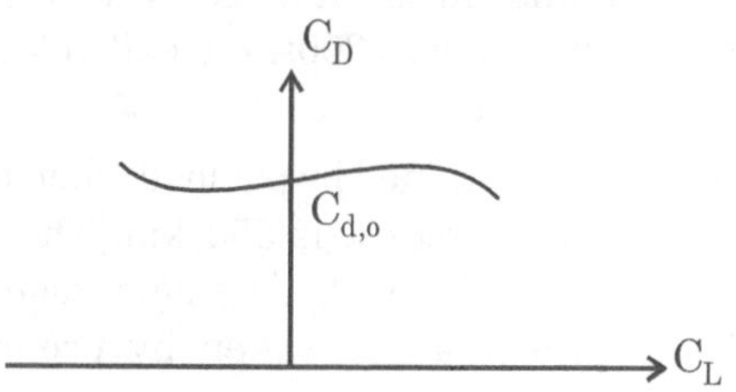

(*d*)

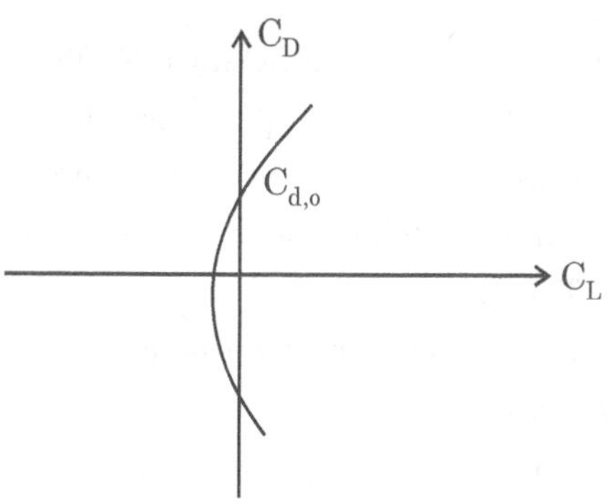

34. For minimum power required of an aircraft, induced drag coefficient should be 0.5. What should be corresponding zero lift drag coefficient?

(*a*) 0.19 (*b*) 0.15

(*c*) 0.167 (*d*) 0.8

35. For a propeller aircraft, engine output efficiency was modified from 0.8 to 0.85 in such a way that there was no effect on aircraft aerodynamics and structure. SFC of engine got increased by 5%. New endurance in terms of older one is E_2 = ______ E_1.

36. For a low speed wind tunnel:

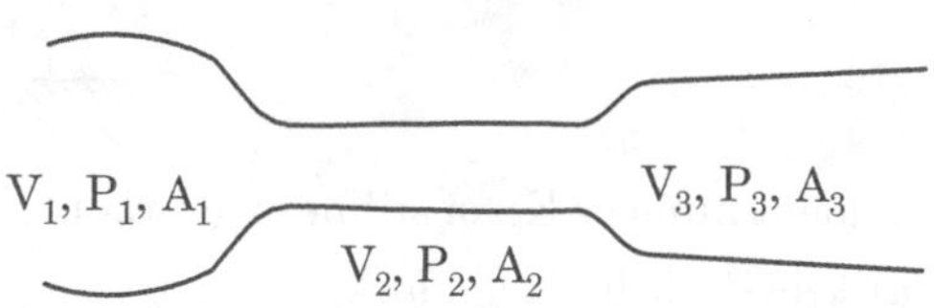

Area ratio $\frac{A_2}{A_1} = \frac{1}{2}$ and inlet and outlet pressure are $P_1 = 0.8 \times 10^4$ Pa, $P_2 = 0.4 \times 10^4$ Pa considering sea level density value, velocity V_2 = ________ m/sec.

37. Indicated airspeed for a aircraft was 50 m/sec at a height where density is 0.8 kg/m³. The equivalent airspeed is V_e = __________ m/sec.

38. Which of the following options holds true for laminar and turbulent boundary layers.

(*a*) $\delta_{turb.} > \delta_{lam}$

(*b*) $\tau_{turb.} > \tau_{lam}$

(*c*) $\left(\frac{dv}{dy}\right)_{y=0,lam} < \left(\frac{dv}{dy}\right)_{y=0,turb}$

(*d*) All of these

39. For dynamically similar flows which of following essentially should be same

(*a*) M_∞, R_e (*b*) V_∞, P_∞

(*c*) T_{01}, P_{01} (*d*) All of these

40. Value of coefficient of lift for a flat plat at 3° angle to a freestream mach number 1.5 is ________

41. Strouhal number for a circular cross section chimney is 0.2. Outer diameter of chimney is 30 cm and first natural frequency of chimney is 12Hz. What should be the velocity of flow to make chimney Iscillate?

V = ________ m/sec

42. For a fixed -pinned column critical buckling load was found to be 4000 N. Column was made up of steel with E = 210 GPa. If length of Column is 40cm, 2nd moment of Area which can resist 4000N load is

(*a*) > 1.54×10^{-10} m^4 (*b*) < 1.54×10^{-10} m^4

(*c*) = 1.54×10^{-10} m^4 (*d*) None of these

43. For figure given below,

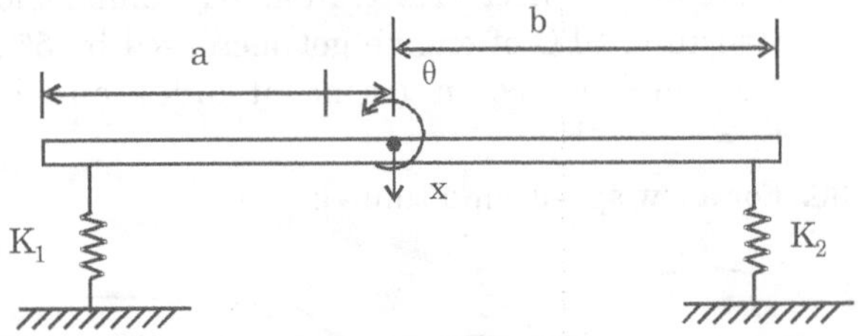

element number K_{22} of stiffness matrix is

(*a*) $k_1 aG^2 - k_2 b^2$ (*b*) $k_1 a^2 + k_2 b^2$

(*c*) $k_1 a^2 \pm k_2 b^2$ (*d*) $k_1 b^2 - k_2 a^2$

44. Amplitude Ratio of 1st peak and 7th peak of single degree of freedom system was found to be 5. Damping ratio of this system is _________%.

45. Torsional Rigidly of a shaft is a product of which of below given options:

(*a*) Shear modulus and Bulk modulus

(*b*) Bulk modulus and Young's modulus

(*c*) 2nd moment of area of shear modulus

(*d*) Polar moment of area and shear modulus

46. Boundary condition for torsional vibration of free-free shaft are:

(*a*) $\theta(o,t) = 0, \frac{\partial\theta}{\partial t}(\ell,t) = 0$

(*b*) $\frac{\partial\theta}{\partial x}(o,t) = 0, \frac{\partial\theta}{\partial x}(\ell,t) = 0$

(*c*) $\frac{\partial\theta}{\partial t}(o,t) = 1, \frac{\partial\theta}{\partial x}(\ell,t) = 2$

(*d*) None of these

47. Airy stress, function for a thin square panel was found as $\phi = 6x^2 + 9xy + 10y^2$.

Ratio of in plane shear stresses produced to the product of normal stresses is _________ $\times 10^{-2}$.

48. Mass flow rate through an turbojet engine is 15 kg/sec. compressor outlet and turbine inlet temperature are 420 K and 1500 K respectively. If mass flow rate of fuel is 0.4 kg/sec. and has a specific heat at constant pressure C_P = 1000 J/kg K. The heating value of fuel is ______ MJ/kg.

49. For a Gas turbine engine:

Total temperature at turbine inlet = 1400 K

Total pressure at turbine exit = 1.9 bar

static temperature at turbine exit = 820 K

Velocity at turbine exit = 180 m/sec

Isentropic efficiency of turbine = 0.9

$\gamma = 1.4$

C_P = 1000 J/KgK

Total pressure for turbine inlet is _________ bar.

50. A liquid rocket engine produces thrust of 3MN with mass flow rate of fuel 50 Kg/sec. If initial mass of rocket was 200 × 10^3 kg and thrust coefficient and characteristic velocity are 1.7 and 200 m/sec. respectively, Then The propellant mass flow rate of rocket is _________ Kg/sec. [propellant = fuel + oxidizer]

51. Effective exhaust velocity of a rocket is 3000 m/s. Total impulse of this Rocket is 150 KNsec. With propellant mass flow rate of 0.3 Kg/sec. The burn time period of rocket is _______ minutes.

52. Orbital period of a satellite is 3 days. Its angular velocity is observed to be 0.002 rad/sec. When it is at 7000 km far from center of planet. What is the area swept by satellite in full orbit?

(*a*) 1.27×10^3 km^2 (*b*) 1.27×10^{11} km^2

(*c*) 1.27×10^9 km^2 (*d*) 1.27×10^{10} km^2

53. A planetary satellite is launched at speed of 300 km/s at a distance 80,000 km from the center of planet of mass M. It is known that orbit is going to be hyperbolic with e = 1.3. The mass M of planet is __________ kg (G = 6.67×10^{-11} m^3 kgs^{-1})

54. A rocket motor has been tested in fully expanded condition and thrust produced was 500 kN and specific impulse was 300s. the exit velocity of jet is _________ m/sec.

55. Max altitude attained by payload when launched vertically on a rocket is 350 km. The Burnout velocity [given that rocket is single stage end 70% of its initial mass is taken by propellant is _________ m/sec.

56. Write the antonym of COMMISSIONED

(*a*) Started (*b*) Finished

(*c*) Closed (*d*) Terminated

57. Which of the following has error:

(*a*) The Indian radio

(*b*) Is free now from the people

(*c*) Both

(*d*) None

58. Choose correct meaning of proverb/Idiom

(*a*) To become hungry

(*b*) To be in good mood

(*c*) To be kind

(*d*) None

59. If $\log_{10}3 = a$, what is the value of $\log_{10}\frac{1}{300}$?

(*a*) $a - 2$ (*b*) $2 - a$

(*c*) $-(a + 2)$ (*d*) $2 \pm a$

60. P is able to do work in 12 days and Q can do same work in 20 days. If they work together For 4 days, what is the fraction of work left?

(*a*) $\frac{7}{15}$ (*b*) $\frac{6}{15}$

(*c*) $\frac{15}{7}$ (*d*) $\frac{15}{8}$

61. $\sqrt{\frac{144}{11}} \times \frac{11}{\sqrt{225}} \times \frac{15}{\sqrt{196}}$ is equal to

(*a*) 0.9 (*b*) 0.15

(*c*) 0.85 (*d*) 0.58

62. A die is rolled twice. What is the probability of getting a sum equal to 9

(*a*) $\frac{1}{9}$ (*b*) $\frac{2}{9}$

(*c*) $\frac{3}{9}$ (*d*) $\frac{4}{9}$

63. There is 70% increase in an amount in 7 years at simple interest. What will be compound interest of R.s. 14,000 after 3 year at same rate?

(*a*) 4635 (*b*) 4634

(*c*) 4636 (*d*) 4633

64. Arrange the pieces of sentences in correct order:

Then

P : It struck me Q : Of course

R : Suitable it was S : How eminently

(*a*) SPQR (*b*) QSRP

(*c*) PSRQ (*d*) QPSR

65. Which option make the sentence grammatically correct: You need not come unless you want to

(*a*) You don't need to come unless you want to

(*b*) You came unless you don't want to

(*c*) Both a & b

(*d*) None of them

OMR SHEET

1. Ⓐ Ⓑ Ⓒ Ⓓ
2. Ⓐ Ⓑ Ⓒ Ⓓ
3. Ⓐ Ⓑ Ⓒ Ⓓ
4. Ⓐ Ⓑ Ⓒ Ⓓ
5. Ⓐ Ⓑ Ⓒ Ⓓ
6. [ ]
7. Ⓐ Ⓑ Ⓒ Ⓓ
8. [ ]
9. [ ]
10. Ⓐ Ⓑ Ⓒ Ⓓ
11. Ⓐ Ⓑ Ⓒ Ⓓ
12. [ ]
13. Ⓐ Ⓑ Ⓒ Ⓓ
14. Ⓐ Ⓑ Ⓒ Ⓓ
15. Ⓐ Ⓑ Ⓒ Ⓓ
16. Ⓐ Ⓑ Ⓒ Ⓓ
17. Ⓐ Ⓑ Ⓒ Ⓓ
18. Ⓐ Ⓑ Ⓒ Ⓓ
19. Ⓐ Ⓑ Ⓒ Ⓓ
20. Ⓐ Ⓑ Ⓒ Ⓓ
21. Ⓐ Ⓑ Ⓒ Ⓓ
22. Ⓐ Ⓑ Ⓒ Ⓓ
23. Ⓐ Ⓑ Ⓒ Ⓓ
24. Ⓐ Ⓑ Ⓒ Ⓓ
25. Ⓐ Ⓑ Ⓒ Ⓓ
26. Ⓐ Ⓑ Ⓒ Ⓓ
27. Ⓐ Ⓑ Ⓒ Ⓓ
28. Ⓐ Ⓑ Ⓒ Ⓓ
29. Ⓐ Ⓑ Ⓒ Ⓓ
30. Ⓐ Ⓑ Ⓒ Ⓓ
31. Ⓐ Ⓑ Ⓒ Ⓓ
32. [ ]
33. Ⓐ Ⓑ Ⓒ Ⓓ
34. Ⓐ Ⓑ Ⓒ Ⓓ
35. [ ]
36. [ ]
37. [ ]
38. Ⓐ Ⓑ Ⓒ Ⓓ
39. Ⓐ Ⓑ Ⓒ Ⓓ
40. [ ]
41. [ ]
42. Ⓐ Ⓑ Ⓒ Ⓓ
43. Ⓐ Ⓑ Ⓒ Ⓓ
44. [ ]
45. Ⓐ Ⓑ Ⓒ Ⓓ
46. Ⓐ Ⓑ Ⓒ Ⓓ
47. [ ]
48. [ ]
49. [ ]
50. [ ]
51. [ ]
52. Ⓐ Ⓑ Ⓒ Ⓓ
53. [ ]
54. [ ]
55. [ ]
56. Ⓐ Ⓑ Ⓒ Ⓓ
57. Ⓐ Ⓑ Ⓒ Ⓓ
58. Ⓐ Ⓑ Ⓒ Ⓓ
59. Ⓐ Ⓑ Ⓒ Ⓓ
60. Ⓐ Ⓑ Ⓒ Ⓓ
61. Ⓐ Ⓑ Ⓒ Ⓓ
62. Ⓐ Ⓑ Ⓒ Ⓓ
63. Ⓐ Ⓑ Ⓒ Ⓓ
64. Ⓐ Ⓑ Ⓒ Ⓓ
65. Ⓐ Ⓑ Ⓒ Ⓓ

ANSWERS

1. (c)	**2.** (a)	**3.** (a)	**4.** (c)	**5.** (b)	**6.** 0.958	**7.** (a)
8. 34.41 m/sec.		**9.** 44.16 m/sec.	**10.** (b)	**11.** (a)	**12.** 2%	**13.** (c)
14. (a)	**15.** (a)	**16.** (b)	**17.** (c)	**18.** (a)	**19.** (c)	**20.** (d)
21. (a)	**22.** (d)	**23.** (c & d)	**24.** (a)	**25.** (a)	**26.** (c)	**27.** (d)
28. (b)	**29.** (c)	**30.** (d)	**31.** (a)	**32.** 4.97 × 10^4 Pa		**33.** (a)
34. (c)	**35.** 1.012	**36.** 93.31 m/sec.		**37.** 40.41 m/sec	**38.** (d)	**39.** (a)
40. 0.187	**41.** 18 m/sec	**42.** (a)	**43.** (b)	**44.** $\zeta = 0.042 = 4.2\%$		**45.** (d)
46. (b)	**47.** 3.75	**48.** 42 MJ/kg	**49.** 15.15 bar	**50.** 882.352 kg/s		
51. 2.78 minutes		**52.** (d)	**53.** M = 8.3 × 10^{28}kg		**54.** 2939.97 m/sec	
55. 2620.4961 m/sec		**56.** (d)	**57.** (b)	**58.** (b)	**59.** (c)	**60.** (a)
61. (c)	**62.** (a)	**63.** (b)	**64.** (c)	**65.** (a)		

EXPLANATIONS

1. $L[y(t)] = y(s)$

$$= \frac{3}{s+1} + \frac{4}{5}\frac{1}{s+3}$$

$$\left[\because L\left[e^{-at}\right] = \frac{1}{s+a}\right]$$

2. If determinant of a matrix is zero, its inverse don't exists

$|A| = 12 - 10 = 2 = +ve > 0$

Inverse of A exist

3. Gradient of $f(x,y) = \nabla f(x,y)$

$$= \frac{\partial f}{\partial x}\hat{i} + \frac{\partial f}{\partial y}\hat{j}$$

$$= \left[4xy - 3y^2\right]\hat{i} + \left[2x^2 - 6yx\right]\hat{j}$$

$$\left[\nabla f(x,y)\right]_{[1,1]} = [4-3]\hat{i} + [2-6]\hat{j}$$

$$= \hat{i} - 4\hat{j}$$

4. In an ordinary differential equation if a unknown variable is multiplied by itself or its derivative then ODE is nonlinear

5. Aerodynamic center

6. $L = \frac{1}{2}\rho V^2 S C_L$

$$\therefore \frac{L_1}{L_2} = \frac{C_{L_1}}{C_{L_2}} = \frac{1.15}{1.2} = 0.958$$

7. Pitching

8. Climb rate $\frac{R}{C} = V_\infty \text{Sin}\theta$

$= 60 \times \text{Sin}(35°)$

$= 34.41$ m/sec.

9. Time period $T = \sqrt{2}\pi\frac{V}{g}$

$$V = \frac{20 \times 9.81 \times \pi}{\sqrt{2}}$$

$= 44.16$ m.sec.

10. $w = \frac{g(n-1)}{v}$

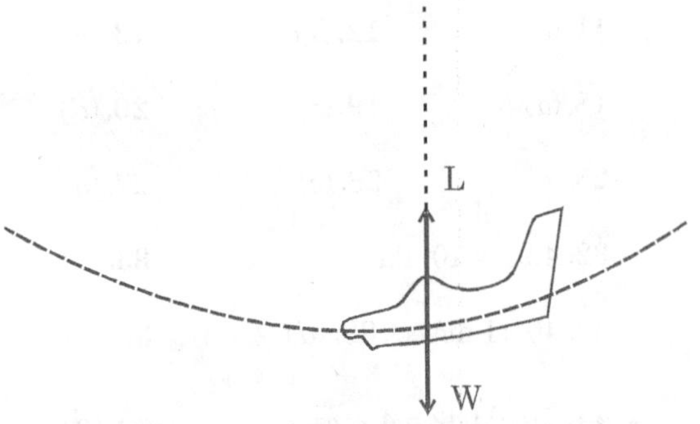

11. Across a normal shock Mach number always reduces to subsonic speeds

12. 1st digit ___ 2% maximum camber

[2nd] located at 40% from leading edge

[3rd – 4th] maximum thickness of 12%

13. Both a and b are form of Bernoulli's equation

14.

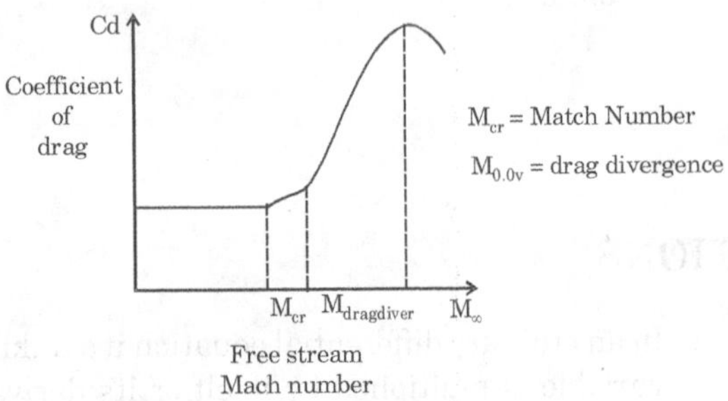

Mcr. = Critical mach number

Mo dragdiver = drag divergence

15. Smaller

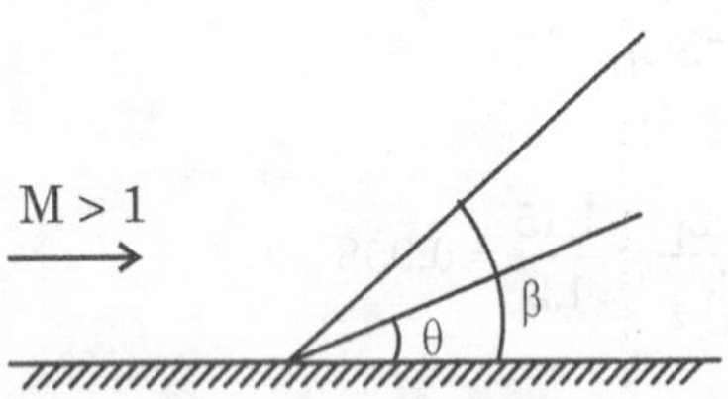

θ = wedge angle

β = shock angle

16. Decreases

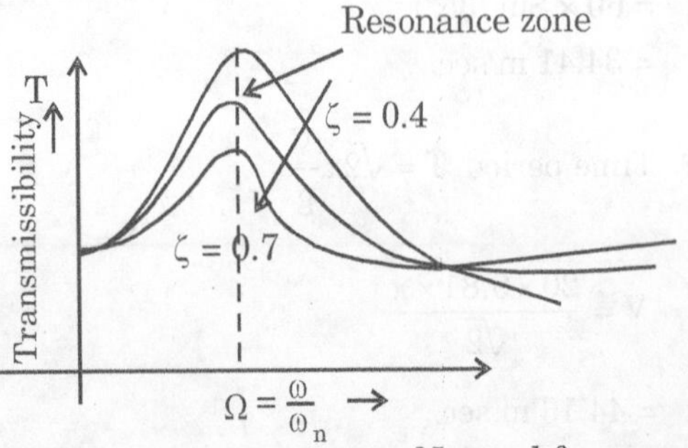

ω_n = Natural frequency

ω = forcing frequency

17. $[P_{cr.}]_{F.P} = \frac{2\pi^2 EI}{l^2}$ fixed - pinned

$[P_{cr.}]_{P.P} = \frac{\pi^2 EI}{l^2}$ pinned – pinned

$\boxed{k = 2}$

18. Natural frequency, $\omega = \frac{\pi^2}{L^2}\sqrt{\frac{EI}{\rho A}}$

$$= \frac{\pi^2}{L^2}\sqrt{\frac{E\frac{\pi d^4}{64}}{\rho\frac{\pi d^2}{4}}}$$

$\therefore \omega \propto d$

19. Because all the stresses are non zero

For plane stress : $\sigma_{zz} = \tau_{xz} = \tau_{yz} = 0$

For plane spain : $\epsilon_{zz} = \epsilon_{xz}, \epsilon_{yz} = 0$

20. At x = 0, $\frac{dy}{dx}, y$ i.e slope and deflection are zero

At x = l, $\frac{d^2y}{dx^2}, \frac{d^3y}{dx^3}$ i.e Bending Moment and shear forces are zero

21. Afterburner is mainly used in fighter aircraft to get sudden increments in thrust.

22. Because of cyclic loading contrifugal compressors are prone to fatigue and vibrations.

23. Ramjet and scramjet both needs high inlet flow velocities to function. Thus cannot be used from take - off.

24. $\eta_n > \eta_d > \eta_t > \eta_c$

nozzle > diffuser > turbine compressor

25. Solid propellant Rockets are called Rocket motors.

26. orthogonal:

(1st column)^T (2nd column) = 0 .....(i)

orthonormal:

with condition (i), magnitude of vectors should be 1

....(ii)

(i) $\left(\frac{1}{\sqrt{2}} \quad \frac{1}{\sqrt{2}}\right)\begin{bmatrix}\frac{1}{\sqrt{2}} \\ \frac{1}{\sqrt{2}}\end{bmatrix} = \frac{1}{2} - \frac{1}{2} = 0$

(ii) Magnitude of

$$\text{Ist} = \sqrt{\left(\frac{1}{\sqrt{2}}\right)^2 + \left(\frac{1}{\sqrt{2}}\right)^2} = 1$$

$$\text{2nd} = \sqrt{\left(\frac{1}{\sqrt{2}}\right)^2 + \left(\frac{1}{\sqrt{2}}\right)^2} = 1$$

27. Determinant of A is zero

Thus inverse of A doesn't exists

$$|A| = 1[-2-1]-1[1+2]-2[1-4]$$

$$= -3-3+6 = 0$$

28. Directional derivative of function g in direction of vector $\vec{v}$

$$f\vec{v} = \left[\left[\frac{4x+3y^2}{\sqrt{6}}\right]\hat{i} + \frac{3xy}{\sqrt{6}}\hat{j}\right] \cdot \left[\frac{3i-4i}{\sqrt{6}}\right]$$

$$\therefore \; \nabla.g.\vec{v} = \frac{1}{6}\left[12x + 9y^2 - 12xy\right]$$

29. As y is a function of time only solution of equation will be in time only not in spatial coordinates. Therefore, this equation cannot be solved in spatial coordinates.

30. $|z_1 + z_2| = |2+3i+3+4i| = |5+7i|$

$$= \sqrt{5^2 + 7^2}$$

$$= \sqrt{25+49} = \sqrt{74} = 8.602$$

31. We know that

$$h = \frac{r}{r+h_G} \times h_G$$

h = Geopotential altitude

$$100 = \frac{r}{(r+h_G)}(h_G)$$

$100\,[r + h_G] = r[h_G]$

$100\,h_G - r\,h_G = -100r$

$h_G[100 - r] = -100\,r$

$$h_G = \frac{-100r}{-[r-100]} = \frac{100r}{(r-100)}$$

$$= \frac{r}{\left[\frac{r}{100} - 1\right]} = \frac{r}{\alpha}\text{Km}$$

32. $P_2 = P_1 e^{\frac{-g_o}{RT}(h_2-h_1)}$

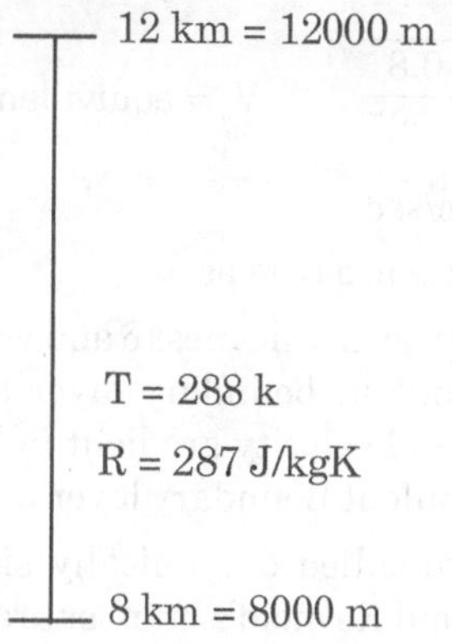

$$= \left(8 \times 10^4\right) \times e^{\frac{-9.81}{287\times288}[12000-8000]} = 4.97 \times 10^4 \text{ Pa}$$

33. Equation of drag polar

$$C_D = C_{d,0} + \frac{{C_L}^2}{\pi e AR}$$

34. For minimum power

$$C_{Do} = \frac{1}{3}C_{Di}$$

$$C_{Do} = \frac{1}{3} \times 0.5 = 0.167$$

35. For propeller plane

$$E = \frac{\eta}{C}\frac{{C_L}^{\frac{3}{2}}}{C_D}(2\rho_\infty s)\left(W_1^{\frac{-1}{2}} - W_0^{\frac{-1}{2}}\right)$$

η and c are changing all other are kept same

$$\frac{E_2}{E_1} = \frac{\left(\frac{\eta}{C}\right)_2}{\left(\frac{\eta}{C}\right)_1} = \frac{\frac{0.85}{1.05C}}{\frac{0.8}{C}} = 1.012$$

$$\boxed{E_2 = 1.012\,E_1}$$

36. For such a wind-tunnel

$$V_2 = \sqrt{\frac{2[P_1 - P_2]}{\rho\left[1 - \left(\frac{A_2}{A_1}\right)^2\right]}} \qquad \left[\rho = 1.225 \text{ kg/m}^3\right]$$

$$= \sqrt{\frac{2\left[0.8 \times 10^4 - 0.4 \times 10^4\right]}{1.2251\left[1 - \left(\frac{1}{2}\right)^2\right]}}$$

= 93.31 m/sec.

37. $V_e = v\sqrt{\frac{\rho}{\rho_s}}$ ρ_s = sea level density

$= 50 \times \sqrt{\frac{0.8}{1.225}}$ V_e = equivalent airspeed

= 40.41 m/sec

38. All of these are true as

Boundary layer-thickness δ and wall shear stress τ_w of turbulent boundary layer are more than laminar and velocity gradient in laminar is less than turbulent Boundary layer near wall.

39. Flows are called dynamically similar if Mach number and Reynold's number are same for both.

40.

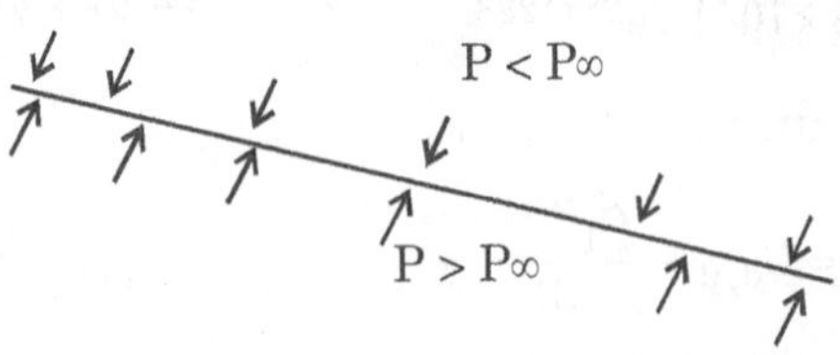

Coefficient of lift

$$C_L = \frac{4\alpha}{\left(M_\infty^2 - 1\right)^{\frac{1}{2}}}$$

angle should be in radian

$$= \frac{4 \times 30 \times \frac{\pi}{180}}{\left(1.5^2 - 1\right)^{\frac{1}{2}}} = 0.187$$

41. Diameter = D 30 cm = 0.03 m

f = 12 Hz

Vortex shedding frequency should be equal to natural frequency to make chimney oscillate

$$S.N = \frac{fD}{v}$$

$$v = \frac{12 \times 0.03}{0.2} = 18 \text{ m/sec}$$

42. For fixed-pinned column

$$P_{cr} = \frac{2\pi^2 EI}{\ell^2}$$

ℓ = 40 cm

= 0.4 m

$$\therefore I = \frac{P_{cr} \times \ell^2}{2\pi^2 E}$$

$$= \frac{4000 \times [0.4]^2}{2\pi^2 \times 210 \times 10^9}$$

= 1.54×10^{-4} m⁴.

∴ I should be greater than this value to resist P_{cr} = 4000 N

43. Kinetic Energy

$$= \frac{1}{2} m\dot{x}^2 + \frac{1}{2} I\dot{\theta}^2$$

Potential Energy

$$= \frac{1}{2} K_1 [x + a\theta]^2 + \frac{1}{2} K_2 [x - b\theta]^2$$

Using Lagrangian equation

Neutral Position

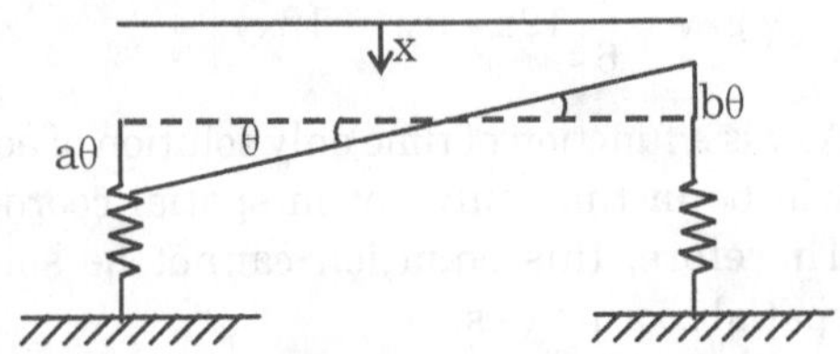

$$\frac{d}{dt}\left(\frac{\partial KE}{\partial \dot{x}}\right) + \frac{\partial (P.E.)}{\partial x} = f(t)$$

x:

$$m\ddot{x} + k_1 x + k_1 a\theta + k_2 x - k_2 b\theta = 0$$

$$m\ddot{x} + (k_1 + k_2)x + [k_1 a - k_2 b]\theta = 0 \quad \text{....(i)}$$

θ :

$$I\ddot{\theta} + k_1 xa + k_1 a^2\theta - k_2 bx + k_2 b^2\theta = 0$$

$$I\ddot{\theta} + (k_1 a - k_2 b)x + \left[k_1 a^2 + k_2 b^2\right]\theta = 0 \quad \text{....(ii)}$$

∴ Stiffness matrix

$$K = \begin{bmatrix} (k_1 + k_2) & (k_1 a - k_2 b) \\ (k_1 a - k_2 b) & (k_1 a^2 + k_2 b^2) \end{bmatrix}$$

$$\therefore k_{22} = k_1 a^2 + k_2 b^2$$

44. $\delta = \ln\left(\frac{A_1}{A_n}\right) = (n-1)\frac{2\pi\zeta}{\sqrt{1-\zeta^2}}$

$$\ln\left[\frac{A_1}{A_2}\right] = [7-1]\frac{2\pi\zeta}{\sqrt{1-\zeta^2}}$$

$$\Rightarrow \ln[5] = 6 \times \frac{2\pi\zeta}{\sqrt{1-\zeta^2}}$$

Solving this

$\boxed{\zeta = 0.042} = 4.2\%$

45. Torsional Rigidity = GJ

G = Shear modulus

J = Polar moment of Area

46.

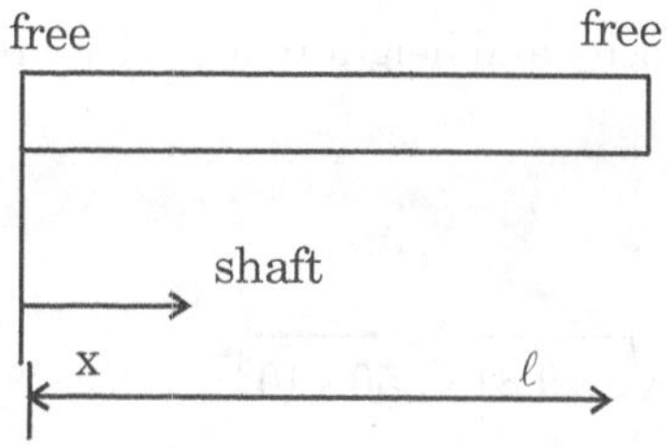

$$\frac{\partial\theta}{\partial x}(o,t) = 0\frac{\partial\theta}{\partial x}(\ell,t) = 0$$

47. In plane shear stresses

$$\tau_{xy} = \frac{\partial^2\phi}{\partial x\, dy}$$

$$= \frac{\partial}{\partial x}\left[\frac{\partial}{\partial y}[\phi]\right]$$

$$= \frac{\partial}{\partial x}[9x + 20y] = 9 \text{ units}$$

Normal stresses

$$\sigma_{xx} = \frac{\partial^2\phi}{\partial x^2} = 12$$

$$\sigma_{yy} = 20$$

$$\sigma_{xx}.\sigma_{yy} = 240 \text{ units}$$

$$\therefore \frac{\tau_{xy}}{\sigma_{xx}.\sigma_{yy}} = \frac{9}{240} = 0.0375$$

$= 3.75 \times 10^{-2}$.

48.

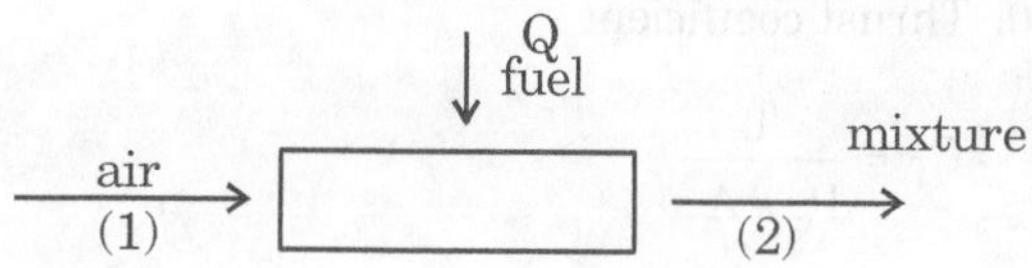

$$\dot{m}_a C_P T_1 + \dot{m}_f Q = \left(\dot{m}_a + \dot{m}_f\right) C_P T_2$$

$$\therefore Q = \frac{\left(\dot{m}_a + \dot{m}_f\right) C_P T_2 - \dot{m}_G\, a\, C_P T_1}{\dot{m}_f}$$

$$= \frac{[15 + 0.4] \times 1000 \times 1500 - 15 \times 1000 \times 420}{0.4}$$

$= 42{,}000000$

$= 42$ MJ/kg

49.

$$T_{o2} = T_2 + \frac{v_2^{\,2}}{2C_P}$$

$$= 820 + \frac{180^2}{2 \times 1000}$$

$= 836.2$ K

$$\text{Now } \frac{P_{o2}}{P_{o1}} = \left[\frac{T_{o\,2}'}{T_{o1}}\right]^{\frac{\gamma}{\gamma-1}}$$

$$\text{and } \eta = \frac{T_{o1} - T_{o2}}{T_{o1} - T_{o2}'}$$

$$0.9 = \frac{14000 - 836.2}{1400 - T_{o2}'} \Rightarrow T_{o2}' = 773.56\text{K}$$

$$\therefore P_{o1} = \frac{P_{o2}}{\left(\frac{T_{o1}'}{T_{o1}}\right)}$$

$$= \frac{1.9}{\left(\frac{773.56}{1400}\right)^{\frac{1.4}{1.4-1}}}$$

$= 15.15$ bar

50. Thrust coefficient

$$C_f = \frac{T}{P_c \times A^*}$$

$$\Rightarrow P_c = \frac{T}{C_f \times A^*} \quad ....(i)$$

Characteristic velocity

$$C^* = \frac{P_c \times A^*}{\dot{M}_P}$$

$$\dot{M}_P = \frac{P_c \times A^*}{C^*}$$

putting value of (i)

$$\Rightarrow \dot{M}_P = \frac{T}{C_f \times C^*} = \frac{3 \times 10^6}{1.7 \times 2000} = 882.352 \text{ kg/s}$$

51. Total impulse = Thrust × time

$$\text{Time} = \frac{\text{Total impulse}}{\text{Thrust}}$$

$$= \frac{\text{Total impulse}}{\dot{m}_P \times v_e} = \frac{150 \times 10^3}{0.3 \times 3000}$$

= 166.66 sec. = 2.78 minutes

52. Area swept

$$\frac{dA}{dt} = \frac{h}{2}$$

h = angular momentum.

On integrating

$$\Rightarrow A = \frac{h}{2}t \Rightarrow A = \frac{\omega R^2}{2} \times t$$

$$= \frac{0.002 \times \left[7000 \times 10^3\right]^2}{2} \times 3 \times 24 \times 60 \times 60$$

$= 1.27 \times 10^{16}$ m² $= 1.27 \times 10^{10}$ km².

53. Velocity of orbit

V = 300 km/s = 300×10^3 m/s

Altitude R = 80000×10^3 m

G = 6.67×10^{-11} m³ kgs⁻¹.

we know that

$$e = \frac{V^2 R}{GM} - 1$$

$$1.3 = \frac{\left[300 \times 10^3\right]^2 \times 80{,}000 \times 10^3}{6.67 \times 10^{-11} \times M}$$

$$\Rightarrow \boxed{M = 8.3 \times 10^{28} \text{ kg}}$$

54. Specific impulse $I_{sp} = \frac{T}{\dot{m}_P \times g}$

$$\Rightarrow \dot{m}_P = \frac{500 \times 10^3}{300 \times 9.8} = 170.07 \text{ kg/s}$$

Now Thrust produced

For fully expanded Nozzle

$$T = \dot{m}_P V_e \Rightarrow V_e = \frac{500 \times 10^3}{170.07} = 2939.97 \text{ m/sec}$$

55. Burnout velocity and height relation for a rocket

$$h = \frac{v^2}{2g}$$

$$v = \sqrt{2gh} = \sqrt{2 \times 9.81 \times 350 \times 10^3}$$

= 2620.4961 m/sec

56. Terminated

57. Is now free from the people

58. To be in good mood

59. $\log_{10}\left[\frac{1}{300}\right]$

$= \log_{10} 1 - \log_{10} 300 = 0 - \log_{10} [3 \times 100]$

$= -[\log_{10} 3 + \log_{10} 100] = -[a + 2]$

60. Amount of work P can do in 1 day $= \frac{1}{12}$

Amount of work Q can do in 1 day $= \frac{1}{20}$

Amount of work both can do in 1 day

$$= \frac{1}{12} + \frac{1}{20} = \frac{2}{15}$$

therfore: Amount of work both can do in 4 days

$$= 4 \times \frac{2}{15} = \frac{8}{15}$$

$$\boxed{\text{Fraction of work left} = 1 - \frac{8}{15} = \frac{7}{15}}$$

61. $\sqrt{\frac{144}{11}} \times \frac{11}{\sqrt{225}} \times \frac{15}{\sqrt{196}}$

$= \frac{12}{11} \times \frac{11}{15} \times \frac{15}{14}$

$= \frac{12}{14} = 0.85$

62. Number of outcomes of a die = 6

n(s) = Total no. of outcomes when die is rolled twice = 6 × 6 = 36

E = Getting sum of 9 = {(3,6), (4,5), (5,4), (6,3)}

Possible no. of element n (E) = 4

$$\therefore P(E) = \frac{n(E)}{n(S)} = \frac{4}{36} = \frac{1}{9}$$

63. Let P = 100

(S.I.) Simple interest = 70 (70% increase due to simple interest)

$$\text{Rate of interest} = \frac{100 \times \text{S.I.}}{\text{P} \times \text{Time}} = \frac{100 \times 70}{100 \times 7} = 10\%$$

Now, compound interest of Rs 14,000 after 3 years at 10%, P = 14,000

Time = 3 years

R = 10%

$$\text{Amount after 5 years} = P\left[1 + \frac{R}{100}\right]^T$$

$$= 14000\left[1 + \frac{10}{100}\right]^3$$

$$= 14000\left[\frac{110}{100}\right]^3 = 14 \times 11^3 = 18634$$

Compound interest = 18634 – 14,000

= Rs. 4634

64. Then it struck me how eminently suitable it was of course.

65. You don't need to come unless you want to.

MOCK TEST – 7

AEROSPACE ENGINEERING

Time Allowed: 3:00 Hours *Maximum Marks: 100*

Read the following instructions carefully before you begin to answer the questions. This Booklet contains questions in English.

INSTRUCTIONS

1. This Booklet contains **65 questions** in total comprising of the following two parts:

   Part -1: Technical Section (85 Marks) *(1-55 Questions)*

   Part -2: General Aptitude Section (15 Marks) *(56-65 Questions)*

2. All questions are compulsory. Q.1-Q.25 & Q.56-Q.60 carry one mark each. Q.26-Q.55 & Q.61-Q.65 carry two marks each.

3. **Types of Questions:**

   (*a*) Multiple Choice Questions(MCQ) carrying 1 or 2 marks each in all papers and sections. These questions are objective in nature, and each will have a choice of four options, out of which the candidate has to mark the correct answer(s).

   (*b*) Numerical Answer Questions of 1 or 2 marks each in all sections. For these questions the answer is a real number, to be entered by the candidate using the virtual keypad. No choices will be shown for these type of questions.

4. For 1-mark multiple-choice questions, 1/3 marks will be deducted for a wrong answer. Likewise, for 2-marks multiple-choice questions, 2/3 marks will be deducted for a wrong answer. There is no negative marking for numerical answer type questions.

MOCK TEST – 7

1. Rate of change of temperature (Temperature lapse rate) is Troposphere is

(a) 6.5 K/km (b) –6.5 K/km

(c) 5 K/km (d) –5 K/km

2. Induce drag coefficient for vertical flight will be

(a) $C_{Di} = 0$ (b) $C_{Di} = 1$

(c) $C_{Di} = \dfrac{C_L^2}{\pi\, eAR}$ (d) $C_{Di} = \infty$

3. Figure given below shown variation of maximum rate of climb with altitude

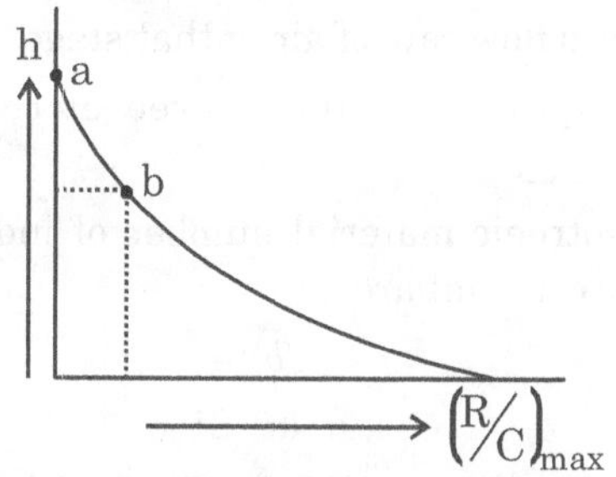

where, point a and b represents

(a) service ceiling, absolute ceiling

(b) absolute ceiling, service ceiling

(c) Maximum rate of climb, absolute ceiling

(d) None of these

4. For straight pointed wing stall occurs at

(a) Leading edge

(b) Trailing edge

(c) Mid of the wing

(d) Depends upon the speed

5. If the root of (5 × 6) metria Q is 4. Which of the following statement is correct?

(a) Q will have four linearly independent rows and four linearly independent columns.

(b) Q will have four linearly independent rows and five linearly independent columns

(c) $Q\,Q^T$ will be in invertiable.

(d) $Q^T Q$ will be in invertiable.

6. For a 3 × 3 matrix the value of determinant $|A| = 5$, then the value of determinant of its $|\text{adj}(\text{adj}\,|A|)|$, is

(a) 625 (b) 125

(c) 25 (d) 225

7. $\lim\limits_{x\to 0} \dfrac{\sin^2 x}{x}$ is equal to ________

8. Given $i = \sqrt{-1}$, what will be the evaluation of definite integral $\int_0^{\pi/2} \dfrac{\cos x + i \sin x}{\cos x - i \sin x}\, dx =$

(a) 0 (b) –1

(c) 1 (d) i

9.

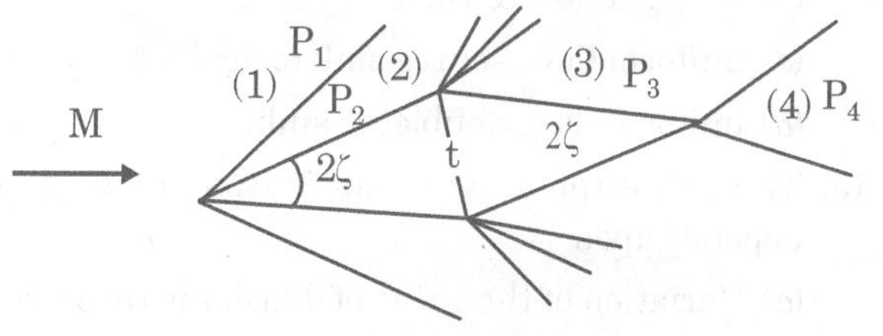

figure (A)

Supersonic flow over diamond wedge airfoil is shown in figure (A). The pressure distribution on the airfoil is

(a)

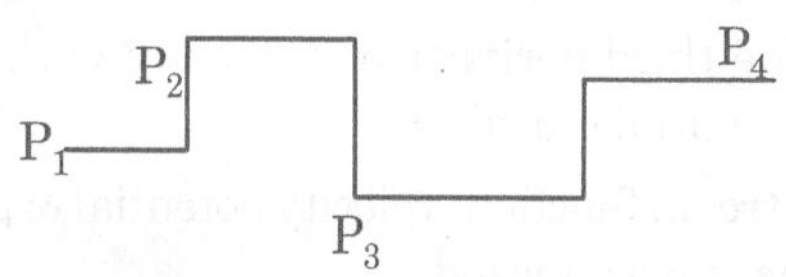

(b)

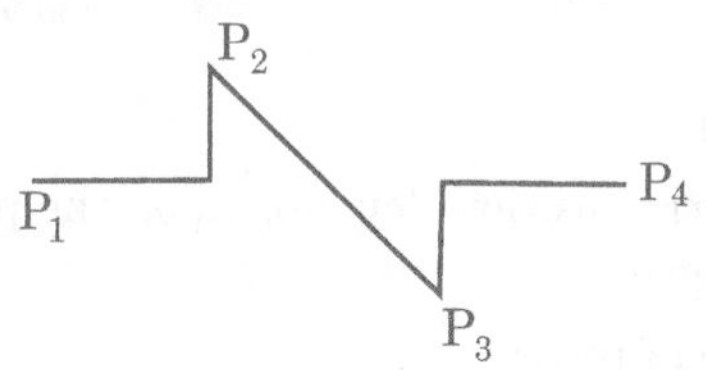

(c)

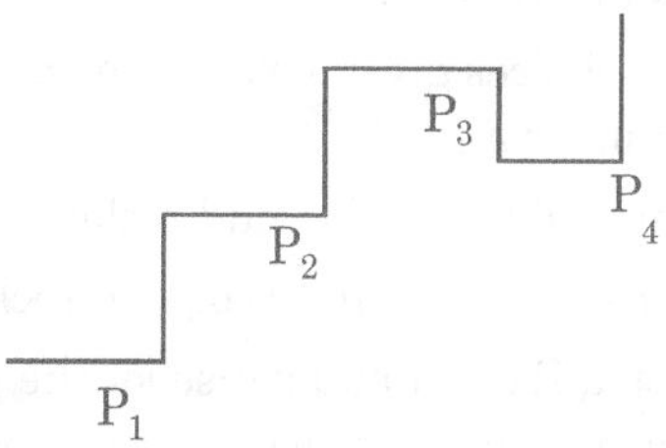

(d)

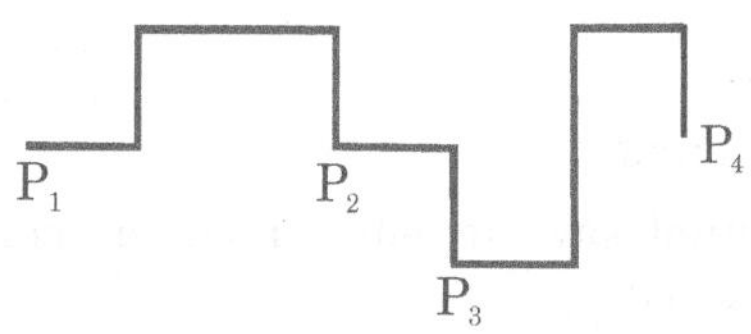

10. Flow of air brought to rest adiabatically of Mach number 3 stagnation pressure will.

(*a*) Decrease or remains constant

(*b*) Increase

(*c*) Decrease

(*d*) Remains constant

11. If M is the Mach number of the flow M* is the characteristics mach number and γ is ratio of specific heats, then as $M \to \infty$. $M^* \to$ ________ take ($\gamma = 1.2$)

12. Rotating flow over the cylinder can be achieved by combination of

(*a*) uniform flow, source-sink doublet & vortex.

(*b*) uniform flow & vortex

(*c*) uniform flow, source-sink & doublet

(*d*) uniform flow, doublet & sink.

13. The schleiren flow visualization technique depends upon

(*a*) Variation of the value of density is the flow

(*b*) The first derivation of density with respect to spatial coordinate.

(*c*) The second derivative of density with respect to spatial coordinate.

(*d*) The third derivative of density with respect to spatial coordinate.

14. The stream function, velocity potential & pressure can be superimposed

(*a*) True

(*b*) False

(*c*) Superimposition depends upon the pressure strength

(*d*) None of these

15. Among the choices given below, specific impulse is maximum for a

(*a*) Ramjet engine (*b*) solid rocket

(*c*) liquid rocket (*d*) cryogenic rocket

16. For Turbojet, Ramjet and pause jet the points of maximum propulsive efficiency and maximum thrust are

(*a*) Same

(*b*) Different

(*c*) Cannot say, depends upon the conditions provided

(*d*) None of these

17. In which of the following cases, a rocket engine would produce maximum thrust

(*a*) Slightly under expanded nozzle

(*b*) optimum expanded nozzle

(*c*) Slightly over expanded nozzle

(*d*) Can't say

18. Burn rate of rocket propellent is strong function of rocket motor chamber pressure (P_C) and given by $r = ap_c^n$ the value of n lies between

(*a*) $0 < n < 0.5$ (*b*) $0.5 < n < 1.0$

(*c*) $1 < n < 1.5$ (*d*) Can't say

19. The pressure ratio in any one stage of jet engine compressor in limited by

(*a*) stagnation temperature at the entry of the stage

(*b*) Mach number at the entry of the stage

(*c*) pressure gradient induced blade stall

(*d*) mass flow rate of air in that stage

20. For impulse turbine degree of reaction is ________.

21. For isotropic material number of independent elastic constant are

(*a*) 1 (*b*) 4

(*c*) 9 (*d*) 21

22. Compatibility and bilinear bihasmosic equations are valid for

(*a*) 2D

(*b*) 3D

(*c*) 2D & 3D

(*d*) Special cases for 2D not valid for 3D

23. A mechanical system is represented by equivalent spring mass damper system with m = 15 kg, K = 4 N/m and C = 1 Ns/m the system is

(*a*) under damped

(*b*) over damped

(*c*) critically damped

(*d*) cannot say

24. 20 Nm external Torque is applied to 1.5 m long hollow square tube with cross section as shown is figure the mean shear stress on the walls is ________ N/cm^2.

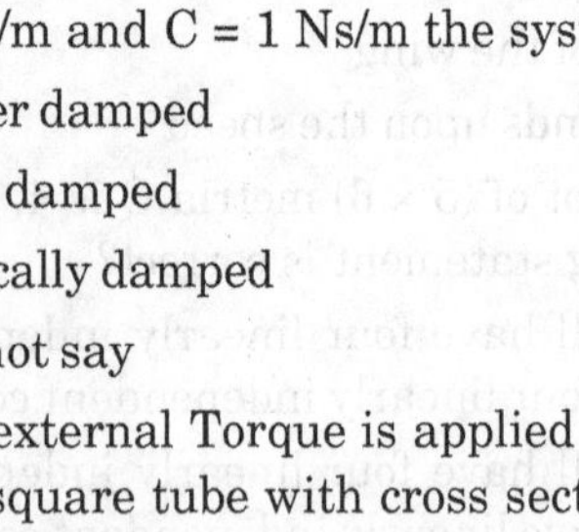

25. For a satellite, a = 6700 Km. e = 0.01 the satellite's altitude at perigue is ______ Km.

26. The Newton- Raphson iteration

$$x_{n+1} = \frac{1}{2}\left(x_n + \frac{R}{x_n}\right)$$ can be used to compute the

(*a*) Square root of R (*b*) Reciprocal of R

(*c*) Square of R (*d*) Logarithm of R

27. For a vibration of mass less simply supported beam having a point mass at the mid of the beam as shown below in figure, the Natural frequency would be

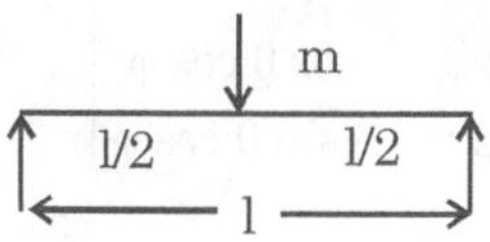

(*a*) $\sqrt{\frac{4IEI}{ml^3}}$ (*b*) $\sqrt{\frac{3\ EI}{ml^3}}$

(*c*) $\sqrt{\frac{384EI}{ml^3}}$ (*d*) $\sqrt{\frac{2EI}{ml^3}}$

28. $\varphi = x^2 y^2 (1 + x + y)$ is valid Airy's stress function?

(*a*) True

(*b*) False

(*c*) Can't say

(*d*) Need numerical data to prove

29. A thin steel plate 4 mm thick and 1000 mm long in bent into a circular arc which substend 45° angle at the centre of arc by a bending moment M_0. The maximum bending stress in the plate is ______ N/mm² (take E = 2 × 10⁵ N/mm²)

30. A cylindrical shell made of mild stell plate having diameter 100 cm is to be subjected to internal pressure of 2 MPa. If the yield stress of the material is 200 MPa. Thickness of the plate on the basis of Von Mises/Distortion Energy Creteria is ______ mm (Assuming Factor of safety = 3)

31. In a gas turbine Engine thermodynamic cycle, heat injection takes place at 11.2 atm and heat rejection takes plate a atmospheric pressure, then efficiency of cycle is ______ % (take γ = 1.2)

32. Consider the ideal jet propulsion cycle in which air enters. The compressor at 1 atm, 15°C. The pressure leaving the compressor is 5 atm. The maximum temperature is 870°C. The air expands in the turbine to such a pressure that turbine work is just equal to compressor work on leading to the turbine. The air expands reversibly and adiabatically is the nozzle at 1 atm. The velocity of air leaving to nozzle is

(*a*) 710.3 m/sec (*b*) 1000 m/sec

(*c*) 1710 m/sec (*d*) 200 m/sec

33. A two stage rocket having the same specific impulse (I_{sp}) of 300 sec for both the stages is designed in such a way that the payload ratio and structural ratio are same for both the stages the second stage of rocket have following mass distribution.

propellant mass = 10208 kg

Structural mass = 1134 kg

Payload mass = 1700 kg

If the rocket is fired from rest and it flies in a zero gravity field and a drag free environment, the maximum height attained by the rocket is

(*a*) 700 km (*b*) 600 km

(*c*) 4500 km (*d*) 3000 km

34. A missile with Ramjet engine flying in air the temperature at inlet and outlet of the combustor are 1200 K and 2500 K respectively. The heating value of fuel is 43 MJ/kg and the burner efficiency is 90%. Considering working fluid to be air (C_p = 1005 J/kg-k and γ = 1.4). The fuel/air $\left(f = \frac{m_f}{m_a}\right)$ for the engine is equal to ______

Common data question

If $\varphi = 3xy$ then

35. velocity and its direction at (1, 3) is

(*a*) 9.5 & 18.5° (*b*) 15 & 18.5°

(*c*) 9.5 & 30° (*d*) –9.5 & 30°

36. Discharge between the stream lines passing through (1, 3) and (3, 3) is

(*a*) –12 units (*b*) –20 units

(*c*) –39 units (*d*) 50 units

37. A wing of small aeroplane is rectangular in plane having a span of 12m and chord of 1.8 m. In a horizontal flight at 200km/hr. The total aerodynamic force acting or the wing is 28 KN. If the lift-drag ratio is 10 this lift coefficient is

(*a*) 0.2 (*b*) 0.9

(*c*) 0.7 (*d*) 1

38. Flow with Mach number 3 encounters a compression corner at angle 8° the shock wave angle would be

(*a*) 8.5° (*b*) 12.5°

(*c*) 17.5° (*d*) 24.5°

39. A convergent-divergent diffuser is to be used at Mach number 3.0. The diffuser has to use a variable throat area so as to shallow the starting shock then the percentage increase in throat area necessary would be

(*a*) 204.6% (*b*) 300%
(*c*) 400% (*d*) 100%

40. A duct in which air flowing a normal shock wave occurs at Mach number of 1.5 the strength of shock wave is ______

41. A projectile is travelling in air having pressure and temperature as 88.3 kN/m² and –2°C. If Mach angle is 40°. the velocity of projectile is ______ m/sec Take γ = 1.4, and R = 287 J/kgk

42. For what value of α and β the following simultaneous equations have a infinite number of solution?

x + y + 2 = 0
x + 3y + 3x = 9
x + 2y + αz = β.

(*a*) 2, 7 (*b*) 3, 8
(*c*) 8, 3 (*d*) 7, 2

43. Given that $\ddot{x} + 3x = 0$ and x(0) = 1, $\dot{x}(0) = 0$ then x (1) = ______

44. For the given function f(x, y) = 4x² + 6y² – 8x – 4y – 8 has a minimum equal to ______

45. The Laplace transformation of function

$$f(t) = \frac{1}{(S^2(S-1))} \text{ the}$$

(*a*) t – t + e^–t^ (*b*) t + 1 + e^–t^
(*c*) – 1 + e^t^ (*d*) 2 t + e^t^

46. A rocket is flying at an airspeed of 1000 m/sec. the angle of attack is 50° and side ship angle is 20°. Sideslip and normal velocity components are

(*a*) 342.02 m/sec, 813.80 m/sec
(*b*) 342.02 m/sec, 100 m/sec
(*c*) 342.05 m/sec, 469.85 m/sec
(*d*) 342.02 m/sec, 1815.00 m/sec

47. The relation between the body fixed angular velocity vector $[p\ q\ r]^T$ and the rate of change of eulers angles $\begin{bmatrix}\varphi\ \theta\ \psi\end{bmatrix}$ is given by

$$\begin{bmatrix} p \\ q \\ r \end{bmatrix} = J^{-1} \begin{bmatrix} \dot{\varphi} \\ \dot{\theta} \\ \dot{\psi} \end{bmatrix}$$

then J^–1^ is equals to

(*a*) $\begin{bmatrix} 1 & 0 & -\cos\theta \\ 0 & \cos\psi & \cos\varphi\sin\theta \\ 0 & \sin\psi & -\sin\varphi - \sin\theta \end{bmatrix}$

(*b*) $\begin{bmatrix} \cos\psi & 0 & 0 \\ 1 & -\sin\theta & \cos\varphi \\ 0 & \cos\theta & -\sin\varphi \end{bmatrix}$

(*c*) $\begin{bmatrix} 1 & -\sin\theta & 0 \\ 0 & \cos\psi & \cos\theta\cos\varphi \\ 0 & -\sin\psi & -\sin\theta\cos\varphi \end{bmatrix}$

(*d*) $\begin{bmatrix} 1 & 0 & -\sin\theta \\ 0 & \cos\varphi & \sin\varphi\cos\theta \\ 0 & -\sin\varphi & \cos\varphi\cos\theta \end{bmatrix}$

Common data questions

A 50% reaction, axial flow compressors runs at a mean blade speed of 250 m/sec. The pressure ratio developed by the machine is 1.9. If the mean flow velocity was 200 m/sec.Condition at inlet are 1 bar and 300 K.

51. The blade angles (β_1, β_2)

(*a*) 10.7°, 21.3° (*b*) 121.3°, 40.7°
(*c*) 507, 21.3° (*d*) 40., 40.7°

52. The air angles (α_1, α_2) are

(*a*) 21.3°, 40.7° (*b*) 40.7°, 21.3°
(*c*) 50.7°, 40.7° (*d*) 40.7°, 121.7°

53. Wing and tail of an aircraft have same aerofoil section the distance between two aerodynamic centre is equal to wing span $\left(\frac{b}{2}\right)$. In addition, following data is applicable

$C_{L_\alpha} = 0.08/\text{degree}, C_{L\delta_{tail}} = 0.04/\text{degree}$

$C_{Mac,wb} = -0.0218$

b = 29.2 m

c = 2.8 m

For tail $\varepsilon_a = 0.0$ $i_t = 2.1°$

$\varepsilon_a = 0.0$

$\varepsilon_a = 0.3$

$s_t = 0.35$

$n_t = 0.9$

the symbols have their usual aerodynamic interpretation.

53. What is value of zero lift moment coefficient C_{Mo} for this aircraft?

(*a*) 0.2410 (*b*) 0.214

(*c*) –0.2410 (*d*) –0.214

54. Which of the following design parameters influence the maximum velocity in steady level flight for a jet propelled airplane?

P. $\frac{(T_A)_{max}}{W}$ (maximum thrust to weight ratio)

Q. $\frac{W}{S}$(wing loading)

R. C_{DO} and K

S. $\left(\frac{L}{D}\right)$(Lift to drag ratio)

(*a*) Q, P and alone

(*b*) P, Q, R_D and S

(*c*) P, Q and S alone

(*d*) Q, R and S alone

55. The natural frequency of an undamped vibrating system is 100 rad/sec. A damper with a damping factor of 0.8 is introduced into system. The frequency of vibration of the damped system, is rad/sec is ______

56. Which of the following options is the closest in the wind indirect

(*a*) Confusing (*b*) circuitous

(*c*) crooked (*d*) confusing

57. If 8 + 12 = 2, and 7 + 13 = 3. then. What is the value of 10 + 18 ?

(*a*) 9 (*b*) 6

(*c*) 6 (*d*) 4

58. If y × W = 5 and u – v = z, is $^r/_X$?

(*a*) 1y (*b*) z

(*c*) S (*d*) X

59. The question below consists of a pair of related words followed by four pairs of words select the pair that best expressess the relation original pair, Grain : Silo

(*a*) pilot : plane (*b*) judge : courtroom

(*c*) water : river (*d*) clock : wall

60. My father was too ______ to open the heavy door.

(i) faint (*b*) feeble

(*c*) timed (*d*) faltering

61. There are two groups A and B consisting of 30 and 70 students respectively. If the average weight of group A is 50 kg and that of group is 70 kg. find the average weight of all students of group A and group B

(*a*) 58 (*b*) 50

(*c*) 40 (*d*) 42

62. Given that $f(y) = \frac{|y|}{y}$ and q is any non-zero real number, the value of | f(q) – f(–q) | is ______

63. In the number system the product of 44 and 11 is 1034. The number 31111 of this system, when converted to the decimal number system, becomes

(*a*) 406 (*b*) 1086

(*c*) 213 (*d*) 691

64. Modern warfare has changed from longer scale clashes of armies to suppression of civilian populations. Chemical agents that do their work silently appear to be suited to such warfare and regretfully, These exit people in unmilitary establishment who think that chemical agents are useful tools for their cause. Which of the following statements best sums of the meaning of the above passage?

(*a*) Modern has resulted in civil strife

(*b*) chemical agents are useful in modern warfare.

(*c*) use of chemical agents in warfare would be undesirable

(*d*) people in miliary establishments time to we chemical agents is war.

65. The two points (5, p) and (0, 9) lie of the straight line having a slope of $\frac{3}{5}$ the value of (p – q) is

OMR SHEET

1. Ⓐ Ⓑ Ⓒ Ⓓ
2. Ⓐ Ⓑ Ⓒ Ⓓ
3. Ⓐ Ⓑ Ⓒ Ⓓ
4. Ⓐ Ⓑ Ⓒ Ⓓ
5. Ⓐ Ⓑ Ⓒ Ⓓ
6. Ⓐ Ⓑ Ⓒ Ⓓ
7.
8. Ⓐ Ⓑ Ⓒ Ⓓ
9. Ⓐ Ⓑ Ⓒ Ⓓ
10. Ⓐ Ⓑ Ⓒ Ⓓ
11.
12. Ⓐ Ⓑ Ⓒ Ⓓ
13. Ⓐ Ⓑ Ⓒ Ⓓ
14. Ⓐ Ⓑ Ⓒ Ⓓ
15. Ⓐ Ⓑ Ⓒ Ⓓ
16. Ⓐ Ⓑ Ⓒ Ⓓ
17. Ⓐ Ⓑ Ⓒ Ⓓ
18. Ⓐ Ⓑ Ⓒ Ⓓ
19. Ⓐ Ⓑ Ⓒ Ⓓ
20.
21. Ⓐ Ⓑ Ⓒ Ⓓ
22. Ⓐ Ⓑ Ⓒ Ⓓ
23. Ⓐ Ⓑ Ⓒ Ⓓ
24.
25.
26. Ⓐ Ⓑ Ⓒ Ⓓ
27. Ⓐ Ⓑ Ⓒ Ⓓ
28. Ⓐ Ⓑ Ⓒ Ⓓ
29.
30.
31.
32. Ⓐ Ⓑ Ⓒ Ⓓ
33. Ⓐ Ⓑ Ⓒ Ⓓ
34.
35. Ⓐ Ⓑ Ⓒ Ⓓ
36. Ⓐ Ⓑ Ⓒ Ⓓ
37. Ⓐ Ⓑ Ⓒ Ⓓ
38. Ⓐ Ⓑ Ⓒ Ⓓ
39. Ⓐ Ⓑ Ⓒ Ⓓ
40.
41.
42. Ⓐ Ⓑ Ⓒ Ⓓ
43.
44.
45. Ⓐ Ⓑ Ⓒ Ⓓ
46. Ⓐ Ⓑ Ⓒ Ⓓ
47. Ⓐ Ⓑ Ⓒ Ⓓ
48.
49. Ⓐ Ⓑ Ⓒ Ⓓ
50. Ⓐ Ⓑ Ⓒ Ⓓ
51. Ⓐ Ⓑ Ⓒ Ⓓ
52. Ⓐ Ⓑ Ⓒ Ⓓ
53. Ⓐ Ⓑ Ⓒ Ⓓ
54. Ⓐ Ⓑ Ⓒ Ⓓ
55.
56. Ⓐ Ⓑ Ⓒ Ⓓ
57. Ⓐ Ⓑ Ⓒ Ⓓ
58. Ⓐ Ⓑ Ⓒ Ⓓ
59. Ⓐ Ⓑ Ⓒ Ⓓ
60. Ⓐ Ⓑ Ⓒ Ⓓ
61. Ⓐ Ⓑ Ⓒ Ⓓ
62. Ⓐ Ⓑ Ⓒ Ⓓ
63.
64. Ⓐ Ⓑ Ⓒ Ⓓ
65. Ⓐ Ⓑ Ⓒ Ⓓ

ANSWERS

1. (*b*)	2. (*a*)	3. (*b*)	4. (*a*)	5. (*a*)	6. (*a*)	7. (0)	8. (*d*)	9. (*a*)	10. (*a*)
11. (3.31)	12. (*a*)	13. (*b*)	14. (*b*)	15. (*a*)	16. (*b*)	17. (*b*)	18. (*b*)	19. (*c*)	20. (0)
21. (*b*)	22. (*a*)	23. (*a*)	24. (2.77)	25. (255)	26. (*a*)	27. (*a*)	28. (*b*)	29. (314)	30. (13)
31. (33%)	32. (*a*)	33. (*c*)	34. (0.0359)	35. (*a*)	36. (*a*)	37. (*c*)	38. (*d*)	39. (*a*)	40. (9.33)
41. (514)	42. (*a*)	43. (0.99)	44. (10/3)	45. (*a*)	46. (*c*)	47. (*d*)	48. (7.2)	49. (*a*)	50. (*a*)
51. (*a*)	52. (*a*)	53. (*b*)	54. (*a*)	55. (60)	56. (*a*)	57. (*d*)	58. (*d*)	59. (*d*)	60. (*b*)
61. (*a*)	62. (*d*)	63. (406)	64. (*d*)	65. (*c*)					

EXPLANATIONS

1. Temperature lapse rate (a) $= \dfrac{dT}{dh}$

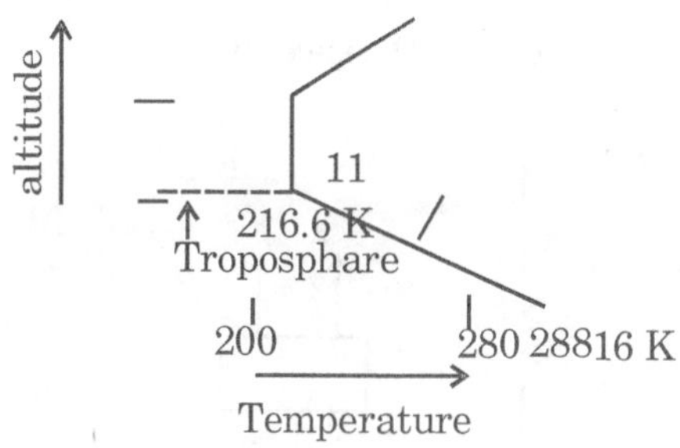

$$= \frac{T - T_1}{h - h_1}$$

$$= \frac{216.16 - 288.16}{11 - 0} = -\,6.5 \text{ k/km}$$

Hence temperature lapse rate of Troposhere is -- 6.5 k/m

2. In case of vertical flight

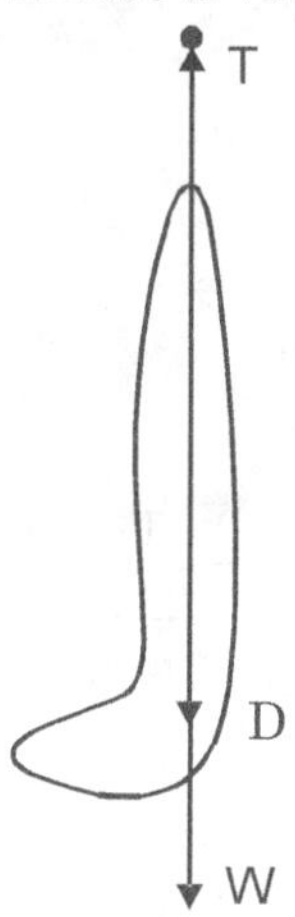

$$\boxed{T = W + D}$$

As lift is zero in vertical flight

$\therefore\ C_{Di} = 0$

$$C_{Di} = \frac{C_L^2}{\pi eAR} = 0$$

$C_{Di} = 0$

3. Absolute ceiling – The altitude where maximum rate of climb is zero is called absolute ceiling

Service ceiling – It is defined as the altitude where $(R/C)_{max}$ = 100 ft/min. Service ceiling represents practical up per limit for steady, level flight.

Heane point a represent absolute ceiling and point b represents Serive ceiling.

4. For Straight pointed wing the flow break down near leading edge hence the stall occurs at leading edge.

5. If rank of matrix is 4, then It surely must have exactly 4 linearly independent rows as well as 4 linearly independent columns, Since rank = row rank = Column rank.

6. $|\text{adj}(\text{adj}(A))| = |A|\,|(n-1)^2$

$$= (5)^{(3-1)^2} = (5)^{2^2} = 5^4 = 625$$

7. $\displaystyle \lim_{x\to 0} \frac{\sin^2 x}{x} = \lim_{x\to 0}\left(\frac{\sin x}{x}\right)^2 \times x = 0$

8. $\displaystyle \int_0^{\frac{\pi}{2}} \frac{\text{Cos}x + i\text{Sin}\,x}{\text{Cos}x - i\text{Sin}x}dx$

$$\int_0^{\frac{\pi}{2}} \frac{e^{ix}}{e^{-ix}}dx = \int_0^{\frac{\pi}{2}} e^{2ix}dx = \left[\frac{e^{i2x}}{2i}\right]_0^{\frac{\pi}{2}}$$

$$= \frac{1}{2i}\left[e^{ih} - e^0\right] = \frac{1}{2i}[-1-1] = \frac{-2}{2i} = \frac{-1}{i}$$

$$= \frac{i^2}{i} = i$$

9. When Supersonic flow occurs on the diamond wedge aerofoil at the leading edge shock wave will form because of flow Compression and at mid section expansion wave takes place and again formation of shock at trailing edge.
As we know across the shock wave pressure rises and across the expansion wave pressure decreases Hence in the given option, (i) is in the order.

10. $$M^* = \sqrt{\frac{(\gamma+1)M^2}{2+(\gamma-1)M^2}}$$

As $M \to \infty$

$$\lim_{M\to\infty} M^* = \lim_{M\to\infty}\sqrt{\frac{(\gamma+1)M^2}{2+(\gamma-1)M^2}}$$

$$= \sqrt{\frac{\gamma+1}{\gamma-1}}$$

$$= \frac{\sqrt{1.2+1}}{1.2-1} = \frac{\sqrt{2.2}}{0.2} = \sqrt{11} = 3.31$$

$$\boxed{M^* = 3.31}$$

13. Schlieren flow visualization technique is optical method for measurement of first derivative of density with respect to spatial coordinate

16. $$\eta_p = \frac{2\alpha}{\alpha+1}, \text{where } \alpha = \frac{C_i}{C_j}$$

The condition for best propulsive efficiency can be obtained by differentiating $\eta_p = \frac{2\alpha}{\alpha+1}$, with respect to α and equating it to zero. which is

$\alpha = 1$ or $c_i = c_j$

Thrust power

$$P_T = \dot{m}_a C_j^2 \alpha(1-\alpha)$$

differentiating P_T with respect to α and equating to zero we get $\alpha = 0.5$, $C_J = 2C_j$.

hence the condition for maximum propulsive efficiency and maximum thrust are different

20. for turbine

Degree of reaction (R)

$$= \frac{\text{Static temperature drop across the rotor of trubine}}{\text{Static temperature drop across the entire stage of turbine}}$$

For impulse turbine, entire temperature drop occurs only across the nozzle vanes, there is no temperature drop occurs across the rotor vanes, Hence degree of reaction is zero, (R = 0)

21. In case of isotropic material it has 2 independent elastic constants, which are E and G.

$E = 2G(1+\nu)$

ν = Poisson's ratio

22. (i) Compatibility equation

$$\frac{\partial^2 \varepsilon_{xx}}{\partial y^2} + \frac{\partial^2 \varepsilon_{xy}}{\partial x^2} = \frac{\partial^2 \gamma_{xy}}{\partial x \partial y}$$

Bilinear harmonic equation

$$\overline{V}^4 \phi = \frac{\partial^4 \phi}{\partial y^4} + \frac{\partial^4 \phi}{\partial x^4} + 2\frac{\partial^2 \partial^2 \phi}{\partial y^2 \partial x^2} = 0$$

both are valid for 2 D case only

23.

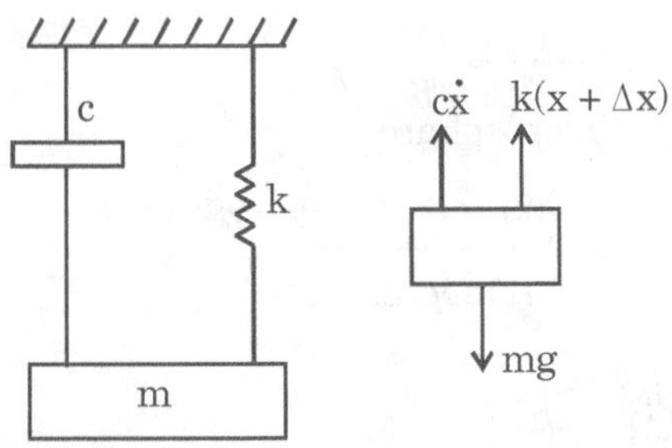

Writing equation for spring mass damper system

$$\boxed{m\ddot{x} + c\dot{x} + kx = 0}$$

Given m = 15 kg, k = 4 N/m, C = 1 Nsec/m

$$15\ddot{x} + 1\dot{x} + 4x = 0$$

$$15\ddot{x} + \dot{x} + 4x = 0$$

$\Delta = B^2 - 4AC$

$= 1^2 - 4 \times 15 \times 4 = < 0$

hence underdamped system.

24. Mean shear stress on the wall $\tau = \frac{T}{2At}$

$$= \frac{20 \times 10^2}{2 \times 19 \times 19 \times \left(\frac{20-18}{2}\right)}$$

$$= \frac{20 \times 10^2}{2 \times 19 \times 19 \times 1} = 2.74 \text{ N/cm}^2$$

25. radius of perigee (r_p) = a (1– e)

= 6700 (1 – 0.01)

= 66.33

radius of earth = 6378.2 km

∴ Altitude of perigee = 6633 – 6378.2

= 255 km

26. $x_n + 1 = \frac{1}{2}\left(x_n + \frac{R}{x_n}\right)$

at convergence

$x_n + 1 = x_n = \alpha$

$\Rightarrow \alpha = \frac{1}{2}\left(\alpha + \frac{R}{\alpha}\right)$

$\Rightarrow 2\alpha^2 = \alpha^2 + R$

$\Rightarrow a^2 = R$

$\Rightarrow a = \sqrt{R}$

So, this iteration will compute square root of R.

27.

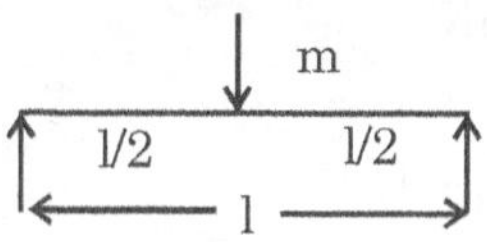

Maximum deflection at centre = $\frac{Wl^3}{48EI}$

Natural frequency = $\sqrt{\frac{k}{m}} = \sqrt{\frac{g}{\delta}}$

$= \frac{\sqrt{\frac{g}{Wl^3}}}{48EI} = \frac{\sqrt{\frac{g}{mgl^3}}}{48EI}$

$\omega_n = \sqrt{\frac{48EI}{ml^3}}$

$$\boxed{\omega_n = \sqrt{\frac{48EI}{ml^3}}}$$

28. $\varphi = x^2y^2(1 + x + y)$

$= x^2y^2 + x^3y^2 + x^2y^3$

$\frac{\partial^4\varphi}{\partial y^4} = 0$

$\frac{\partial^4\varphi}{\partial x^4} = 0$

$\frac{2\partial^2}{\partial y^2}\frac{\partial^2\varphi}{\partial x^2} = 2(4 + 12x + 12y)$

Hence $\overline{V^4}\varphi = \frac{\partial^4\varphi}{\partial x^4} + \frac{\partial^4\varphi}{\partial y^4} + \frac{2\partial^2\partial^2\varphi}{\partial y^2\partial x^2} \neq 0$

Hence Airy stress function is not valid

29.

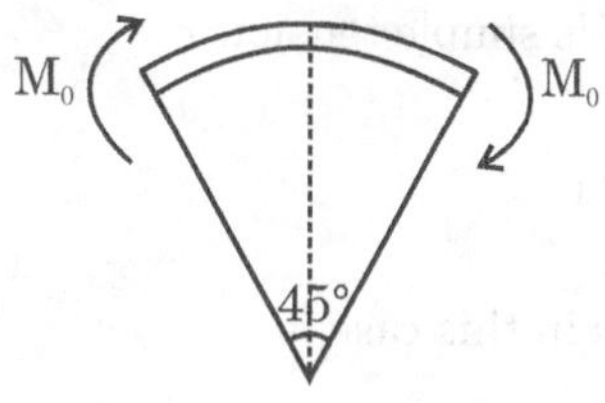

$I = \frac{bt^3}{12} = \frac{b \times (4)^3}{12}$

We have

$\frac{M}{I} = \frac{E}{R} = \frac{\sigma}{Y}$

$\frac{E}{R} = \frac{\sigma}{Y}$

$\Delta L = Y\theta$

$\Delta L = R\theta$

$1000 = R \times \frac{\pi}{4}$

R = 1274 mm.

Here

$y = \frac{4}{2} = 2$

$\frac{E}{R} = \frac{6}{y}$

$\Rightarrow \frac{2 \times 10^5}{1.274} = \frac{\sigma}{2}$

$\Rightarrow \sigma = 314 \frac{N}{mm^2}$

30. Hoop stress. $\sigma_\theta = \frac{pd}{2h} = \frac{2\times1}{2h} = \frac{1}{h}$ MPa

Longitudinal stress $\sigma_l = \frac{pd}{4h} = \frac{2\times1}{4\times h}$

$= \frac{2\times1}{2\times h} = \frac{1}{h}$ MPa

Allowable stress is simple tension σ

$= \frac{200}{3} = 66.6$MPa

Von mises stress in this case is

$$\sigma_y = \sqrt{\sigma_\theta^2 + \sigma_l^2 - \sigma_\theta.\sigma_l}$$

$$66.66 = \sqrt{\left(\frac{1}{h}\right)^2 + \left(\frac{0.5}{h}\right)^2 - \frac{1}{h}\times\frac{0.5}{h}}$$

$$\Rightarrow 66.66^2 = \frac{1}{h^2}[1+0.25-0.5]$$

$$h^2 = \frac{0.75}{66.66^2}$$

$$h = \frac{\sqrt{0.75}}{66.66} = 0.01299\text{m}$$

= 13 mm.

31. $P_2 = 11.2$ atm

$P_1 = 1$atm

$$r_P = \frac{P_2}{P_1} = 11.2$$

Efficiency of Brayton cycle (n) $= 1 - \frac{1}{(r_P)^{\frac{\gamma-1}{\gamma}}}$

$$= 1 - \frac{1}{(11)^{\frac{0.2}{1.2}}}$$

= 0.3294

= 32.94%

$\boxed{\eta = 33\%}$

32.

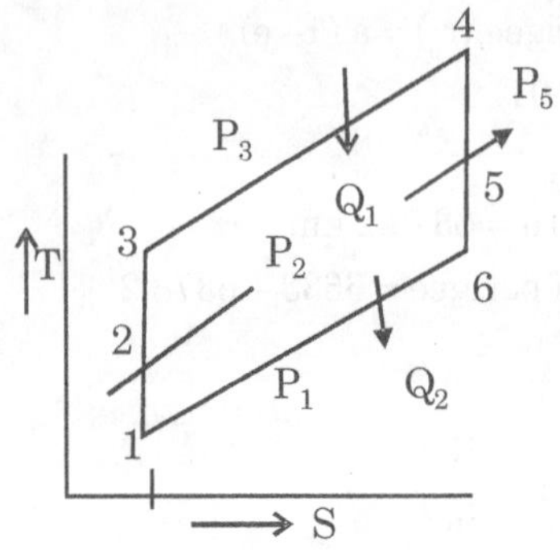

Given

$P_1 = 1$ atm

$T_1 = 288$ K

$P_3 = 5$ atm

$T_1 = 870 + 273 = 1143$K

$$\left(\frac{T_3}{T_1}\right) = \left(\frac{P_3}{P_1}\right)^{\frac{r-1}{r}} = (5)\frac{1.4-1}{1-4} = 1.58$$

$T_3 = 1.58 \times 288 = 455$K

Since $W_T = W_c$

$\therefore C_p (T_3 - T_1) = C_p (T_4 - T_5)$

$455 - 288 = 1143 - T5$

$\boxed{T_5 = 976K}$

$$\frac{T_4}{T_6} = \left(\frac{P_4}{P_6}\right)^{\frac{r-1}{r}} = (5)^{\frac{0.4}{1.4}} = 1.58$$

$$T_6 = \frac{1143}{1.58} = 725K$$

Assuming the velocity of air leaving to the nozzle is very small. then

$$h_5 = h_6 + \frac{V_6^2}{2}$$

$$V_6 = \sqrt{2(h_5 - h_6)}$$

$$= \sqrt{2C_p(T_{05} - T_{06})}$$

$$= \sqrt{2\times1005(976-720)}$$

= 710.3 m/sec.

$\boxed{V_6 = 710\text{m/sec}}$

33. Given

$m_p = 10208$ kg

$m_s = 1134$ kg

$m_l = 1700$ kg

$I_{SP} = 300$ sec

$$\text{payload ratio } (\pi) = \frac{m_L}{m_p + m_s}$$

$$= \frac{1700}{10208+1134} = 0.14$$

$$\text{Structural ratio} = (\varepsilon) = \frac{m_s}{m_p + m_s}$$

$$= \frac{1134}{1020+1134} = 0.099$$

Number of stages = 2,

$$\text{burnout velocity} = \eta I_{sp} g_0 \ln\left(\frac{1+\lambda}{\varepsilon+\lambda}\right)$$

$$= \eta V_e \ln\left(\frac{1+\lambda}{\varepsilon+\lambda}\right)$$

$$= 2 \times 2940 \ln\left(\frac{1+0.14}{0.099+0.14}\right)$$

= 9187 m/sec.

$V_e = Isp\ g_o = 300 \times 9.8$

= 2940 m/sec

Maximum height attained by rocket

$$h = \frac{V_b^2}{2g}$$

$$= \frac{(9187)^2}{2 \times 9.8} = 4305 \text{ km}$$

34. $Q = mC_p \Delta T \Rightarrow$

$$\dot{m}_f \Delta H_c = (\dot{m}_a + \dot{m}_f) c_p \Delta T)$$

$$0.9\ \Delta H_c = \left(1+\frac{1}{f}\right) \times 1005 \times 130$$

$\boxed{f = 0.0359}$

35. $\varphi = 3xy$

$$\therefore u = \frac{-\partial \varphi}{\partial x}, v = \frac{-\partial \varphi}{\partial y} = \frac{-\partial(3xy)}{\partial y} = -3x$$

$$= \frac{\partial}{\partial x}(3xy)$$

$= -3y$

$u = -3y$, $v = -3x$

Velocity Component at (1, 3) is

$u = -3 \times 3 = -9$

$v = -3 \times 1 = -3$

$$\therefore v = \sqrt{u^2 + v^2}$$

$$\therefore v = \sqrt{(-3)^2 + (-9)^2} = \sqrt{9+81}$$

$$= \sqrt{90}$$

= 9.5 m/sec

Its directions has a slope

$$\frac{\partial y}{\partial x} = \frac{v}{u} = \frac{-3}{-9} = \frac{1}{3}$$

$$\tan\theta = \frac{1}{3}$$

$$\theta = \tan^{-1}\left(\frac{1}{3}\right) = 18.5°$$

36. Discharge between two stream lines

Total derivative ψ may be written as

$$d\psi = \frac{\partial \psi}{\partial x}dx + \frac{\partial \psi}{\partial y}dy$$

$= -vdx + udy$

$= -(-3x)\,dx + (-3y)\,dy = 3xdx - 3ydy$

$$\psi = \frac{3}{2}x^2 - \frac{3}{2}y^2 + C$$

where C = Constant of Integration Discharge between two streamlines passing through (1, 3) & (3, 3)

$$= \psi_{1\times3} - \psi_{3\times3} = \frac{3}{2}(1-9) - \frac{3}{2}(9-9)$$

$$= \frac{3x-8}{2}$$

$= -12$ units

37.

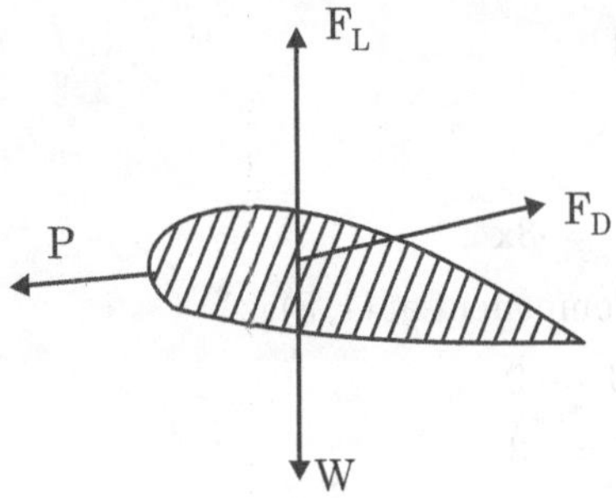

Span (l) = 12m

Chord (c) = 1.8m

$$V = 200\text{km/h} = \frac{200 \times 1000}{60 \times 60}$$

= 55.57 m/sec

Acodynamic force. F_L = 28 kN

For horizontal flight

$$F_L \uparrow = W \downarrow \qquad F_D = C_D \times \frac{1}{2} eAV^2$$

$$\text{given } \frac{F_L}{F_D} = 10$$

$$\Rightarrow \frac{28 \times 1000}{C_D \times \frac{1}{2} \times 1 \times 2(12 \times 1.8) \times 55.5^2} = 10$$

$$\therefore C_D = \frac{28 \times 1000 \times 2}{10 \times 12 \times 1.8 \times 1.2 \times 55.5^2} = 0.07$$

$$\therefore \frac{C_L}{C_D} = 10$$

$\therefore C_L = 10 \times C_D = 0.7$

38. Given M = 3

Mach angle

$$\mu = \sin^{-1}\left(\frac{1}{3}\right)$$

= 19.5°

Shock wave angle always greater then Mach angle, hence from given option (iv) is correct

39. For diffuser relation between throat area and pressure is

$$\frac{A_2^*}{A_1^*} = \frac{P_{01}}{P_{02}} \qquad ...(1)$$

A_2^* = Second throat area

A_1^* = first throat ϕ area

P_{01} = Total pressure before shock

P_{02} = Total pressure after shock wave

$$\therefore \frac{P_{01}}{P_{02}} = \frac{P_{01}}{P_1} \times \frac{P_1}{P_2} \times \frac{P_2}{P_{02}}$$

$$= \left(1 + \frac{\gamma - 1}{2} M_1^2\right)^{\frac{\gamma}{\gamma-1}} \times \left(1 + \frac{2\gamma}{\gamma+1}\left(M_1^2 - 1\right)\right)^{-1} \times \left(1 + \frac{\gamma-1}{2} M_2^2\right)^{\frac{-\gamma}{\gamma-1}} \qquad ...(2)$$

$M_1 = 3, \ \gamma = 1.2$

$$M_2 = \sqrt{\frac{1 + \frac{\gamma-1}{2} M_1^2}{\gamma M_1^2 - \frac{\gamma-1}{2}}}$$

$$= \sqrt{\frac{1 + 0.2 \times 3^2}{1.4 \times 3^2 - 0.2}}$$

$$= \sqrt{\frac{2.8}{12.4}}$$

$= \sqrt{0.225}$

$M_2 = 0.475$

$$\frac{Po_1}{Po_2} = \left(1 + 0.2 \times 3^2\right)^{3.5} \times \left(1 + \frac{2.8}{2.4}\left(3^2 - 1\right)\right) \times \left(1 + 0.2 \times 0.475^2\right)$$

= 3.046

$$\therefore \frac{A_L^*}{A_I^*} = 3.046$$

$\therefore$ percentage increase in area required to swallow the starting shock wave

$$\therefore \frac{A_2^* - A_I^*}{A_I^*} \times 100 = 2046\%$$

40. Static pressure ratio across shock wave in given by

$$\frac{P_2}{P_1} = 1 + \frac{2\gamma}{\gamma+1}\left(M_1^2 - 1\right)$$

$$\frac{P_2}{P_1} - 1 = \frac{2\gamma}{\gamma+1}\left(M_1^2 - 1\right)$$

$$\frac{P_2 - P_1}{P_1} = \text{strength of shock wave}$$

$$\therefore \text{ Strength of shock wave } = \frac{2\times 1.4}{1.4+1}\left(3^2+1\right)$$

$$= \frac{2.8}{2.4}(8)$$

$$= 9.33$$

41. Given P = 88.3 KN/m^2

T = – 2 + 273 = 271 K

M = 40°

γ = 1.4 R = 287 J/KgK

Sonic velocity $C = \sqrt{\gamma RT}$

$$C = \sqrt{1.4 \times 287 \times 271} = 330 \text{m/sec}$$

Mach anglem μ is given by

$$\mu = \sin - 1\frac{1}{M}$$

$$\text{Sin}\mu = \frac{1}{M}$$

$$= \frac{1}{\frac{V}{C}} = \frac{C}{V}$$

$$V = \frac{C}{\text{Sin}\mu} = \frac{330}{\text{Sin}40} = 513.7 \text{m/sec}$$

$$\boxed{V = 514 \text{m/sec}}$$

42. Applying Determinant method to solve the equation for any system of equation

D = 0, D_1, D_2, D_3 $\neq$ 0 the system has no solution

D = 0, $D_1 = D_2 = D_3 = 0$ the system has infinite many solution

$D \neq 0, D_1 = D_2 = D_3 = 0$ has only one trival solution

$\therefore$ Determinant of system of equations

$$= \begin{vmatrix} 1 & 1 & 1 \\ 1 & 3 & 3 \\ 1 & 2 & \alpha \end{vmatrix}$$

putting D = 0

$\alpha = 2$

$$D_1 = \begin{vmatrix} 1 & 3 & 1 \\ 1 & 9 & 3 \\ 1 & \beta & 2 \end{vmatrix}$$

putting $D_1 = 0$

$\beta = 7$

43. Given

$$\ddot{x} + 3x = 0, \ x(0) = 1, \ \dot{x}(0) = 1$$

Auxiliary equation

$(D^2 + 3) = 0$

$$D = \pm\sqrt{3}i$$

$$x = A\cos\sqrt{3}t + B\sin\sqrt{3}t$$

t = 0, x = 1

1 = A cos 0° + B sin 0°

A = 1

$$\dot{x} = \left(A\cos\sqrt{3}t - B\sin\sqrt{3}t\right)\times\sqrt{3}$$

At t = 0, $\dot{x}$ = 0

B = 0

$$x = \text{Cos}\sqrt{3}\times 1$$

$$x(1) = \text{Cos}\sqrt{3}\times 1$$

$$= \text{Cos}\sqrt{3} = 0.99.$$

44. $F(x, y) = 4x^2 + 6y^2 - 8x - 4y - 8$

$$\frac{\partial f}{\partial x} = 8x - 8$$

$$\frac{\partial f}{\partial y} = 12y - 4$$

putting $\frac{\partial f}{\partial x} = 0, \frac{\partial f}{\partial x} = 0$, we get

$$x = 1 \; \& \; y = \frac{1}{3}$$

$\left(1, \frac{1}{3}\right)$ is stationary point

$$r = \left(\frac{\partial^2 f}{\partial x^2}\right)\left(1, \frac{1}{3}\right) = 8$$

$$5 = \left(\frac{\partial^2 f}{\partial x \partial y}\right)_{\left(1, \frac{1}{3}\right)} = 0$$

$$t = \left(\frac{\partial^2 f}{\partial y^2}\right)_{\left(1, \frac{1}{3}\right)} = 12$$

$rt - S^2 = 48 - 0 > 0$

$r > 0$ the function f (x,y) has minimum value at this point

∴ minimum value is =

$$4 \times 1^2 + 6 \times \frac{1}{3^2} - 8 \times 1 -$$

$$4 \times \frac{1}{3} + 8 = \frac{10}{3}$$

Hence the optimum value of f (x,y) is minimum to $\frac{10}{3}$

45. Equation $\frac{1}{S^2(S+1)}$ Can be written as

$$\frac{A}{S^2} + \frac{B}{S} + \frac{C}{S+1} = \frac{1}{S^2(S+1)} \left(\begin{matrix}\text{By partial fraction} \\ \text{method}\end{matrix}\right)$$

where A, B, C are constants.

$\Rightarrow A(S+1) + B(S+1)(S) + C(S^2) = 1$

Let S = 0

A = 1

Let S = – 1

C = 1

Let S = 1

B = – 1

$$\text{So} \frac{1}{S^2(S+1)} + \frac{1}{S^2} + \frac{(-1)}{S} + \frac{1}{(S+1)}$$

$$= \frac{1}{S^2} - \frac{1}{S} + \frac{1}{(S+1)}$$

$$L^{-1}\left(\frac{1}{S^2(S+1)}\right) L^{-1}\left(\frac{1}{S^2} - \frac{1}{S} + \frac{1}{(S+1)}\right)$$

$= t - 1 + e^{-t}$

46. Sideslip velocity Compoaent

$V_{yb} = V \sin \beta e$

= 1000 Sin 20° = 1000 (0.3420) = 342.02 m/sec

Normal velocity Compoaent

$V_{2b} = V \sin\alpha \cos \beta e$

= 1000 Sin 30° . Sin co°

= 469.85 m/sec

47. We resolve the eulers rate into body fixed co-ordinate frame. such we obtain

$$\begin{bmatrix} p \\ q \\ r \end{bmatrix} = \begin{bmatrix} \phi \\ 0 \\ 0 \end{bmatrix} + [\varphi_1]\begin{bmatrix} 0 \\ \theta \\ 0 \end{bmatrix} + [\varphi_1][\varphi_2]\begin{bmatrix} 0 \\ 0 \\ \dot{\psi} \end{bmatrix}$$

$$\begin{bmatrix} \dot{\varphi} \\ 0 \\ 0 \end{bmatrix} + \begin{bmatrix} 1 & 0 & 0 \\ 0 & \cos\varphi & \sin\varphi \\ 0 & -\sin\varphi & \cos\varphi \end{bmatrix}\begin{bmatrix} 0 \\ \dot{\theta} \\ 0 \end{bmatrix} +$$

$$\begin{bmatrix} 1 & 0 & 0 \\ 0 & \cos\varphi & \sin\varphi \\ 0 & -\sin\varphi & \cos\varphi \end{bmatrix}$$

$$\begin{bmatrix} \cos\theta & 0 & -\sin\theta \\ 0 & 1 & 0 \\ \sin\theta & 0 & \cos\theta \end{bmatrix}\begin{bmatrix} 0 \\ 0 \\ \dot{\psi} \end{bmatrix} \times$$

$$\begin{bmatrix} \dot{\varphi} - \dot{\psi}\,\mathrm{Sin}\theta \\ \dot{\theta}\,\mathrm{Cos}\varphi + \mathrm{Sin}\varphi\mathrm{Cos}\theta\,\dot{\psi} \\ -\dot{\theta}\,\mathrm{Sin}\varphi + \mathrm{Cos}\varphi\mathrm{Cos}\theta\,\dot{\psi} \end{bmatrix} = \begin{bmatrix} \dot{\varphi} \\ \dot{\theta} \\ \dot{\psi} \end{bmatrix}$$

$$\begin{bmatrix} 1 & 0 & -\mathrm{Sin}\theta \\ 0 & \mathrm{Cos}\varphi & \mathrm{Cos}\theta\mathrm{Sin}\varphi \\ 0 & -\mathrm{Sin}\varphi & \mathrm{Cos}\theta\mathrm{Cos}\varphi \end{bmatrix}$$

$$= J^{-1}\begin{bmatrix} \dot{\varphi} \\ \dot{\theta} \\ \dot{\psi} \end{bmatrix}$$

48. Velocity impulse magnitude required for a manoever with V_i as initial velocity and V_f as final velocity such as $\overrightarrow{V_f}$ makes as angle α with $\overrightarrow{V_i}$ is

$$\Delta V = \sqrt{V_i^2 + V_f^2 - 2V_iV_f\mathrm{Cos}\alpha}$$

In a place change manoener without any modification of speed or flight – path angle at a point of impulse then $V_i = V_f$

$$\Delta V = \sqrt{V_j^2 + V_j^2 - 2V_jV_j\mathrm{Cos}\alpha}$$

$$= \sqrt{2V_i^2 - (1 - \mathrm{Cos}\alpha)}$$

$$= \sqrt{2V_i^2 \times 2\mathrm{Sin}^2\frac{\alpha}{2}} = 2V_i^2\mathrm{Sin}\frac{\alpha}{2}$$

$$= 2 \times 7.2 \times 10^3 \times \mathrm{Sin}\frac{60}{2}$$

$$= 2 \times 7.2 \times 10^3 \times \mathrm{Sin}30°$$

$= 7.2 \times 10^3$

$= 7.2$ km/sec

49. Required Impulsive force = $m\Delta V$

$$= m\left[(\Delta V)_{\text{periqee}} - (\Delta V)_{\text{Circulan}}\right]$$

$$= m\left[\sqrt{\frac{\mu}{a}\left(\frac{1+e}{1-e}\right)} - \sqrt{\frac{\mu}{a}}\right]$$

$$= m\sqrt{\frac{\mu}{a}}\left[\sqrt{\frac{(1+e)}{(1-e)}} - 1\right]$$

50. Given

excess power = 300000

V = 100m/sec

W = 15000 N

$$\frac{dh}{dt} = 10\text{m/sec}$$

$$\text{Specific power} = \frac{\text{excess power}}{\text{weight}} = \frac{3 \times 10^5}{15000}$$

$= 20\text{W/N}$

$$P_S = \frac{dh}{dt} + \frac{V}{g}\frac{dv}{dt}$$

$$20 = 10 + \frac{100}{10}\frac{dv}{dt}$$

$$\frac{dv}{dt} = \frac{(20-10)}{10}$$

$= 1\text{m/sec}^2$

51.

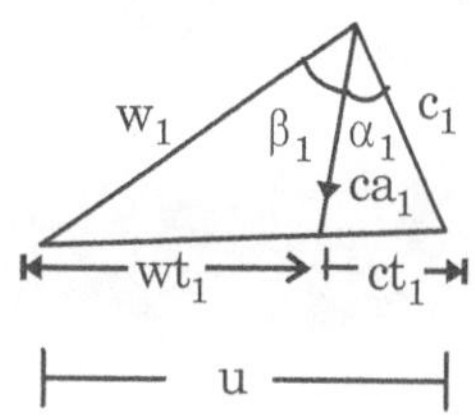

$$\Delta T = \left(r_p^{\frac{\gamma-1}{\gamma}} - 1\right) = \left(1.3^{0.286} - 1\right) \times 300$$

= 23.38 K

Degree of reaction

$$R = \frac{Ca}{2u}\left[\tan\beta_1 + \tan\beta_2\right]$$

$$0.5 = \frac{200}{2 \times 250}(\tan\beta_1 + \tan\beta_2)$$

$\tan\beta_1 + \tan\beta_2 = 1.25$ ...(1)

temperature rise per stage (DT)

$$\Delta T = \frac{VCa}{Cp}(\tan\beta_1 + \tan\beta_2)$$

$\tan \beta_1 - \tan \beta_2 = \frac{\Delta TCp}{VCa} = \frac{23.38 \times 1005}{250 \times 200}$

$= 0.47$

$\tan \beta_1 - \tan\beta_2 = 0.47$ ...(2)

Adding this tan equation, we get

$2 \tan\beta_1 = 1.25 + 0.47$

$\tan\beta_1 = 0.86$

$\beta_1 = \tan + (0.86) = 40.7°$

from equation (1)

$\tan \beta_2/m = 1.25 - \tan\beta_1$

$= 1.25 - \tan 40.7$

$= 1.25 - 0.86$

$\beta_2 = 21.3°$

β_1 and β_2 are 40.7 & 21.3°

52. for 50% reaction stage

$\alpha_1 = \beta_2, \alpha_2 = \beta_2$

$$\left.\begin{aligned}\alpha_1 &= 21.3^0 \\ \alpha_2 &= 40.7^0\end{aligned}\right\}$$

53. $V_H = \frac{S_t \, l_t}{SC}$

$$= \frac{0.3 \times \frac{29.2}{2}}{2.8}$$

$= 1.56$

S_t = tail surface area

S = wing surface area

l_t = distance of the tail aerodyamic centre from centre of gravity

C = chord

$$\therefore C_{m_o} = C_{m_{ocwb}} + V_H \eta_t at\left(\dot{i}_t + \varepsilon_o\right)$$

$= -0.0218 + 1.56 \times 0.9 \times 0.08\ (2.1 + 0)$

$= 0.21472$

54. V max for steady level flight is given by

$$V_{max} = \left\{\frac{\left(\frac{(T_A)_{max}}{W}\right) \times \left(\frac{W}{S}\right) + \left(\frac{W}{S}\right)\sqrt{\left(\frac{(T_A)_{max}}{W}\right)^2 - 4C_{DoK}}}{e_\infty C_{D,o}}\right\}^{\frac{1}{2}}$$

$$V_{max} = \left\{\frac{\left(\frac{(T_A)_{max}}{W}\right)\left(\frac{W}{S}\right) + \left(\frac{W}{S}\right)\sqrt{\left(\frac{(T_A)_{max}}{W}\right)^2 - 4C_{DoK}}}{e_\infty C_{Do}}\right\}^{\frac{1}{2}}$$

Hence, V max is influcrce by

$\frac{(T_A)_{max}}{W}, \frac{W}{S} C_{DO}$ & K

It is not influenced by $\frac{L}{D}$.

55. $\omega_n = 100$ rad/sec

Damping factor (ζ) = 0.8

frequency of vibration of damped

system is given by $\omega_d = \omega_n\sqrt{1-\zeta^2}$

$= 100\sqrt{1-0.8^2} = 60 \text{rad/sec}$

56. Circuitous is marked by obliqueness or indirection in speech or Conduct.

57. $\frac{12-8}{2} = 2$, and $\frac{15-7}{2} = 3$ to code employed

here is special In x + y '+' stands for $\frac{y-x}{2}$

$\therefore 10 + 18 = \frac{18-10}{2} = 4$

58. If we Count backward from z, we get y as the 2nd letter, W as the 4th letter, and S as the 8th letter. In this order u is 6th, V is 5th and Z the 1st letter of the code. So

2 × 4 = 8 and 6 – 5 = 1

Hence r = 9 and x = 3

$\frac{r}{x} = \frac{9}{3} = 3$

59. A silo is a place where we would find grain

A plane is place where you would find a pilot

A Courtroom is place where would find writer

A Clock is place where you would find time not wall.

60. The sense in sentence in that due to physical weakness my father was unable to open the heavy door. So, feeble is the best word to complete the sentence

61. Averge weight of group A = 30kg
Total weight of group A = 30 × 30 = 900kg
Averge weight of group B = 70 kg
Total weight of group B = 70 × 70 = 4900 kg
Averge weight of group A and group

$$B = \frac{900+4900}{30+70} = \frac{5800}{100} = 58kg$$

62. $F(y) : \frac{|y|}{y} \quad \begin{matrix} |y| = y \; y > 0 \\ -y \; y < 0 \\ 0 \; y \; : 0 \end{matrix}$

y + he

$$|f(q) - f(-q)| = \frac{y}{y} - \frac{(-y)}{y}$$

$$= \frac{y}{y} + \frac{y}{y} = 2$$

63.

$$\begin{array}{r} 44 \\ \times 11 \\ \hline 44 \\ 44 \\ \hline 1034 \end{array}$$

From the above Calculator it is clear that the base of given number is 5. $(3111)_5 = 3 \times 5^3 + 1 \times 5^2 + 1 \times 5^1 + 1 \times 5^\circ = 375 + 25 + 5 + 1 = 406$

64. From the Statement of the passage. It is clear that people is military establishments thing that chemical agents are useful tool for their Causes. It is not given is the passage that chemical agents are useful or undesirable in modern warfare.

65. Shape of lire is given by

$$\tan\theta = \frac{y_2 - y_1}{x_2 - x_1}$$

$$\frac{9-P}{0-5} = \frac{9-P}{-5} = \frac{P-9}{5}$$

$$\text{Given } \tan\theta = \frac{3}{5}$$

$$\therefore = \frac{3}{5} = \frac{P-9}{5}$$

$$\boxed{P - 9 = 3}$$

MOCK TEST – 8

AEROSPACE ENGINEERING

Time Allowed: 3:00 Hours *Maximum Marks: 100*

Read the following instructions carefully before you begin to answer the questions. This Booklet contains questions in English.

INSTRUCTIONS

1. This Booklet contains **65 questions** in total comprising of the following two parts:

   Part -1: Technical Section (85 Marks) *(1-55 Questions)*

   Part -2: General Aptitude Section (15 Marks) *(56-65 Questions)*

2. All questions are compulsory. Q.1-Q.25 & Q.56-Q.60 carry one mark each. Q.26-Q.55 & Q.61-Q.65 carry two marks each.

3. **Types of Questions:**

   (*a*) Multiple Choice Questions(MCQ) carrying 1 or 2 marks each in all papers and sections. These questions are objective in nature, and each will have a choice of four options, out of which the candidate has to mark the correct answer(s).

   (*b*) Numerical Answer Questions of 1 or 2 marks each in all sections. For these questions the answer is a real number, to be entered by the candidate using the virtual keypad. No choices will be shown for these type of questions.

4. For 1-mark multiple-choice questions, 1/3 marks will be deducted for a wrong answer. Likewise, for 2-marks multiple-choice questions, 2/3 marks will be deducted for a wrong answer. There is no negative marking for numerical answer type questions.

MOCK TEST – 8

1. The lift of an aircraft in steady climbing flight at an given forward speed is

(*a*) Inversely proportional to climb angle

(*b*) higher than lift in steady level flight at a given same forward speed.

(*c*) Lower than lift in steady level flight at same forward speed.

(*d*) Independent of climb angle.

2. To reduce the aerodynamic heating of nose which type of nose is preferable

(*a*) sharp nose (*b*) Blunt nose

(*c*) both (*a*) and (*b*) (*d*) None of these

3. Which of the following statement is correct?

(*a*) velocity potential is defined only for irrational flow, inconstrast, the stream function is usd for rotational or irrotational flows flames.

(*b*) velocity potential is defined only for rotational flow, inconstrast. The stream function is used for irrotation flow only.

(*c*) velocity potential and stream function both are valid for irrotational flow only.

(*d*) None of the above.

4. In Infinite wing induce drag coefficient (c_{Di}) is

(*a*) 1 (*b*) ∞

(*c*) 0 (*d*) cannot say

5. Which of the following statement incorrect about sweep back wing?

(*a*) sweep backing the wing increase the drag divergence Mach number to higher Mach numbers.

(*b*) sweep backing the wing increase the critical Mach number

(*c*) Making wing sweepback decrease C_L and $\dfrac{C_l}{C_d}$ ratio

(*d*) increasing sweepback angle decrease is $\dfrac{C_l}{C_d}$ ratio, this happens mainly due to C_R increase.

6. A static fluid can have.

(*a*) Non-zero normal stress and shear stress

(*b*) Negative normal stress and zero shear stress

(*c*) Positive normal stres and zero shear stres

(*d*) zero normal stress and zero shear stress

7. SI unit of kinematic viscosity is

(*a*) m^2/sec (*b*) kg/m^2

(*c*) m/sec^2 (*d*) m^3/sec^2

8. Consider an incompressible laminar boundary layer flow over a flat plate of length l, alinged with a direction of an oncoming uniform free flow. If f is the ratio of drag force on the front half on the plate to drag force on the rear half then

(*a*) $f < \dfrac{1}{2}$ (*b*) $f = \dfrac{1}{2}$

(*c*) $f = 1$ (*d*) $f > 1$

9. Which of the following features of an aircraft do not support in lateral stability?

(*a*) High wing

(*b*) Lower position of C.G.

(*c*) wing sweepback & dihedral

(*d*) wing taper

10. For which of these applications in the turboshaft engine most suited?

(*a*) High speed combat aircraft

(*b*) Low speed fixed-wing aircraft

(*c*) High speed fighter aircraft

(*d*) Helicopters

11. The main objective of by-pass is turbofan engine is to

(*a*) mass flow rate through engine inlet

(*b*) Turbine inlet temperature

(*c*) mass flow rate through exhaust nozzle

(*d*) Compressor pressure ratio

12. Which of the following can provide lateral control of the airplane?

(*a*) Spoilers

(*b*) Different differential stabizers

(*c*) Ailerons

(*d*) All of them

13. The state of plane stress of point in an isotropic material is given by

$$[\sigma] = \begin{bmatrix} -200 & 100 \\ 100 & 100 \end{bmatrix} \text{ in MPa.}$$

the maximum shear stress is MPa is

(*a*) 111.8 (*b*) 150.1

(*c*) 180.3 (*d*) 225.6

14. A column has rectangular cross section of 100 mm 200 mm and a length of 1 m the slenderness ratio of column is close to ________

15. A simply supported beam PQ is load by moment of 1 N-m at the mid span of the beam as shown in figure the reaction forces R_P and R_Q at the supports P and Q respectively are

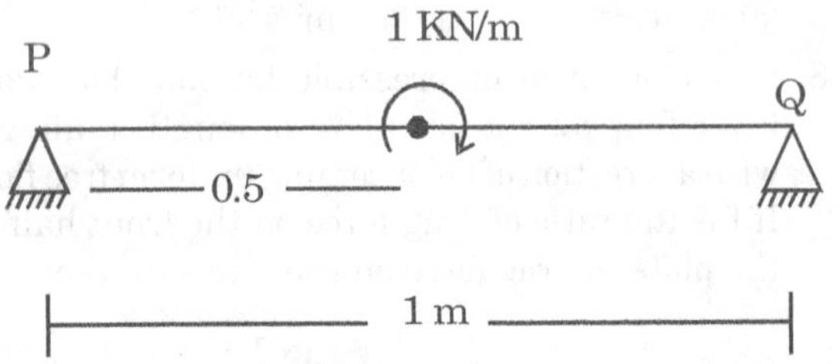

(*a*) 1 kN downward

(*b*) 1 kN upward

(*c*) 0.5 kN downward 0.5 kN upward

(*d*) 1 kN upward, 1 kN upward

16. A length of cantilever beam is halved the natural frequency of man M at the end of this cantilever beam of negligible mass is increased by a factor of

(*a*) 2 (*b*) 4

(*c*) $\sqrt{8}$ (*d*) 8

17. The natural frequency of spring-mass system on earth is w_n. the natural frequency of the system or the mas $\left(g_{moon} = \frac{g_{earth}}{6}\right)$ is ________.

18. A rectangular beam 100 mm wide and 250 mm deep is subjected to a maximum shear force of 50 KN. maximum shear stress is ________ N/mm^2.

19. A solid rocket motor operating at an I_{sp} of 210 sec. The speed of exhaust gases the nozzle exit is ______ m/sec.

20. Which of the following satellite quantities are constant is elliptical orbit of a satellite?

(*a*) Kinetic Energy

(*b*) Angular velocity

(*c*) Angular momentum

(*d*) All of the above

21. A satellite is revolving around the earth is a circular orbit with a speed of 5.9 m/sec. How much additional velocity is required for satellite so that it escapes from earths gravitational field?

(*a*) 2.44 m/sec (*b*) 5.05 m/sec

(*c*) 5.88 m/sec (*d*) data is insufficient

22. If I_1, I_2 are integrating factors of the equations $xy^1 + 2y = 1$ $xy^1 - 2y = 1$ then

(*a*) $I_1 = -I_2$ (*b*) $I_1 I_2 = x^2$

(*c*) $I_1 = x^2 I_2$ (*d*) $I_1 I_2 = 1$

23. If $f(x) = |x|$ then

(*a*) $f'(0) = 0$

(*b*) f(x) is maximum at x = 0

(*c*) f(x) is minimum at x = 0

(*d*) None of these

24. If C of the men value theorem for the function $f(x) = x(x-2)$, where $a = 0$, $b = \frac{3}{2}$ is ______

25. The value of integral $\int_{-2}^{2} \frac{dx}{x^2}$ is ______

26. If $\varphi(x) = \int_0^{x^2} \sqrt{t}\, dt$ then $\frac{d\varphi}{dx}$ is

(*a*) $2x^2$ (*b*) $\sqrt{x}$

(*c*) 10 (*d*) 1

27. The value of y as $t \to \infty$ for an initial value of y(1) = 0 for the differential equations.

$$\left(4t^2 + 1\right)\frac{dy}{dt} + 8yt - t = 0 \text{ is}$$

(*a*) 1 (*b*) $\frac{1}{2}$

(*c*) $\frac{1}{\infty}$ (*d*) $\frac{1}{8}$

28. Given matrix $A = \begin{bmatrix} 3 & 0 & 0 \\ 5 & 4 & 0 \\ 3 & 6 & 1 \end{bmatrix}$ the eigen vector corresponding to smallest eigen value is

(*a*) $\begin{bmatrix} 0 \\ 0 \\ 1 \end{bmatrix}$ (*b*) $\begin{bmatrix} 1 \\ 0 \\ 0 \end{bmatrix}$

(*d*) $\begin{bmatrix} 0 \\ 1 \\ 0 \end{bmatrix}$ (*d*) None of these

29. The rank of metrix $\begin{bmatrix} 1 & 3 & 4 \\ 1 & 3 & 5 \\ 1 & 3 & 4 \end{bmatrix}$ is

(*a*) 0 (*b*) 1

(*c*) 2 (*d*) 3

30. Using trapezoidal rule, and the table given below 5.2

$\int_{4}^{5.2} \ln x\, dx$ will be

x	4	4.2	4.6	4.8	5	5.2
l_{nx}	1.39	1.44	1.4	1.37	1.61	1.65

(*a*) 1.8277 (*b*) 1.9284
(*c*) 1.64 (*d*) 1.9127

31. Match the right pairs:

(i) Dutch roll made	P. Coupled roll-yaw oscillation
(ii) phupoid mode	Q. couple roll-pitch oscillation
(iii) short period mode	R. Angle of attack remains cosntant
	S. Speed remains constant

(*a*) I - R, II - Q, III - P
(*b*) I - P, II - R, III - S
(*c*) I - P, II - Q, III - R
(*d*) I - R, II - R, III - S

32. Air enters an ideal turbine of the turbojet engine at total temperature of 800 K and leaves at axial speed 300 m/sec temperature of 312 K the work extracted per unit mass flow rate by turbine (Assume exhaust = 1.005 KJ/kg – K)

(*a*) 357 KJ/kg (*b*) 400 KJ/kg
(*c*) 446 KJ/kg (*d*) 600 KJ/kg

33. A rocket engine nozzle has a throat area of 3.8 m^2 and equilibrium chamber pressure 35 bar. The specific impulse I_{sp}. is 189 sec and rate of flow of propellant gasses through nozzle is 1.12 Kg/sec the thrust coefficient is equal to ________.

34. Absolute ceiling of aircraft is 20,500 m above mean sea level (SSL). If rate of climb of aircraft at MSL is 210 m/sec and assumed to decrease linearly with allitude, how much time would it take to climb to its cruise altitude of 11 mm from MSL?

(*a*) 30 sec (*b*) 1 min 27 sec
(*c*) 1 min 35 sec (*d*) 2 min 20 sec

35. for flow over half - damond wedge as shown in figure given below. Pressure coefficient is region (2) is

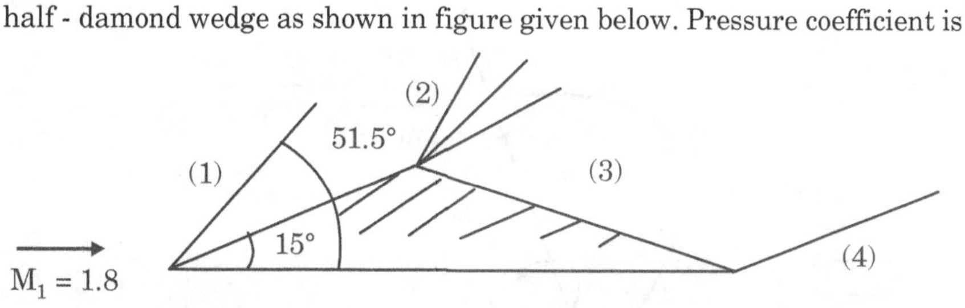

(i) 0.506 (*b*) 0.806 (*c*) 1.21 (*d*) 2.149

36. A flat plate is 15° angle of attack to a supersonic stream at Mach 2.7. the value of lift coefficient and wave drag coefficient are ________.

37. An oblique shock wave is formed at the leading edge of 10° wedge. The shock wave angle is 28. 900 and upstream velocity 925 m/sec. The density ratio across the shock wave.

(*a*) 1.06 (*b*) 2.5
(*c*) 1.081 (*d*) 3.5

38. Consider the rocket engine burning hydrogen and oxygen the total mass Flow rate of propellant and oxidiser is 287.2 Kg/sec. The combustion chamber temperature is 3600 K. assume that combustion chamber is low-velocity reservoir for the rocket engine. If the throat area of rocket engine is 0.2 m^2, the combution chamber (reservoir) pressure is? Assume that the gas that flows through the engine has a ratio of specific heats $\gamma = 1.2$ and a molecular weight at 16 is

(*a*) 100 bar (*b*) 200 bar
(*c*) 135 bar (*d*) 50 bar

39. If load factor n at steady pullout is 9 and radios of curvature of flight path during pull out is 200 m, the velocity of aero plane during pull out is

(*a*) 127 m/sec (*b*) 125 m/sec
(*c*) 120 m/sec (*d*) 152 m/sec

40. An airplane flying straight and level at a speed of 110 m/sec at height of 11 km is disturbed by a symmetric vertical just resulting a phugoid oscillation the approximate frequency and damping ratio of phugoid oscillation are.

$$\left[\text{if } \frac{x_u}{m} = -0.0446 \text{ sec}^{-1}\right]$$

(*a*) 2 sec^{-1} (*b*) 0.44
(*c*) 0.126 sec^{-1}, 0.18 (*d*) 0.4 sec^{-1}, 0.18

41. Assuming lift curve slope for symmetric aero foil is given by this aerofoil theory, percentage change in C_L Value at $\alpha = 6$ for an untwisted elliptical wing of aspect ratio 8 as compared to infinite wing is ______%

42. Consider an ideal ramjet flying at match 1.2 at ambient temperature 280 K. The exhaust gasses are optimallyexpanded and leave the engine at an temperature of 1160 K. Assuming the specific heat ratio constant, what will be specific thrust obtained by engine? (Rair = 287 J/Kg-K)

(*a*) 350 N (*b*) 320 N

(*c*) 380 N (*d*) 500 n

43. Combination of doublet of strength 'K' uniform flow with velocity V and a vertex of strength T produces lifting flow over a cylinder of radius R. If strength of vortex is 4 times which keeping all other parameters unchanged what would happen to the radius of cylinder?

(*a*) Remains same (*b*) becomes double

(*c*) becomes half (*d*) None of these

44. An aircraft has an approach speed of 40 m/ec with an decent angle of 6.6°. If the aircraft load factor is 1.2 and constant declination of touchdown is 0.25 g . If the loading strips was at an altituddude 2900 Km above mean sea level (air density = 0.9026 Kg/m³), how much distance would be required to l and ?

(*a*) 1000 m (*b*) 500

(*c*) 550 m (*d*) 445.5 m

45. Consider a potential flow over a finite wing the following circulation distribution

$$\Gamma(y) = 100\sqrt{1-\left(\frac{2y}{4}\right)^2} \ \text{m}^2/\text{sec}.$$

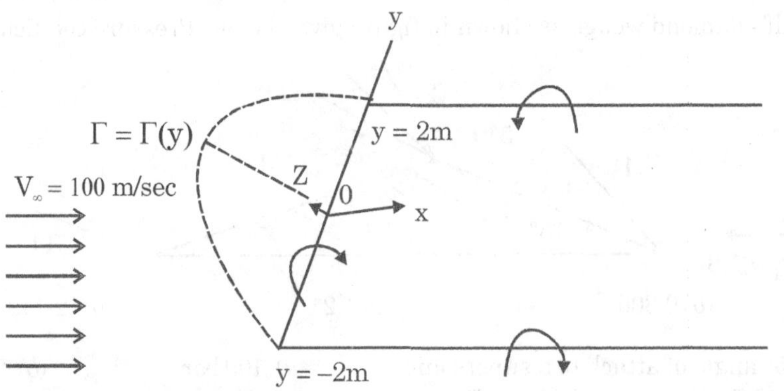

If the free steam velocity is 100 m/sec, the downwash is

(*a*) 12.5 m/sec (*b*) –25 m/sec (*c*) $25\sqrt{1-\left(\frac{2y}{4}\right)^2}$ m/sec (d) $-2\sqrt{1-\left(\frac{2y}{4}\right)^2}$ m/sec

46. In Hohmann transfer, a satellite in one circular orbit is transperred to another circular orbit via a transfer orbit which in elliptical in nature. The perigee of transfer orbit lies lower circular orbit and appogee in higher circular orbit.

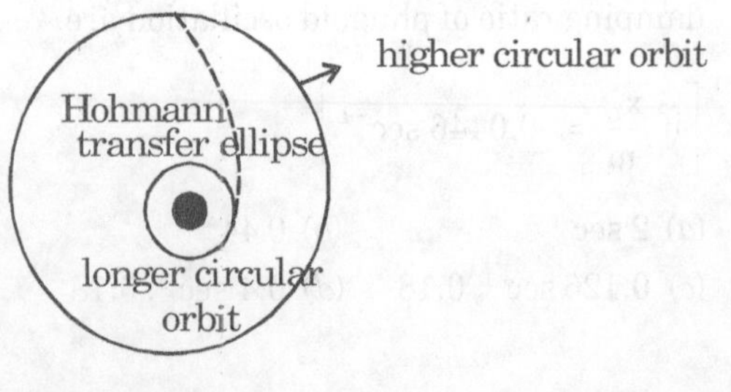

An earth satllite in transfrred from a 400 km circular orbit to 1200 Km circular orbit using hohmaan's transfer methodology, how much velocity incremental required in lower orbit to put the satellite is transfer ellipse?

(*a*) 11.8 m/sec (*b*) 210 m/sec

(*c*) 340 m/sec (*d*) 426 m/sec

47. Air enters compressor of a turbojet engine of 260 K temperature and leaves at 700 K the pressure rating of the compressor is ________ (assuming adiabatic efficiency to be 0.88, $\gamma = 1.35$)

48. A cantilever ABC is fixed at A and rigidly propped at C and is loaded as shown in figure reaction at C is ______ KN.

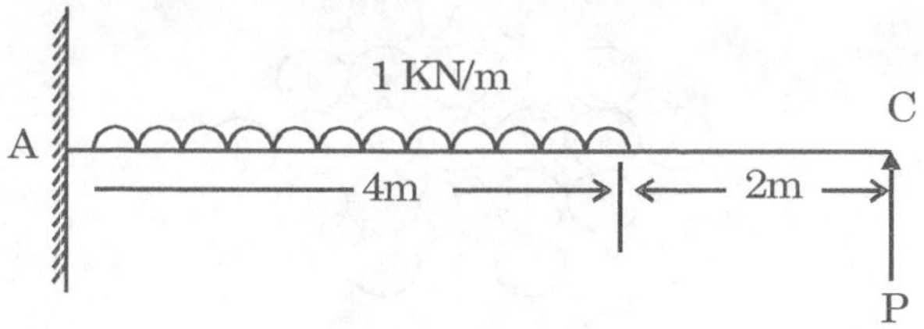

49. Stell cantilever beam of length 1000 mm having a cross section 25 mm × 25 mm. assume mass of cantilever beam is negligible and E_{steel} = 200 m/kg. If the lateral virabtion of this system is critically damped using a viscous damper, the damping constant of the the damper is

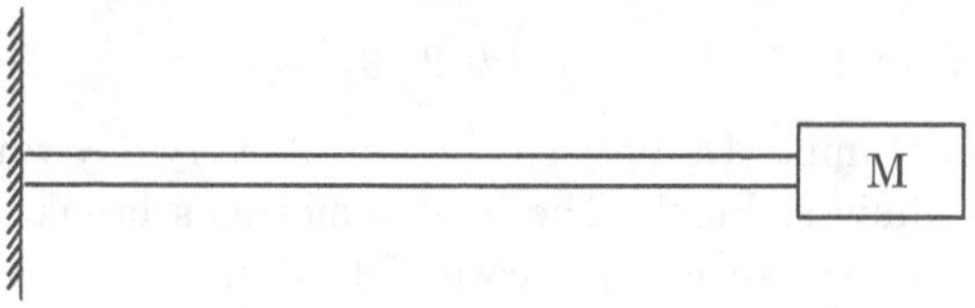

(*a*) 1200 N sec/m (*b*) 625 N sec/m
(*c*) 312.50 N sec/m (*d*) 15.25 N sec/m

50. Damper mass-spring system as shown in figure has m = 10 kg, stiffness coefficient K = 100 N/m. damping factor c = 300 N/m displacement of mass is function of time is given by

(*a*) $x(t) = A_1e^{3.18t} + A_2e^{-26.18t}$

(*b*) $x(t) = A_1e^{-6.28t} + A_2e^{-13.09t}$

(*c*) $x(t) = A_1e^{3.18t} + A_2e^{26.18t}$

(*d*) $x(t) = (A_1 + A_2t)e^{-3.18t}$

51. Consider linear spring mass system having damping ratio of 0.02 the fractional loss in total energy per cycle is

(*a*) 0.0004 (*b*) 0.98
(*c*) 0.78 (*d*) 0.22

Common data question

Data given for jet - powered aircraft weight = 30 KN. Wing Span area, S = 30 m². Maximum lift coefficient $0.015 + 0.027\ C_L^2$ (assume air density at sea level)

52. At what velocity it should fly to maximise its range?

(*a*) 45 m/sec (*b*) 50 m/sec
(*c*) 60 m/sec (*d*) 40 m/sec

53. What is minimum speed that airplane can fly?

(*a*) 50 m/sec (*b*) 35 m/sec
(*c*) 26 m/sec (*d*) 100 m/sec

54. Which of the following is true for 50% reaction stage of axial flow compressor.

P. Halt of the enthalpy rise takes place is rotor and half in stator.

Q. Half of the pressure rise takes place in rotor and half in the stator

R. It has symmetrical blades is its rotor and stator.

S. Full enthalphy rise takes place is rotor no enthalpy rise takes place is stator.

(*a*) only P (*b*) P, Q & R
(*c*) P, Q, R & S (*d*) Q R & S

55. An aircraft with weight 15000 Kg takes off from mean sea level with forward speed of 100 m/sec and starts to climb with a climb angle of 10°. The rate of climb and excess thrust available at the start of climb are ______ m/sec and ______ KN.

56. Choose the most appropriate word from the options given below to complete the following sentence.

The principal presented the chief guest with a ________ as token of appreciation.

(*a*) momento (*b*) memento
(*c*) momentum (*d*) moment

57. Choose the appropriate word/phrase, out of the four options given below, to complete the following sentence:

Frogs __________.

(*a*) croak (*b*) roar
(*c*) hiss (*d*) patter

58. Choose the word most similar in meaning to the given word:

Educe

(*a*) Exert (*b*) Educate
(*c*) Extract (*d*) Extend

59. Operators □ , ◇ and → are defined by:

$a \square b = \frac{a-b}{a+b}$; $a \diamond b = \frac{a+b}{a-b}$; $a \to b = ab$.

Find the value of (66 □ 6) → (66 ◇ 6).

(*a*) – 2 (*b*) – 1
(*c*) 1 (*d*) 2

60. If $\log_x (5/7) = -1/3$, then the value of *x* is

(*a*) 343/125 (*b*) 125/343
(*c*) – 25/49 (*d*) – 49/25

61. The following question presents a sentence, part of which is underlined. Beneath the sentence you find four ways of phrasing the underlined part. Following the requirements of the standard written English, select the answer that produces the most effective sentence.

Tuberculosis, together with its effects, <u>ranks one of the leading causes of death</u> in India.

(*a*) ranks as one of the leading causes of death
(*b*) rank as one of the leading causes of death
(*c*) has the rank of one of the leading causes of death
(*d*) are one of the leading causes of death

62. Read the following paragraph and choose the correct statement.

Climate change has reduced human security and threatened human well being. An ignored reality of human progress is that human security largely depends upon environmental security. But on the contrary, human progress seems contradictory to environmental security. To keep up both at the required level is a challenge to be addressed by one and all. One of the ways to curb the climate change may be suitable scientific mnnovations, while the other may be the Gandhian perspective on small scale progress with focus on sustainability.

(*a*) Human progress and security are positively associated with environmental security.
(*b*) Human progress is contradictory to environmental security.
(*c*) Human security is contradictory to environmental security.
(*d*) Human progress depends upon environmental security.

63. Fill in the missing value

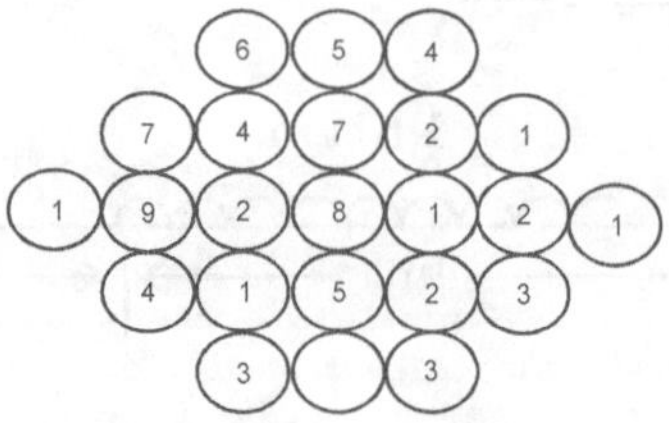

64. A cube of side 3 units is formed using a set of smaller cubes of side 1 unit. Find the proportion of the number of faces of the smaller cubes visible to those which are NOT visible.

(*a*) 1 : 4 (*b*) 1 : 3
(*c*) 1 : 2 (*d*) 2 : 3

65. Humpty Dumpty sits on a wall every day while having lunch. The wall sometimes breaks. A person sitting on the wall falls if the wall breaks.

Which one of the statements below is logically valid and can be inferred from the above sentences?

(*a*) Humpty Dumpty always falls while having lunch
(*b*) Humpty Dumpty does not fall sometimes while having lunch
(*c*) Humpty Dumpty never falls during dinner
(*d*) When Humpty Dumpty does not sit on the wall, the wall does not break

OMR SHEET

1. (A) (B) (C) (D)	18.	35. (A) (B) (C) (D)	52. (A) (B) (C) (D)
2. (A) (B) (C) (D)	19.	36.	53. (A) (B) (C) (D)
3. (A) (B) (C) (D)	20. (A) (B) (C) (D)	37. (A) (B) (C) (D)	54. (A) (B) (C) (D)
4. (A) (B) (C) (D)	21. (A) (B) (C) (D)	38. (A) (B) (C) (D)	55.
5. (A) (B) (C) (D)	22. (A) (B) (C) (D)	39. (A) (B) (C) (D)	56. (A) (B) (C) (D)
6. (A) (B) (C) (D)	23. (A) (B) (C) (D)	40. (A) (B) (C) (D)	57. (A) (B) (C) (D)
7. (A) (B) (C) (D)	24.	41.	58. (A) (B) (C) (D)
8. (A) (B) (C) (D)	25.	42. (A) (B) (C) (D)	59. (A) (B) (C) (D)
9. (A) (B) (C) (D)	26. (A) (B) (C) (D)	43. (A) (B) (C) (D)	60. (A) (B) (C) (D)
10. (A) (B) (C) (D)	27. (A) (B) (C) (D)	44. (A) (B) (C) (D)	61. (A) (B) (C) (D)
11. (A) (B) (C) (D)	28. (A) (B) (C) (D)	45. (A) (B) (C) (D)	62. (A) (B) (C) (D)
12. (A) (B) (C) (D)	29. (A) (B) (C) (D)	46. (A) (B) (C) (D)	63.
13. (A) (B) (C) (D)	30. (A) (B) (C) (D)	47.	64. (A) (B) (C) (D)
14.	31. (A) (B) (C) (D)	48.	65. (A) (B) (C) (D)
15. (A) (B) (C) (D)	32. (A) (B) (C) (D)	49. (A) (B) (C) (D)	
16. (A) (B) (C) (D)	33.	50. (A) (B) (C) (D)	
17.	34. (A) (B) (C) (D)	51. (A) (B) (C) (D)	

ANSWERS

1. (c)	**2.** (b)	**3.** (a)	**4.** (c)	**5.** (d)	**6.** (c)	**7.** (b)	**8.** (d)	**9.** (d)	**10.** (d)
11. (c)	**12.** (d)	**13.** (c)	**14.** 346.4	**15.** (d)	**16.** (c)	**17.** (w_n)	**18.** (2)	**19.** (2058)	**20.** (c)
21. (a)	**22.** (d)	**23.** (d)	**24.** –1/2	**25.** 1	**26.** (a)	**27.** (d)	**28.** (a)	**29.** (c)	**30.** (a)
31. (b)	**32.** (c)	**33.** 1.53	**34.** (b)	**35.** (a)	**36.** 0.1229	**37.** (*)	**38.** (d)	**39.** (c)	**40.** (d)
41. 20%	**42.** (c)	**43.** (c)	**44.** (d)	**45.** (a)	**46.** (b)	**47.** 24	**48.** 0.174	**49.** (a)	**50.** (a)
51. (d)	**52.** (c)	**53.** (c)	**54.** (c)	**55.** 17.36 & 2.604		**56.** (b)	**57.** (a)	**58.** (c)	**59.** (c)
60. (a)	**61.** (a)	**62.** (b)	**63.** (3)	**64.** (c)	**65.** (b)				

EXPLANATIONS

1.

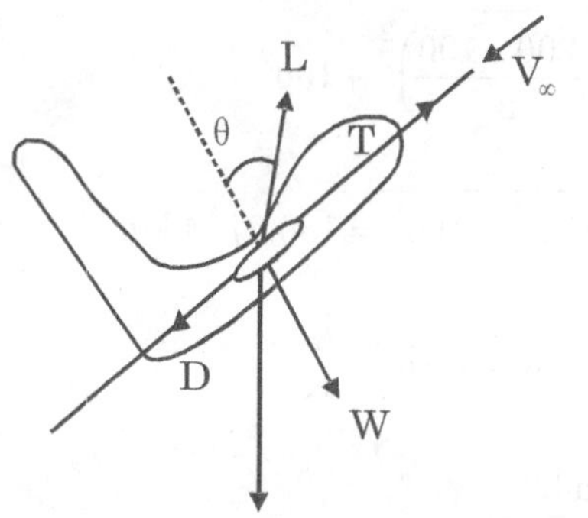

Consider force equilibrium perpendicular to free stream

$$L = W\cos\theta = \frac{1}{2}\rho\, v_\infty^2 S C_L$$

$$(C_L)_{C_{Limb}} = \frac{2W}{\rho v^2 s}\cos\theta \quad ....(1)$$

In steady level flight

$$(C_L)_{level} = \frac{2W}{\rho v^2 s} \quad ...(2)$$

$$(C_L)_{level} > (C_L)_{climb}$$

2. $\dot{q}_w \propto \frac{1}{\sqrt{R}}, R$

R = radius of nose

Stagnation point heating varies inversely with square root of the nose radius hence to reduce the heating increase the nose radius this is the region the nose and the lead edge region of hypersonic vehicles are blunt.

4. Induce drag coefficient $(C_{Di}) = \frac{C_L^2}{\pi\, eAR}$

where C_L = lift coefficient

e = oswal efficiency factor

AR = Aspect ratio

For infinite wing

$AR \to \infty$

$C_{Di} \to 0$

6.

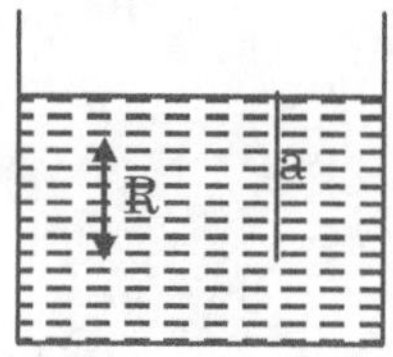

In a static fluid

Considering a area a,

$$\text{Normal stress} = \frac{\text{force}}{\text{Area}}$$

$$\frac{\text{Weight of water column above area a}}{\text{Area}}$$

$$= \frac{\rho\, gh}{a} > 0$$

thus, normal stress = +ve

As there is no flow, hence shear stress is zero because shear stress is stress exerted by one layer of floating fluid on the other layer which moves with different velocity in the boundary layer.

In static fluid, no flow, hence no boundary layer and no shear stress.

∴ positive normal stress

zero shear stress.

7. Kinematics viscosity = $\frac{\text{Dynamic viscosity}}{\text{Density}}$

Now according to newton's law of viscosity

$$T = \frac{\mu dv}{dy}$$

$$\mu = \frac{T}{\frac{du}{dy}} = \frac{N/m^2}{\frac{m/sec}{m}}$$

$$= \frac{Nm}{m^2 x\, m/sec} = \frac{Kg}{m^2} \times \frac{m}{sec^2} \times sec = \frac{kg}{m-sec}$$

8. Let L = length of the plate
V = velocity of the fluid
x = kinematic viscosity of the fluid
Average coefficient of drag for total

$$\text{plate} = \left(\frac{1.328}{\sqrt{\frac{VL}{\nu}}}\right)$$

Drag coefficient for Front half

$$C_{D_1} = \frac{1.328}{\sqrt{\frac{VL}{2\nu}}}$$

$$= \frac{1.328\sqrt{2}}{\sqrt{\frac{VL}{\nu}}} = \frac{1.878}{\sqrt{VL\nu}}$$

Frictional drag for total plate = $C_D \times \frac{1}{2}\rho\, Av^2$

$$C_D \times \frac{1}{2}\rho\, LBv^2$$

$$= \frac{1.328}{\sqrt{\frac{VL}{\nu}}} \times \frac{1}{2}\rho\, LBV^2$$

$$= \frac{0.664}{\sqrt{\frac{VL}{\nu}}} \times \rho L\, BV^2$$

$$= \frac{0.4695}{\sqrt{\frac{VL}{\nu}}} \times \rho LBV^2$$

¥ frictional drag as rear half $F_{D_2} = F_D - F_{D_1}$

$$= \frac{0.669}{\sqrt{\frac{VL}{\nu}}} \times \rho\, L\, B\, v^2 - \frac{0.4895}{\sqrt{\frac{VL}{\nu}}} \times \rho\, L\, B\, v^2$$

$$= \frac{0.1945}{\sqrt{\frac{VL}{\nu}}} \times \rho\, L\, B\, V^2$$

$$\therefore = \frac{F_{D_1}}{F_{D_2}} = \frac{\frac{04695}{\sqrt{\frac{VL}{x}}} \times \rho\, LBV^2}{\frac{04695}{\sqrt{\frac{VL}{\nu}}} \times \rho\, LBV^2} = 2.414$$

13. Given
$6_{xx} = -200$ mpa
$6_{xx} = 100$ m pa
$T_{xy} = 100$ m pa

$$\therefore T_{max} = \frac{6_1 - 6_2}{2}$$

$$= \sqrt{\left(\frac{6_x - 6_y}{2}\right)^2 + T_{xy}{}^2}$$

$$= \sqrt{\left(\frac{-200 - 100}{2}\right)^2 + 100^2}$$

$$= \sqrt{(.150)^2 + 100^2} = 180.27 \text{ M kg}$$

14. slenderness ratio = $\frac{l_e}{K}$

$$I = \frac{bd^3}{12}$$

$$K = \sqrt{\frac{I}{A}}$$

$$= \sqrt{\frac{bd^3}{12 \times A}}$$

$$= \sqrt{\frac{20 \times 10^3}{12 \times 10 \times 20}} = 2.886$$

$$\lambda = \frac{1 \times 10^3}{2.886} = 346.4$$

15.

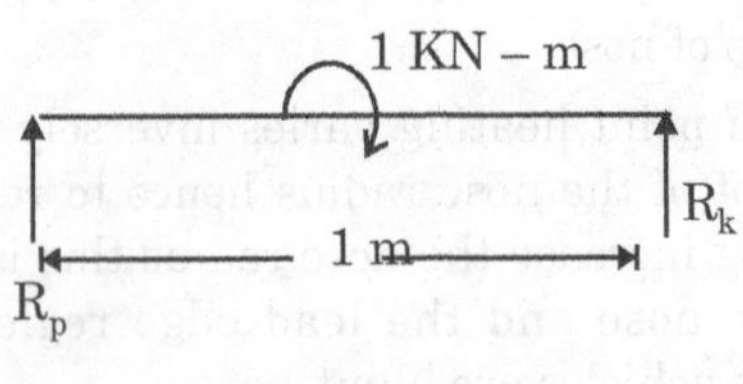

taking moment about Q
$R_p \times 1 + 1 = 0$
$R_p = -1$ KN.
taking moment about P.
$1 + (-R_Q \times 1) = 0$
$R_Q = 1$ KN
1 KN ↑ 1 KN ↑

16. For a cantilever loaded by mass M at the free end and of length L,

Deflection $\delta = \frac{MgL^3}{3\,EI}$

$$W_n = \sqrt{\frac{g}{b}} = \sqrt{\frac{g}{MgL^3}} \times 3E\,I = \sqrt{\frac{3EI}{M\,L^3}}$$

$$W_n \propto \frac{1}{L^{\frac{3}{2}}}$$

$$\frac{W_i}{W_f} = \frac{L_f^{\frac{3}{2}}}{L_i^{\frac{3}{2}}} = \frac{\left(\frac{L_i}{2}\right)^{\frac{3}{2}}}{(L_i)^{\frac{3}{2}}} = \frac{1}{2^{\frac{3}{2}}} = \frac{1}{\sqrt{8}}$$

$$\boxed{w_f = \sqrt{8}\,W_i}$$

17. $W_n = \sqrt{\frac{g}{\delta}} = \sqrt{\frac{K}{m}}$

Natural frequency is independent at g.

18. $\frac{2N}{mm^2}$

maximum shear stress (T_{max}) $= \frac{3}{2} T_{av}$

$= 1.5 \frac{F}{Area}$

$= \frac{1.5 \times 50 \times 10^3}{100 \times 250} = \left(\frac{2N}{mm^2}\right)$

19.2058 m/sec

I_{sp} = 210 sec

I_{sp} = 210sec

$$I_{sp} = \frac{v_e}{g_o}$$

$V_e = I_{sp}\, g_o$. = 210 × 9.8 = 2 km/sec

21. relation between escape velocity and circular velocity is given by

$v_{esc} = \sqrt{2}\, V_{circular} = \sqrt{2} \times 5.9 = 8.34$ m/sec

Additional velocity required = 8.34 – 5.9

= 2.44 m/sec

22. Given

$xy^1 + 2y^1 = 1$ ....(1)

$xy^1 - 2y = 1 - (2)$

Both differential equation are linear

I.F of (1),

$$I_1 = e^{\int \frac{2}{x}\,dx} = e^2 \log x = e\log x^2 = x^2.$$

I.F of (2),

$$I_2 = e^{\int \frac{-2}{x}\,dx} = e^{\log x^{-2}} = \frac{1}{x^2}$$

multiplying I.F of (1) and I.fF (2), we get

$$I_1 I_2 = x^2 \times \frac{1}{x^2} = 1$$

23. F(x) = |x|

= x x > 0 = 0 x = 0 = – x x < 0

$f'(0) \neq 0$ f(x) is not maximum at x = 0

24. Since $f(x) = x(x-2) = x^2 - 2x$

f(x) is continuous is intevoil $\left[0, \frac{3}{2}\right]$ and differentiable in interval $\left(0, \frac{3}{2}\right)$ therefore mean value theorem applicable in $\left[0, \frac{3}{2}\right]$

f(a) = f(0) = 0

$$f(b) = f\left(\frac{3}{2}\right) = \left(\frac{3}{2}\right)^2 - 2 \times \frac{3}{2}$$

$$= \frac{9}{4} - 3 = -\frac{3}{4}$$

By langrange's mean value theorem, there exit a

CE$\left(0, \frac{3}{2}\right)$ such that

$$f'(c) = \frac{f\left(\frac{3}{2}\right) - f(0)}{\frac{3}{2} - 0}$$

$$(2c - 2) = \frac{-\frac{3}{4} - 0}{\frac{3}{2}} = -\frac{1}{2}$$

25. $\int_{-2}^{2} \frac{dx}{x^2} = \int_{-2}^{2} \frac{1}{x^2}\,dx$

$f(x) = \frac{1}{x^2}$ = even function.

for even function

$$\int_{-2}^{2} f(x)\,dx = 2\int_{-2}^{2} \frac{1}{x^2}\,dx = -\left[-\frac{1}{2}\right]_0^2 = 1$$

26. $\varphi(x) = \int_0^{x^2} \sqrt{t}\, dt = \left| \frac{2}{3} t^{\frac{2}{3}} \right|_0^{x^2} = \frac{2}{3}x^2$

$$\frac{d\varphi(x)}{dx} = \frac{2}{3} \times 3x^2 = 2x^2$$

27. $(t^2+1)\frac{dy}{dx} + 8y - t = 0$

$$\Rightarrow \frac{dy}{dx} + 8y\frac{t}{(4t^2+1)} = \frac{t}{(4t^2+1)}$$

$$\Rightarrow \frac{dy}{dx} + y\frac{8t}{(4t^2+1)} = \frac{t}{(4t^2+1)}$$

$$\text{I.F} = e\int \frac{8t}{4t^2+1}dt = (4t^2+1)$$

Hence, the solution of this equation

$$y(4t^2+1) = \int \frac{1}{(4t^2+1)}(4t^2+1)dt + c$$

$$= \int t dt + c = \frac{t^2}{2} + C$$

$$y = \frac{t^2}{2(4t^2+1)} + \frac{C}{(4t^2+1)} \quad ....(ii)$$

But y (1) = 0

$$0 = \frac{1}{10} + \frac{C}{5}$$

$$C = -\frac{1}{2}$$

From the equation (ii), we get

$$y = \frac{t^2}{2(4t^2+1)} - \frac{1}{2t(4t^2+1)}$$

$$\frac{y}{t+\alpha} = \underset{t+\infty}{Lt} \frac{1}{2\left(4+\frac{1}{t^2}\right)} - \frac{1}{2(4t^2+1)}$$

$$= \frac{1}{2(4+0)} - = \frac{1}{8}$$

28. $A = \begin{bmatrix} 3 & 0 & 0 \\ 5 & 4 & 0 \\ 3 & 6 & 1 \end{bmatrix}$

Single lower triangle is zero

Here the eigen value of matrix are 3, 4, 1

$$A - \lambda I = \begin{bmatrix} 3 & 0 & 0 \\ 5 & 4 & 0 \\ 3 & 6 & 1 \end{bmatrix} - \begin{bmatrix} 1 & 0 & 0 \\ 0 & 1 & 0 \\ 0 & 0 & 1 \end{bmatrix}$$

$$= \begin{bmatrix} 2 & 0 & 0 \\ 5 & 3 & 0 \\ 3 & 6 & 0 \end{bmatrix}$$

Rank of coefficient matrix = Number of non-zero diagonal elements = 2 <3

Hence non - zero solution vector = 3 – 2 = 1

$$\therefore \begin{bmatrix} 2 & 0 & 0 \\ 5 & 3 & 0 \\ 3 & 6 & 0 \end{bmatrix} = \begin{bmatrix} x_1 \\ x_2 \\ x_3 \end{bmatrix}$$

These equations can be re Written as

$2x_1 = 0$

$5x_1 + 3x_2 = 0$

$3x_1 + 6x_2 = 0$

From the equations

$x_1 = 0, \quad x_2 = 0$

$$x = \begin{bmatrix} x_1 \\ x_2 \\ x_3 \end{bmatrix} = \begin{bmatrix} 0 \\ 0 \\ x_3 \end{bmatrix} = x_3 \begin{bmatrix} 0 \\ 0 \\ 1 \end{bmatrix}$$

Hence $\begin{bmatrix} 0 \\ 0 \\ 1 \end{bmatrix}$ is required eigen vectors.

29. $A = \begin{bmatrix} 1 & 3 & 4 \\ 1 & 3 & 5 \\ 1 & 3 & 4 \end{bmatrix}$

To check rank

$$\left.\begin{array}{c} \Delta \neq 0 \\ \text{for any } \Delta_{(2\times2)} \neq 0 \end{array}\right\} \text{rank will be 3}$$

$$\left.\begin{array}{c} \text{every } \Delta = 0 \\ \text{for } \Delta_{(2\times2)} = 0 \text{ from} \end{array}\right\} \text{rank will be 1}$$

every $\Delta = 0$
for $\Delta_{(2\times2)} \neq 0$ from } rank will be 2

$$\Delta = \begin{bmatrix} 1 & 3 & 4 \\ 1 & 3 & 5 \\ 1 & 3 & 4 \end{bmatrix}$$

$= 1\,(12 - 15) + 3\,(5 - 4\,) + 4\,(3 - 3)$
$= -3 + 3 \times 1$
$= 0$

$$\Delta_{2\times2} = \begin{bmatrix} 1 & 3 \\ 1 & 3 \end{bmatrix}$$

$= 3 \times 1 - 3 \times 1 = 0$

$$\left(\Delta_{2\times2}\right)_{2^{nd}} = \begin{bmatrix} 3 & 4 \\ 3 & 5 \end{bmatrix}$$

$= 15 - 12$
$\neq 0$, rank $= 2$

30. By trapedoidal role

$$\int_4^{5.2} y\,dx = \frac{0.2}{2}$$

$$= \left[1.39 + 1.65 + 2\left[1.44 + 1.48 + 151 + 161\right]\right]$$

$= 1.8277$

31. Phugoid mode - It persists after short period mode has died down and influences the changes in velocity and pitch angle in periodic manner angle of attack remains constant damping of this mode is very light. Short period mode:- It is heavily damped oscillation, where angle of attack, pitch angle changes, velocity remains constant.
Dutch roll mode: - It coupled with roll and yaw motion

32. given

$T_{o_3} = 800K$
$T_4 = 312$ K
$v = 300$ m/sec

$\therefore$ Work extracted per unit mass $= h_{o_1} - h_{o_2}$

$$= C_p T_{o_1} - C_p T_{o_2}$$

$$= C_p T_{o_1} - \left(C_p\left(T_2 + \frac{U_2^2}{2}\right)\right)$$

$$C_p\left(T_{o_1} - \left(T_2 + \frac{U_2^2}{2\,C_p}\right)\right)$$

$$= 1005\left(800 - \left(312 + \frac{300^2}{2 \times 1005}\right)\right)$$

$$= 1005\left(800 - (356.77)\right)$$

$= 445.45 \times 10^3$
$= 446$ J/kg

33. mass flow rate (m) $= \dfrac{P_C At}{C*}$

$$C^* = \frac{P_C At}{m} \qquad ....(1)$$

Specific impulse $\left(I_{sp}\right) = \dfrac{C^*\ Cf}{go}$

$$C_f = \frac{I_{sp} go}{C*}$$

$$= \frac{I_{sp} go}{\frac{P_c A_t}{m}}$$

[putting value of C^* from eqn (1)

$$= \frac{I_{sp} go\ \dot{m}}{\frac{P_c A_t}{m}} = \frac{185.2 \times 9.8 \times 1.12}{35 \times 10^5 \times 3.8 \times 10^{-4}} = 1.53$$

34. Given

$\left(R/C\right)_{sea} = 210 m/sec$ at $h = 0$

at $h = 20.500$ m $\left(R/C\right) = 0$

Rate of climb of aircraft can be assumed to decrease linearly over altitude.
Therefore assuming

$\left(R/C\right) = a + bh,$

where

$a = \left(R/C\right)$ of sea level

$$b = \frac{\left(R/C\right)_{altitude} - \left(\frac{R}{C}\right)_{sea}}{h_a - h_{sea}}$$

$$= \frac{0 - 210}{20.500 - 0} = -0.01$$

$\therefore\ \left(R/C\right) = 210 - 0.01h$

we have

$$\left(R/C\right) = \frac{dh}{dt}$$

$$dt = \frac{dh}{\left(R/C\right)} = \int_{t_1}^{t_2} dt = \int_{t_1}^{t_2} \frac{dh}{\left(R/C\right)}$$

$$\int_{h_1}^{h_2} \frac{dh}{(a+bh)}$$

here $\left(R/C\right)$ from sea level $h_1 = 0$ $h_2 = h$

$$\Delta t = \int_0^h \frac{dh}{(a+bh)}$$

$$= \frac{1}{b}\left|\ln(a+bh) - \ln a\right|$$

$$= \frac{1}{b}\left|\ln\left(\frac{a+bh}{a}\right)\right|$$

$$= \frac{1}{0.01}\left|\ln\left(\frac{210 - 0.01\times 11000}{210}\right)\right|$$

$$= \frac{1}{0.01}\left|\ln\left(\frac{210-110}{210}\right)\right|$$

$$= \frac{1}{0.01}\left|\ln\left(\frac{100}{210}\right)\right| = \frac{-0.7419}{-0.01} = 74.19 \text{ sec}$$

= 1 min 14 sec.

35. Pressure ratio between (1) and (2)

$$\frac{P_2}{P_1} = 1 + \frac{2\gamma}{\gamma+1}\left(M_{\eta_1}^2 - 1\right)$$

$$= 1 + \frac{2\gamma}{\gamma+1}\left((M_1 \sin\beta)^2 - 1\right)$$

$$= 1 + \frac{2\times 1.4}{2.8}\left((1.8 \sin 51.5)^2 - 1\right) = 2.149$$

Pressure coefficient in region (2)

$$C_{P_2} = \frac{P_2 - P_1}{q_2} = \frac{2\left(P_2/P_1 - 1\right)}{\gamma M_1^2}$$

$$= \frac{2(2.149-1)}{1.4\times 1.8^2} = 0.506$$

36. Lift coefficient is supersonic flow

$$\frac{4\alpha}{\sqrt{M_\infty^2 - 1}}$$

$$= \frac{4\times\frac{15\pi}{290}}{\sqrt{2.4^2 - 1}} = 0.47$$

wane drag coefficient

$$\left(C_{L_w}\right) = \frac{4\alpha^2}{\sqrt{M_\infty^2 - 1}} = C_l \times \alpha$$

$$= 0.4797 \times \frac{15\pi}{180}$$

$$= 0.1229$$

38. (iv) Given

m = 287.2 Kg/sec

$T_1 = 3600$ K

$A_t = 0.2$ m^2

Molecular weight (M) = 16

$\gamma = 1.2$

$P_0 = ?$

$$\overline{R} = \frac{R}{\text{Molecular weight}}$$

$$= \frac{8314}{16} = 519.62 \text{ J/kg}$$

Now, mass flow rate $\left(\dot{m}\right) = \frac{P_0 A^*}{\sqrt{T_0}}\sqrt{\frac{\gamma}{R}\left(\frac{2}{\gamma+1}\right)^{\frac{\gamma+1}{\gamma-1}}}$

$$= \frac{P_0 A^*}{\sqrt{T_0 R}}\sqrt{\gamma\left(\frac{2}{\gamma-1}\right)^{\frac{\gamma-1}{\gamma-1}}}$$

$$= \frac{P_0 A^*}{\sqrt{T_0 R}} \times 0.3931$$

$$\Rightarrow P_0 = \frac{m\sqrt{T_0 R}}{A^* \times 0.3931}$$

$$= \frac{287\times\sqrt{360}\times 520}{0.2\times 0.3931} = 50 \text{ bar}$$

39. During steady pull out balancing the forces in the radial direction, we obtain

$$n = \frac{v^2}{Rg} - \cos\theta,$$

where $\frac{L}{W} = n$ and $w = mg$

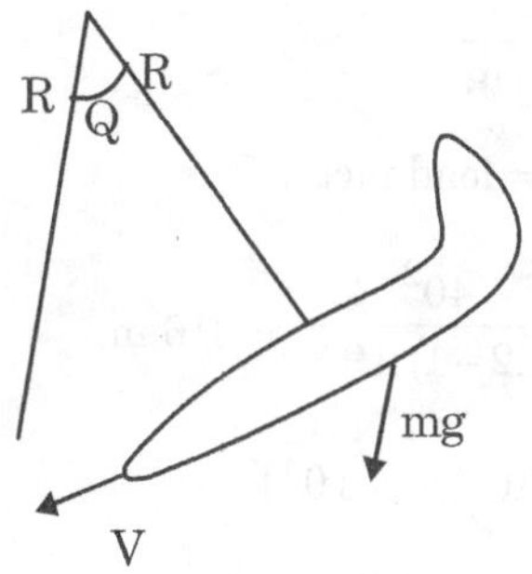

at lowest point of pull out, $\theta = 0°$

$$v = \sqrt{(n-1)gR}$$

$$= \sqrt{(9-1) \times 9.8 \times 200}$$

$$= \sqrt{8 \times 9.8 \times 200}$$

= 125.21 m/sec

40. $g/h_{=11\ km} = g_e\left(\frac{r}{R}\right)^2$

$$= g_e\left(\frac{R+h}{R}\right)^2$$

$$= 9.8\left(\frac{6400+11}{6400}\right)^2$$

$$= 9.8\left(\frac{6411}{64\ m}\right)^2 = 9.8\left(\frac{6411}{6400}\right)^2 = 9.833\ \text{m/sec}^2.$$

Lanchester formula for frequency of phugoid oscillation

$$W_n = \frac{\sqrt{2}g}{u_0}$$

$$= \sqrt{2}\frac{9.833}{110}$$

$= 0.126\ \text{sec}^{-1}$.

damping ratio of phugoid mode is given by

$$\zeta = \frac{-\frac{X_u}{m}}{2W_n}$$

$$-\frac{-(0.0446)}{2 \times 0.126}$$

$= 0.1769$

$= 0.177$

$= 0.18$

41. $a = \dfrac{a_\infty}{1 + \dfrac{a_\infty}{\pi e A\ R}}$

e = 1 for elliptical wing

$$= \frac{a_\infty}{1 + \frac{a_\infty}{\pi A\ R}}$$

$$= \frac{a_\infty}{1 + \frac{a_\infty}{\pi \times 8}}$$

$$= \frac{2\pi}{1 + \frac{2\pi}{8\pi}}$$

$$= \frac{2\pi}{1 + \frac{1}{4}} = \frac{2\pi \times 4}{5}$$

$$= \frac{8\pi}{5}$$

$$\alpha = \frac{8\pi}{5}$$

$$\frac{d\,C_L}{d\propto} = \frac{8\pi}{5}$$

$$C_l = \frac{8\pi}{5} \times 6 \times \frac{\pi}{180}$$

$$= \frac{8\pi}{5} \times \frac{\pi}{30}$$

$$= \frac{8\pi^2}{150} = \frac{4\pi^2}{75}$$

for infinite wing

$$a_o = \frac{d\,C_L}{d\,\alpha}$$

$$C_L = a_o\alpha$$

$$= 2\pi \propto$$

$$= 2\pi \times \frac{6\pi}{180} = \frac{\pi^2}{15}$$

$$\Delta C_L = \frac{\pi^2}{15} - \frac{l\pi^2}{75}$$

$$= \frac{\pi^2}{75}$$

percentage charge is lift coefficient

$$= \frac{\frac{\pi^2}{75}}{\frac{\pi^2}{15}} \times 100 = 20\%$$

42. Specific thrust $= \frac{\dot{m}(u_e - u_i)}{\dot{m}}$

$$= u_e - u_i$$

$$= u_i\left(\frac{u_e}{u_i} - 1\right)$$

$$= M_1\sqrt{\gamma RT_i}\left(\sqrt{\frac{T_e}{T_i}} - 1\right)$$

$$= 1.2\sqrt{1.2 \times 287 \times 280}\left(\sqrt{\frac{1160}{280}} - 1\right)$$

= 373 (1.03)

= 386.20 N

43. Radius of cylinder $= \sqrt{\frac{\Gamma}{2\pi v_\infty}}$

According to question

When $\Gamma' = 4\Gamma$,

$$R' = \sqrt{\frac{4\Gamma}{2\pi\, v_\infty}}$$

$$= 2\sqrt{\frac{\Gamma}{2\pi\, v_\infty}} = 2R$$

44. Given v = 40 m/sec,

n =1.2, Qa = 6.6°, a = 0.25 g

Total landing distance

$$= \frac{15 - h_f}{\tan\theta_a} + R\sin\theta_a + \frac{v^2}{2a}$$

where, $h_f = R(1 - \cos\theta_a)$

$$R = \frac{v^2}{(n-1)g}$$

where n = load factor

$$\therefore\ R = \frac{40^2}{(1.2-1)\times 9.8} = 816 \text{ m.}$$

$$\therefore h_f = 816(1 - \cos\theta_a)$$

$$= 816(1 - \cos 6.6) = 5.40\text{m}$$

$\therefore$ Total landing distance

$$= \frac{15 - 5.4}{\tan 6.6°} + 816\sin 6.6° + \frac{40^2}{2 \times 0.25g} = 326.2 \text{ m}$$

landing distance (s) $\propto \frac{1}{\rho}$

$$\therefore\ S_L\rho_5 = S_a\rho_a$$

$$S_a = S_L\frac{\rho_5}{\rho_a} = 326.2 \times \frac{1.2b}{0.9026} = 445.5\text{m}$$

45. $\Gamma_{(y)} = 100\sqrt{1 - \left(\frac{2y}{4}\right)^2}$

Circulation distribution over finite wing is given by

$$\Gamma_{(y)} = \Gamma_0\sqrt{1 - \left(\frac{2y}{b}\right)^2} \quad ....(2)$$

comparing equation (1) and equation (2), we get

$\Gamma_0 = 100$

b = 4

$$\therefore\ \text{Downwash (w)} = \frac{\Gamma_0}{2b}$$

$$= \frac{100}{2\times 4} = \frac{100}{8} = 12.5\text{m/sec}$$

46. $r_{min} = 6800$ Km

$r_{max} = 7600$ Km

velocity incremental required is lower orbit

$$= \sqrt{\frac{2\mu\, r_{max}}{(r_{max}+r_{min})r_{min}}} - \sqrt{\frac{\mu}{r_{min}}}$$

$$= \sqrt{\frac{\mu}{r_{min}}}\left[\sqrt{\frac{2\, r_{max}}{r_{max}+r_{min}}} - 1\right]$$

$$= \sqrt{\frac{4\times10^{14}}{6.8\times10^{6}}}\left[\sqrt{\frac{2\times7600}{1\times400}} - 1\right]$$

$$= \sqrt{\frac{4\times10^{8}}{6.8}}[1.0274-1]$$

$= 7.6 \times 10^3 \times 0.02740$

$= 208$ m/sec

47. Given $T_{o_1} = 260$ K

$T_{o_2} = 700$ K

$\eta_C = 0.88$

$\gamma = 1.35$

∴ Temperature rise across the compressor is given by

$$T_{o_2} - T_{o_1} = \frac{T_{o_1}}{\eta_C}\left(r_p^{\frac{\gamma-1}{\gamma}} - 1\right)$$

$$T_{oo} - 260 = \frac{260}{0.88}\left(r_p^{\,0.285} - 1\right)$$

$$r_p^{\,0.285} = \frac{(T00-260)\times0.88}{260} + 1$$

$= 2.18$

$r_p = (2.48)^{3.5}$

$r_p = 24.$

∴ pressure rating of compressor = 24

48. Down ward defection at C due to u.d.l on length AB is given by.

$$y = \frac{Wl_1^4}{8\,EI} + \frac{Wl_1^{\,3}}{6EI}(L-l_1)$$

$$= \frac{1\times4^4}{8EI} + \frac{1\pi4^3}{6EI}\times(6-4)$$

$$= \frac{160}{3EI} \qquad ...(i)$$

upward deflection at point C due to prop reaction

$$= \frac{Pl^3}{3EI} = \frac{72P}{EI} \qquad ...(ii)$$

Equating deflection from equation (1) and equation (ii) we get

$P = 0.791$ KN

49. moment of Inertia

$$(I) = \frac{25\times25^3}{12}$$

$= 3.255 \times 10^4$ mm^4. $= 3.255 \times 10^{-8}$ m^4.

Due to mass attached at end of cantilever

$$\text{Deflection } (\delta) = \frac{Wl^3}{3EI}$$

$$= \frac{20\times9.81\times1^3}{3\times2\times10^{11}\times3.255\times10^{-8}}$$

$= 0.010045$M

$= 10$ mm

Natural frequency of vibration

$$\omega_n = \sqrt{K/m} = \sqrt{\frac{g}{\delta}}$$

$$= \sqrt{\frac{9.81}{0.010045}} = 31.25 \text{ rad/sec}$$

∴ damping constant $= 2\sqrt{Km}$

$= 2m\, W_n = 2 \times 20 \times 31.25 = 1250$ N-sec/m

50. m = 10 kg

K = 1000 N/m

C = 300 N sec/m

$$W_n = \sqrt{K/m} = \sqrt{\frac{1000}{10}} = 10\text{rad/se}$$

$$\zeta = \frac{C}{C_C} = \frac{C}{2mW}$$

$$= \frac{C}{2m\sqrt{K/m}}$$

$$= \frac{C}{2\sqrt{KM}} = \frac{300}{2\sqrt{10\times100}} = \frac{300}{2\times100} = 1.5$$

$\therefore P_1 = \zeta W_n + W_n\sqrt{\zeta^2 - 1}$

$= -1.5 \times 10 + 10\sqrt{1.5^2 - 1}$

$= -15 + 11.18$

$= -3.18$

$P_2 = \xi W_n - W_n\sqrt{\xi^2 - 1}$

$= -1.5 \times 10 - 10\sqrt{1.5^2 - 1}$

Solution of the system can be written as

$x(t) = A_1 e^{P_1 t} + A_2 e^{P_2 t}$

$= A_1 e^{-3.18t} + A_2 e^{-26.18t}$

$\boxed{x(t) = A_1 e^{-3.18t} + A_2 e^{-26.18t}}$

51. Given

$\zeta = 0.02$

$\therefore$ logarithmic decrement $(\delta) = \dfrac{2\pi\zeta}{\sqrt{1-\zeta^2}}$

$= \dfrac{2\pi \times 0.02}{\sqrt{1-0.02^2}} = 0.1256$

$\boxed{\delta = 0.1256}$

specific energy loss $\left(\dfrac{\Delta V}{V_i}\right)$

$= 1 - e^{-2\delta} = 1 - e^{-2\times 0.1256} = 0.22$

52. (iii) W = 30 KN

$S = 30\ m^2$

$C_{L\,max} = 2.5$

$C_D = C_{Do} + KC_L^2$

$C_{Do} = 0.015$

$K = 0.025$

Velocity for maximum range for jet propelled aircraft occurs at velocity for maximum

$$\left(\frac{C_L^{\frac{1}{2}}}{C_D}\right)_{max}$$

$$\therefore V_{\left(\frac{C_L^{1/2}}{C_D}\right)_{max}} = \left(\frac{2}{e_\infty}\sqrt{\frac{3K}{C_{Do}}} \times \frac{W}{S}\right)^{\frac{1}{2}}$$

$$= \left(\frac{2}{1.225}\sqrt{\frac{3\times 0.025}{0.015}} \times \frac{30\times 10^3}{30}\right)^{\frac{1}{2}}$$

$$= \left(\frac{2000}{1.225}\sqrt{\frac{30.025}{0.015}}\right)^{\frac{1}{2}}$$

$= 40.40 \times 1.48$

$= 60.41$ m/sec

$= 60$ m/sec

53. $V_{min} = \sqrt{\dfrac{2W}{\rho_\infty S C_{L\,max}}}$

$= \sqrt{\dfrac{2\times 30\times 10^3}{1.225\times 30\times 2.5}}$

$= \sqrt{\dfrac{2000}{1.225\times 2.5}}$

$= 25.55$

$= 26$ m/sec

55.

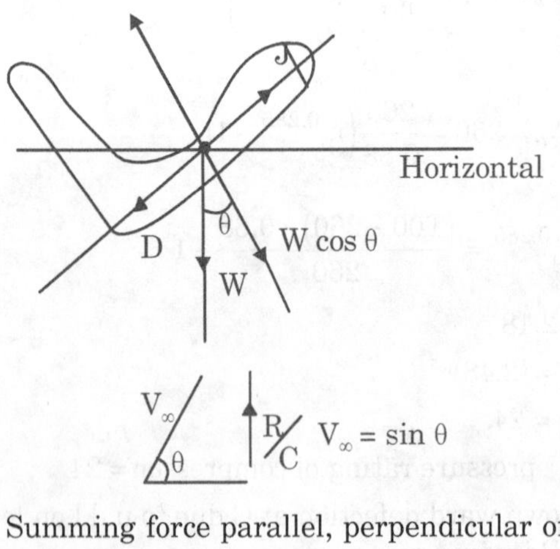

Summing force parallel, perpendicular of flight path

$T = D + W\sin\theta$ ...(1)

$L = W\cos\theta$ ...(2)

multiplying equations (1) by V_∞, we get

$TV_\infty = DV_\infty + WV_\infty \sin\theta$

$$\frac{TV_\infty - DV_\infty}{W} = V_\infty \sin\theta \quad ...(3)$$

$V_\infty \sin\theta$ is airplane vertical velocity called as rate of climb $\left(R/C\right)$

$$\left(R/C\right) = \frac{TV_\infty - DV_\infty}{W}$$

$$= \frac{(T - D)V_\infty}{W}$$

$$= \frac{(\text{Excess thrust})V_\infty}{W} \quad ...(4)$$

$$\therefore \; R/C = V_\infty \sin\theta$$

= 100 sin 10°

= 17.36 m/sec

from equation (4)

$$17.36 = \frac{(\text{Excess thrust}) \times 100}{15000}$$

$$\Rightarrow \text{Excess Thrust} = \frac{17.36 \times 15000}{100}$$

= 2604 N

= 2.604 KN

56. 'Memento' means an object kept as a reminder of a person or event.

57. The sound that frogs make is known as 'croaking'.

58. 'Educe' means to infer.

59. $(66 \,\square\, 6) \longrightarrow (66 \,\lozenge\, 6)$

$$\Rightarrow \left(\frac{66-6}{66+6}\right)\left(\frac{66+6}{66-6}\right) = 1$$

60. $\log_x\left(\frac{5}{7}\right) = -\frac{1}{3}$

$$\log\frac{5}{7} = (\log x)\left(-\frac{1}{3}\right)$$

$$\log\frac{5}{7} = \log x^{\frac{-1}{3}}$$

$$\Rightarrow \left(\frac{5}{7}\right)^{-3} = x$$

$$\Rightarrow \boxed{\frac{343}{125} = x}$$

61. Only A is written correctly. In B 'rank' cause subject-verb disagreement; in C there is tense error.

62. Only B can follow from the given paragraph.

63. Middle value first row $= \frac{6+4}{2} = 5$

Middle value of 2nd row $= \frac{7+4+2+1}{2} = 7$

Similarly, for last row $= \frac{3+3}{2} = 3$ is the missing value

64. Small cubes are of 1 unit, therefore,

total cubes = 9 × 3 = 27

total faces = 27 × 6 = 162

total faces visible = 9 × 6 = 54

faces that are not visible = 162 – 54 = 108

$$\therefore \text{Required Ratio} = \frac{54}{108} = \frac{1}{2}$$

65. Only B follows logically from the given premises.

MOCK TEST – 9

AEROSPACE ENGINEERING

Time Allowed: 3:00 Hours *Maximum Marks: 100*

Read the following instructions carefully before you begin to answer the questions. This Booklet contains questions in English.

INSTRUCTIONS

1. This Booklet contains **65 questions** in total comprising of the following two parts:

   Part -1: Technical Section (85 Marks) *(1-55 Questions)*

   Part -2: General Aptitude Section (15 Marks) *(56-65 Questions)*

2. All questions are compulsory. Q.1-Q.25 & Q.56-Q.60 carry one mark each. Q.26-Q.55 & Q.61-Q.65 carry two marks each.

3. **Types of Questions:**

   (*a*) Multiple Choice Questions(MCQ) carrying 1 or 2 marks each in all papers and sections. These questions are objective in nature, and each will have a choice of four options, out of which the candidate has to mark the correct answer(s).

   (*b*) Numerical Answer Questions of 1 or 2 marks each in all sections. For these questions the answer is a real number, to be entered by the candidate using the virtual keypad. No choices will be shown for these type of questions.

4. For 1-mark multiple-choice questions, 1/3 marks will be deducted for a wrong answer. Likewise, for 2-marks multiple-choice questions, 2/3 marks will be deducted for a wrong answer. There is no negative marking for numerical answer type questions.

MOCK TEST – 9

1. A uniformly simply supported beam of span (l) carries a uniformly distributed load w per unit length over the whole span, the downward deflection at centre will be

   (*a*) $\frac{wl^3}{8EI}$ (*b*) $\frac{wl^3}{3EI}$

   (*c*) $\frac{5wl^3}{384EI}$ (*d*) $\frac{wl^3}{48EI}$

2. When stress (s) and young's modulus of elasticity (E) remains constant, the energy absorbing capacity of part subjected to dynamic forces, is a function of its

   (*a*) length (*b*) cross-section

   (*c*) volume (*d*) none of these

3. The Torque transmitted by shaft of diameter 40 mm if the shear stress is not to exceed 400 N/cm^2, would be

   (*a*) 1.6 × p N – m (*b*) 16 × p N – m

   (*c*) 0.8 × p N – m (*d*) 0.4 × p N – m

4. The ratio of circumferential stress to longitudinal stress for a thin cylindrical shell will be ________

5. Shear centre is the point (in or outside a section) through which the applied shear force produces

   (*a*) Infinite Torsion of the member

   (*b*) Finite Torsion of the member

   (*c*) no Torsion, no twisting of the member

   (*d*) Bending and finite Torsion of the member.

6. The beam shown in figure is

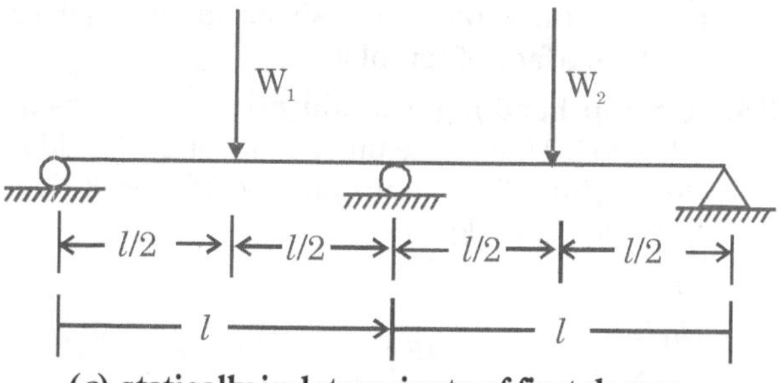

   (*a*) statically indeterminate of first degree

   (*b*) statically indeterminate of infinite degree

   (*c*) statically indeterminate of second degree

   (*d*) statically determinate

7. Work done factor of a axial flow compressor varies with number of stages?

   (*a*) Increasing with increasing number of stages

   (*b*) Decreasing with increasing number of stages.

   (*c*) Decreasing with decreasing number of stages.

   (*d*) Independent of number of stages.

8. For propeller driven aircraft, maximum range occurs at which of the following conditions?

   (*a*) Max $\frac{C_L}{C_D}$ (*b*) Max $\frac{C_L^{\frac{3}{2}}}{C_D}$

   (*c*) Max $\frac{C_L^{\frac{1}{2}}}{C_D}$ (*d*) None of these

9. Which of the following engines used in a aircraft has maximum efficiency at low supersonic speeds.

   (*a*) Ramjet (*b*) Scramjet

   (*c*) Turbofan (*d*) Turboprop

10. Arrange the component of conventional turbojet engine in order.

    (*a*) Inlet - High pressure compressor - Low pressure compressor-combustor-high pressure turbine - low pressure turbine - Afterburner - Nozzle

    (*b*) Inlet - Low pressure compressor - high pressure compressor - combustor - high pressure turbine - low pressure turbine - afterburner - nozzle

    (*c*) Inlet - low pressure compressor - high pressure - combustor - low pressure turbine - high pressure turbine - Afterburners - Nozzle

    (*d*) None of these

11. A rocket thrust chamber contains a propellant of molecular weight 35 and specific heat ratio γ = 1.2, the value of specific heat at constant pressure (C_P) is ________ J/Kg-k (assume universal gas constant = 8314 J/kg-K)

12. Nozzle used in the rocket motor at mean sea level condition is?

    (*a*) Over expanded

    (*b*) optimum expanded

    (*c*) under expanded

(*d*) may be over expanded or optimum expanded

13. Which of the following remains approximately constant for all stages of an axial compressor?

(*a*) Average air velocity

(*b*) relative air velocity

(*c*) pressure & temperature

(*d*) temperature & density

14. Select the correct statement about impulse stage of an axial turbine

(*a*) Half of the pressure drop occurs in rotor and half in stator

(*b*) The entire pressor drop occurs in rotor

(*c*) The entire pressure drop occurs in stator

(*d*) Pressure remains constant in the stage

15. The system of linear equations

$4x + 2y = 7$

$2x + y = 6$

(*a*) A unique solution

(*b*) No solution

(*c*) Infinite number of solution

(*d*) Exactly two solutions

16. The number of linearly independent eigen vectors $\begin{bmatrix} 2 & 1 \\ 0 & 2 \end{bmatrix}$ is

(*a*) 0 (*b*) 1

(*c*) 2 (*d*) Infinite

17. The Blasius equation $\frac{d^3f}{dx^3} + \frac{f}{2}\frac{d^2f}{dx^2} = 0$ is a

(*a*) Second order non-linear ordinary differential equation

(*b*) Third order non-linear ordinary differential equation

(*c*) Third order linear ordinary differential equation

(*d*) Mixed ordinary non-linear ordinary differential equation.

18. The function $y = (A + Bx)\, e^{3x}$ is solution of differential equation

(*a*) $\frac{d^2y}{dx^2} + 6\frac{dy}{dx} + 9y = 0$

(*b*) $\frac{d^2y}{dx^2} - 6\frac{dy}{dx} + 9y = 0$

(*c*) $\frac{d^2y}{dx^2} - 6\frac{dy}{dx} - 9y = 0$

(*d*) $\frac{d^2y}{dx^2} + 9\frac{dy}{dx} - 6y = 0$

19. Water (Prandtl number = 6) flows over a flat plate, which is heated over a entire length. Which one of the following relationships between the hydrodynamic boundary layer thickness (d) and thermal boundary layer thickness (δ_T) is true?

(*a*) $d_T > d$ (*b*) $d_T < d$

(*c*) $d_T = d$ (*d*) can't predicted

20. For compressible fluid, sonic velocity is

(*a*) property of fluid

(*b*) always given by $(\gamma RT)^{\frac{1}{2}}$, where γ, R, T are respectively ratio of specific heat gas constant and temperature in Kelvin

(*c*) Always given by $\left(\frac{dP}{d\rho}\right)^{\frac{1}{2}}$, where P, r and S are respectively pressure, density and entropy.

(*d*) always greater than the velocity of fluid at any location.

21. For the continuity equation given $\nabla.\overline{V} = 0$ to be valid, where $\overline{V}$ is the velocity vector. Which one of the following is a necessary condition?

(*a*) Steady flow (*b*) Irrotational flow

(*c*) Inviscid flow (*d*) Incompressible flow

22. For incompressible fluid (kinematic viscosity = 7.4×10^{-7} m²/s, specific gravity = 0.88) is held between two parallel plates. It the top plate is moved with velocity of 0.5 m/sec while the bottom one is held stationary, the fluid attains the linear velocity profile in the gap of 0.5 mm between these plates the shear stress in pascal on the surface of top plate is ________

23. A pump handling a liquid raises its pressure 1 bar to 30 bar. Take the density of the liquid as 900 kg/m³. The Isentropic specific work done by pump in kJ/kg is

(*a*) 0.1

(*b*) 0.3

(*c*) 0.25

(*d*) 2.93

24. The maximum velocity of a one dimensional incompressible fully developed viscous flow, between two fixed parallel plates is 6 m/sec. The mean velocity (in m/sec) of the mean velocity (in m/sec) of the flow is ________

25. The primary function of fin in vertical tail of the aircraft is to provide

(*a*) yaw control (*b*) yaw stability

(*c*) Roll damping (*d*) Roll stability

26. Satellites used for telecommunication relay are kept in a geostationary orbit. A satellite is said to be in such an orbit when

a. Orbit is geosynchronous

b. The orbit is circular

c. The orbit lies in plane of earth's equator

d. The orbit is at an altitude of 22236 KM. Select the correct answer using the codes given below.

(*a*) a, b, and c (*b*) a, b and d

(*c*) b and d (*d*) a, b, c and d

27. A satellite is revolving around earth in 400 km × 1200 km elliptical orbit, the time period of satellite in second is ________

28. A satellite two stage rocket with initial mass of 16000 kg, having mass of structure 2100 kg carrying a payload of 1000 kg, is fired in vertical trajectory from surface of the earth. Both the stage of rocket have same specific impulse of 300 sec and same payload ratio of 0.066. The acceleration due to gravity is 9.8 m/sec². Neglecting drag and gravity effects and considering both the stages of with same payload ratio the terminal velocity attained by payload in m/sec is ________

29. The contribution of horizontal tail to the pitching moment coefficient about the centre of gravity $\left(C_{m_{CG}}\right)$ of an aircraft is given by $C_{m_{tail}} = 0.3 + 0.0115\,a$. Where a is the angle of attack of the aircraft. The contribution of tail to the aircraft longitudinal stability

(*a*) is stabilizing

(*b*) is destabilizing

(*c*) is nil

(*d*) cannot be determined from the given information

30. The trim curves of an aircraft are of the from $C_m = 0.04 - 0.18 + 0.2C_L$. Where elevator deflection angle, d in a radians the static margin of the aircraft is ________.

31. A pilot probe on a aircraft in a steady level flight records a pressure of 55,000 N/m². The static pressure and density are 45,280 N/m² and 0.6 Kg/m³ respectively equivalent airspeed of aircraft in m/sec is ________.

32. For air flow over a flat plate, velocity (u) and boundary layer thickness (d) can be expressed respectively as

$$\frac{u}{u_\infty} = \frac{3}{2}\left(\frac{y}{\delta}\right) - \frac{1}{2}\left(\frac{y}{\delta}\right)^3$$

$\delta = \frac{4.65}{\sqrt{R_{e_x}}}$, if the free stream velocity is 2m/sec, and air has kinematic viscosity of 1.5×10^{-5} m²/sec and density of 1.23 kg/m³. Wall shear stress of x = 1m, is

(*a*) 2.36×10^{2} N/m² (*b*) 43.6×10^{-3} N/m²

(*c*) 4.36×10^{-3} N/m² (*d*) 2.18×10^{-3} N/m²

33. A U-tube mercury manometer with a small quantity of mercury is used to measure to the static pressure difference between two locations A and B in a conical section through which compressible fluid flows. At a particular flow rate, the mercury column appears as shown in figure the density of mercury is 13600 kg/m³ and g = 9.81 m/sec². Which of the following is correct?

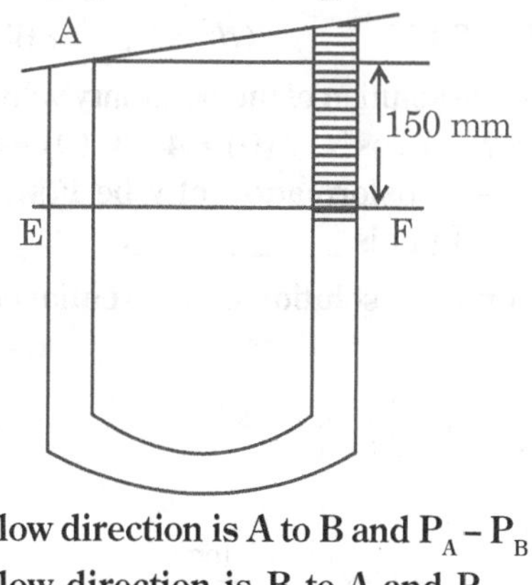

(*a*) Flow direction is A to B and $P_A - P_B$ = 20 kPa

(*b*) Flow direction is B to A and $P_A - P_B$ = 1.4 kPa

(*c*) Flow direction is A to B and $P_B - P_A$ = 20 kPa

(*d*) Flow direction is B to A and $P_B - P_A$ = 1.4 kPa

34. The properties of mercury at 300K and density 13529 kg/m³, specific heat at constant pressure = 0.139 kJ/kg-K, dynamic viscosity = 0.1523×10^{-2} N-sec/m² and thermal conductivity = 8.540 W/m–K. The prandtl number of mercury at 300 K is ______

35. A flow properties upstream of the normal shock wave are 550 m/sec, 100 kPa, and 300 K the characteristics Mach number down steam of shock wave is ________.

36. Which of the following is correct about normal shock wave, When $M_1 \to \infty$, $\gamma = 1.4$

P. $\lim_{M_1 \to \infty} M_2 = 0.378$

Q. $\lim_{M_1 \to \infty} \left(\frac{\rho_2}{\rho_1}\right) = 6$

R. $\lim_{M_1 \to \infty} \frac{P_2}{P_1} = \infty$

S. $M_2^* M_1^* = 1$

Where, subscript 1, 2 denotes the shock wave property upstream and downstream of the shock wave and * denotes characteristics property.

(*a*) only R (*b*) only P

(*c*) P, Q and R (*d*) P, Q, R, and S

37. For prandtl-Meyer expansion wave, the upstream Mach number is 2 and the pressure ratio across the fan is 0.5 the angles of the front and end Mach lines of expansion for relative to free stream is

(*a*) 30°, 128° (*b*) 30°, 12.86°

(*c*) 60°, 245° (*d*) 60°, 12.86°

38. If y be the solution of the boundary value problem y¢¢ + y = 4 cos²x, y (0) = 4, y¢ (0) = 2, Let the Laplace transformation of y be F(s), Then the value of F(1) is __________.

39. The complete solution of differential equation

$$x^2\frac{d^2y}{dx^2} - 2x\frac{dy}{dx} - 4y = x^4 \text{is}$$

(*a*) $y = \frac{C_1}{x} + C_2x^4 + \frac{x^4 \log x}{5}$

(*b*) $y = \frac{C_1}{x} + C_2x^4 - \frac{x^4 \log x}{5}$

(*c*) $y = \frac{C_1}{x} + C_2x^3 - \frac{x^4 \log x}{5}$

(*d*) $y = \frac{C_1}{x} + C_2x^3 - \frac{x^4 \log x}{5}$

40. The value of integral $\oint(xy\,dy - y^2dx)$. Where C is square cut from the first gradient by the lines x = 1 & y = 1 will be (use Green's Theorem to change the integral into double integral)

(*a*) $\frac{1}{2}$ (*b*) 1

(*c*) $\frac{3}{2}$ (*d*) $\frac{5}{3}$

41. Given the orthogonal matrix

$$A = \begin{bmatrix} 1 & 1 & 1 & 1 \\ 1 & 1 & -1 & -1 \\ 1 & -1 & 0 & 0 \\ 0 & 0 & 1 & -1 \end{bmatrix} \text{ then } \left[AA^T\right]^{-1} \text{ is}$$

(*a*) $\begin{bmatrix} \frac{1}{4} & 0 & 0 & 0 \\ 0 & \frac{1}{4} & 0 & 0 \\ 0 & 0 & \frac{1}{4} & 0 \\ 0 & 0 & 0 & \frac{1}{4} \end{bmatrix}$ (*b*) $\begin{bmatrix} \frac{1}{2} & 0 & 0 & 0 \\ 0 & \frac{1}{2} & 0 & 0 \\ 0 & 0 & \frac{1}{2} & 0 \\ 0 & 0 & 0 & \frac{1}{2} \end{bmatrix}$

(*c*) $\begin{bmatrix} 1 & 0 & 0 & 0 \\ 0 & 1 & 0 & 0 \\ 0 & 0 & 1 & 0 \\ 0 & 0 & 0 & 1 \end{bmatrix}$ (*d*) $\begin{bmatrix} \frac{1}{2} & 0 & 0 & 0 \\ 0 & \frac{1}{2} & 0 & 0 \\ 0 & 0 & \frac{1}{4} & 0 \\ 0 & 0 & 0 & \frac{1}{4} \end{bmatrix}$

42. The value of ordinary differential equation $\frac{dy}{dx} = x + y$ and y = 1, when x = 0 the value of y when x = 0.2 using runge - Kutta fourth order method is ______

43. Consider an airplane with acceleration of 8 ft/sec^2. At existing flight condition the specific excess power is 300 ft/sec and instantaneous velocity of 800 ft/sec. Instantaneous maximum rate of climb that can be obtained as these alleviated flight conditions is _______ ft/sec.

44. Neutral point represents

P. Aerodynamic centre of full aircraft

Q. Most aft location of center of gravity for stable aircraft.

R. Location of centre of gravity such that $C_{m.cg}$ does not change with angle of attack.

S. Most forward location of Centre of gravity for stable aircraft.

Which of the above statement is correct.

(*a*) Only S

(*b*) Q, R and S

(*c*) P, Q and R

(*d*) P, Q, R and S

45. A metallic alloy is compressed under plain strain condition as shown in figure

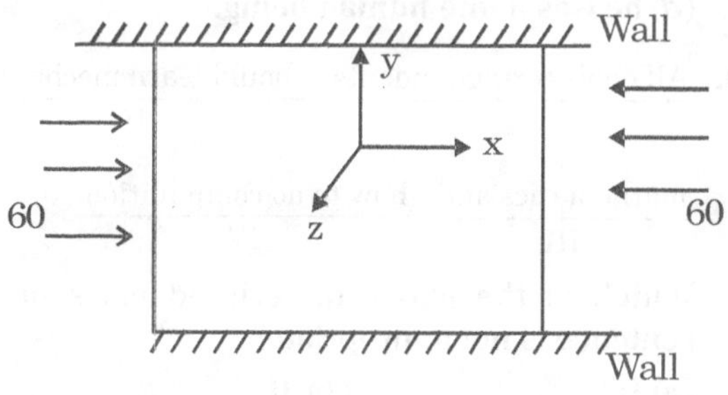

Given s_{xx} = – 60, E = 90 GPa, v = 0.3 and $s_{xx} = K\hat{e}_{xx}$.

Value of K is?

(*a*) 90 GPa

(*b*) 200 GPa

(*c*) 120 GPa

(*d*) 250 GPa

46. In the above question no. 45 metallic allow is compressed under plain stress condition the value of K is

(*a*) 200 GPa

(*b*) 120 GPa

(*c*) 99 GPa

(*d*) 250 GPa

47. In the given below thin walled open and closed section are given having average radius 0.5 and thickness t = 0.05m.

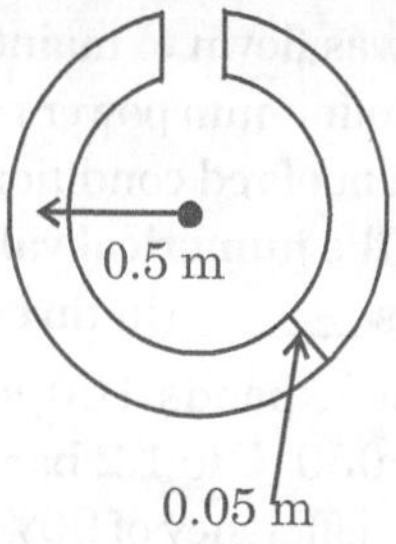

The ratio of torsional constant of open and closed section is ________.

48.

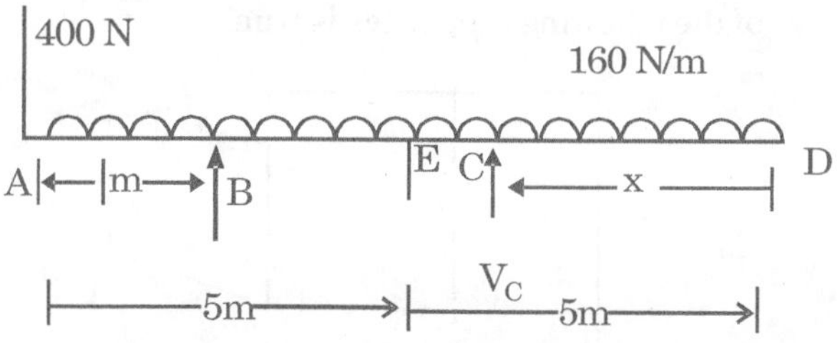

For horizontal beam AD, 10 m long carries a uniformly distributed load of 160 N/m together with concentrated load of 400 N at the left end A. The beam is supported at point B which is 1 m from A and at C which is at right hand, half of the beam and x meters from end D. If mid point is point of contraflexure, the value of x in meter.

(*a*) 3 (*b*) 5

(*c*) 3.7 (*d*) 6

49. During an under damped oscillation of a single degree of freedom system in the time displacement plot the third peak is of magnitude 100 and tenth peak of magnitude 10 the damping ratio is approximately _______.

50. Consider following statement about longitudinal static stability.

P. $C_{M,cg}$ at zero lift must be negative

Q. $C_{M,cg}$ at zero lift must be positive

R. Slope of C_L versus $\propto_\alpha$ must be negative

S. $\dfrac{\partial C_{m,cg}}{\partial \propto_\alpha}$ must be negative ($\propto_\alpha$ is absolute angle of attack) of aircraft is true

(*a*) only Q (*b*) only P

(*c*) Q and S (*d*) P, Q and S

51. A conventional low speed aircraft had the following Aerodynamics characteristics:

$C_{DO} = 0.2$, e = 1 (oswald efficiency)

The aircraft was flown to maintain a steady level flight and for minimum power required condition, at lift coefficient of red condition, at lift coefficient of $C_L = 0.8$, the numerical value of aspect ratio of the wing is ______ (in three decimal places)

52. A gas turbine expands 100 kg/sec of air from 12 bar and 1050° C to 1.2 bar adiabatically with an isentropic efficiency of 90% the power output of engine is ______ hp (take, $\gamma = 1.4$, C_P = 1005 J/kgK)

53. For ideal ramjet engine as shown in figure which of the following properties is true?

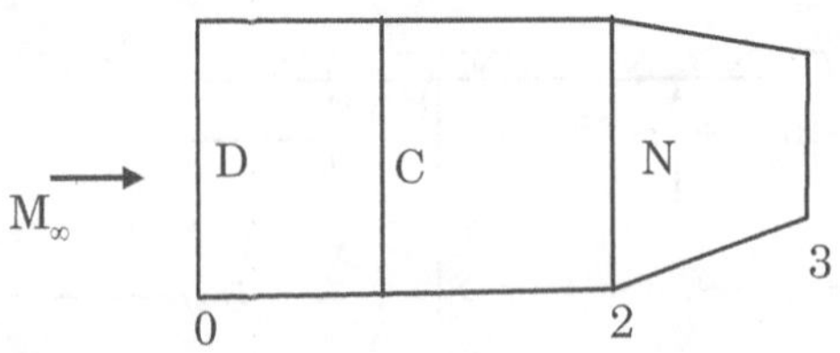

D - Diffuser

C - Combustion chamber

N - Nozzle

P. $\left(T_{0_1} = T_{0_2}\right) \neq \left(T_{0_1} = T_{0_\infty}\right)$

Q. $T_{0_3} = T_{0_2} = T_{0_1} = T_{0_\infty}$

R. $P_{0_3} = P_{0_2} \neq P_{0_1} = P_{0_\infty}$

S. $P_{0_3} = P_{0_2} = P_{0_1} = P_{0_\infty}$

(*a*) only Q (*b*) P and R

(*c*) P and S (*d*) Q and S

54. Consider a rocket engine burning hydrogen and oxygen the combustion chamber pressure and temperature are 25 atm and 3517 K respectively. The area of rocket Nozzle throat is 0.1 m². The area of exit is designed so that exit pressure exactly equals to ambient pressure at the standard altitude of 30 km. For this mixture assume that γ = 1.22 and molecular weight $\overline{M} = 16$. at the standard 30 km altitude. Specific impulse of the rocket engine is?

(*a*) 400 sec

(*b*) 320 sec

(*c*) 500 sec

(*d*) 450 sec

55. A rocket engine using hydrogen - oxygen as the fuel-oxidizer combination has a specific impulse of 360 sec. The ratio of propellant mass of initial mass required to achieve a burnout equal to escape velocity from the earth surface is ______.

56. If 3 £ X £ 5 and 8 £ Y £ 11 then which of the following options is TRUE?

(*a*) $\frac{3}{5} \le \frac{X}{Y} \le \frac{8}{5}$ (*b*) $\frac{3}{11} \le \frac{X}{Y} \le \frac{5}{8}$

(*c*) $\frac{3}{11} \le \frac{X}{Y} \le \frac{8}{5}$ (*d*) $\frac{3}{5} \le \frac{X}{Y} \le \frac{8}{11}$

57. The Headmaster __________ to speak to you. Which of the following options is incorrect to complete the above sentence?

(*a*) is wanting (*b*) wants

(*c*) want (*d*) was wanting

58. Mahatama Gandhi was known for his humility as

(*a*) he played an important role in humiliating exit of British from India.

(*b*) he worked for humanitarian causes.

(*c*) he displayed modesty in his interactions.

(*d*) he was a fine human being.

59. All engineering students (I) should learn mechanics, (II) mathematics and (III) how to do computation. (IV)

Which of the above underlined parts of the sentence is not appropriate?

(*a*) I (*b*) II

(*c*) III (*d*) IV

60. Select the pair that best expresses a relationship similar to that expressed in the pair :

water : pipe::

(*a*) cart : road

(*b*) electricity : wire

(*c*) sea : beach

(*d*) music : instrument

61. Velocity of an object fired directly in upward direction is given by V = 80 – 32t, where t (time) is in seconds. When will the velocity be between 32 m/sec and 64 m/sec ?

(*a*) (1, 3/2)

(*b*) (1/2, 1)

(*c*) (1/2, 3/2)

(*d*) (1, 3)

62. In a factory, two machines M1 and M2 manufacture 60% and 40% of the autocomponents respectively. Out of the total production, 2% of M1 and 3% of M2 are found to be defective. If a randomly drawn autocomponent from the combined lot is found defective, what is the probability that it was manufactured by M2 ?

(*a*) 0.35 (*b*) 0.45

(*c*) 0.5 (*d*) 0.4

63. Following table gives data on tourists from different countries visiting India in the year 2011.

Country	Number of Tourists
USA	2000
England	3500
Germany	1200
Italy	1100
Japan	2400
Australia	2300
France	1000

Which two countries contributed to the one third of the total number of tourists who visited India in 2011 ?

(*a*) USA and Japan

(*b*) USA and Australia

(*c*) England and France

(*d*) Japan and Australia

64. If $| -2X + 9 | = 3$ then the possible value of $| -X | - X^2$ would be :

(*a*) 30 (*b*) – 30

(*c*) – 42 (*d*) 42

65. All professors are researchers

Some scientists are professors

Which of the given conclusions is logically valid and is inferred from the above arguments :

(*a*) All scientists are researchers

(*b*) All professors are scientists

(*c*) Some researchers are scientists

(*d*) No conclusion follows

OMR SHEET

1.	(A) (B) (C) (D)
2.	(A) (B) (C) (D)
3.	(A) (B) (C) (D)
4.	
5.	(A) (B) (C) (D)
6.	(A) (B) (C) (D)
7.	(A) (B) (C) (D)
8.	(A) (B) (C) (D)
9.	(A) (B) (C) (D)
10.	(A) (B) (C) (D)
11.	
12.	(A) (B) (C) (D)
13.	(A) (B) (C) (D)
14.	(A) (B) (C) (D)
15.	(A) (B) (C) (D)
16.	(A) (B) (C) (D)
17.	(A) (B) (C) (D)
18.	(A) (B) (C) (D)
19.	(A) (B) (C) (D)
20.	(A) (B) (C) (D)
21.	(A) (B) (C) (D)
22.	(A) (B) (C) (D)
23.	(A) (B) (C) (D)
24.	
25.	(A) (B) (C) (D)
26.	(A) (B) (C) (D)
27.	
28.	
29.	(A) (B) (C) (D)
30.	
31.	
32.	(A) (B) (C) (D)
33.	(A) (B) (C) (D)
34.	
35.	
36.	(A) (B) (C) (D)
37.	(A) (B) (C) (D)
38.	
39.	(A) (B) (C) (D)
40.	(A) (B) (C) (D)
41.	(A) (B) (C) (D)
42.	
43.	
44.	(A) (B) (C) (D)
45.	(A) (B) (C) (D)
46.	(A) (B) (C) (D)
47.	
48.	(A) (B) (C) (D)
49.	
50.	(A) (B) (C) (D)
51.	
52.	
53.	(A) (B) (C) (D)
54.	(A) (B) (C) (D)
55.	
56.	(A) (B) (C) (D)
57.	(A) (B) (C) (D)
58.	(A) (B) (C) (D)
59.	(A) (B) (C) (D)
60.	(A) (B) (C) (D)
61.	(A) (B) (C) (D)
62.	(A) (B) (C) (D)
63.	(A) (B) (C) (D)
64.	(A) (B) (C) (D)
65.	(A) (B) (C) (D)

ANSWERS

1. (*c*) **2.** (*c*) **3.** (*a*) **4.** (2) **5.** (*c*) **6.** (*a*) **7.** (*b*) **8.** (*a*)

9. (*c*) **10.** (*b*) **11.** (1428.J/kg – K) **12.** (*a*) **13.** (*a*) **14.** (*c*) **15.** (*b*)

16. (*b*) **17.** (*b*) **18.** (*b*) **19.** (*b*) **20.** (*b*) **21.** (*d*) **22.** (*a*) **23.** (*d*)

24. (4 m/sec) **25.** (*b*) **26.** (*a*) **27.** 6077 sec. **28.** (9666 m/sec.) **29.** (*b*)

30. (– 0.2) **31.** (Vequivalent = 89 m/sec) **32.** (*c*) **33.** (*a*) **34.** (0.0248)

35. ($M_2^* = 0.70$) **36.** (*d*) **37.** (*d*) **38.** 3.4 **39.** (*a*) **40.** (*c*) **41.** (*c*)

42. (1.24266) **43.** (101 ft/sec.) **44.** (*c*) **45.** (*c*) **46.** (*c*)

47. (0.003312) **48.** (*a*) **49.** (0.050) **50.** (*d*) **51.** (0.33) **52.** (772) **53.** (*c*)

54. (*a*) **55.** (24) **56.** (*b*) **57.** (*b*) **58.** (*c*) **59.** (*d*) **60.** (*b*) **61.** (*c*)

62. (*c*) **63.** (*c*) **64.** (*b*) **65.** (*c*)

EXPLANATIONS

1. (c) Cantilever beam

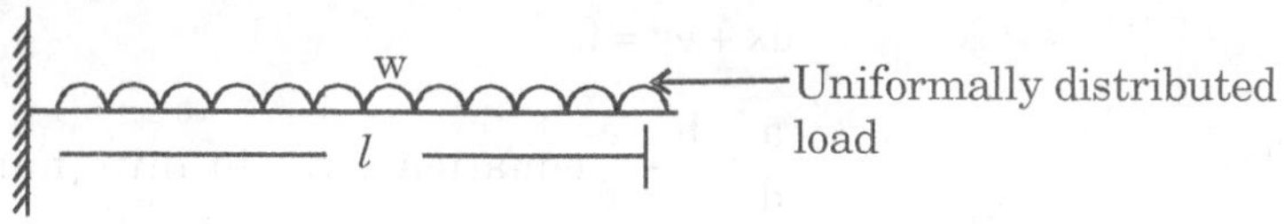

$$\delta = \frac{Wl^3}{8EI}$$

(b)

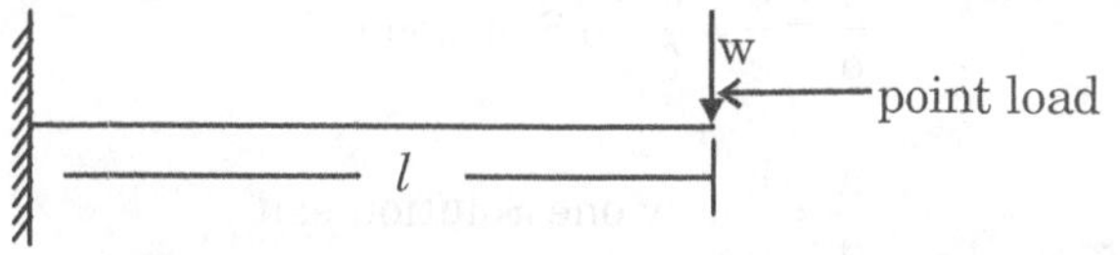

$$\delta = \frac{Wl^3}{3EI}$$

Simply Supported beam

(a)

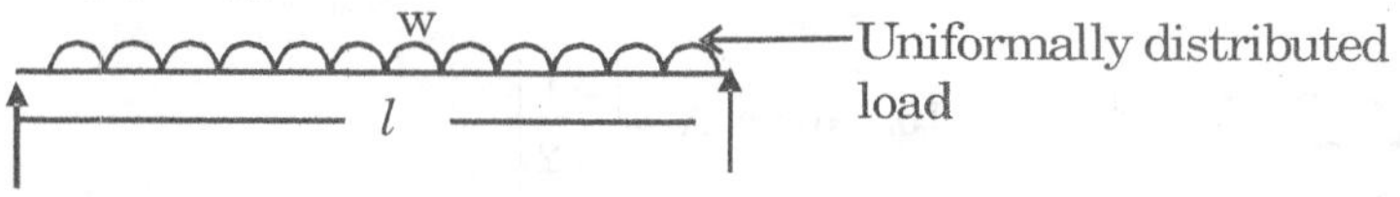

$$\delta = \frac{5Wl^3}{384EI}$$

(b)

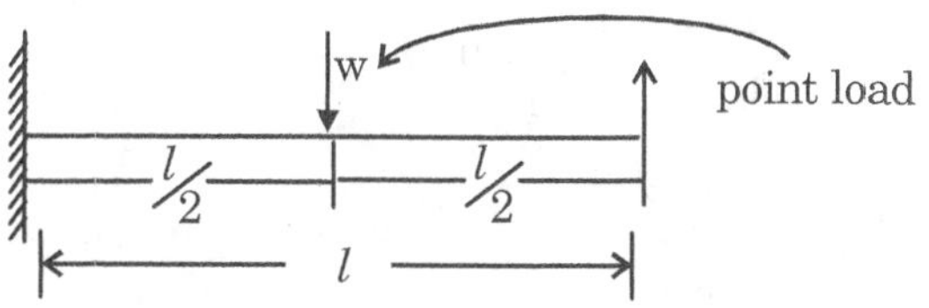

$$\delta = \frac{Wl^3}{48EI}$$

2. Energy absorbed by a part subject to dynamic force is given by $V = \frac{\sigma^2}{2E} \times \text{Volume}$

since σ, E are constants
$V \propto \text{Volume}$

3. $\frac{\tau}{r} = \frac{T}{J} = \frac{C\theta}{l}$

$\frac{\tau}{r} = \frac{T}{J}$

$\tau = \frac{TJ}{r}$

$= \frac{T \times \pi d^3}{32}$

$= \frac{400 \times (10^{-2})^2 \times \pi \times (40 \times 10^{-3})^3}{32}$

$= 1.6 \times \pi \text{ N} - \text{m}$

4. For thin cylindrical shell

$\sigma_h = \frac{pd}{2t}, \sigma_l = \frac{pd}{4t}$

$\frac{\sigma_h}{\sigma_l} = \frac{pd}{2t} = 2$

5. By definition of shear centre

6.

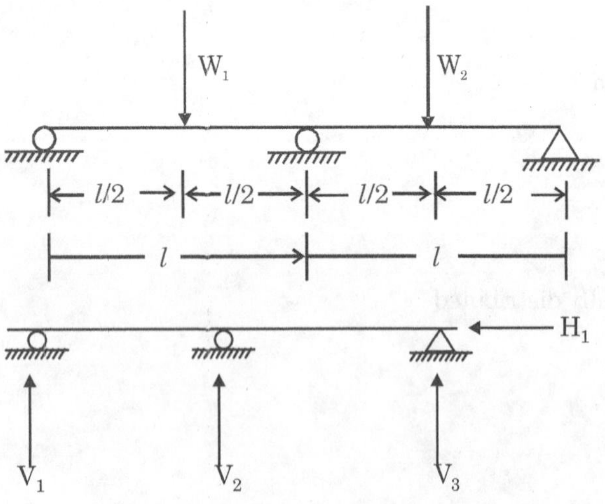

Here, there are 4 joints force variables which can be, Calculated by applying 3 equilibrium equations. So degree of Static indeterminacy
= 4 – 3 = 1

9. Ramjet & Scramjet is used for high super sonic or greater than the velocity of supersonic. Turboprop is used to subsonic speed.

11. M = 35

$\bar{R} = \frac{R}{M} = \frac{8314}{35} = 238 \text{J/kg} - \text{K}$

$C_p = \frac{\gamma R}{\gamma - 1} = \frac{1.2 \times 238}{0.2}$

$= 1428.\text{J/kg} - \text{K}$

12. $P_{atm} > P_e$ Over expanded
$P_{atm} = P_e$ Optimum expanded
$P_{atm} < P_e$ under expanded
P_{atm} = Atmospheric pressure
P_e = pressure at exit.

13. For all Stages of axial Compressor, average air velocity remains constant. Temperature pressure, density changes.

14. For Impulse turbine

Degree of reaction is zero. Entire pressure drop occurs only across the Nozzle vanes called as stator.

15. 4x + 2y = 7
2x + y = 6
For ax + by = c
dx + ey = f

$\frac{a}{d} = \frac{b}{e} = \frac{c}{f}$ equation have infinite number of solution

$\frac{a}{d} = \frac{b}{e} \neq \frac{c}{f}$ no Solution exit.

$\frac{a}{d} \neq \frac{b}{e}$ only one Solution exit.

here $\frac{4}{2} = \frac{2}{1} \neq \frac{7}{6}$ no Solution.

16. Given $A = \begin{bmatrix} 2 & 1 \\ 0 & 2 \end{bmatrix}$

Characteristics equation $|A - \lambda I| = 0$

$\begin{bmatrix} 2-\lambda & 1 \\ 0 & 2-\lambda \end{bmatrix} = 0$

$[2 - \lambda]^2 = 0$

$\lambda = 2, 2$

Now, Consider eigen value problem

$[A - \lambda I]x = 0$

$$\begin{bmatrix} 0 & 1 \\ 0 & 0 \end{bmatrix}\begin{bmatrix} x_1 \\ x_2 \end{bmatrix} = 0$$

$x_2 = 0$

The Solution is therefore. $x_2 = 0$, x_1 = anything therefore it has only one independent eigen vector.

17. The Blasius equations

$$\left[\frac{\partial^3 f}{\partial x^5} + \frac{f}{2}\frac{\partial^2 f}{\partial x^2} = 0\right]$$

The equation is 3rd order because highest differential is $\frac{d^3 f}{dx^3}$ and it is non – linear

Since the product $f \times \frac{d^2 f}{dx^2}$ is not allowed to be linear differential equation.

The equation is ordinary differential equation Since differential coefficient with respect to single independent Variable

18. Since $y = (A + Bx)e^x$ is solution of differential equation and this form is given when roots are equal.

Now, A.E of 2nd option

$(\Delta^2 - 6\Delta + 9) = 0$

$(\Delta - 3)^2 = 0$

$\Delta = 3, 3$

$y = (A + Bx)^{3x}$

19. $\frac{\delta}{\delta_T} = (Pr)^n$

where $Pr = \frac{\mu C_p}{K}$

$\mu = \Delta$ viscosity coefficient

C_p = Specific heat

K = Thermal Conductivity

here Pr = 6 > 1

$\therefore \frac{\delta}{\delta_T} > 1$

$\delta > \delta_T$

20. For Compressible fluid sonic velocity is given by

$V = \sqrt{\gamma RT}$

$$\gamma = \frac{C_p}{C_v} = \frac{\text{Specific heat at Constant pressure}}{\text{Specific heat at Constant Volume}}$$

R = Universal gas Constant

T = absolute temperature in Kelvin.

21. Continuity equation

$\nabla . \vec{V} = 0$

$$\frac{\partial u}{\partial x} + \frac{\partial v}{\partial y} + \frac{\partial w}{\partial z} = 0$$

multiplying by density on the both side, we get

$$\rho\left(\frac{\partial u}{\partial x} + \frac{\partial v}{\partial y} + \frac{\partial w}{\partial z}\right) = 0$$

This is the equation for incompressible flow.

22. Kinematic viscosity (ν) = 7.4×10^{-7} m^1/sec

$\therefore$ dynamic viscosity (μ) = $\nu\rho$

$= 7.4 \times 10^{-7} \times 0.88$

$\therefore$ shear stress (τ) = $\mu\frac{du}{dy}$

$$= 7.4 \times 10^{-7} \times 0.88 \times \frac{0.5}{0.5 \times 10^{-3}}$$

$= 0.651 \times 10^{-3}$ Pa

23. P_1 = 1 bar

P_2 = 30 bar

Specific work done (W) = Vdp

$$= \frac{m}{\rho}dp$$

$$\frac{W}{m} = \frac{dp}{\rho}$$

$$= \frac{(30-1) \times 10^5}{990 \times 1000}$$

$\boxed{= 2.93 \times 10}$

= 2.93 KJ/kg

24. In case of fully developed flow between two fixed parallel plates

$$V_{max} = \frac{3}{2}$$

$$\therefore V_{avg} = \frac{2}{3} \times 6 = 4\text{m/sec}$$

27. Given

$R_a = 1200$ km

$R_p = 400$ km

Let, Radius of earth, r = 6400 km

$G = 6.67 \times 10^{-11}$

$M = 5.98 \times 10^{11}$ kg

∴ As per formula

$R_p = a(1 - e)$ (1)

and $e = \dfrac{R_a - R_p}{R_a + R_p}$

$= \dfrac{7600 - 6800}{7600 + 6800}$

e = 0.055

putting value of e in equation (1) we get

$a = \dfrac{6800 \times 10^3}{1 - 0.055}$

$= 7195 \times 10^3$m

$V = \sqrt{\dfrac{GM}{a}}$

$= \dfrac{\sqrt{6.67 \times 10^{-11} \times 5.97 \times 10^{24}}}{7195.76 \times 10^3}$

= 7435.85 m/sec

$T = \dfrac{2\pi a}{V} = \dfrac{2 \times \pi \times 7195376 \times 10^3}{7438.95}$

= 6077 sec.

28. $m_j = 16000$ kg

$m_L = 1000$ kg

$m_s = 2100$ kg

$I_{sp} = 300$ sec

payload ratio (λ) = 0.066

∴ $m_j = m_p + m_s + m_L$

$\Rightarrow m_j - m_L = m_p + m_s$

$16000 - 1000 = m_p + m_5$

$\Rightarrow m_p + m_s = 15000$ kg

Structural ratio $(\varepsilon) = \dfrac{m_s}{m_p + m_s} = \dfrac{2100}{15000} = 0.14$

Terminal velocity attained by this payload

$= \eta I_{sp}$ go $\ln\left(\dfrac{1+\lambda}{\varepsilon+\lambda}\right)$

$= 2 \times 300 \times 9.8 \ln\left(\dfrac{1+0.066}{0.14+0.066}\right)$

= 9666 m/sec.

29. $C_{m_{CG}} = 0.3 + 00.115\alpha$

$\dfrac{\partial C_{m_{CG}}}{\partial \alpha} = 0.0115$

= positive value

= destabilizing

30. Given

$C_m = 0.04 - 0.18 + 0.2C_L$

$\dfrac{\partial C_m}{\partial C_L} = 0.2$

$\dfrac{-\partial C_m}{\partial C_L} = -0.2$

Static margin of aircraft is – 0.2

31. $P_o = 55{,}000$ N/m²

$P_s = 45{,}280$ N/m²

$\rho_a = 0.6$ kg/m³

$V_{equivalent} = ?$

Applying Bernoulli's theorem

$P_o = P_s + P_d = P_s + \dfrac{1}{2}\rho_a V_T^2$

$\Rightarrow \sqrt{\dfrac{2(P_o - P_s)}{\rho_a}} = V_T$

$\Rightarrow \sqrt{\dfrac{2(55000 - 45280)}{0.6}} = V_T$

$\Rightarrow V_T = 127$ m/sec

∴ Vequivent $= V_T\sqrt{\dfrac{\rho_a}{\rho_s}}$

Vequivent $= \sqrt{\dfrac{\rho_a}{\rho_s}} \times V_T$

$= \sqrt{\dfrac{0.6}{1.225}} \times 127$

= 88.88

Vequivalent = 89m/sec

32. $\frac{u}{u_\infty} = \frac{3}{2}\left(\frac{y}{\delta}\right) - \frac{1}{2}\left(\frac{y}{\delta}\right)^3$

$$R_{e_x} = \frac{ux}{\nu} = \frac{2 \times 1}{1.5 \times 10^{-5}} = 1.33 \times 10^5$$

$$8 = \frac{4.64x}{\sqrt{R_{ex}}} = \frac{4.64 \times 1}{\sqrt{1.33 \times 10^5}} = 0.0127$$

$$\frac{du}{dy} = u\left[\frac{3}{2} \times \frac{y}{\delta} - \frac{3}{2}\left(\frac{y^2}{\delta^3}\right)\right]$$

$$\frac{du}{dy} = \frac{3u}{2\delta}$$

$$\tau = \mu\left(\frac{du}{dy}\right)_{y=0} = \mu \times \frac{3u}{2\delta} = \rho \times \nu \times \frac{3\mu}{2\delta}$$

$= 4.36 \times 10^{-3} N/m^2$

33.

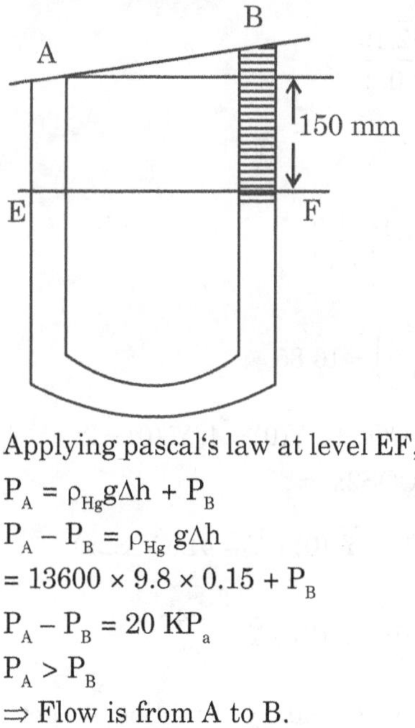

Applying pascal's law at level EF,

$P_A = \rho_{Hg} g \Delta h + P_B$

$P_A - P_B = \rho_{Hg}\, g\Delta h$

$= 13600 \times 9.8 \times 0.15 + P_B$

$P_A - P_B = 20\ KP_a$

$P_A > P_B$

$\Rightarrow$ Flow is from A to B.

34. Prandial number $(P_r) = \frac{\mu C_p}{K}$

$\mu = 0.1523 \times 10 -2$ N – sec/m^2

$C_p = 0.1393$ KJ/kg – K

Thermal Conductivity (K) = 8.540 W/m – K

$$\text{Prandtl number} = \frac{0.1523 \times 10^{-2} \times 0.1393 \times 10^3}{8.54}$$

$= 0.0248$

35. Given

V = 550 m/sec

P = 100 KPa

T = 300 K

$$M = \frac{v}{a} = \frac{v}{\sqrt{\gamma RT}} = \frac{500}{\sqrt{1.4 \times 287 \times 300}}$$

$= 1.58$

Normal Shock wave

M_1	M_2
P_1	P_2
e_1	e_2
M_1^*	M_2^*

relation between M and M* is given by

$$M^{*2} = \frac{(\gamma + 1)M^2}{(\gamma - 1)M^2 + 2}$$

$$M_1^{*2} = \frac{(1.4 + 1)1.58^2}{(1.4 - 1)1.58^2 + 2}$$

$$M_1^* = \sqrt{\frac{2.4 \times 1.58^2}{0.4 \times 1.58^2 + 2}}$$

$M_1^* = 1.41$

We know by prandtl relation

$M^*_2 M^*_1 = 1$

$$M_2^* = \frac{1}{M_1^*}$$

$$= \frac{1}{1.41}$$

$= 0.70$

$\boxed{M_2^* = 0.70}$

36.

$$M_2 = \sqrt{\frac{1 + \frac{\gamma - 1}{2} M_1^2}{\gamma M_1^2 - \frac{\gamma - 1}{2}}}$$

When

$$\lim_{M_1 \to \infty} M_2 = \sqrt{\frac{\gamma - 1}{2r}} = 0.378$$

$$\frac{\rho_2}{\rho_1} = \frac{(\gamma+1)M_1^2}{(\gamma-1)M_1^2}$$

$$\lim_{M_1 \to \infty} \frac{\rho_2}{\rho_1} = \frac{\gamma+1}{\gamma-1} = \frac{2.4}{0.4} = 6$$

$$\lim_{M_1 \to \infty} \frac{P_2}{P_1} = \lim_{M_1 \to \infty}\left[1 + \frac{2\gamma}{\gamma+1}\left(M_1^2 - 1\right)\right]$$

$= \infty$

By prandtl relation

$$\boxed{M_1^* M_2^* = 1}$$

37.

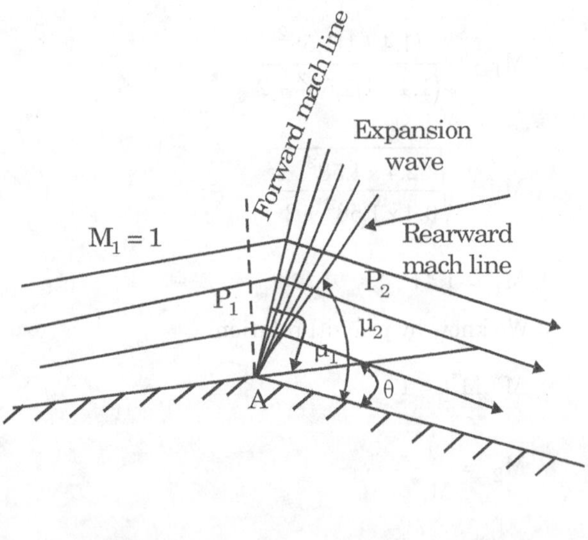

$M_1 = 2$

angle of front

$$\mu_1 = \text{Sin}^{-1}\frac{1}{M_1} = \text{Sin}^{-1}\frac{1}{2}$$

$= 30°$

Pressure ratio across the fan $\left(\frac{P_2}{P_1}\right) = 0.5$

$$\frac{P_2}{P_1} = 0.5$$

$$\Rightarrow \left(\frac{1 + \frac{\gamma-1}{2}M_1^2}{1 + \frac{\gamma-1}{2}M_2^2}\right)^{\frac{r}{r-1}} = 0.5$$

$$\Rightarrow \left(\frac{1 + 0.2M_1^2}{1 + 0.2M_2^2}\right)^{3.5} = 0.5$$

$$\Rightarrow \frac{1 + 0.2 \times 2^2}{1 + 0.2M_2^2} = 0.5^{0.285}$$

$$\Rightarrow 1 + 0.2M_2^2 = \frac{1 + 0.2 \times 2^2}{0.5^{0.285}}$$

$$\Rightarrow 1 + 0.2M_2^2 = \frac{1.8}{0.82}$$

$$\Rightarrow 0.2M_2^2 = 2.19$$

$$\Rightarrow M_2 = \sqrt{\frac{2.19}{0.2}}$$

$M_2 = 3.3$

$$\therefore \mu_2 = \text{Sin}^{-1}\left(\frac{1}{\mu_2}\right)$$

$$= \text{Sin}^{-1}\left(\frac{1}{3.3}\right) = 16.85$$

38. Y" + Y = 4 COS2x, Y(0) = 4, Y' (0) = 2

Y" + Y = 4COS2x

$$S^2\overline{Y} - SY(0) - Y'(0) + \overline{Y} = 4L(COS2x)$$

$$S^2\overline{Y} - SY(0) - Y'(0) + \overline{Y} = \frac{4S}{S^2+4}$$

Putting given boundary Condition, we get

$$\left(S^2 + 1\right)\overline{Y} - SY(0) - Y'(0) = \frac{4S}{S^2+4}$$

$$\left(S^2 + 1\right)\overline{Y} - 4S - 2 = \frac{4S}{5^2+4}$$

$$\Rightarrow \left(S^2 + 1\right)\overline{Y} = \frac{4S}{S^2+4} + 4S + 2$$

$$\overline{Y} = \frac{4S}{(S^2+4)(S^2+1)} + \frac{4S+2}{(S^2+1)}$$

$$\therefore \overline{Y} = F(S)$$

$$\therefore F(S) = \frac{4S}{(S^2+4)(S^2+1)} + \frac{4S+2}{(S^2+1)}$$

$$F(1) = \frac{4\times 1}{(1^2+4)(1^2+1)} + \frac{4\times 1+2}{(1^2+1)}$$

$$= \frac{4}{5\times 2} + \frac{6}{2} = \frac{2}{5} + 3 = \frac{17}{5}$$

$= 3.4$

39. $x^2\frac{d^2y}{dx^2} - 2x\frac{dy}{dx} - 4y = x^4$

Let $x = e^z$

$$\Delta = \frac{\partial}{\partial z}$$

$$\Delta y = x\frac{\partial y}{\partial x}$$

$$= x^2\frac{d^2y}{dx^2} = \Delta(\Delta-1)y$$

$$\Delta(\Delta-1)y - 24y = e^{4z}$$

$$(\Delta^2 - \Delta - 2\Delta - 4)y = e^{4z}$$

$$(\Delta^2 - 3\Delta - 4)y = e^{4z}$$

$\therefore$ Auxiliary equation

$$\Delta^2 - 3\Delta - 4 = 0$$

$$(\Delta+1)(\Delta-4) = 0$$

$\Delta = -1, 4$

C.F $= C_1e^{-z} + C_2\,e^{4z}$

$$\text{P.I} = \frac{e^{4z}}{\Delta^2 - 3\Delta - 4} = \frac{e^{4z}\bullet z}{2\Delta - 3} = \frac{1}{(2\times 4-3)}e^{4z}\bullet z$$

$$= \frac{ze^{4z}}{5}$$

Complete Solution of this differential equation

$$\text{C.F} = \text{P.I} = C_1e^{-z} + C_2e^{4z} + \frac{ze^{4z}}{5}$$

$$= \frac{C_1}{e^{-\log x}} + C_2e^{4\log x} + \log x.\frac{e^{4\log x}}{5}$$

$$= \frac{C_1}{x} + C_2{}^{x^4} + .\frac{x^4}{5}\log x$$

40. $I = \oint(xy\,dy - y^2dx)$

Applying Green's Theorem

$$\oint_C \varphi dx + \psi dy \iint_R \left(\frac{\partial\psi}{\partial x} - \frac{\partial\varphi}{\partial y}\right) dxdy$$

$$\therefore \oint_C (xydy - y^2dx) = \oint(-y^2)dx + (2y)dy$$

$$\varphi = -y^2, \psi = xy$$

$$\frac{\partial\psi}{\partial y} = x, = \frac{\partial\psi}{\partial x} = y$$

Substituting in the green theorem we get

$$I = \int_{y=0}^{1}\int_{x=0}^{1} [y-(-2y)]dxdy$$

$$= \int_{y=0}^{1}\int_{x=0}^{1} 3ydxdy$$

$$= \int_{y=0}^{1} [3xy]_{x=0}^{1}\, dy$$

$$= \int_{y=0}^{1} 3ydy$$

$$= \frac{3\{Y^2\}_0^1}{2}$$

$$= \frac{3}{2}$$

41. $A = \begin{bmatrix} 1 & 1 & 1 & 1 \\ 1 & 1 & -1 & -1 \\ 1 & -1 & 0 & 0 \\ 0 & 0 & 1 & -1 \end{bmatrix}$

For orthogonal matrix $AA^T = I$; I = Identity matrix

$[AA^T]^{-1} = I^{-1} = I$

For Identity matrix diagonal element are always one.

42. $\frac{dy}{dx} = x + y$

$x_0 = 0,\ y_0 = 1,\ h = 0.2$

$f(x_0, y_0) = 0 + 1 = 1$

$K_1 = hf(x_0, y_0) = 0.2 \times 1 = 0.2$

$$K_2 = hf\left(x_0 + \frac{h}{2}, y_0 + \frac{K_1}{2}\right)$$

$$= 0.2f\left(0 + \frac{0.2}{2}, 1 + \frac{0.2}{2}\right)$$

$= 0.2\ f(0.1, 1.1)$

$= 0.2 \times 1.2 = 0.24$

$$K_3 = hf\left(x_0 + \frac{h}{2}, y_0 \frac{K_2}{2}\right)$$

$= 0.2\ f(0.1, 1.12)$

$= 0.2 \times 1.22 \simeq 0.244$

$K_4 = h\ f(x_0 + h, y_0 + K_3)$

$= 0.2 \times f(0.2, 1.244)$

$$\therefore K = \frac{1}{6}[K_1 + 2K_2 + 2K_3 + K_4]$$

$$= \frac{1}{6}[0.2 + 2 \times 0.24 + 2 \times 0.244 + 0.2888]$$

$$= \frac{1.456}{6} = 0.2426$$

$\therefore y_1 = 1 + 0.242666$

$= 1.24266.$

43. Instantaneous Rate of C limb $\frac{dh}{dt} = P_s - \frac{V\infty}{g}A$

A = Instantaneous acceleration

P_s = Specific excess power

V_∞ = Instantaneous Velocity

$$\frac{dh}{dt} = 300 - \frac{800}{32.2} \times 8$$

= 101 ft/sec.

45. Given

$\sigma_{xx} = -\sigma_0$

E = 90 G Pa

$\nu = 0.3$

For plain strain Condition

$\varepsilon_{zz} = 0$

$$\varepsilon_{zz} = \frac{\sigma_{zz}}{E} - \frac{\nu(\sigma_{xx} + \sigma_{yy})}{E}$$

$\sigma_{zz} = \nu(\sigma_{xx} + \sigma_{yy})$ ...(1)

According to Question

$\varepsilon_{yy} = 0$ (As fixed by wall)

$$\varepsilon_{yy} = \frac{\sigma_{yy}}{E} - \nu\frac{(\sigma_{xx} + \sigma_{zz})}{E}$$

$$= \frac{\sigma_{yy} - \nu(\sigma_{xx} + \sigma_{zz})}{E}$$

$$\Rightarrow 0 = \sigma_{yy} - \nu(\sigma_{xx} + \sigma zz)$$

$$\boxed{\sigma_{yy} = \nu(\sigma_{xx} + \sigma_{zz})} \quad ...(2)$$

$\sigma_{yy} = \nu(\sigma_{xx} + \nu(\sigma_{xx} + \sigma_{yy}))$

$\nu\sigma_{xx} + \nu^2\sigma_{xx} + \nu^2\sigma_{yy}$

$$\Rightarrow \sigma_{yy} = \frac{\nu(1+\nu)\sigma_{xx}}{1-\nu^2} = \frac{\nu(1+\nu)\sigma_{xx}}{(1+\nu)(1-\nu)} = \frac{\nu}{1-\nu}\sigma_{xx}$$

$$\boxed{\sigma_{yy} = \frac{\nu}{1-\nu}\sigma_{xx}} \text{ and } 6zz = V\left(1 + \frac{\nu}{1-\nu}\right)\sigma_{xx}$$

$$= \left(\frac{\nu}{1-\nu}\right)\sigma_{xx}$$

we have

$$\varepsilon_{xx} = \frac{\sigma_{xx}}{E} - \nu\frac{(\sigma_{yy} + \sigma_{zz})}{E}$$

$$= \frac{\sigma_{xx}}{E} - \frac{V}{E}\left(\frac{\nu}{1-\nu}\sigma_{xx} + \frac{\nu}{1-\nu}\sigma_{xx}\right)$$

$$= \frac{\sigma_{xx}}{E}\left(1 - \left(\frac{\upsilon^2}{1-\upsilon} + \frac{\upsilon^2}{1-\upsilon}\right)\right)$$

$$\varepsilon_{xx} = \frac{\sigma_{xx}}{E}\left(1 - \left(\frac{2\nu^2}{1-\nu}\right)\right) \quad ....(3)$$

According to problem

$\sigma_{xx} = K\varepsilon_{xx}$. ....(4)

Equating equations (3) and equations (4), we get

$$\frac{\sigma_{xx}}{K} = \frac{\sigma_{xx}}{E}\left(1 - \frac{2V^2}{1-\nu}\right)$$

$$\Rightarrow E = \frac{E}{1 - \frac{2v^2}{1-\nu}} = \frac{90 \times 101}{-\frac{2 \times 0.32}{1-0.3}} = 121 \text{ GPa}$$

46. Given

$\sigma_{xx} = -\sigma$

E = 90 GPa

ν = 0.3

$\sigma_{xx} = K\varepsilon_{xx}$

ε_{yy} = 0 (as fixed by wall)

For plain stress Condition

$\sigma_{zz} = 0$

$$\varepsilon_{yy} = \frac{\sigma_{yy}}{E} - \frac{\nu\sigma_{xx}}{E}$$

$\Rightarrow 0 = \sigma_{yy} - \nu\sigma_{xx}$

$\Rightarrow \sigma_{yy} = \nu 6_{xx}$ ...(1)

Now

$$\varepsilon_{xx} = \frac{\sigma_{xx} - \nu\sigma_{yy}}{E} = \frac{\sigma_{xx}}{E} - \frac{\nu}{E}(\nu\sigma_{xx})$$

$$= \frac{\sigma_{xx} - \nu^2\sigma_{xx}}{E}$$

$$\varepsilon_{xx} = \frac{\sigma_{xx} - \nu^2\sigma_{xx}}{E}$$

$$= \frac{\sigma_{xx}}{E}(1 - \nu^2) \quad ...(2)$$

Given boundary condition

$\sigma_{xx} = K\varepsilon_{xx}$

$$\varepsilon_{xx} = \frac{\sigma_{xx}}{K} \quad ...(3)$$

from equation (2) and equation (3), we get

$$\frac{\sigma_{xx}}{E}(1 - \nu^2) = \frac{\sigma_{xx}}{K}$$

$$K = \frac{E}{1-\nu^2} = \frac{90}{1-0.3^2} = \frac{90}{0.91} = 99\text{GPa}$$

$$\boxed{K = 99\text{GPa}}$$

47. For open section:

$$J_1 = \frac{bt^3}{3} = \frac{2\pi rt^3}{3}$$

$$= \frac{2\pi \times 0.5 \times (0.05)^3}{3}$$

$J_1 = 13.083 \times 10^{-5}$ m^4

For closed section

$$J_2 = \frac{4\bar{A}^2}{\oint \frac{dS}{t}} = \frac{4 \times (\pi r^2)^2}{\frac{2\pi(r)}{t}}$$

$$= 2\pi r^3 t$$

$$= 2\pi \times 0.5^3 \times 0.05$$

$= 3925 \times 10^{-5}$ m^4

$$\therefore \frac{J_1}{J_2} = \frac{13.083 \times 10^{-5}}{3925 \times 10^{-5}}$$

= 0.003312

48.

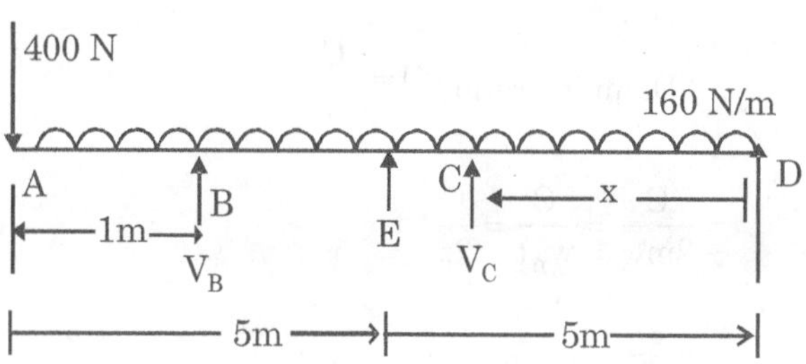

Let V_B and V_c are the reaction at support B and C resolving the forces on the beam vertically

$V_B + V_C = 400 + 160 \times 10$

= 2000 N

According to question bending moment at point E = 0, Taking moment about E of the forces on L.H.S of E.

$$V_B \times 4 - 400 \times 5 - 160 \times \frac{5^2}{2} = 0$$

$V_B \times 4 - 2000 - 2000 = 0$

$V_B = 1000$ N

$\therefore V_B + V_C = 2000$

$\therefore V_C = 2000 - 1000 = 1000$ N

Taking moments about E of the forces on the R.H.S of E.

$$1000(5-x)-160\times\frac{5^2}{2}=0$$

$\boxed{x = 3m}$

49. We have

$A_P = A_o e^{-PC}$.

P = number of peak

$\therefore A_3 = A_o e^{-3C}$.

$100 = A_o e^{-3C}$ ....(1)

and

$A_{10} = = A_o e^{-10C}$.

$10 = = A_o e^{-10C}$ ....(2)

$\therefore$ from equation (1) and equation (2), we get

$$\frac{100}{10}=\frac{e^{-3C}}{e^{-10C}}=e^{(-3+10)C}=e^{7C}$$

$\Rightarrow 10 = e^{7C}$

log 10 = 7C

$$\Rightarrow \frac{\ln 10}{7}=C$$

C = 0.32

$$\therefore \text{ Damping ratio } (\zeta)=\frac{C}{C_m}$$

$$=\frac{C}{2mw}=\frac{C}{w_n t}=\frac{C}{2\pi}$$

$$=\frac{0.32}{6.28}=0.050$$

50. for longitudinal static stability

(i) C_{mo} must be positive

(ii) $\frac{\partial C_{m_{cg}}}{\partial \alpha}$ must be negative

51. $C_{DO} = 0.2$, e = 1

$C_L = 0.8$, AR = ?

Condition of or minimum power required for aircraft is aircraft should fly at

$$\left(\frac{C_L^{\frac{3}{2}}}{C_D}\right)_{max}, \text{for } \left(\frac{C_L^{\frac{3}{2}}}{C_D}\right)_{max}, C_{DO}=\frac{C_{Di}}{3}=\frac{C_L^2}{3\pi\, eAR}$$

$$\Rightarrow AR=\frac{C_L^2}{3\pi e C_{DO}}$$

$$=\frac{(0.8^2)}{3\times3.14\times1\times0.2}=0.33$$

$\boxed{AR = 0.33}$

52. Given

$\eta_T = 0.9$

$P_3 = 12$

$P_4 = 1.2$

$\gamma = 1.4$

$C_P = 1005$ J/kg–K

m = 1000 Kg/sec.

$T_{O3} = 1050 + 273 = 1323$ K

Power output of the turbine = $mC_P (T_{O3} - T_{O4})$

$$mC_P\eta_T T_{o3}\left(1-\frac{1}{(r_t)^{\frac{1.4-1}{1.4}}}\right)$$

= 575829.61 watt

= 772 .

54. Given

$P_o = 25$ atm

$T_o = 3517$ K

$A_t = 0.1\ m^2$.

$\gamma = 1.22$,

$\overline{M} = 16$

$$R=\frac{8314}{16}=520\ J/kg-K$$

$$\text{Specific impuse }(I_{sp})=\frac{\sqrt{\frac{2\gamma RT_0}{(\gamma-1)}\left(1-\left(\frac{P_e}{P_0}\right)^{\frac{\gamma-1}{\gamma}}\right)}}{g_0}$$

$$=\frac{\sqrt{\frac{2\times1.22\times520}{(1.22-1)}\left(1-\left(\frac{1.74\times10^{-2}}{25}\right)\right)^{0.18}}}{9.8}$$

= 397.9 sec.

55. Given

$I_{sp} = 360$ sec

escape velocity from earth surface = 11.2 Km/sec

$\therefore$ Burnout velocity $(V_b)\gamma$ = Isp go $\ln\frac{m_i}{m_f}$

According to question Burnout velocity is equal to escape velocity

$\therefore$ V_b = 112.2 Km/sec

$\therefore$ $11.2 \times 10^3 = 360 \times 9.8 \ln\frac{m_i}{m_f}$

$$\frac{11.2\times10^3}{360\times9.8} = \ln\frac{m_i}{m_f}$$

$$\Rightarrow 3.17 = \ln\frac{m_i}{m_f}$$

$$\Rightarrow \frac{m_i}{m_f} = e^{3.17}$$

$$\Rightarrow \frac{m_i}{m_f} = 23.80$$

$$\boxed{\frac{m_i}{m_f} = 24}$$

56. $3 \le x \le 5$

$8 \le y \le 11$

$$\Rightarrow \frac{1}{8} \ge \frac{1}{y} \ge \frac{1}{11}$$

$$\Rightarrow \frac{1}{11} \le \frac{1}{y} \le \frac{1}{8}$$

$$\therefore \boxed{\frac{3}{11} \le \frac{x}{y} \le \frac{5}{8}}$$

57. The sentence is in present tense and the blank should maintain the tense structure. Additionally 'principal' is in singular form the verb should also be in the singular form. Hence, (b) is the correct answer.

58. Mahatma Gandhi was known for his humility as he displayed modesty in his interactions.

59. In part IV, the 'how to do' is redundant.

60. Water flows through a pipe and similarly electricity flows through a wire.

61. V = 80 – 32t

$$V\ 32\text{m/sec} \Rightarrow t = \frac{8}{2}$$

$$V = 64\ \text{m/sec} \Rightarrow t = \frac{1}{2}$$

$$\therefore \boxed{t \in \left(\frac{1}{2}, \frac{3}{2}\right)}$$

62. Let P(M) denote the probability that the component is manufactured by Machine Mi & P (def) denote the probability that the component is defective.

$$P(M_2 \mid def) = \frac{P(M_2 \cap def)}{P(def)}$$

$$= \frac{0.4 \times 0.03}{(0.4 \times 0.3) + (0.6 \times .02)}$$

$= 0.5$

63. Total No. of tourists = 13500.

$$\frac{1}{3} \times 13500 = 4500$$

Option (c) England (3500) + France(1000)

= 4500

64. $|-2x + 9| = 3$

$\Rightarrow +(2x + 9) = 3$

$\Rightarrow x = 3$

$-(-2x + 9) = 3$

$\Rightarrow x = 6$

$x = 3$

$\Rightarrow |-x| - x^2$

$\Rightarrow |-3| - (3)^2$.

$\Rightarrow 3 - 9 = -6$

$x = 6$

$\Rightarrow |-6| - 6^2$

$\Rightarrow 6 - 36 = -30$

Option (b)

65. From clear logical deduction, (c) is the correct choice.

MOCK TEST – 10

AEROSPACE ENGINEERING

Time Allowed: 3:00 Hours *Maximum Marks: 100*

Read the following instructions carefully before you begin to answer the questions. This Booklet contains questions in English.

INSTRUCTIONS

1. This Booklet contains **65 questions** in total comprising of the following two parts:

   Part -1: Technical Section (85 Marks) *(1-55 Questions)*

   Part -2: General Aptitude Section (15 Marks) *(56-65 Questions)*

2. All questions are compulsory. Q.1-Q.25 & Q.56-Q.60 carry one mark each. Q.26-Q.55 & Q.61-Q.65 carry two marks each.

3. **Types of Questions:**

   (*a*) Multiple Choice Questions(MCQ) carrying 1 or 2 marks each in all papers and sections. These questions are objective in nature, and each will have a choice of four options, out of which the candidate has to mark the correct answer(s).

   (*b*) Numerical Answer Questions of 1 or 2 marks each in all sections. For these questions the answer is a real number, to be entered by the candidate using the virtual keypad. No choices will be shown for these type of questions.

4. For 1-mark multiple-choice questions, 1/3 marks will be deducted for a wrong answer. Likewise, for 2-marks multiple-choice questions, 2/3 marks will be deducted for a wrong answer. There is no negative marking for numerical answer type questions.

MOCK TEST – 10

1. The rank of the matrix $\begin{bmatrix} 1 & 1 & 1 \\ 2 & 0 & 1 \\ 2 & 1 & 1 \end{bmatrix}$ is

(*a*) 0 (*b*) 1
(*c*) 2 (*d*) 3

2. The value of integral $\int_{-1}^{1} \frac{dx}{x^2+1}$ is ______.

3. Which option correctly relate the geometric altitude (h_G) and geo-potential altitude (h) :

(*a*) $\frac{1}{h} = \frac{1}{h_G} + \frac{1}{r}$ (*b*) $\frac{1}{h_G} = \frac{1}{h} - \frac{1}{r}$

(*c*) $\frac{1}{h_G} = \frac{1}{h} + \frac{1}{r}$ (*d*) $\frac{1}{h} = \frac{1}{h_G} - \frac{1}{r}$

'r' denotes the radius of earth.

4. Ratio of temperature at 3 km altitude to the temperature at Sea level is ______ .

(T_{SL} = 2.88.15. k , Lapse rate = 0.0065 k/m)

5. For a Supersonic flow entering into the duct which option represents the correct configuration of nozzle (Assume isentropic flow through the nozzle)

(*a*) 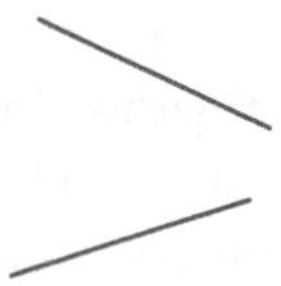 (*b*) 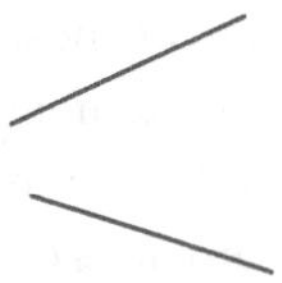

(*c*) 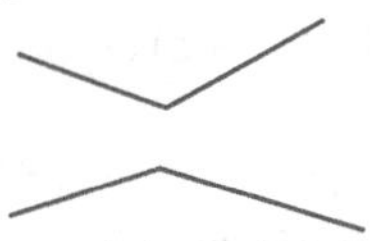 (*d*) (*a*) & (*c*)

6. Supercritical airfoil is used to increase
(*a*) Critical Mach Number
(*b*) Super Critical Mach Number
(*c*) Drag-Divergence Mach Number
(*d*) Stall angle

7. Thin airfoil theory is simulated by placing the vortex sheet on
(*a*) Chord line
(*b*) Camber line
(*c*) Upper Surface of airfoil
(*d*) Lower Surface of airfoil

8. Figure Shows flow over the wedges with same freestream Mach number but different wedge angle

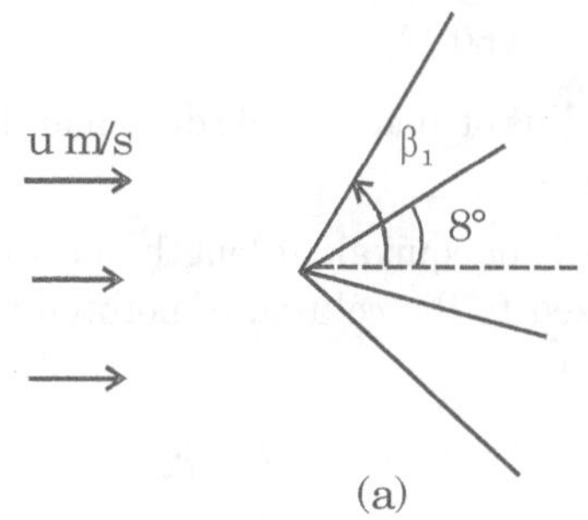

(a)

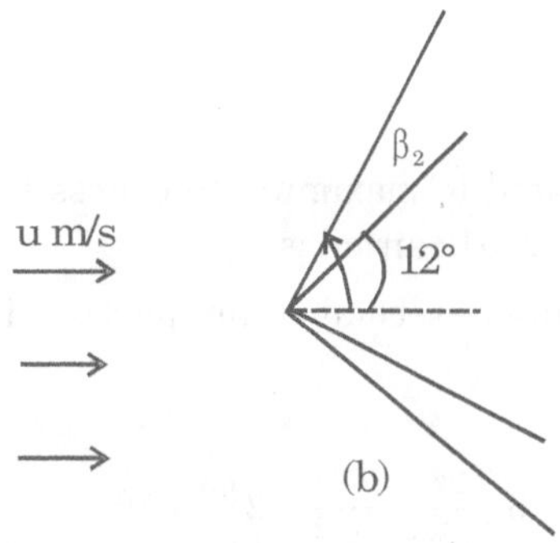

(b)

'β_1' and 'β_2' represents shock angles. Then the relationship between 'β_1' and 'β_2' is for weak shock wave solution is

(*a*) $\beta_1 = \beta_2$ (*b*) $\beta_2 > \beta_1$
(*c*) $\beta_1 \neq \beta_2$ (*d*) (*b*) & c

9. Which option is correct for homogeneous material?
(*a*) A material possessing the same properties at all points.
(*b*) A material possessing the same properties in diagonal.
(*c*) A material which is isotropic.
(*d*) A material possessing the same properties at corner points.

10. The Relationship between Young's Modulus of Elasticity (E), Poission's ratio (ν) and Modulus of rigidity (G) is represented by

(*a*) $\frac{E}{2} = G(1-\nu)$ (*b*) $\frac{E}{3} = G(1+\nu)$

(*c*) $\frac{E}{2} = G(1+\nu)$ (*d*) $\frac{E}{3} = G(1-\nu)$

11. The Stresses at a point on two planes perpendicular to each other are shown below

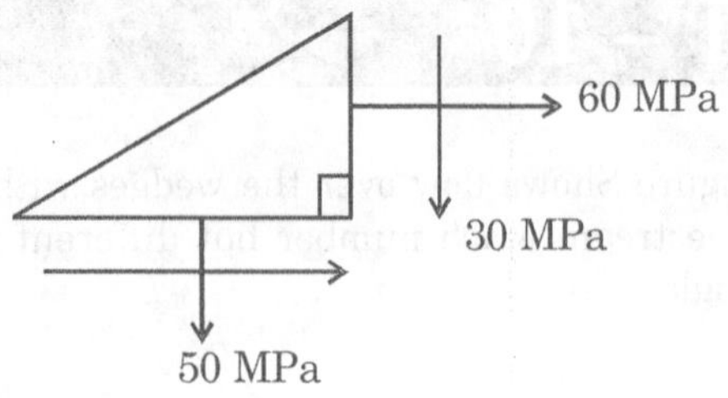

The radius of mohr's circle drawn for this system is (in pa) ______ .

12. The ratio of equivalent length of column of both end fixed to the column of both ends hinged is given by

(*a*) 2L (*b*) $\sqrt{2}L$

(*c*) $\frac{L}{2}$ (*d*) $\frac{L}{\sqrt{2}}$

13. The chord to maximum thickness ratio for the NACA 24012 airfoil is ______ .

14. The linear second order partial differential equation

$$4\frac{\partial^2 z}{\partial x^2} - 6\frac{\partial^2 z}{\partial x \partial y} + \frac{\partial^2 z}{\partial y^2} + 2\frac{\partial z}{\partial x} + 16 = 0 \text{ is}$$

(*a*) Parabolic (*b*) Elliptic
(*c*) Circular (*d*) Hyperbolic

15. Lift on an aircraft climbing vertically up is

(*a*) Zero (*b*) equal to weight
(*c*) equal to drag (*d*) equal to thrust

16. With increase in altitude of propeller-driven airplane, Rate of climb of airplane is

(*a*) remains constant
(*b*) First increase then decrease
(*c*) Increases
(*d*) Decreases

17. In an impulse stage of turbine

(*a*) Static pressure remains constant in stator
(*b*) Static pressure remains constant in rotor
(*c*) Static pressure increases in both rotor and stator
(*d*) Static pressure decreases in both rotor and stator.

18. Consider the Statement related to isentropic process:

(i) reversible and adiabatic process
(ii) reversible and non-adiabatic process
(iii) Entropy constant process

Which statements is correct

(*a*) (i) and (ii) (*b*) (i), (ii) and (iii)
(*c*) (i) and (iii) (*d*) (ii) and (iii)

19. An aircraft is flying with a speed of 300 KPh. The value of Co-efficient of pressure at a location on the wing is –2.0. Then velocity of air passing through the point is

(*a*) 48 m/s (*b*) 144 m/s
(*c*) 168 m/s (*d*) 190 m/s

20. An aircraft is flying at geometric altitude of 2 km. The force by which it is pull down due to the gravity is. (g_0 = 10 m/s², r_{earth} = 6400km)

(*a*) 9.81 $\frac{N}{kg}$ (*b*) 10 $\frac{N}{kg}$

(*c*) 9.89 $\frac{N}{kg}$ (*d*) 9.99 $\frac{N}{kg}$

21. For longitudinal Static Stability of aircraft

(*a*) N.P. should lies ahead of C.G.
(*b*) N.P. should lies behind of C.G.
(*c*) N.P should lies exactly at the C.G.
(*d*) N.P should lies aerodynamic centre.

22. A swept back wing

(*a*) Contributes negatively in directional stability
(*b*) Contributes positively in directional stability
(*c*) Not Contributes in directional stability
(*d*) None of these

23. Which statements is true for ventral fin?

(i) It is located on the lower surface of fuselage
(ii) It provides directional stability of aircraft
(iii) It does not provide directional stability to the aircraft.
(iv) It is located on the upper surface of fuselage

(*a*) (i) & (iii) (*b*) (ii) & (iii)
(*c*) (i) & (ii) (*d*) (i) & (iv)

24. With initial condition x (1) = 1, the solution of the differential equation $y\frac{dx}{dy} + x = y$ is

(*a*) $xy = \frac{y^2}{2} + \frac{1}{2}$ (*b*) $xy = \frac{y^2}{2} + \frac{1}{6}$

(*c*) $xy = \frac{y^2}{2} + \frac{1}{4}$ (*d*) $x = \frac{xy^2}{2} + \frac{1}{2}$

25. The period of revolution of the earth about the Sun is 365 days. The semimajor axis of the earth's orbit is 1.445278×10^{11} m. The period of revolution of mars about the sun is, if its semimajor axis is 2.278×10^{11} m ______ .

26. The propellant in a single state sounding rocket occupies 70% of its initial mass. It all of it is expanded instantaneously at an equivalent exhaust velocity of 3000 m/s, the altitude attained by the payload when launched vertically is (Neglect drag, g = 9.81 m/s²)

(*a*) 565 km (*b*) 585 km

(*c*) 665 km (*d*) 685 km

27. The Euler's energy equation for rotating machines are given by

(*a*) $E = C_{t1} U_1 - C_{t2} U_2$

(*b*) $E = C_{t1} U_2 - C_{t2} U_1$

(*c*) $E = W_{t1} U_1 - W_{t2} U_2$

(*d*) $E = W_{t1} U_2 - W_{t2} U_1$

28. For a single degree of freedom system shown in figure (K = 20 KN/m, m = 100 kg and $\zeta = 0.4$) The value of damping co-efficient 'C' is ______ .

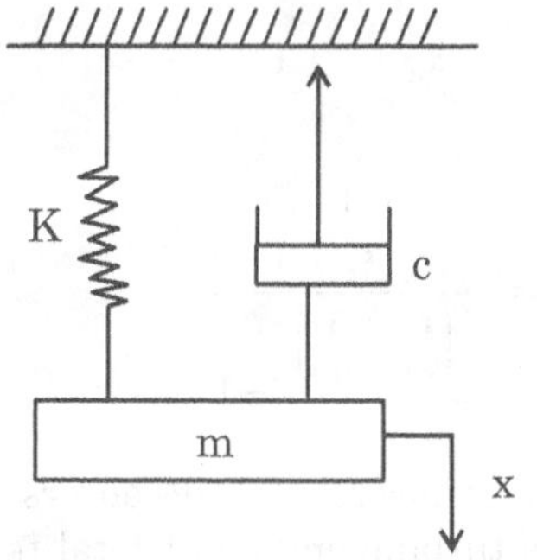

29. A Cantiliver of rectangular section 48 mm wide and 36 mm deep, of length 250 mm carries a UDL. The maximum value of intensity of UDL if the maximum deflection in the cantiliver is 1 mm. (Take E = 70 GPa) ______ .

30. A horizontal bar AB is supported by identical pin jointed column CD and EF of length 'l' as shown

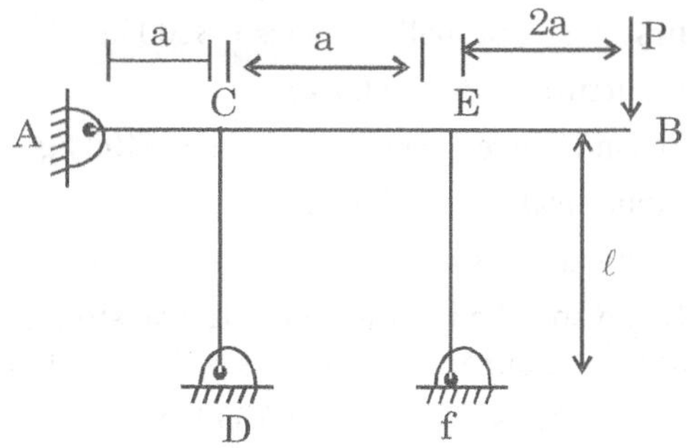

The load 'P' at which system collapse by buckling is given by

(*a*) $\frac{\pi^2 EI}{l^2}$ (*b*) $\frac{1}{2}\frac{\pi^2 EI}{l^2}$

(*c*) $\frac{3}{4}\frac{\pi^2 EI}{l^2}$ (*d*) $\frac{1}{\sqrt{2}}\frac{\pi^2 EI}{l^2}$

31. The advantage of semimonocoque type of fuselage over monocoque type is

(*a*) Higher weight to strength ratio

(*b*) Lighter weight

(*c*) Higher strength to weight ratio

(*d*) Suitable for small as well as large aircraft

32. Figure shows the resultant streamlines which formed after the superposition of uniform flow and doublet flow

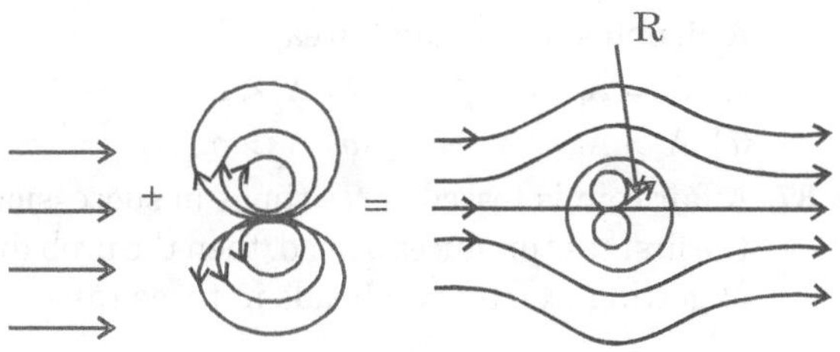

The equation of streamline of uniform and doublet flow is given by

$\psi_{\text{Uniform}} = 10\, r \sin\theta$,

$\psi_{\text{Doublet}} = \frac{-100}{2\pi}\frac{\sin\theta}{r}$

The radius 'R' of the dividing circle is given by (In meter) ______ .

33. Consider a vortex sheet of strength 10 m/s, the local jump of tangential component of velocity is shown in figure.

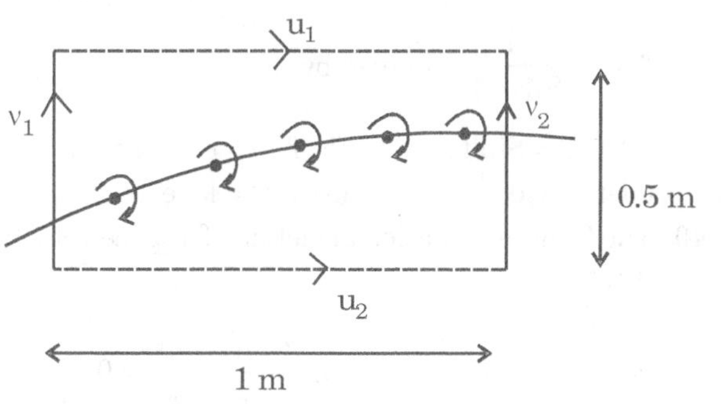

If u_1, u_2, v_1 and v_2 is 2 m/s, x m/s, 10 m/s and 5 m/s respectively, then value of 'x' is given by

(*a*) 2.5 m/s (*b*) 3.5 m/s

(*c*) – 8 m/s (*d*) 8 m/s

34. The asymptotic value of velocity ratio across the normal shock wave is ______ (Take r = 1.4,

If $\frac{f_1}{f_2} = \frac{(r+1)m_1^2}{2+(r-1)m_1^2}$

Where f_1 and f_2 is density of air ahead and behind of the normal shock respectively.

35. For a 50% reaction axial-flow compressor, the inlet and exit blade angles are 45° and 15° respectively. Axial velocity is maintain constant to 200 m/s. The tangential force required to produce power of 300 kw, if the inlet temperature and pressure is 300 k and l bar respectively.

(*a*) 1183 N (*b*) 3712N

(*c*) –1085 N (*d*) 3568 N

36. For the low speed operation of subsonic intake, which option is correct.

If A_c denotes capture area and

A_i denotes intake inlet area

(*a*) $A_c = A_i$ (*b*) $A_c < A_i$

(*c*) $A_c > A_i$ (*d*) $A_c > o$

37. A fair coin is tossed three times in succession. If the first toss produces a head. then the probability of getting exactly two heads in three tosses is

(*a*) $\frac{1}{8}$ (*b*) $\frac{1}{2}$

(*c*) $\frac{3}{8}$ (*d*) $\frac{3}{4}$

38. The Complete solution of the ordinary differential equation $\frac{d^2y}{dx^2} + P\frac{dy}{dx} + qy = 0$

is $y = c_1 e^{-x} + c_2 e^{-3x}$, then the value of P and q are

(*a*) p = 3, q = 3 (*b*) p = 3, q = 4

(*c*) p = 4, q = 3 (*d*) p = 4, q = 4

39. The inverse Laplace transform of the function $f(s) = \frac{1}{S(S+1)}$ is given by

(*a*) f (t) = Sin t (*b*) $f(t) = e^{-t}$ Sin t

(*c*) $f(t) = e^{-t}$ (*d*) $f(t) = 1 - e^{-t}$

40. The General equation of motion of a space vehicle is given by

(*a*) $\ddot{r} - \frac{h^2}{r^3} + \frac{k^2}{r^3} = 0$ (*b*) $\dot{r} - \frac{h^2}{r^3} + \frac{k^2}{r^3} = 0$

(*c*) $\dot{r} - \frac{h^2}{r^3} + \frac{k^2}{r^2} = 0$ (*d*) $\ddot{r} - \frac{h^2}{r^3} + \frac{k^2}{r^2} = 0$

Where 'h' and 'k' is constant

41. A Jet propelled plane is flying at an altitude where pressure and temperature are 0.55 bar and 255 k respectively at mach 0.6. mass flow rate of air is 18.2 kg/s and fuel- air ratio is 0.05 for maximum propulsive efficiency the thrust and thrust power produced by the engine are respectively

(*a*) 12080 N, 120 kw (*b*) 12088 N, 125 kw

(*c*) 11000 N, 120 kw (*d*) 0N, 0 kw

42. Isentropic efficiency and polytropic efficiency of an axial compressor stage are related by

(*a*) $\eta_{is} = \frac{\pi_c^{\frac{r-1}{r}}}{\pi_c^{\frac{r-1}{r} \times \frac{1}{\eta p}} - 1}$

(*b*) $\eta_{is} = \frac{\pi_c^{\frac{r-1}{r} \times \frac{1}{\eta p}} - 1}{\pi_c^{\frac{r-1}{r}} - 1}$

(*c*) $\eta_{is} = \frac{\pi_c^{\left(\frac{r-1}{r} \times \frac{1}{\eta p}\right)} + 1}{\pi_c^{\frac{r-1}{r}} + 1}$

(*d*) $\eta_{is} = \frac{\pi_c^{\frac{r-1}{r}} - 1}{\pi_c^{\left(\frac{r-1}{r} \times \frac{1}{\eta p}\right)} - 1}$

43. Air of mass flow rate 80 kg/s enters the combustor of a gas turbine engine at total temperature of 400 k and leaves at 1300 k. If specific heat of air and combustion product is 1005 J/kg-k and 1147 J/kg-k respectively and heating value of fuel is 44 MJ/kg. It efficiency of the combustor is 98% then mass flow rate of fuel is (kg/s) ______ .

44. Consider a two-stage rocket with the following design characteristics

first stage: propellant mass = 8200 kg

Structural mass = 800 kg

Second Stage: propellant mass = 5400 kg

Structural mass = 500 kg

Payload mass = 80 kg

It the specific impulse for both the stages is 275 sec, then burnout velocity is (take g = 10 m/s²)

(*a*) 6416 m/s (*b*) 7590 m/s

(*c*) 8596 m/s (*d*) 6580 m/s

45. Characteristics roots of equation for longitudinal dynamic stability is given by
$\lambda_{1,2} = -2.508 \pm i\,2.587$
$\lambda_{3,4} = -0.017 \pm i\,0.2125$
Which option is correct regarding the equation?
(a) $\lambda_{1,2}$ represent phugoid mode and $\lambda_{3,4}$ represent short period oscillation
(b) $\lambda_{1,2}$ represent short period oscillation and $\lambda_{3,4}$ represent phugoid mode
(c) $\lambda_{1,2}$ and $\lambda_{3,4}$ both represents pure divergence characteristics
(d) $\lambda_{1,2}$ and $\lambda_{3,4}$ represents spiral mode

46. Following equation represents the motion of phugoid mode of an aircraft

$$\frac{d^2h}{dt^2} + \frac{4}{3}h = 0$$

Then time required for one oscillation is (s) _______.

47. A wing root cross-section is idealized using booms as shown below

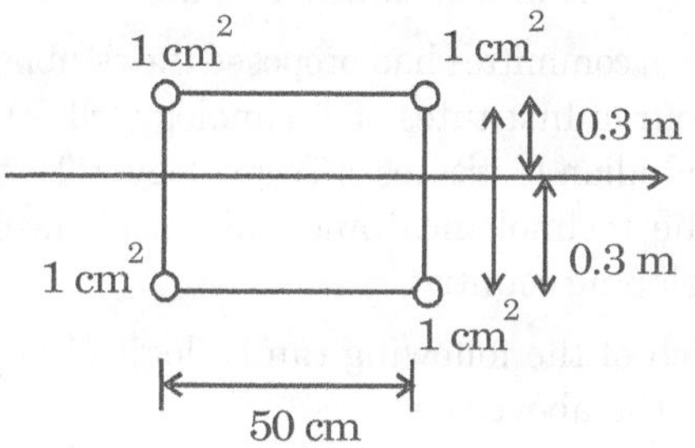

The wing root bending moment in steady stage level flight is my = 10 N – m. If the airplane flies at a load factor n = 3.5, the maximum bending stress at the root is
(a) 10^6 N/m²   (b) 70×10^5 N/m²
(c) 35×10^5 N/m²   (d) 28×10^4 N/m²

48. An aircraft with a mass of 4000 kg takes off from sea level with a forward speed of 60 m/s. It climbs with a climbing angle of 15°. The excess thrust available at the start of the climb is (in KN) ______.

49. If 'H_N' represents the static margin of an aircraft flying in longitudinal stick fixed stable condition. The which option is correct representation of this
(a) $H_N = 0$ (b) $H_N > O$
(c) $H_N < 0$ (d) $H_N > -1$

50. A wing-body model is tested in a subsonic wind tunnel. The lift is found to be zero at a geometric angle of attack $\alpha = -1.5°$. At $\alpha = 5°$, the lift co-efficient is measured as 0.52. Also at $\alpha = 1.0°$ and 8°, the moment coefficient about the C.G are measured as –0.01 and 0.05 respectively. The C.G. is located as 0.35C. Then the location of the aerodynamic center is (from the leading edge) ________.

51. Consider the function $f(x) = 1 - x^2$. The Newton-Raphson formula to find the root of the function in K[th] iteration is
(a) $\frac{x_K^2 + 1}{2x_K}$ (b) $\frac{-2x_K + 1}{x_K^2}$
(c) $\frac{x_K^2 + 1}{-2x_K}$ (d) $\frac{2x_K - 1}{x_K^2}$

52. The direction of vector A is radially outward from the origin, with $|A| = kr^n$ where $r^2 = x^2 + y^2 + z^2$ and K is a constant.
The value of n for which $\nabla.A = 0$ is _______.

53. A satellite is in elliptic orbit whose length of semi-major axis is 'a' around the earth. The velocity of satellite at apogee is
(a) $V_a = \sqrt{\frac{\mu}{a}\left(\frac{1+e}{1-e}\right)}$ (b) $V_a = \sqrt{\frac{\mu}{a}\left(\frac{1-e}{1+e}\right)}$
(c) $V_a = \frac{\mu}{a}\sqrt{\left(\frac{1+e}{1-e}\right)}$ (d) $V_a = \frac{\mu}{a}\sqrt{\left(\frac{1-e}{1+e}\right)}$
Where, $\mu = Gm_e m$
m_e = mass of earth
m = mass of satellite

54. Airy's stress function are represented by
(a) $\sigma_x = \frac{\partial^2\phi}{\partial y^2}, \sigma_y = -\frac{\partial^2\phi}{\partial x^2}, \tau_{xy} = \frac{\partial^2\phi}{\partial x\alpha y}$
(b) $\sigma_x = \frac{\partial^2\phi}{\partial y^2}, \sigma_y = -\frac{\partial^2\phi}{\partial y^2}, \tau_{xy} = -\frac{\partial^2\phi}{\partial x\alpha y}$
(c) $\sigma_x = \frac{\partial^2\phi}{\partial y^2}, \sigma_y = -\frac{\partial^2\phi}{\partial x^2}, \tau_{xy} = \frac{-\partial^2\phi}{\partial x\alpha y}$
(d) $\sigma_x = \frac{\partial^2\phi}{\partial x^2}, \sigma_y = \frac{-\partial^2\phi}{\partial x^2}, \tau_{xy} = \frac{\partial^2\phi}{\partial x\alpha y}$

55. The radial and tangential velocity component of flow field is given by
(a) $V_r = \frac{1}{r}\frac{\partial\psi}{\partial\theta}, V_\theta = \frac{\partial\psi}{\partial r}$
(b) $V_r = \frac{\partial\psi}{\partial r}, V_\theta = \frac{1}{r}\frac{\partial\psi}{\partial\theta}$
(c) $V_r = \frac{\partial\psi}{\partial r}, V_\theta = -\frac{1}{r}\frac{\partial\psi}{\partial\theta}$
(d) $V_r = \frac{1}{r}\frac{\partial\psi}{\partial\theta}, V_\theta = \frac{\partial\psi}{\partial r}$
Where, Vr = Radial Component
Vθ = Tangential Component
Ψ = Stream function

56. The chairman requested the aggrieved shareholders to ______________ him.

(*a*) bare with (*b*) bore with

(*c*) bear with (*d*) bare

57. Identify the correct spelling out of the given options:

(*a*) Managable (*b*) Manageable

(*c*) Mangaeble (*d*) Managible

58. Pick the odd one out in the following:

13, 23, 33, 43, 53

(*a*) 23 (*b*) 33

(*c*) 43 (*d*) 53

59. R2D2 is a robot. R2D2 can repair aeroplanes. No other robot can repair aeroplanes.

Which of the following can be logically inferred from the above statements?

(*a*) R2D2 is a robot which can only repair aeroplanes.

(*b*) R2D2 is the only robot which can repair aeroplanes.

(*c*) R2D2 is a robot which can repair only aeroplanes.

(*d*) Only R2D2 is a robot.

60. If $|9y - 6| = 3$, then $y^2 - 4y/3$ is _____.

(*a*) 0 (*b*) + 1/3

(*c*) – 1/3 (*d*) undefined

61. The following graph represents the installed capacity for cement production (in tonnes) and the actual production (in tonnes) of nine cement plants of a cement company. Capacity utilization of a plant is defined as ratio of actual production of cement to installed capacity. A plant with installed capacity of at least 200 tonnes is called a large plant and a plant with lesser capacity is called a small plant. The difference between total production of large plants and small plants, in tonnes is __________.

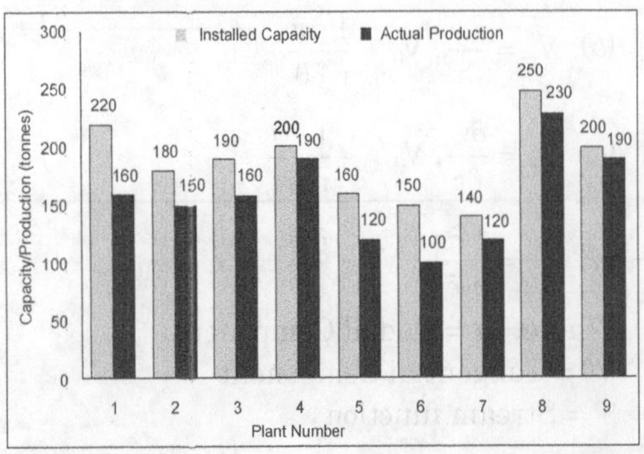

62. A poll of students appearing for masters in engineering indicated that 60 % of the students believed that mechanical engineering is a profession unsuitable for women. A research study on women with masters or higher degrees in mechanical engineering found that 99 % of such women were successful in their professions.

Which of the following can be logically inferred from the above paragraph?

(*a*) Many students have misconceptions regarding various engineering disciplines.

(*b*) Men with advanced degrees in mechanical engineering believe women are well suited to be mechanical engineers.

(*c*) Mechanical engineering is a profession well suited for women with masters or higher degrees in mechanical engineering.

(*d*) The number of women pursuing higher degrees in mechanical engineering is small.

63. Sourya committee had proposed the establishment of Sourya Institutes of Technology (SITs) in line with Indian Institutes of Technology (IITs) to cater to the technological and industrial needs of a developing country.

Which of the following can be logically inferred from the above sentence?

Based on the proposal,

(i) In the initial years, SIT students will get degrees from IIT.

(ii) SITs will have a distinct national objective.

(iii) SIT like institutions can only be established in consultation with IIT.

(iv) SITs will serve technological needs of a developing country.

(*a*) (iii) and (iv) only

(*b*) (i) and (iv) only

(*c*) (ii) and (iv) only

(*d*) (ii) and (iii) only

64. Shaquille O' Neal is a 60% career free throw shooter, meaning that he successfully makes 60 free throws out of 100 attempts on average. What is the probability that he will successfully make <u>exactly</u> 6 free throws in 10 attempts?

(*a*) 0.2508 (*b*) 0.2816

(*c*) 0.2934 (*d*) 0.6000

65. The numeral in the units position of $211^{870} + 146^{127} \times 3^{424}$ is _____.

OMR SHEET

1. (A) (B) (C) (D)	18. (A) (B) (C) (D)	35. (A) (B) (C) (D)	52.
2. (A) (B) (C) (D)	19. (A) (B) (C) (D)	36. (A) (B) (C) (D)	53. (A) (B) (C) (D)
3. (A) (B) (C) (D)	20. (A) (B) (C) (D)	37. (A) (B) (C) (D)	54. (A) (B) (C) (D)
4.	21. (A) (B) (C) (D)	38. (A) (B) (C) (D)	55. (A) (B) (C) (D)
5. (A) (B) (C) (D)	22. (A) (B) (C) (D)	39. (A) (B) (C) (D)	56. (A) (B) (C) (D)
6. (A) (B) (C) (D)	23. (A) (B) (C) (D)	40. (A) (B) (C) (D)	57. (A) (B) (C) (D)
7. (A) (B) (C) (D)	24. (A) (B) (C) (D)	41. (A) (B) (C) (D)	58. (A) (B) (C) (D)
8. (A) (B) (C) (D)	25.	42. (A) (B) (C) (D)	59. (A) (B) (C) (D)
9. (A) (B) (C) (D)	26. (A) (B) (C) (D)	43.	60. (A) (B) (C) (D)
10. (A) (B) (C) (D)	27. (A) (B) (C) (D)	44. (A) (B) (C) (D)	61.
11.	28.	45. (A) (B) (C) (D)	62. (A) (B) (C) (D)
12. (A) (B) (C) (D)	29.	46.	63. (A) (B) (C) (D)
13.	30. (A) (B) (C) (D)	47. (A) (B) (C) (D)	64. (A) (B) (C) (D)
14. (A) (B) (C) (D)	31. (A) (B) (C) (D)	48.	65.
15. (A) (B) (C) (D)	32.	49. (A) (B) (C) (D)	
16. (A) (B) (C) (D)	33. (A) (B) (C) (D)	50.	
17. (A) (B) (C) (D)	34.	51. (A) (B) (C) (D)	

ANSWERS

1. (*a*) 2. (1.57) 3. (*b*) 4. (0.93) 5. (*b*) 6. (*c*) 7. (*b*)

8. (*b*) 9. (*a*) 10. (*c*) 11. (30.414×10^6) 12. (*d*) 13. (8.33)

14. (*d*) 15. (*a*) 16. (*d*) 17. (*b*) 18. (*c*) 19. (*b*) 20. (*d*)

21. (*b*) 22. (*b*) 23. (*c*) 24. (*a*) 25. (686.35) 26. (*c*) 27. (*a*)

28. (1131.37) 29. (26.753) 30. (*c*) 31. (*c*) 32. (1.262) 33. (*c*)

34. (0.1667) 35. (*a*) 36. (*c*) 37. (*b*) 38. (*c*) 39. (*d*)

40. (*d*) 41. (*d*) 42. (*d*) 43. (2.0496) 44. (*c*) 45. (*b*) 46. (5.44)

47. (*c*) 48. (10.16) 49. (*b*) 50. (0.24) 51. (*a*) 52. (–2) 53. (*b*)

54. (*b*) 55. (*d*) 56. (*c*) 57. (*b*) 58. (*b*) 59. (*b*) 60. (*c*)

61. 120 62. (*c*) 63. (*c*) 64. (*a*) 65. (7)

EXPLANATIONS

1. $$A = \begin{bmatrix} 1 & 1 & 1 \\ 2 & 0 & 1 \\ 2 & 1 & 1 \end{bmatrix} \begin{matrix} R_1 \\ R_2 \\ R_3 \end{matrix}$$

$R_2 \to R_2 - R_1,\ R_3 \to R_3 - R_1$

$$a = \begin{bmatrix} 1 & 1 & 1 \\ 1 & -1 & 0 \\ 1 & 0 & 0 \end{bmatrix} \begin{matrix} R_1' \\ R_2' \\ R_3' \end{matrix}$$

Rank = number of non - zero rows = 0

2. $$\int_{-1}^{1} \frac{dx}{x^2+1}$$

$$= \left[\tan^{-1}(x)\right]_{-1}^{1}$$

$$= \left[\tan^{-1}(1) - \tan^{-1}(-1)\right] = 1.57$$

3. We know that, $h = \left(\dfrac{r}{r+h_G}\right) h_G$

$$hr + hh_G = h_G r$$

$$hr + h_G = (r - h)$$

$$h_G = \frac{hr}{r-h} = \frac{1}{\frac{1}{h} - \frac{1}{r}}$$

$$\frac{1}{hG} = \frac{1}{h} - \frac{1}{r}$$

4. At 3 km altitude in standard atmosphere

$$T_{3km} = T_o - Lh$$

$$T = 288.15 - 0.0065 \times 3000$$

$$T_{3Km} = 268.65 \text{ K}$$

$$\frac{T_{3km}}{T_{SL}} = \frac{268.65}{288.15} = 0.93$$

5. From area - velocity relationship

$$\frac{dA}{A} = \frac{dV}{V}\left(M^2 - 1\right)$$

$$\text{for } M > 1,\ \frac{dA}{A} = \frac{dv}{V}$$

means increase in area results in increase in velocity. Function of nozzle is to increase the velocity of flow so option (b) is correct.

6. Beyond the Drag-divergence mach number, drag will suddenly increase due to the formation of terminal normal shock on the upper surface of airfoil. The terminal normal shock is formed in the high mach number region, due to which pressure behind the shock is large. This strong adverse pressure gradient leads to boundary layer separation.

Super critical airfoil has relatively flatter upper surface which forces the normal shock further towards the trailing edge. In other words in low mach number zone.

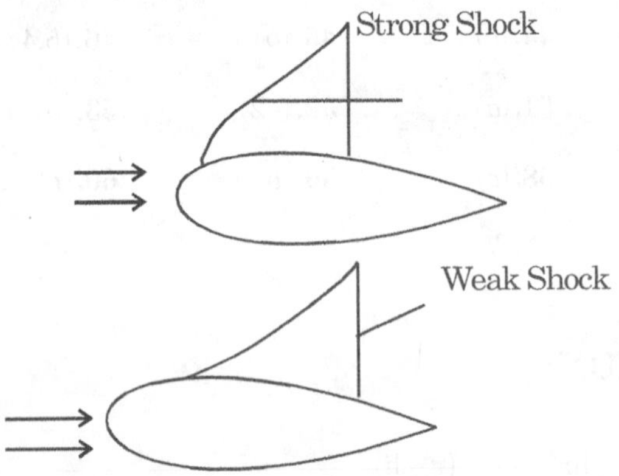

8. From this graph, we can see that, for a particular mach number with increase in 'θ' the shock angle increases.

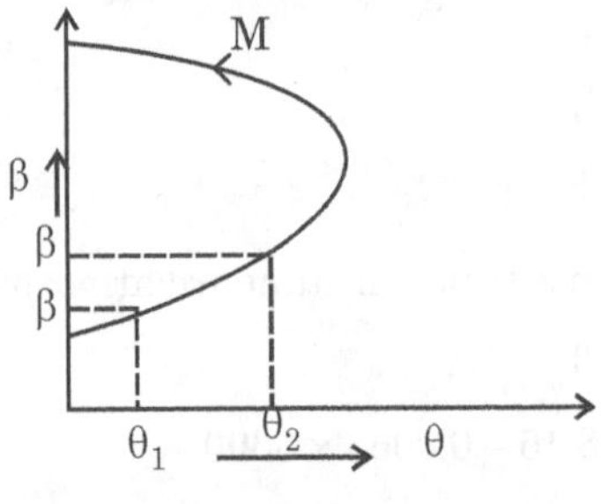

10. We know that, $E = 2G(1 + v)$

$$\text{so} \frac{E}{2} = G(1+v)$$

11. 30.414×10^6 Pa

Radius of mohr's circle

$$= \sqrt{\left(\frac{\sigma_1 - \sigma_2}{2}\right)^2 + \tau^2}$$

$$= \sqrt{\left(\frac{60-50}{2}\right)^2 + (30)^2}$$

= 30.414 MPa

= 30.414×10^6 Pa

12. For both end fixed, $\ell e_1 = \frac{\ell}{\sqrt{2}}$

for both end Hinged, $\ell e2 = \ell$

$$\frac{\ell e_1}{\ell e_2} = \frac{\ell}{\sqrt{2} \times \ell} = \frac{1}{\sqrt{2}}$$

13. for NACA-5 digit airfoil

$$(t)_{max} = \text{last two digit} \times \frac{C}{100}$$

$$\frac{C}{(t)_{max}} = \frac{100}{12} = 8.33$$

14. A = 4, B = –6, C = 1

$B^2 - 4AC = (-6)^2 - 4 \times 4 \times 1$

= 36 – 16

= 20

So, $B^2 - 4AC > 0$

If $B^2 - 4AC > 0 \rightarrow$ then it is hyperbola

15.

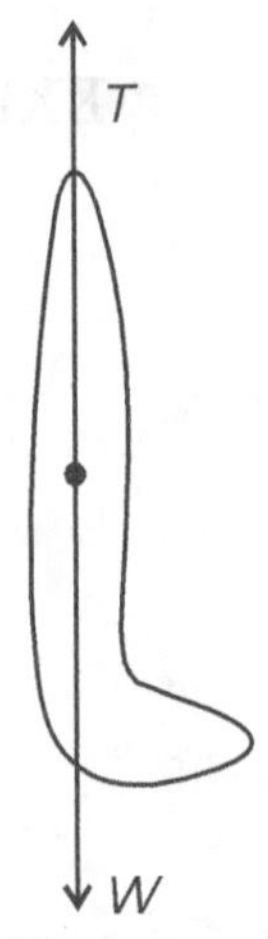

Thrust produced by the engine will balance the weight of aircraft.

Hence, T > W

16.

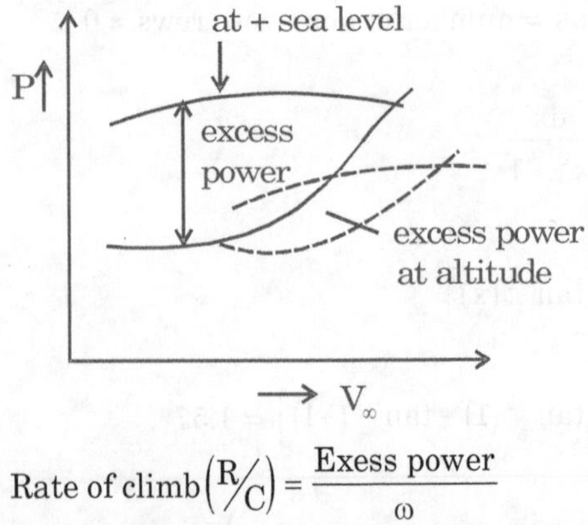

$$\text{Rate of climb} \left(R/C\right) = \frac{\text{Exess power}}{\omega}$$

with increase in altitude the magnitude of excess power reduces hence R/C decreases.

19. $(V_\infty) = 300 \times \dfrac{1000}{3600} = 83.333\ m/s$

$$C_p = 1 - \left(\frac{V}{V_\infty}\right)^2$$

$$-2 = 1 - \left(\frac{V}{V_\infty}\right)^2$$

$$\left(\frac{V}{V_\infty}\right)^2 = 3$$

$V = \sqrt{3} \times V_\infty = \sqrt{3} \times 83.333$

V = 144.34 m/s

20. We know that, $g = go\left(\dfrac{r}{r + h_G}\right)^2$

$$g = 10 \times \left(\frac{6400}{6400+2}\right)^2 = 9.99 m/s^2$$

$$\frac{g}{m} = 9.99 \frac{N}{Kg}$$

24. $\dfrac{dx}{dy} + \dfrac{x}{y} = 1$

It is in the form of $\dfrac{dx}{dy} + Px = Q$

$$P = \frac{1}{y}$$

I.F. = $e^{\int P\, dy}$

$= e^{\int \frac{1}{y} dy}$

$= e^{\log y} = y$

Solution, $x \times (I.F.) = \int Q \times (I.F.)\, dy + C$

$x \times y = \int 1.y\ dy + C$

$$xy = \frac{y^2}{2} + C$$

$$x = \frac{y}{2} + \frac{C}{y}$$

x(1) = 1

$$1 = \frac{1}{2} + \frac{C}{1}$$

$$\Rightarrow C = \frac{1}{2} = 0.5$$

25. from Kepler's 3rd law:-

$$\left(\frac{T_1}{T_2}\right)^2 = \left(\frac{a_1}{a_2}\right)^3$$

$$\left(\frac{365}{T_2}\right)^2 = \left(\frac{1.49527 \times 10^{11}}{2.278 \times 10^{11}}\right)^3$$

$$(T_2)^2 = (365)^2 \times \left(\frac{2.278}{1.49527}\right)^3$$

T_2 = 686.35 days

26. m_p = 70%, m_i = 0.7m

$m_f = m_i - m_p = m_i - 0.7m_i$

$= 0.3m_i$

$$V_b = Ve\, \ell n \frac{m_i}{m_f}$$

$$= 3000 \times \ell n\left(\frac{m_i}{0.3\ m_i}\right)$$

= 3611.918 m/s

$$h = \frac{(V_b)^2}{2g} = \frac{(3611.918)^2}{2 \times 9.81}$$

= 664931 m

= 664.931 km

28. $C_C = 2\sqrt{km} = 2\sqrt{20 \times 1000 \times 100}$

= 2828.427 kg/s

$$\zeta = \frac{C}{C_C}$$

C = 2828.427 × 0.4 = 1131.37

29.

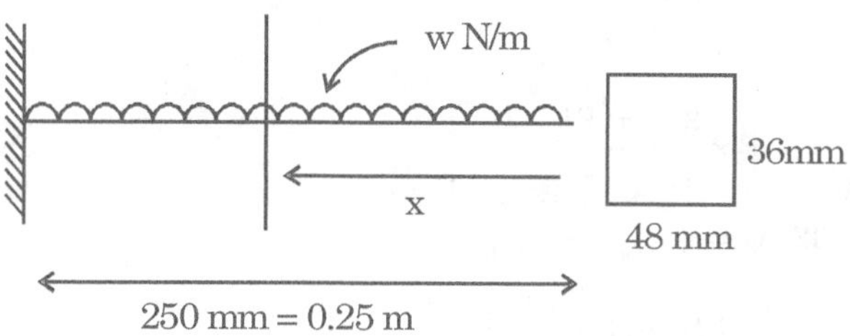

$$m_x = -wx \times \frac{x}{2}$$

$$EI \frac{d^2y}{dx^2} = \frac{-wx^2}{2}$$

$$EI\frac{dy}{dx} = -\frac{wx^3}{6} + C_1$$

$$EJy = -\frac{wx^4}{24} + C_1x + C_2$$

at x = 0.25m, y = 0

$$0 = -\frac{w \times (0.25)^4}{24} + C_1 \times 0.25 + C_2$$

$0.25C_1 + C_2 = 1.6276 \times 10^{-4}$ w ...(1)

at x = 0.25 m, $\frac{dy}{dx} = 0$

$$0 = -w \times \frac{(0.25)^3}{6} + C_1$$

$C_1 = 2.6042 \times 10^{-3} w$

$C_2 = 1.6276 \times 10^{-4}$ w – 0.25 × 2.6042 × 10^{-3} w

$C_2 = -4.8829 \times 10^{-4}$ w

Maximum deflection occurs at free end (x = 0)

$EIy_{max} = C_2$

$EIy_{max} = C_2$

$70 \times 10^9 \times 1.86624 \times 10^{-7} \times 1 \times 10^{-3}$

$= -4.8829 \times 10^{-4}$ w

w = 26.753 kN/m

30. Let the forces on member 'CD' and 'EF' is Euler's buckling load,

$$P_{cr} = \frac{\pi^2 EI}{\ell^2}$$

Taking moment about 'A'

$$P \times 4a = P_{cr} \times a + P_{cr} \times 2a$$

$$P \times 4 = \frac{\pi^2 EI}{\ell^2} + \frac{\pi^2 EI}{\ell^2} \times 2$$

$$P \times 4 = \frac{3\pi^2 EI}{\ell^2}$$

$$P \times \frac{3}{4} = \frac{\pi^2 EI}{\ell^2}$$

32. $\psi = \psi_{uniform} + \psi_{doublet}$

$$= 10r\sin\theta - \frac{100}{2\pi}\frac{\sin\theta}{r}$$

$$= 10\ r\sin\theta - \left(1 - \frac{100}{2\pi \times 10r^2}\right)$$

$$= 10\ r\sin\theta - \left(1 - \frac{R^2}{r^2}\right)$$

$$R^2 = \frac{100}{2\pi \times 10}$$

$$R = \sqrt{\frac{100}{2\pi \times 10}} = 1.262 \text{ m}$$

33. T = – (V_1dn – V_2 dn + u_1 dS – u_2 dS)

rds = (V_2 – V_1) dn + (u_2 – u_1) ds

10 × 1 = (5 – 10) × 0.5 + (2 – x)1

10 + 5 × 0.5 = 2 – x

x = –10.5 m/s

34. $\rho_1 u_1 = \rho_2 u_2$.

$$\frac{u_2}{u_1} = \frac{\rho_1}{\rho_2} = \frac{2 + (\gamma - 1)M_1^2}{(\gamma + 1)M_1^2}$$

$$\frac{u_2}{u_1} = \underset{M_1 \to \infty}{Lt} \frac{M_1^2\left[\frac{2}{M_1 2} + (\gamma - 1)\right]}{(\gamma + 1)M_1^2}$$

$$\frac{u_2}{u_1} = \frac{\gamma - 1}{\gamma + 1} = \frac{1.4 - 1}{1.4 + 1}$$

$$\frac{u_2}{u_1} = 0.1667$$

35. for 50% reaction stage

$\alpha_1 = \beta_2 = 15°$

$\beta_1 = \alpha_2 = 45°$

$$R = \frac{C_a}{2u}(\tan\beta_1 + \tan\beta_2)$$

$$\frac{1}{2} = \frac{C_a}{2u}(\tan\beta_1 + \tan\beta_2)$$

$$\frac{1}{2} = \frac{200}{2u}(\tan 45° + \tan 15°)$$

u = 253.589 m/s

$$p = \dot{m}w = \dot{m}u(C_{t2} - C_{t1}) = \dot{m}uC_a(\tan\alpha_2 - \tan\alpha_1)$$

$$\dot{m} = \frac{300 \times 10^3}{253.589 \times 200 \times (\tan 45 - \tan 15)}$$

$$\dot{m} = 8.08 \text{kg/s}$$

Tangential force, $f = \dot{m}(C_{t2} - C_{t1})$

$$= \dot{m}C_a(\tan\alpha_2 - \tan\alpha_1)$$

$= 8.08 \times (\tan 45° - \tan 15°)$

36. For low speed operation to produce high thrust large mass flow rate of air is required.

So capture area will be greater than the intake inlet area and flow will accelerates from freestream to the intake inlet area and flow will accelerates from freestream to the intake.

37. $P\left(\frac{\text{2 heads in 3 tosses}}{\text{first toss is head}}\right)$

$$= \frac{\begin{pmatrix}\text{2 heads in 3 tosses} \\ \text{and first toss is head}\end{pmatrix}}{P(\text{first toss is head})}$$

$$= \frac{P(HHT) + P(HTH)}{P(H)}$$

$$= \frac{1}{2} \times \frac{1}{2} \times \frac{1}{2} + \frac{1}{2} \times \frac{1}{2} \times \frac{1}{2}$$

38. $\frac{d^2y}{dx^2} + P\frac{dy}{dx} + qy = 0$

$$\left(D^2 + PD + q\right)y = 0$$

$$D^2 + PD + q = 0 \quad ...(1)$$

Solution of equation (1) is

$y = C_1 e^{-x} + C_2 e^{-3x}$.

Roots are –1 and –3

$\alpha = -1$ and $\beta = -3$

$D^2 + PD + q = D^2 - (\alpha + \beta) D + \alpha.\beta = 0$

$-(\alpha + \beta) = P$

$P = -(-1-3) = 4$

$\alpha.\beta = q$

$q = (-1)(-3) = 3$

39. $\frac{1}{s(s+1)} = \frac{1}{s} - \frac{1}{s+1}$

$$L^{-1}\left[\frac{1}{s(s+1)}\right] = L^{-1}\left(\frac{1}{s}\right) - L^{-1}\left(\frac{1}{s+1}\right)$$

$= 1 - e^{-t}$.

41. $\eta_{propulsive} = \frac{\text{Thrust power}}{\text{propulsive power}}$

$$= \frac{T \times V_i}{\frac{1}{2}\left(\dot{m}_a + \dot{m}_f\right)V_e^2 - \frac{1}{2}\dot{m}_a v_i^2}$$

$$= \frac{\left[\left(\dot{m}_a + \dot{m}_f\right)V_e - \dot{m}_a V_i\right]V_i}{\frac{1}{2}\left(\dot{m}_a + \dot{m}_f\right)V_e^2 - \frac{1}{2}\dot{m}_a V_i^2}$$

for $\dot{m}_f << \dot{m}_a$

$$\eta_{propulsive} = \frac{\dot{m}_a\left(V_e - V_i\right)V_i}{\frac{1}{2}\dot{m}_a\left(V_e - V_i\right)\left(V_e + V_i\right)}$$

$$= \frac{2V_i}{V_e + V_i}$$

So propulsive efficiency is maximum when $V_e = V_i$.

$\eta_{prop} = 100\%$

So for $V_e = v_i$ both thrust and thrust power is zero

42. $\eta_{is} = \frac{T_{02} - T_{01}}{T_{02} - T_{01}} = \frac{\gamma_p^{\frac{\gamma-1}{\gamma}} - 1}{\gamma_p^{\frac{n-1}{n}} - 1}$

$$\frac{n-1}{n} = \frac{\gamma-1}{\gamma} \times \frac{1}{\eta_P}$$

$$\eta_{is} = \frac{\pi_c^{\frac{\gamma-1}{\gamma}} - 1}{\pi_c^{\frac{\gamma-1}{\gamma} \times \frac{1}{\eta_p}} - 1}$$

43. 2.09 kg/s

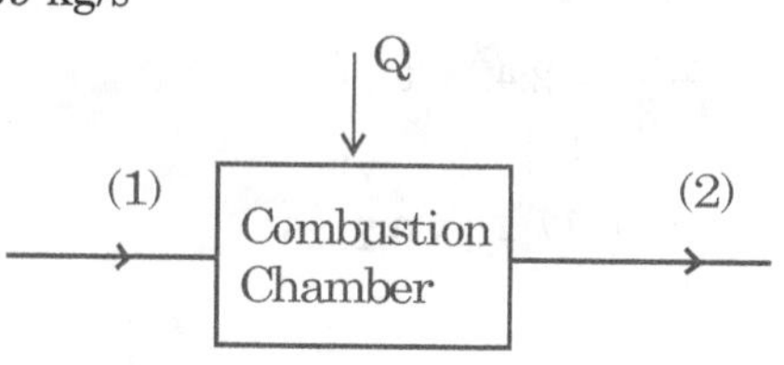

$$\eta_{cc} = \frac{h_{o2} - h_{oi}}{\dot{m}f \times HV}$$

$$\eta_{cc}\,\dot{m}_f \times HV = \left(\dot{m}_a + \dot{m}_f\right)C_{pg}T_{02} - \dot{m}_a C_{pa}T_{01}$$

$$\dot{m}_f\left(\eta_{cc} \times HV - C_{pg}T_{02}\right)\dot{m}_a\left(C_{pg}T_{02} - C_{pa}T_{01}\right)$$

$$\dot{m}_f = \frac{\dot{m}_a\left(C_{pg}T_{02} - C_{pa}T_{01}\right)}{\eta_{cc}HV - C_{pg}T_{02}}$$

$$= \frac{80 \times (1005 \times 400 - 1147 \times 1300)}{1147 \times 1300 - 44 \times 10^6} = 2.0496 \text{ kg/s}$$

44. For first stage

$m_{i1} = 8200 + 800 + 5400 + 500 + 80$

$= 14980$ kg

$m_{f1} = 14980 - m_{p1}$.

$= 14980 - 8200 = 6780$ kg

for second stage:

$m_i = m_{f1} - m_{s1}$

$= 6780 - 850 = 5980$ kg

$m_{f2} = 5980 - 5400 = 580$ kg

$$\Delta V_b = (\Delta V_b)_1 + (\Delta V_b)_2$$

$$= gI_{sp} \ell n \frac{m_{i1}}{mf_1} \times \frac{m_{i2}}{m_{f2}}$$

$$= 10 \times 275 \times \ell n\left(\frac{14980 \times 5980}{6780 \times 580}\right)$$

$= 8596$ m/s

45. Time period $(T) = \frac{2\pi\hat{t}}{|\beta|}$

β = Imaginary parts of root

$\hat{t}$ = Airplane density factor

for $\lambda_{1,2} = -2.508 \pm i\, 2.587$

$$T_1 = \frac{2\pi\hat{t}}{2.587} = 2.4287\hat{t}$$

for $\lambda_{3,4} = -0.17 \pm \hat{\ell} \times 0.2125$

$$T_2 = \frac{2\pi\hat{t}}{0.2125} = 29.568\hat{t}$$

Since time period of $\lambda_{1,2}$ is small so its represent short period oscillation and $\lambda_{1,2}$ is small so its represent short period oscillation and $\lambda_{3,4}$ represents phugoid mode.

46. General equation of motion for phugoid

$$\frac{d^2h}{dt^2} + 2\frac{g^2}{V^2}h = 0$$

Given equation, $\frac{d^2h}{dt^2} + \frac{4}{3}h = 0$

$$T = \frac{2\pi}{\omega_n} \text{ and } \omega_n = \sqrt{\frac{2g^2}{V^2}}$$

$$\omega_n = \sqrt{\frac{4}{3}} = 1.155$$

$$T = \frac{2\pi}{1.155} = 5.44 \text{ sec.}$$

47. $A = 1 \text{ cm}^2 = 10^{-4} \text{ m}^2$

$I_{yy} = 4 \times A \times (0.03)^2$

$= 4 \times 10^{-4} \times (0.03)^2$

$$\frac{M}{I} = \frac{\sigma}{y}$$

$$\frac{M}{I} = \frac{\sigma}{y}$$

$$= \frac{10 \times 0.03}{4 \times 10^{-4} \times (0.03)^2}$$

$= 833333.33 \text{ N/m}^2$

this is σ_{max} at level flight i.e. n = 1

At n = 3.5, the load on the aircraft increases by n times on the aircraft structure

$\sigma_{max}|_{n=3.5} = 3.5 \times 833333.33$

$= 2.916 \times 10^6 \text{ N/m}^2$

$\simeq 30 \times 10^5 \text{ N/m}^2$

48. $R/C = V_\infty \sin\theta = 60 \times \sin 15° = 15.529$ m/s

$$R/C = \frac{\text{Excess thrust} \times V_\infty}{w}$$

$$\text{Excess thrust} = \frac{R/C \times w}{V_\infty}$$

$$= \frac{15.529 \times 4000 \times 9.8j}{60}$$

$= 10.156$ KN

49. we know that

$$\frac{dC_{mg}}{dC_L} = -H_N$$

For static stability, $\frac{dC_{mg}}{dC_L} < 0$

Hence > 0

50. $a = \frac{dC_L}{d\alpha} = \frac{0.52-0}{5-(-1.5)} = 0.08$ per deg ree

$C_{mcg} = C_{mac} + a\,\alpha\,(h - hac)$

$-0.01 = Cmac + 0.08\,(1 + 1.5)\,(h - hac)$ ...(1)

at $\alpha = 7.88°$

$0.05 = C_{mac} + 0.08\,(7.88 + 1.5)\,(h - hac)$

subtracting (2) and (1)

$-0.06 = -0.55\,(h - hac)$

$h - hac = 0.11$

$hac = 0.35 - 0.11 = 0.24$

51. $f(x) = 1 - x^2$

$f'(x) = -2x$

By Newton's Raphson formula

$$x_{k+1} = x_k - \frac{f(x_k)}{f'(x_k)}$$

$$= x_k - \frac{(1-x_k^2)}{-2x_k}$$

$$= \frac{2x_k^2 + 1 - x_k^2}{2x_k}$$

$$= \frac{x_k^2 + 1}{2x_k}$$

52. $\vec{A} = Kr^n\hat{i}$

$$\nabla\vec{A} = \frac{1}{r^2}\frac{\partial}{\partial r}(r^2kr^n)$$

$$= \frac{1}{r^2}\frac{\partial}{\partial r}(kr^{n+2})$$

$$= \frac{1}{r^2} \times (n+2)k.r^{n+1} = (n+2)K.r^{n-1}$$

$\nabla.\vec{A} = 0$

$n + 2 = 0$

$n = -2$

56. "Bear with" means "to have patience with."

57. The other spellings are wrong.

58. All are prime numbers except 33.33 is divisible by 3 & 11.

59. Since no other robot except R2D2 robot can repair aeroplanes. Therefore, option (b) is correct.

60. $|9y - 6| = 3$

Possible values of y are $1 \& \frac{1}{3}$.

For y = 1 : $y^2 - \frac{4y}{3}$

$\Rightarrow 1 - \frac{4}{3} = -\frac{1}{3}$

For $y = \frac{1}{3}$: $y^2 - \frac{4y}{3}$

$\frac{1}{9} - \frac{4}{3} \times \frac{1}{3} = \frac{-1}{3}$

61. Large plant production = 160 + 190 + 230 + 190 = 770 ...(i)

Small plant production = 650 ...(ii)

Difference = (i) – (ii) = 770 – 550 = 120

62. From the last line, option C can be inferred.

63. SITs will have a national character like IITs. Statement iv can be inferred from what is stated in the given premise.

64. Here we have,

n = number of trials,

p = success probability = 0.6,

q = failure probability = 0.4 and

k = number of shots hit = 6

p (6 hits in 10 attempts) = $C(10, 6)p^6q^4$

$= \frac{10!}{6!4!}(0.6)^6(0.4)^4$

$= 210 \times .046656 \times .0256$

$= .2508$

65. $211^{870} + 146^{127} \times 3^{424}$ unit digits

$= 1^{870} + 6^{127} + 3^{424}$ units digit

$= 1 + 6 \times 1 = 7$

MOCK TEST – 11

AEROSPACE ENGINEERING

Time Allowed: 3:00 Hours *Maximum Marks: 100*

Read the following instructions carefully before you begin to answer the questions. This Booklet contains questions in English.

INSTRUCTIONS

1. This Booklet contains **65 questions** in total comprising of the following two parts:

   Part -1: Technical Section (85 Marks) *(1-55 Questions)*

   Part -2: General Aptitude Section (15 Marks) *(56-65 Questions)*

2. All questions are compulsory. Q.1-Q.25 & Q.56-Q.60 carry one mark each. Q.26-Q.55 & Q.61-Q.65 carry two marks each.

3. **Types of Questions:**

   (*a*) Multiple Choice Questions(MCQ) carrying 1 or 2 marks each in all papers and sections. These questions are objective in nature, and each will have a choice of four options, out of which the candidate has to mark the correct answer(s).

   (*b*) Numerical Answer Questions of 1 or 2 marks each in all sections. For these questions the answer is a real number, to be entered by the candidate using the virtual keypad. No choices will be shown for these type of questions.

4. For 1-mark multiple-choice questions, 1/3 marks will be deducted for a wrong answer. Likewise, for 2-marks multiple-choice questions, 2/3 marks will be deducted for a wrong answer. There is no negative marking for numerical answer type questions.

MOCK TEST – 11

1. Which of the following is incorrect for adverse yaw
   (a) It is yawing moment in opposite direction of roll.
   (b) It's caused partly by the change in drag between left and right wings.
   (c) An unwanted side effect of ailerons operation is adverse yaw.
   (d) During adverse yaw rising wing generates decreased lift which causes increased induce drag.
2. A slip is a aerodynamic state where an aircraft
   (a) moves somewhat sideways as well as forward relative to the coming on airflow or relative wind.
   (b) moves what some ways side as well as back ward relative to coming on airflow or relative wind.
   (c) moves some what sideways and flow is coming parallel to longitudinal axis.
   (d) None of these
3. Radius of Mohr circle for the stress state shown in figure is

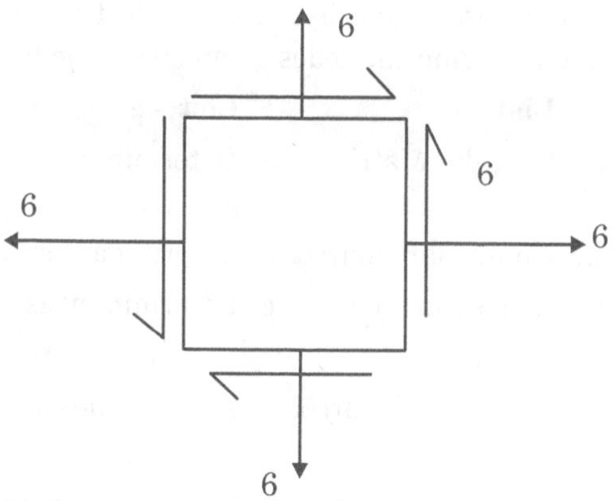

   (a) 0 (b) 26
   (c) 36 (d) 6
4. The ratio of Eulers buckling loads of Columns with the same parameters having (i) both ends fixed, and (b) both ends hinged is ______.
5. In a steady flow through Nozzle, the flow velocity on the Nozzle axis given by $V = V_0\left(1+\frac{3x}{L}\right)$, where x is the distance along the axis of nozzle from the inlet plane and L is the length of the nozzle the time required for the fluid particle on the axis of travel from inlet to exit plane of nozzle for L = 2 m and V_0 = 5 m/sec is
   (a) 0.1848 sec (b) 1.8 sec
   (c) 0.25 sec (d) 1.4 sec
6. DB denotes double base solid propellant LOX–RP1 denotes liquid oxygen - kerosene combination LOX-LH_2 denotes liquid oxygen-hy-drogen combination.
   The correct order of increasing specific impulse is
   (a) DB < LOX – RP1 < LOX – RP2
   (b) LOX – RP1 < DB < Lox – LH_2
   (c) LOX – LH_2 < DB < LOX – RP1
   (d) DB < LOX – LH_2 < LOX – RP1
7. The maximum possible value of C_P pressure coefficient for compressible flow is.
   (a) 1 (b) 0.5
   (c) 2 (d) None of these
8. The parameters that remains constant in cruise climb for an aircraft are
   (a) Equivalent airspeed and lift coefficient
   (b) Altitude and lift coefficient
   (c) Equivalent airspeed and altitude
   (d) lift coefficient and aircraft mass
9. In half Rankine oval condition the distance between stagnation point and source (If m strength of vortex and V_∞ = free stream velocity) is
   (a) $\frac{m}{2\pi V_\infty}$ (b) $\frac{m}{4\pi V_\infty}$
   (c) $\frac{m}{\pi V_\infty}$ (d) $\frac{m}{5\pi V_\infty}$
10. The equation of motion for vibrating rod is given by $\frac{\partial^2 u}{\partial x^2} = \frac{1}{c^2}\frac{\partial^2 u}{\partial t^2}$, u = displacement along the rod and is function of both position x and time t. The natural frequerries ω of the fixed-fixed condition is (ℓ = length of road)
   (a) $\cos\left(\frac{w\ell}{c}\right) = 0$ (b) $\sin\left(\frac{w\ell}{c}\right) = 0$
   (c) $\cos\left(\frac{wc}{\ell}\right) = 0$ (d) $\sin\left(\frac{wc}{\ell}\right) = 0$

11. Drag vidergenle mach number is usually

(*a*) Close to and always greater than the critical mach number

(*b*) equal to critical mach number

(*c*) Less than the critical mach number

(*d*) None of these

12. Centre of pressure is chord wise location about which the

(*a*) pitching moment is varying with angle of attack

(*b*) pitching moment is zero

(*c*) pitching moment is constant

(*d*) pitching moment is infinity

13. Interferometer is technique which measures

(*a*) pressure (*b*) Temperature

(*c*) Density (*d*) Velocity

14. A beam is made up of two identical bars AB and BC by hinging them together at B. The end A is built-in (Cantilever) and the end C is simply supported, with P acting as shown in figure, the bending moment at point A is.

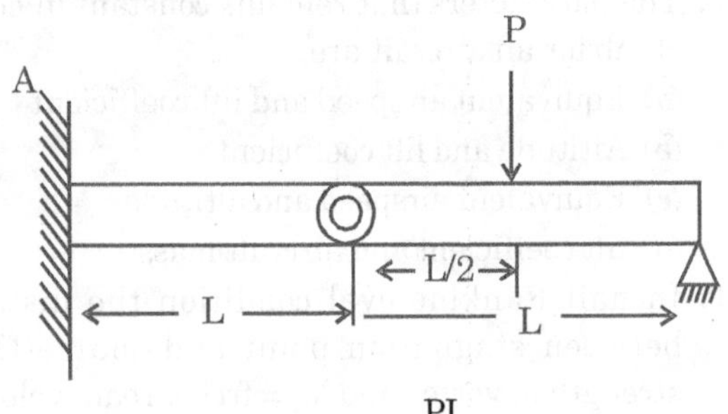

(*a*) zero (*b*) $\frac{PL}{2}$

(*c*) $\frac{3PL}{2}$ (*d*) Indeterminate

15. For forced vibration system. When damping ratio is zero and frequency ratio 0.5 the magnification factor for damped system is ______

16. An elliptic wing has

(*a*) elliptic lift distribution, elliptic downwash

(*b*) elliptic lift distribution, constant downwash

(*c*) Constant lift distribution, elliptic downwash

(*d*) None of the above.

17. Combustor efficiency of the aircraft engine

(*a*) decrease with altitude

(*b*) remains constant with altitude

(*c*) Increases with altitude

(*d*) may increase or decrease.

18. If the amplitude of oscillation of a system reduces by $\frac{1}{4}$ after 12 oscillation, then damping ratio ______.

19. For solid propellant rocket engine, when the area ratio decreases with time is called.

(*a*) Regressive burning

(*b*) Progressive burning

(*c*) Neutral burning

(*d*) Neutral and regressive

20. Match the following

P. Isocoric process M. $W = \frac{R(T_2 - T_1)}{r - 1}$

Q. Isobaric process N. $W = 0$

R. Isothermal process O. $W = RT_1 \ell n \frac{e_1^{(roh)}}{e_2}$

S. Adiabatic process K. $W = P(V_2 - V_1)$

(*a*) P - M, Q - N, R - K, S - O

(*b*) P - N, Q - K, R - O, S - M

(*c*) P - N, Q - N, R - O, S - M

(*d*) P - M, Q - N, R - K, S - M

21. Match list - 1 and list - 2 and select the correct answer using the codes given below the lists

List - 1	**List - 2**
A. Singular Matrix	a. Determinant is not defined
B. Non-square matrix	b. Determinant is always
C. Real symmetric matix	c. Determinant is zero.
D. Orthogonal matrix	d. Eigen values are always real
	e. Eigen values are not defined

Codes

(*a*) A -c, B - a, C- d, D - b

(*b*) A - a, B - b, C - d, D - c

(*c*) A - b, B - c, C - d, D - c

(*d*) A - c, B - d, C - a, D - e

22. The eigen values of the following matrix are $\begin{bmatrix} -1 & 3 & 5 \\ -3 & -1 & 6 \\ 0 & 0 & 3 \end{bmatrix}$

(a) 3, 3 + 5j, 6 – j (b) –6 + 5j, 3 + j, 3 – j

(c) 3 + j, 3 – j, 5 + j (d) 3, – 1 + 3j, –1 – 3j

23. The area enclosed between the curves $y^2 = 4x$ and $x^2 = 4y$ is

(a) $\frac{16}{3}$ (b) 8

(c) $\frac{32}{3}$ (d) 16

24. The solution of differential equation $x(f) + 2x(t) = 8(t)$ with initial condition $x(0) = 0$ is

(a) $e^{-2t}u(t)$ (b) $e^{2t}u(t)$

(c) $e^{-t}u(t)$ (d) $e^{t}u(t)$

25. Match the list 1 and list 2 to select. The correct answer.

P. Newton - Rapson method	K. Solving non-linear equations
Q. Range - Kutla method	L. Numerical integration
R. Simpson's rule	M. Solving ordinary differential equation
S. Gauss elimination method	N. Solving simultaneous equations

(a) P - K, Q - L, R - M, S - N

(b) P - K, Q - M, R - L, S - N

(c) P - M, Q - P, R - L, S - N

(d) P - M, Q - L, R - P, S - N

26. Let P, Q, R, S be n × n matrices, each with non - zero determinant, if PQRS = 1, then Q^{-1} is

(a) $S^{-1} R^{-1} P^{-1}$

(b) RSP

(c) PSR

(d) does not necessarily exist

27. Consider points A and B in the x – y plane, with A = (1, 0) and B = (0,1). The line integral

$2\int_A^B (xdx + ydy)$ along the semicircle with the line segment AB as its diameter is _____ .

28. For a thin walled closed circular cross - section of radius 'r' and thickness 't' the applied twisting moment is T. If for the same circular cross - section there is a small slit (as shown in fig), then the ratio of maximum shear stress in first to second is

(a) $\frac{3t}{r}$ (b) $\frac{t}{3r}$

(c) $\frac{3t}{2r}$ (d) $\frac{2t}{3r}$

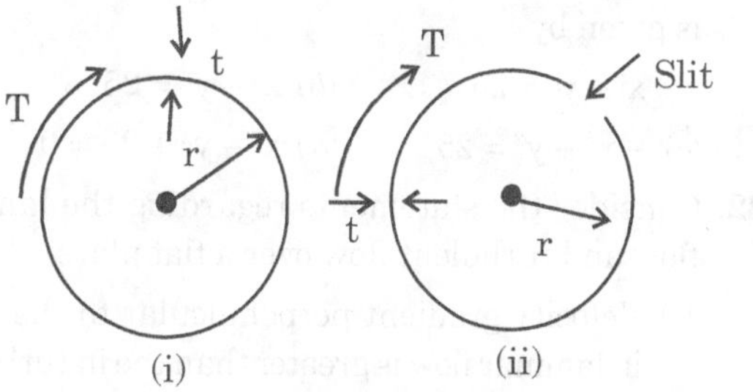

29. A thin walled two - cell closed section (shown in fig) is subjected to twisting moment of magnitude 50 N - m. If 'q_1' and 'q_2' is shear flow in cell '1' and 2 then magnitude of shear stress in member 'BE' is.

(Assume thickness of wall is uniform = 1 mm)

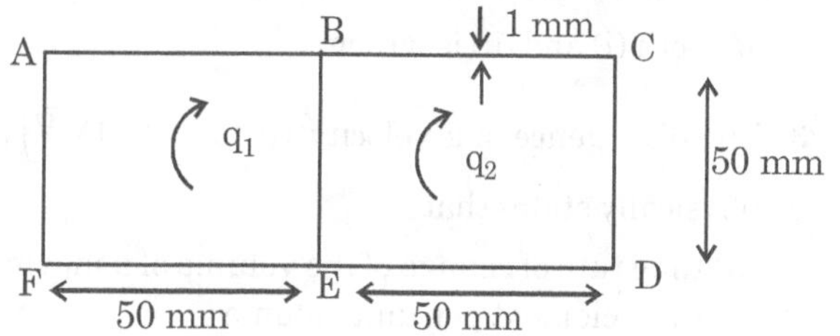

(a) 5 N/mm (b) 10 N/mm

(c) 7 N/mm (d) 0 N/mm

30. A Structure is idealized by using massboom concept as shown in fig, with area of each boom is 100 mm². The Shear centre of system is lies at a distance of e_x mm from point '0' Then the magnitude of e_x is given by

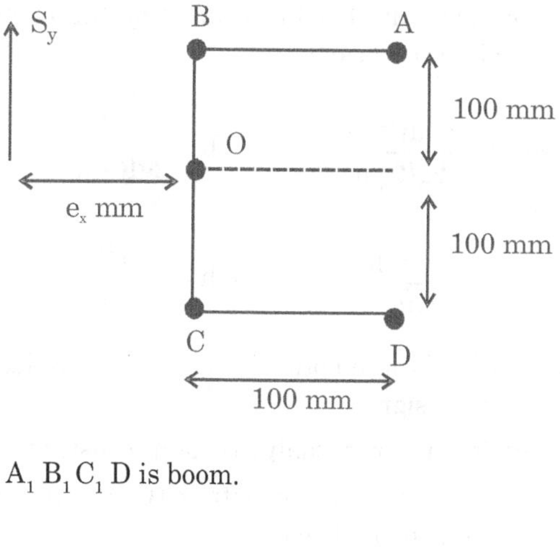

A_1 B_1 C_1 D is boom.

31. The velocity field is given by the equation

$$U = \frac{y}{x^2 + y^2} \text{ and } V = \frac{-x}{x^2 + y^2}$$

Then the equation of streamline for the given velocity field and passing through the point (O, 5) is given by

(a) $x^2 + y^2 = 25$ (b) $x^2 - y^2 = 25$

(c) $-x^2 + y^2 = 25$ (d) $x^2 - y^2 + 25 = 0$

32. Consider the statements regarding the laminar flow and turbulent flow over a flat plate

(i) Velocity gradient perpendicular to the plate in laminar flow is greater than the in turbulent flow.

(ii) Shear stress in laminar flow is greater than the turbulent flow which option is correctly matched these statement

(a) (i) is true,(ii) is wrong

(b) (i) is wrong,(ii) is true

(c) both (i) and (ii) is true

(d) both (i) and (ii) is wrong

33. The divergence of a velocity vector '$\vec{V}$' $(\nabla.\vec{V})$ is physically states that

(a) time rate of change of the volume of a moving fluid element per unit volume

(b) time rate of change of volume of a stationary fluid element per unit volume.

(c) time rate of change of velocity of a moving fluid element, per unit volume

(d) time rate of change of velocity of a moving fluid element.

34. A rocket of mass 'm' is lanced vertically with initial speed u Assuming negligible air resistance. Then the maximum height attained by the rocket is (R = Radius of earth)

(a) $h = \frac{uR^2}{2gR - u}$ (b) $h = \frac{uR^2}{2gR + u}$

(c) $h = \frac{u^2R}{2gR - u^2}$ (d) $h = \frac{u^2R}{2gR + u^2}$

35. Which of these options are correct for the free vortex design

(a) Stagnation enthalpy remains constant

(b) $C_t \times r$ remains constant (C_t is tangential component of velocity)

(c) Entropy remains constant

(d) All of the above

36. If η_d = Efficiency of diffuser, η_c = Efficiency of compressor, η_b = Efficiency of combustion chamber, η_t = Efficiency of turbine and η_n = Efficiency of nozzle of a turbojet engine then which option is correct

(a) $\eta_n > \eta_t > \eta_b > \eta_c > \eta_d$

(b) $\eta_n < \eta_t < \eta_b < \eta_c < \eta_d$

(c) $\eta_b < \eta_n < \eta_t < \eta_c < \eta_d$

(d) $\eta_b > \eta_n > \eta_t > \eta_c > \eta_d$

37. An ideal ramjet is flying with a velocity of 1000 m/s at an altitude where ambient pressure and temperature is 0.25 bar and 225 k respectively. The exhaust gases from the engine are optimally expanded and leave the nozzle at 1000 k . If the the specific heat ratio (r) remain's constant, then the specific thrust developed by the engine is given by (in N – S /kg) _____ .

38. The mach number at the inlet of a nozzle is 1.2 then the shape of the nozzle should be

(a) Converging

(b) Diverging

(c) Converging - Diverging

(d) Diverging - Converging

39. Which one of the given diagram correctly represents the ideal gas?

(a) P_V

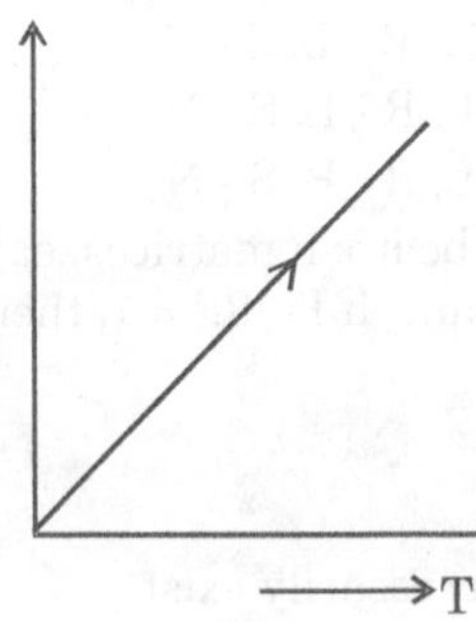

(b) P_V

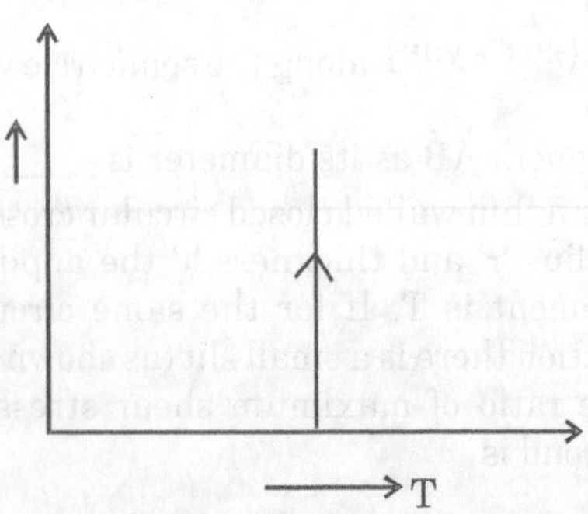

(c) 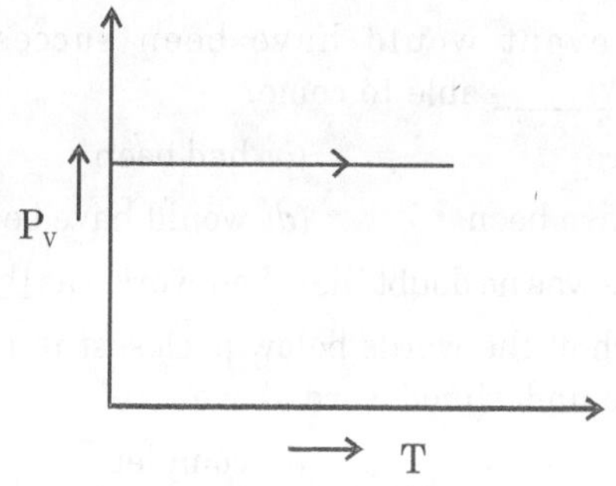

(d) 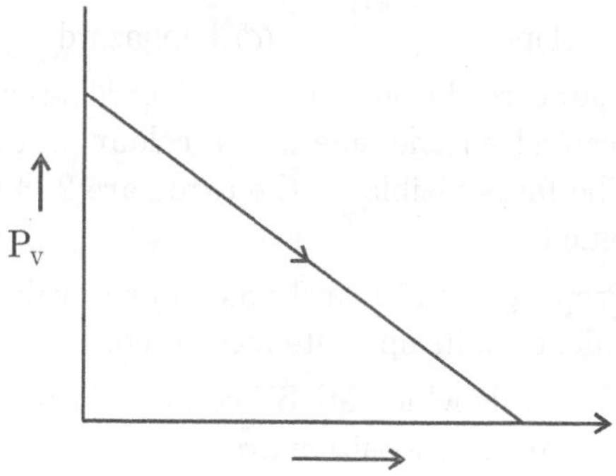

40. The Stagnation temperature at Compressor inlet is increased by two times and stagnation pressure is reduced by 0.5 times of its value. In order to maintain the same pressure ratio across the compressor, the mass flow rate will be reduced by the factor

(a) 0.4 (b) 0.35

(c) 0.3 (d) 0.45

41. An aircraft weighing 30000 N having wing surface area 10 m², Zero lift drag Co-efficient 0.015, oswald efficiency factor 0.6 and aspect ratio 6 is flying at a speed of 100 m/s at an altitude where air density is 1 kg/m³. The drag force acts on the aircraft is (in KN) _____ .

42. Read the statements given regarding to ventral fin

(i) Ventral fin provide positive static stability in directional mode.

(ii) Ventral fin provide Negative static stability in lateral stability.

(iii) Ventral fin provide proper ground clearance during take - off.

Which statements are true.

(a) (i), (ii) & (iii) (b) (ii) & (iii)

(c) (i) (d) (i) & (ii)

43. The Characteristics equation of longitudinal dynamic stability of an airplane is given by

$$\lambda^4 + 6.05\lambda^3 + 13\lambda^2 + 0.6\lambda + 0.6 = 0$$

Which option is correct statement of the above equation.

(a) Airplane will be dynamically Stable.

(b) Airplane will be dynamically Neutral.

(c) Airplane will be dynamically Unstable

(d) Can't decide about dynamic Stability.

44. The geometric & aerodynamic characteristics of a glider are given as follows

Wing : (NACA 23012, AR 8, $a_0 = 0.104$, $\alpha_{oL} = -1.2°$)

Horizontal tail (NACA 0009, AR = 4, $a_0 = 0.110$)

Tail volume ratio = 0.6 , Tail efficiency = 0.9

Location of Aerodynamic centre = 0.24 C

$\left(\frac{dc_m}{dc_L}\right)_{fuserage}$ 0.08, Elevator Area ratio = 0.35 and elevator effectiveness = 0.55, $\frac{d\varepsilon}{d\alpha} = 0.5$ at what location the stick fixed

Neutral point of glider will lie (In % age of C)

(a) 49 % (b) 39 %

(c) 59 % (d) 42 %

45. The Neutral point for stick free operation of glider in Question (44) lies at, if

Floating tendency = – 0.003

Restoring tendency = – 0.0055

(a) 38 % of chord (b) 48 % of chord

(c) 30 % of chord (d) 40 % of chord

46. In a rocket engine.40 kg/s of propellant is pumped into the combustion chamber, where the temperature after combustion is 3335 K. The combustion products have mixture values of R = 402 J /kg – K and r = 1.21 . If the throat area is 465 cm², then pressure in combustion chamber is _____ . (In KPa).

47. An aircraft of weight 'w', wing area 's' and thrust produced by power plant is 'T' require 1.5 km ground roll to lift - off from the ground. The ground roll distance required if the weight of aircraft is increased by 25% of initial weight and having same maximum lift co-efficient _____ (in km)

48. Torque exerted on a flywheel over a cycle is listed in the tabel flywheel energy (in J per unit cycle) using Simpson's rule is

Angle (degree)	0	60	120	180	240	300	360
Torque (N-m)	0	1066	-323	0	323	-355	0

49. For lateral and directional stability of the aircraft. Which of the following conditions is true?

(a) $l_\beta < 0, N_\beta < 0$ (b) $l_\beta > 0, N_\beta > 0$

(c) $l_\beta < 0, N_\beta > 0$ (d) $l_\beta > 0, N_\beta < 0$

50. A Rocket in flying at air speed at 1000 m/sec angle of attack 30° and side flip angle is 20°. Then the axial, side flip and normal velocity components are

(a) 813.80 m/sec, 342 m/sec, 469 m/sec

(b) 820 m/sec, 342 m/sec, 470 m/sec

(c) 825 m/sec, 350 m/sec, 880 m/sec

(d) None of the above

51. Equation of the line normal to function

$f(x) = (x-8)^{\frac{2}{3}} + 1$ at P (0, 5) is

(a) $y = 3x - 5$ (b) $y = 3x + 5$

(c) $3y = x + 18$ (d) $3y = x - 18$

52. A Planet is observed to be at its slowest when it is at a distance r_1 from the sun and at its fastest when it is at a distance r_2 from the sun. The eccentricity e of the planet's orbit is given by

(a) $e = \frac{r_1 - r_2}{r_1 + r_2}$ (b) $e = \frac{r_1 + r_2}{r_1 - r_2}$

(c) $e = \frac{r_1}{r_2}$ (d) $e = \frac{r_2}{r_1}$

53. The lite co-efficient for a high - aspect - ratio straight wing with a thin symmetrical airfoil at an angle of attack of 6° in a supersonic flow in mach 2.5 is _____ .

54. The ratio of turn radius of an aircraft in level turn to the pull - up maneuvers with same speed V_∞ and load factor of 'n' is given by

(a) $\sqrt{n+1}$ (b) $\sqrt{n-1}$

(c) $\sqrt{\frac{n+1}{n-1}}$ (d) $\sqrt{\frac{n-1}{n+1}}$

55. Prandtl's lifting line theory states that

(a) Effective angle of attack is equal to geometric angle of attack plus induced angle of attack.

(b) Geometric angle of attack equal to effective angle of attack plus induced angle of attack

(c) Absolute angle of attack equals to effective angle of attack plus induced angle of attack

(d) None

56. The event would have been successful if you _______ able to come.

(a) are (b) had been

(c) have been (d) would have been

57. There was no doubt that their work was thorough.

Which of the words below is closest in meaning to the underlined word above?

(a) pretty (b) complete

(c) sloppy (d) haphazard

58. Four cards lie on a table. Each card has a number printed on one side and a colour on the other. The faces visible on the cards are 2, 3 red, and blue.

Proposition: If a card has an even value on one side, then its opposite face is red.

The cards which MUST be turned over to verify the above proposition are

(a) 2, red (b) 2, 3 red

(c) 2, blue (d) 2, red, blue

59. What is the value of x when

$$81 \times \left(\frac{16}{25}\right)^{x+2} \div \left(\frac{3}{5}\right)^{2x+4} = 144?$$

(a) 1 (b) –1

(c) –2 (d) Cannot be determined

60. Two dice are thrown simultaneously. The probability that the product of the numbers appearing on the top faces of the dice is a perfect square is

(a) 1/9 (b) 2/9

(c) 1/3 (d) 4/9

61. Bhaichung was observing the pattern of people entering and leaving a car service centre. There was a single window where customers were being served. He saw that people inevitably came out of the centre in the order that they went in. However, the time they spent inside seemed to vary a lot:

some people came out in a matter of minutes while for others it took much longer.

From this, what can one could?

(a) The centre operates on a first-come-first-served basis, but with variable service times, depending on specific customer needs.

(b) Customers were served in an arbitrary order, since they took varying amount of time for service completion in the centre.

(c) Since some people came out within a few minutes of entering the centre, the system is likely to operate on a last-come-first-served basis.

(d) Entering the centre early ensured that one would have shorter service times and most people attempted to do this.

62. A map shows the elevations of Darjeeling, Gangtok, Kalimpong, Pelling, and Siliguri. Kalimpong is at a lower elevation than Gangtok. Pelling is at a lower elevation than Gangtok. Pelling is at a higher elevation than Siliguri. Darjeeling is at a higher elevation than Gangtok.

Which of the following statements can be inferred from the paragraph above?

i. Pelling is at a higher elevation than Kalimpong
ii. Kalimpong is at a lower elevation than Darjeeling
iii. Kalimpong is at a higher elevation than Siliguri
iv. Siliguri is at a lower elevation than Gangtok

(a) Only ii (b) Only ii and iii
(c) Only ii and iv (d) Only iii and iv

63. P, Q, R, S, T and U are seated around a circular table. R is seated two places to the right of Q. P is seated three places to the left of R. S is seated opposite U. If P and U now switch seats, which of the following must necessarily be true?

(a) P is immediately to the right of R
(b) T is immediately to the left of P
(c) T is immediately to the left of P or P is immediately to the right of Q
(d) U is immediately to the right of R or P is immediately to the left of T

64. Budhan covers a distance of 19 km is 2 hours cycling one fourth of the time and walking the rest. The next day he cycles (at the same speed as before) for half the time and walks the rest (at the same speed as before) and covers 26 km in 2 hours. The speed in km/h at which Budhan walks is

(a) 1 (b) 4
(c) 5 (d) 6

65. The points in the graph below represent the halts of a lift for durations of 1 minute, over a period of 1 hour.

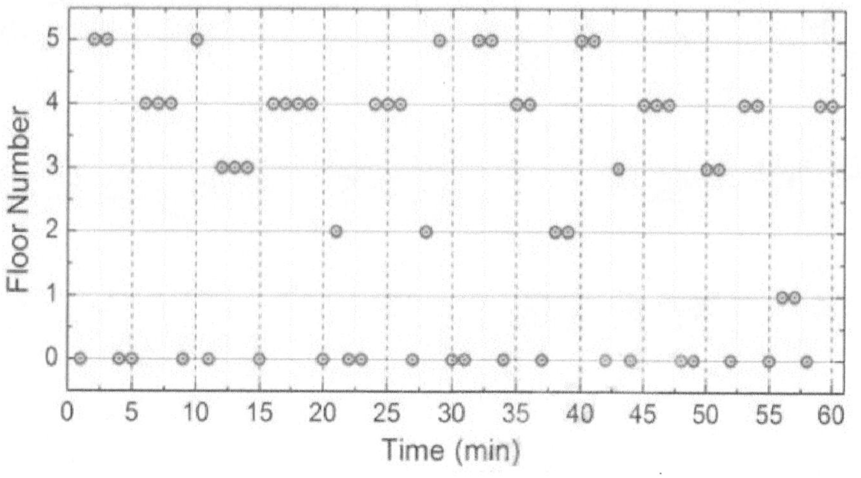

Which of the following statements are correct?

i. The elevator never moves directly from any non-ground floor to another non-ground floor over the one hour period.
ii. The elevator stays on the fourth floor for the longest duration over the one hour period

(a) Only i (b) Only ii (c) Both i and ii (d) Neither i nor ii

OMR SHEET

1. Ⓐ Ⓑ Ⓒ Ⓓ	18.	35. Ⓐ Ⓑ Ⓒ Ⓓ	51. Ⓐ Ⓑ Ⓒ Ⓓ
2. Ⓐ Ⓑ Ⓒ Ⓓ	19. Ⓐ Ⓑ Ⓒ Ⓓ	36. Ⓐ Ⓑ Ⓒ Ⓓ	52. Ⓐ Ⓑ Ⓒ Ⓓ
3. Ⓐ Ⓑ Ⓒ Ⓓ	20. Ⓐ Ⓑ Ⓒ Ⓓ	37.	53.
4. Ⓐ Ⓑ Ⓒ Ⓓ	21. Ⓐ Ⓑ Ⓒ Ⓓ	38. Ⓐ Ⓑ Ⓒ Ⓓ	54. Ⓐ Ⓑ Ⓒ Ⓓ
5.	22. Ⓐ Ⓑ Ⓒ Ⓓ	39. Ⓐ Ⓑ Ⓒ Ⓓ	55. Ⓐ Ⓑ Ⓒ Ⓓ
6. Ⓐ Ⓑ Ⓒ Ⓓ	23.	40. Ⓐ Ⓑ Ⓒ Ⓓ	56. Ⓐ Ⓑ Ⓒ Ⓓ
7. Ⓐ Ⓑ Ⓒ Ⓓ	24.	41.	57. Ⓐ Ⓑ Ⓒ Ⓓ
8. Ⓐ Ⓑ Ⓒ Ⓓ	25. Ⓐ Ⓑ Ⓒ Ⓓ	42. Ⓐ Ⓑ Ⓒ Ⓓ	58. Ⓐ Ⓑ Ⓒ Ⓓ
9. Ⓐ Ⓑ Ⓒ Ⓓ	26. Ⓐ Ⓑ Ⓒ Ⓓ	43. Ⓐ Ⓑ Ⓒ Ⓓ	59. Ⓐ Ⓑ Ⓒ Ⓓ
10. Ⓐ Ⓑ Ⓒ Ⓓ	27.	44. Ⓐ Ⓑ Ⓒ Ⓓ	60. Ⓐ Ⓑ Ⓒ Ⓓ
11. Ⓐ Ⓑ Ⓒ Ⓓ	28. Ⓐ Ⓑ Ⓒ Ⓓ	45. Ⓐ Ⓑ Ⓒ Ⓓ	61. Ⓐ Ⓑ Ⓒ Ⓓ
12. Ⓐ Ⓑ Ⓒ Ⓓ	29. Ⓐ Ⓑ Ⓒ Ⓓ	46.	62. Ⓐ Ⓑ Ⓒ Ⓓ
13. Ⓐ Ⓑ Ⓒ Ⓓ	30.	47.	63. Ⓐ Ⓑ Ⓒ Ⓓ
14. Ⓐ Ⓑ Ⓒ Ⓓ	31. Ⓐ Ⓑ Ⓒ Ⓓ	48.	64. Ⓐ Ⓑ Ⓒ Ⓓ
15.	32. Ⓐ Ⓑ Ⓒ Ⓓ	49. Ⓐ Ⓑ Ⓒ Ⓓ	65. Ⓐ Ⓑ Ⓒ Ⓓ
16. Ⓐ Ⓑ Ⓒ Ⓓ	33. Ⓐ Ⓑ Ⓒ Ⓓ	50. Ⓐ Ⓑ Ⓒ Ⓓ	
17. Ⓐ Ⓑ Ⓒ Ⓓ	34. Ⓐ Ⓑ Ⓒ Ⓓ		

ANSWERS

1. (*d*)	2. (*a*)	3. (*d*)	4. (*d*)	5. (0.1848)	6. (*b*)	7. (*c*)	8. (*a*)
9. (*a*)	10. (*b*)	11. (*a*)	12. (*b*)	13. (*c*)	14. (*b*)	15. (1.33)	16. (*)
17. (*a*)	18. (0.0184)	19. (*a*)	20. (*b*)	21. (*a*)	22. (*d*)	23. ($8\sqrt{y}$)	24. ($e^{-2t}u(t)$)
25. (*b*)	26. (*c*)	27. (0)	28. (*b*)	29. (*d*)	30. (50)	31. (*b*)	32. (*c*)
33. (*a*)	34. (*)	35. (*d*)	36. (*d*)	37. (1108.19)		38. (*b*)	39. (*a*)
40. (*b*)	41. (2350)	42. (*c*)	43. (*a*)	44. (*b*)	45. (*c*)	46. (1692.02)	47. (2.34)
48. (993)	49. (*c*)	50. (*a*)	51. (*b*)	52. (*a*)	53. (0.18)	54. (*c*)	55. (*b*)
56. (*b*)	57. (*b*)	58. (*c*)	59. (*b*)	60. (*b*)	61. (*a*)	62. (*c*)	63. (*c*)
64. (*d*)	65. (*d*)						

EXPLANATIONS

1. During adverse yaw operation rising wing generates increased lift. Which causes increased induce drag. The decending wing generates reduce lift which causes reduced induce drag.

2.

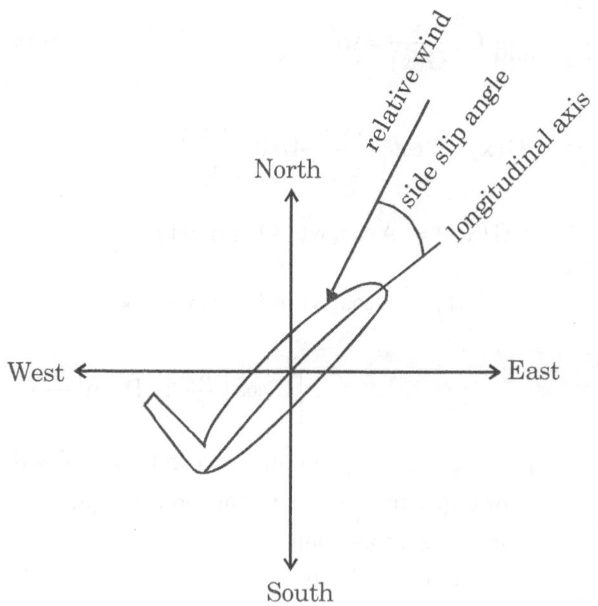

3. $6_{xx} = 6$

$6_{yy} = 6$

$T_{xy} = 6$

$\therefore$ Radius of mohr's circle

$$= \sqrt{\left(\frac{x_{xx} - 6_{yy}}{2}\right)^2 + T_{xy}{}^2}$$

$$= \sqrt{\left(\frac{6-6}{2}\right)^2 + 6^2}$$

$$\sqrt{6^2} = 6$$

4. (i) For both ends fixed lequivalent $= \frac{\ell}{2}$

(ii) For both ends hinged lequivalent $= \ell$.

Euler's buckling load.

$$P_{cr} = \frac{\pi^2 EI}{L^2_{equivant}}$$

$$\frac{(P_{cr})_1}{(P_{cr})_{11}} = \frac{\pi^2 EI}{\left(\frac{\ell}{2}\right)^2} \times \frac{(\ell)^2}{\pi^2 EI} = 4$$

5.

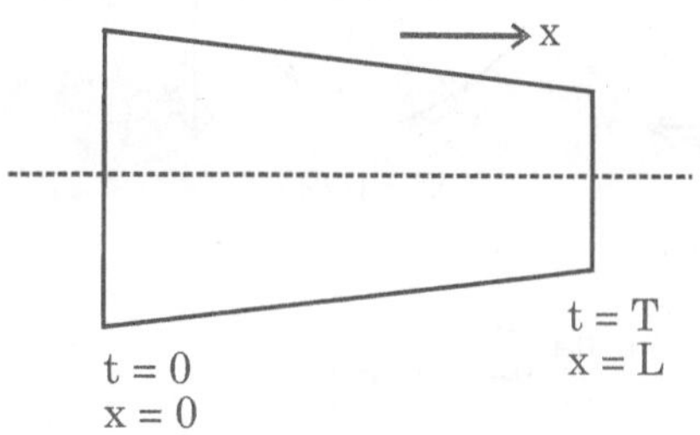

$$V = V_0\left(1\frac{3x}{L}\right)$$

$$\frac{dx}{dt} = V_0\left(1 + \frac{3x}{L}\right)$$

$$\int_0^L \frac{dx}{\left(1+\frac{3x}{L}\right)} = V_0 \int_0^T dt$$

$$\frac{L}{3V_0}\left[\ell n\left(1 + \frac{3x}{L}\right)\right]_0^L = T$$

$$T = \frac{L}{3V_o}\ell n4$$

at L = 2m, V_o = 5m/sec

$$T = \frac{2}{3\times5}\ell n4 = 0.1848.$$

6. I_{sp} (Specific impulse) is lower for solid propellants compared to liquid propellants. Also hydrocarbons have lower I_{SP} then liquid hydrogen.

7. For compressible flow.

$$C_p = 2\ C_{P_{in}}\ \sin^2\theta.$$

$$\therefore\ C_{P_{max}} = 2$$

8. During cruise climb, Altitude increases and mass decreases due to fuel combination the aircraft flies at $\left(\frac{C_L}{C_D}\right)_{max}$ so that range is maximised.

9. Half Rankine oval condition is combination of uniform flow and source

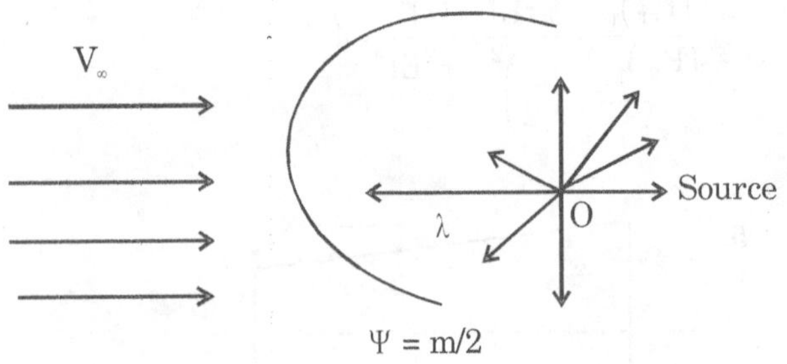

$$\psi = V_\infty r\sin\theta + \frac{m}{2\pi}\theta$$

At stagnation point V = 0 i.e, radial and tangential velocity both zero.

$$\therefore\ V_r = \frac{1}{r}\frac{\partial\psi}{\partial\theta}$$

$$= V_\infty\cos\theta + \frac{m}{2\pi r}$$

$$\therefore\ V_r = 0, \theta = 180°$$

$$\therefore\ 0 = V_\infty - \frac{m}{2\pi\lambda}$$

$$\therefore\ \boxed{\lambda = \frac{m}{2\pi V_\infty}}$$

Where λ = distance between stagnation point and source.

10. $\frac{\partial^2 u}{\partial x^2} = \frac{1}{C^2}\frac{\partial^2 u}{\partial t^2}$

Let u (x, t) = F(x) G(t)

$$C^2\left(\frac{\partial u}{\partial x^2}\right) = \left(\frac{\partial^2 u}{\partial t^2}\right)$$

$$C^2\left(F(t), G''(x)\right) = F''(t)G(x)$$

$$C^2\frac{G''(x)}{G(x)} - \frac{F''(t)}{F(t)} = -K^2$$

$$\therefore\ \frac{F''}{F(t)} = -w^2$$

$$F''(t) + w^2F(t) = 0$$

F(t) = A cos (wt) + B sin (wt)

and $C^2\frac{G''}{G(x)} = w^2$

$$G(x) = C\cos\left(\frac{wx}{c}\right) + D\sin\left(\frac{wx}{c}\right)$$

$$\therefore G(x,t) = \left[A\cos(wt) + B\sin(wt)\right]$$

$$\therefore G(x,t) = \left[A\cos(wt) + B\sin(wt)\right] \times \left[C\cos\left(\frac{wx}{C}\right) + D\sin\left(\frac{wx}{C}\right)\right]$$

putting boundary condition we get A = 0, C = 0

Now applying boundary conditions we get

fixed – fixed condition

u(0, t) = 0 u (ℓ, t) = 0

Frequency equation

$$\sin\left(\frac{w\ell}{C}\right) = 0$$

(ii) for fixed free condition

u (0, t) = 0

$$\frac{\partial u}{\partial t}(\ell, t) = 0$$

Frequency equations

$$\cos\left(\frac{w\ell}{C}\right) = 0$$

(iii) free – free condition

$$\frac{\partial u}{\partial t}(0,t)=0,\ \frac{\partial u}{\partial t}(\ell,t)=0$$

frequency equation,

$$\sin\left(\frac{w\ell}{C}\right)=0$$

11.

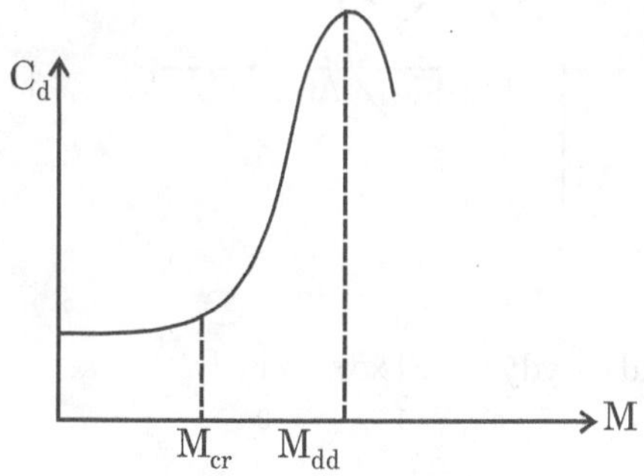

$m_{cr} < m_{dd} < 1$

Cr → critical

dd → drag di-vergence

14.

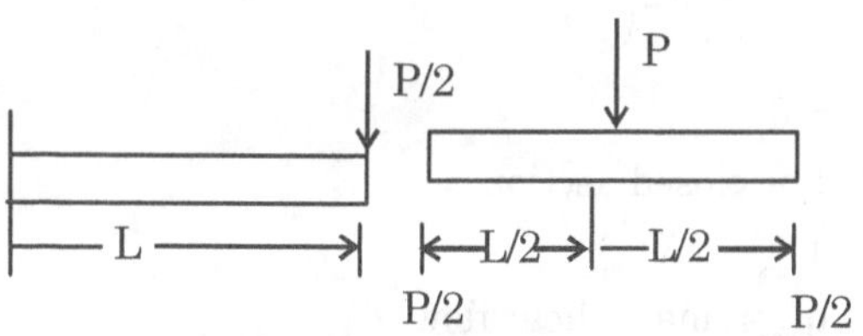

∴ Bending moment at $A=\frac{P}{2}\times L$

$$=\frac{PL}{2}$$

15. Magnification factor (β) for forced vibration system is given by

$$\beta=\frac{1}{\sqrt{(1-r^2)+(2r\xi)^2}}$$

r = frequency ratio = $\frac{w_f}{w}$

w_f = forced vibration frequency.

$\xi = 0$

$$\beta=\frac{1}{1-r^2}=\frac{1}{1-0.5^2}=\frac{1}{0.75}$$

= 1.33

16. For elliptic wing

down wash $w-\frac{T_o}{26}$ = constant

Lift distribution

$$(L)=e_\infty V_\infty T_o\sqrt{1-\left(\frac{2y}{b}\right)^2}$$

= elliptic

17. Combustor efficiency of the aircraft engine decreases with altitude since the pressure is reduced at altitudes and so rate of reaction of burning component decreases, the time of reaction increases.

18. Given $\frac{x_o}{x_n}=\frac{1}{4}, n=12$

∴ logarithmic decrement

$$(\delta)=\frac{1}{n}\ell n\frac{x_o}{x_n}=\frac{1}{12}\ell n\frac{1}{4}$$

$\delta = 0.1155$

$$\xi=\frac{1}{\sqrt{1+\left(\frac{2\pi}{\delta}\right)^2}}$$

$$=\frac{1}{\sqrt{1+\left(\frac{6.28}{0.1155}\right)^2}}$$

$$=\frac{1}{\sqrt{2957.34}}=\frac{1}{54.37}$$

= 0.0184

20. $W_{adiabatic}=\frac{R(T_1-T_2)}{r-1}$

$W_{isothermal}=RT_1\ell n\frac{e_2}{e_1}$

$W_{isocoric}=0$

$W_{isobaric}=P(V_2-V_1)$

y = 4 and y = 0

∴ x = 4, and x = 0

Area unclosed between curve 1 and 2

$$= \int_{x_1}^{x_2} y_1 dx - \int_{x_1}^{x_2} y_2 dx$$

$$\int_0^4 2\sqrt{x}dx - \int_0^4 \frac{x^2}{4}dx$$

$$= \frac{16}{3}$$

21. A. Singular matrix → Determinant is zero

B. Non - square matrix → Determinant is not defined

C. Real symmetric matrix → Eigen values are always real.

D. Orthogonal matrix → Determinant is always one.

22. Sum of eigen values

Tr (A) = –1 + –1 + 3 = 1

$\sum \lambda = 1$

from the given option only (d), satisfy this condition

23. Curve 1, $y^2 = 4x$

Curve 2, $x^2 = 4y$

intersection points of curve 1 and 2

$$y^2 = 4x = 4\sqrt{4y} = 8\sqrt{y}$$

24. Taking Laplace transformation

we get

5 × (5) – × (0) + 2 × (5) = 1

X(5) [S + 2] = 1

$$X(5) = \frac{1}{S+2}$$

$$X(t) = e^{-2t}u(t)$$

25. Simpsons rule - Numerical integration

Newton - Raphson - solving non-linear equation

Runge - Kutta Method - solving ordinary differential equations.

Gauss - Elemination method-

Solving simultaneous linear equations.

26. PQRS = 1

$PQRSS^{-1} R^{-1} = S^{-1} R^{-1}$

$PQ = S^{-1} R^{-1}$

$Q = P^{-1} S^{-1} R^{-1}$

$Q-1 = (P^{-1} S^{-1} R^{-1})^{-1}$

$= (P^{-1}).(S^{-1})^{-1}.(R^{-1})^{-1}$

$Q^{-1} = PSR$

27.

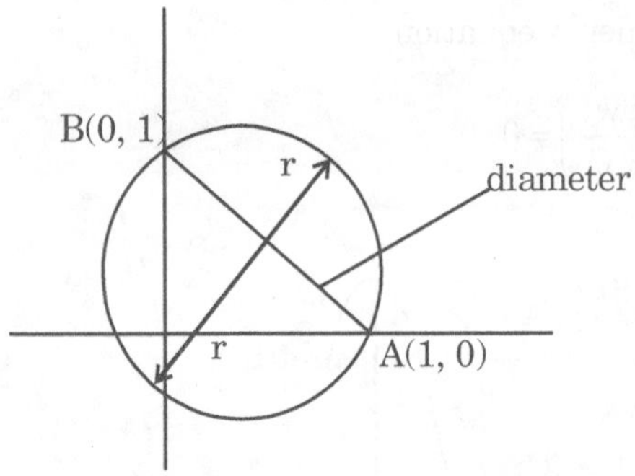

$$2\int_A^B (xdx + ydy) = 2\int_1^0 xdx + 2\int_0^1 ydy$$

$$= 2\left[\frac{x^2}{2}\right]_1^0 + 2\left[\frac{y^2}{2}\right]_0^1$$

$$2\left[\frac{-1}{2}\right] + 2\left(\frac{1}{2}\right)$$

$$= 0$$

28. For closed section.

$T_{max} = 2A\, q_{max}$

q_{max} = max. shear flow

$= T_{max} \times T_{min}$

$T_{max} = 2A\, t_{max}\, t_{min}$

$$\tau_{max} = \frac{T_{max}}{2A\, t_{min}}$$

$$\text{so } (\tau_{max})_1 = \frac{T}{2\pi r^2 t}$$

for open section (slit)

$$\frac{T}{J} = G\frac{d\theta}{d2} = \frac{T}{t}$$

$$J = \frac{st^3}{3}$$

$$(\tau_{max})_2 = \frac{T \times t}{J}$$

$$= \frac{T \times t}{\frac{1}{3} \times 2\pi r \times t^3}$$

$$= \frac{3T}{2\pi rt^2}$$

$$\frac{(\tau_{max})_1}{(\tau_{max})_2} = \frac{T}{2\pi r^2 t} \times \frac{2\pi rt^2}{3T}$$

$$= \frac{t}{3r}$$

29. $T = 2A_1 q_1 + 2A_2 q_2$.

$50 \times 10^3 = 2 \times 2500\, q_1 + 2 \times 2500\, q_2$.

$q_1 + q_2 = 10$ ...(1)

$$\frac{d\theta}{dz} = \frac{1}{2AG}\sum \frac{qs}{t}$$

$$\left(\frac{d\theta}{dz}\right)_1 = \frac{1}{2A_1 G}\left[\frac{q_1 \times 50}{t} + \frac{q_1 \times 50}{t} + \frac{(q_1 - q_2) \times 50}{t} + \frac{q_1 \times 50}{t}\right]$$

$$= \frac{50}{2A_1 tG}(4q_1 - q_2)$$

$$\left(\frac{d\theta}{dz}\right)_2 = \frac{1}{2A_2 G}\left[\frac{q_2 \times 50}{t} + \frac{q_2 \times 50}{t} + \frac{q_2 \times 50}{t} - \frac{(q_1 - q_2) \times 50}{t}\right]$$

$$= \frac{50}{2A_2 Gt}(4q_2 - q_1)$$

$$\left(\frac{d\theta}{dz}\right)_1 = \left(\frac{d\theta}{dz}\right)_2$$

$4q_1 - q_2 = 4q_2 - q_1$.

$q_1 = q_2$

Shear flow in 'BF'

$= q_1 - q_2 = 0$ N/m

$$\text{Shear stress} = \frac{q_1 - q_2}{t} = 0\ \text{N/m}^2$$

30. $I_{xy} = 0$, $S_x = 0$

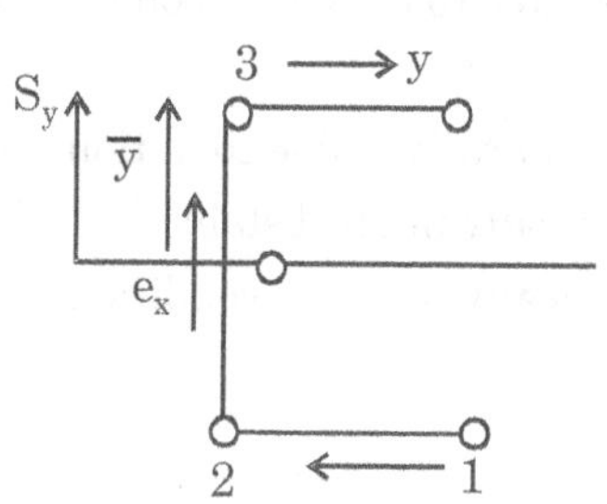

$$q = \frac{-S_y}{I_{xx}}\sum A_r y_r$$

$\bar{y} = 100$mm

$I_{xx} \sum A\bar{y}^2$

$= 100 \times (100)^2 \times 4$

$= 4 \times 10^6$ mm^4.

$$q_{1-2} = \frac{-S_y}{I_{xx}}(A_1 y_1)$$

$$= \frac{-S_y}{4 \times 10^6} \times 100 \times (-100)$$

$q_{1-2} = 2.5 \times 10^{-3}\ S_y$ N/mm

$$q_{2-3} = q_{1-2} - \frac{S_y}{I_{xx}}(A_2 y_2)$$

$$= 2.5 \times 10^{-3} S_y - \frac{S_y}{4 \times 10^{+6}}(100) \times (-100)$$

$q_{2-3} = 5 \times 10^{-3}$ Sy N/mm

$$q_{3-y} = q_{2-3} - \frac{S_y}{I_{xx}}(A_3 y_3)$$

$$= 5 \times 10^{-3} S_y - \frac{S_y}{y \times 10^6}(100 \times 100)$$

$q_{3-4} = 2.5 \times 10^{-3}\ S_y$.

Equating clockwise moment about 'O'

$S_y e_x = q_{1-2} \times \ell_{1-2} \times 100 + q_{3-y} \times \ell_{3-y} \times 100$

$S_y e_x = 2 \times 2.5 \times 10^{-3} \times 5y \times 100 \times 100$

$e_x = 50$ mm

31. $$\frac{dy}{dx} = \frac{-v}{u} = \frac{x}{y}$$

$$\int y dy = \int x dx$$

$$\frac{y^2}{2} = \frac{x^2}{2} + c$$

$y^2 = x^2 + c_1$

at (0, 5) $25 = c_1$

$x^2 - y^2 = 25$

34. * (Total energy at h) = (Total energy during launch)

$(P.E)_n = (P.E)_L + (K.E)_L$

$$\frac{-Gmm}{R+h} = \frac{-Gmm}{R} + \frac{1}{2}mu^2$$

Let g = Gm

$$\frac{-g}{R+h} = \frac{-g}{R} + \frac{1}{2}u^2$$

$$-g\left(\frac{1}{R+h} - \frac{1}{R}\right) = \frac{1}{2}u^2$$

$$-g\left(\frac{R-R-h}{R(R+h)}\right) = \frac{1}{2}u^2$$

36. $0.97 < \eta_b < 0.99$, $0.95 < \eta_n < 0.98$

$0.9 < \eta_t < 0.95$, $0.85 < \eta_c < 0.9$

$0.7 < \eta_d < 0.9$

37. For ideal Ramjet, $M_i = M_e$

$$\frac{v_i}{\sqrt{rRT_i}} = \frac{v_e}{\sqrt{rRT_e}}$$

$$v_e = v_i\sqrt{\frac{T_e}{T_i}}$$

$$= 1000 \times \sqrt{\frac{1000}{225}} = 2108.185 \text{ m/s}$$

$$T = \dot{m}_a(v_e - v_i)$$

$$\frac{T}{\dot{m}_a} = v_e - v_i = 2108.185 - 1000$$

$$= 1108.185 \frac{N-S}{kg}$$

38. from $\frac{dA}{A} = \frac{dv}{v}(m^2 - 1)$

m > 1, dA = + dv

means increases in area results in increase in velocity.

39. for ideal gas, Pv = nRT

Since nR is constant

so we can write $Pv \propto T$

40. Non dimensional form of compressor is

$$\frac{\dot{m}\sqrt{T_{01}}}{P_{01}}$$

$$\text{Initial condition} = \frac{\dot{m}_i\sqrt{T_{01}}}{P_{01}}$$

$$\text{final condition} = \frac{\dot{m}_f\sqrt{2T_{01}}}{0.5P_{01}}$$

To maintain the same pressure ratio

$$\frac{\dot{m}_f\sqrt{2T_{01}}}{0.5P_{01}} = \dot{m}_i \times \frac{\sqrt{T_{01}}}{P_{01}}$$

$$\dot{m}_f = \frac{0.5}{\sqrt{2}}\dot{m}_i$$

$$\dot{m}_f = 0.35\,\dot{m}_i$$

41. $K = \frac{1}{\pi A\,Re}$

$$\frac{1}{\pi \times 6 \times 0.6} = 0.088$$

$$C_L = \frac{2L}{\rho v^2 S}$$

$$= \frac{2 \times 30000}{1 \times (100)^2} = 0.6$$

$$C_D = C_{Do} + KC_L^2 = 0.015 + 0.088 \times 0.6^2$$

$C_D = 0.047$

$$D = C_D \times \frac{1}{2}\rho v^2 s$$

$$= 0.047 \times \frac{1}{2} \times (100)^2 \times 10$$

= 2350 N

42. Only statement (i) is correct

43. General form of characteristic equation.

$a\lambda^4 + B\lambda^3 + C\lambda^2 + D\lambda + E = 0$

Routh's Discriminant

so, $R = BCD - D^2A - B^2E$

$R = 6.05 \times 13 \times 0.6 - (0.6)^2 \times 1 - (6.05)^2 \times 0.6$

= 24.8685

If, $R > 0 \rightarrow$ Dynamically stable condition

$R = 0 \rightarrow$ Dynamically neutral stable

$R < 0 \rightarrow$ Dynamically unstable condition

44. $a_{o_w} = 0.104$ per degree $= 0.104 \times \dfrac{180}{\pi}$

$= 5.9588$ per rad.

$a_{o_t} = 0.110$ per degree

$= 0.110 \times \dfrac{180}{\pi}$

$= 6.3025$ per rad

$$a = a_w = \frac{a_{ow}}{1 + \dfrac{a_{ow}}{\pi AR}}$$

$$= \frac{5.9588}{1 + \dfrac{5.9588}{\pi \times 8}}$$

$a = 4.8167$ per radian

$$a_1 = a_t = \frac{6.3025}{1 + \dfrac{6.3025}{\pi \times 4}}$$

$= 4.1994$ per radian

$$\frac{dC_m}{dC_L} = \frac{dC_f}{dC_L} + \left(\frac{x_{cg} - x_{ac}}{c}\right) - \eta_t \bar{v}\frac{a_1}{a}\left(1 - \frac{d\varepsilon}{d\propto}\right)$$

at Neutral point, $\dfrac{dC_m}{C_L} = 0$ and $x_{cg} = x_N$

$$0 = \frac{dC_{mf}}{dC_L} + \left(\frac{x_N - x_a}{\tau}\right) - n + \bar{v}\frac{a_1}{C}\left(1 - \frac{d\varepsilon}{d\propto}\right)$$

$$0 = 0.08 + \frac{x_N}{C} - 0.24 - 0.9 \times 0.6 \times \left(\frac{4.1974}{4.8167}\right) \times (1 - 0.5)$$

$x_N = 0.3951$ C

$x_N = 39\%$ of C

45. $$0 = \frac{d\,C_m t}{dC_2} + \left(\frac{xN'}{\bar{C}} - \frac{x_{ac}}{\bar{C}}\right) - \eta_t \bar{v}\frac{a_1}{a}\left(1 - \frac{d\varepsilon}{d\propto}\right)\left(1 - \frac{a_2 b_1}{a_1 b_2}\right)$$

a_2 = elevator effectiveness = 0.55

$$0 = 0.08 + \left(\frac{x_{N'}}{c} - 0.24\right) - 0.9 \times 0.6 \times \left(\frac{4.1974}{4.8167}\right)$$

$$(1 - 0.5) \times \left[\frac{1 - 0.55 \times (-0.003)}{4.1974 \times (-0.0055)}\right]$$

$$\frac{x_{N'}}{C} = 0.3785$$

$x_{N'} = 37\%$ of C

46. Mass flow rate through throat $\dot{m}_p = 40\text{kg/s}$

$$\dot{m}_p = \rho^* A^* v^*$$

$$= \rho^* \times \sqrt{rRT^*} \times A^*$$

$$= \frac{\rho^*}{\rho o} \times \sqrt{\frac{T^*}{To}} \times \sqrt{To} \times \sqrt{rR} \times A^* \times \rho_o$$

$$= \left(\frac{2}{r+1}\right)^{\frac{r}{(r-1)}} \times \left(\frac{2}{r+1}\right)^{\frac{1}{2}} \times \sqrt{rRT_o} \times \frac{P_o}{RT_o} \times A^*$$

$$\dot{m}_p = \left(\frac{2}{r+1}\right)^{\frac{3r-1}{2(r-1)}} \times \sqrt{\frac{r}{RT_o}} \times P_o \times A^*$$

$$40 = \left[\frac{2}{(1.21+1)}\right]^{\frac{3\times1.21-1}{2(1.21-1)}} \times \sqrt{\frac{1.21}{402 \times 3335}} \times 0.0465 \times P_o$$

Po = 1692021.02 Pa = 1692.02 kpa

47. Ground Roll distance $(S_{Lo})_1 = \dfrac{1.44w^2}{g\rho_\infty S\, C_{L\max} T}$

$w_2 = 1.25w$

$$(S_{Lo})_2 = \frac{1.44 \times (1.25)^2 \times w^2}{g\rho_\infty S\, C_{L\max} T}$$

$$\frac{(S_{Lo})_2}{(S_{Lo})_1} = \frac{1.44 \times (1.25)^2 \times w^2}{g\,\rho_\infty\, S\, C_{L\max} T} \times \frac{g\,\rho_\infty\, S\, C_{L\max} T}{1.44\, w^2}$$

$$\frac{(S_{Lo})_2}{(S_{Lo})_1} = (1.25)^2$$

$(SLo)_2 = (1.25)^2 \times 1.5 = 2.34$ Km

48. flywheel energy = $\int_0^{2\pi} T(\theta)\, d\theta$

from simpson's rule

$$J = \frac{h}{3}\left[(y_o + y_6) + 4(y_1 + y_3 + y_5) + (y_2 + y_4)\right]$$

$$h = 60° = 60 \times \frac{\pi}{180} = \frac{\pi}{3} \text{ rad}$$

$$I = \frac{\pi}{3}\left[(0+0) + 4(1066 + 0 - 355) + 2(-323 + 323)\right]$$

= 993

49. For lateral and directional stability of aircraft

$l_\beta < 0$

$N_\beta > 0$

β = side slip angle, L = rolling moment

N = yawing moment

for the position side ship the aircraft should rate over list an should yaw towards the right.

50. We have the velocity component along axial, side ship and normal direction are

$V_{xb} = V \cos\alpha \cos\beta$

= 1000 cos 36° cos 20° = 813.80 m/sec

$V_{yb} = V \sin\beta = 1000 \sin 20° = 342$ m/sec

$V_{zb} = V \sin\alpha \cos\beta = 1000 \sin 36°, \cos 0$

= 46.9 85 m/h

51. $f(x) = (x-8)^{\frac{2}{3}} + 1$

$$f'(x) = \frac{2}{3}(x-8)^{-\frac{1}{3}}$$

$$m_1 = f'(x)\big|_{(0,5)}$$

$$= \frac{2}{3} \times (-8)^{\frac{-1}{3}}$$

$$= \frac{2}{3} \times \frac{1}{(-2)} = \frac{-1}{3}$$

Slope of line normal to given function is m_2.

$m_1 \cdot m_2 = -1$

$$-\frac{1}{3} . m_2 = -1$$

$$\Rightarrow m_2 = 3$$

Equation of normal

y − 5 = 3 (x)

3x + 5 = y

52. $r_1 = \dfrac{P}{1-e}$

$$r_2 = \frac{P}{1+e}$$

$$\frac{r_1}{r_2} = \frac{1+e}{1-e}$$

$$\frac{r_1 + r_2}{r_1 - r_2} = \frac{1+e+1-e}{1+e-1+e} = \frac{1}{e}$$

$$e = \frac{r_1 - r_2}{r_1 + r_2}$$

53. Flow is compressible

$$C_{\ell} = \frac{4\alpha}{\sqrt{m_\infty^2 - 1}}$$

$$\frac{dC\ell}{d\alpha} = a_{comp} = \frac{4}{\sqrt{m_\infty^2 - 1}}$$

$$a_{comp} = \frac{4}{\sqrt{(2.5)^2 - 1}} = 1.746 \text{ per rad.}$$

$$C_L = a_{comp} \times \alpha.$$

$$= 1.746 \times \frac{6}{180} \times \pi = 0.1827$$

54. Turn - Radius for level turn,

$$R_1 = \frac{V_\infty^2}{g\sqrt{n^2 - 1}}$$

Turn radius for pull-up,

$$R_2 = \frac{V_\infty^2}{g(n+1)}$$

$$\frac{R_1}{R_2} = \frac{V_\infty^2}{g\sqrt{n^2-1}} \times \frac{g(n+1)}{V_\infty^2}$$

$$= \frac{n+1}{\sqrt{(n+1)(n-1)}}$$

$$\frac{R_1}{R_2} = \sqrt{\frac{n+1}{n-1}}$$

56. The past perfect tense should be used in the sentence.

57. "Thorough" means complete.

58. The faces visible on the cards are 2, 3, red and blue

If the card has an even value on one side, then its opposite face is red.

$\therefore$ The cards which must be turned over to verify the above proposition are 2, blue.

59. $81 \times \left(\frac{16}{25}\right)^{x+2} \div \left(\frac{3}{5}\right)^{2x+4} = 144$

$$81 \times \frac{2^{4x+8}}{5^{2x+4}} \times \frac{5^{2x+4}}{3^{2x+4}} = 144$$

$$3^4 \times \frac{2^{4x+8}}{3^{2x+4}} = 2^4 \times 3^2$$

$$\frac{2^{4x+8}}{3^{2x+4}} = \frac{2^4}{3^2}$$

$$\frac{2^{4x+4}}{3^{2x+2}} = 1$$

$$\left(\frac{4}{3}\right)^{2x+2} = \left(\frac{4}{3}\right)^0$$

$\Rightarrow 2x + 2 = 0$

$x = -1$

60. **11** 21 31 41 51 61
12 **22** 32 42 52 62
13 23 **33** 43 53 63
14 24 34 **44** 54 64
15 25 35 45 **55** 65
16 26 36 46 56 **66**

No. of favourable outcomes = 8

Total outcomes = 36

Total outcomes = 36

$\therefore$ Probability that the product if the two numbers on the dice is a perfect square is

$$\frac{8}{36} = \frac{2}{9}$$

61. Since the customers were addressed to on a first come first serve basis, option A is the most logical.

62. The other options are wrong.

Alternate:

Darjeeling > Gangtok > Pelling > Siliguri

and Gangtok > Kalimpong (in terms of elevation)

$\Rightarrow$ The elevation of Kalimpong can be greater than that of feeling or between feeling and Siliguri or lower than that of Siliguri

Only (ii) and (iv) statements can be inferred.

63.

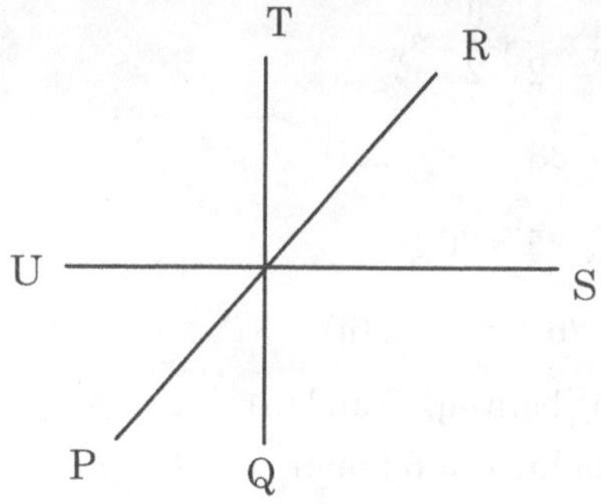

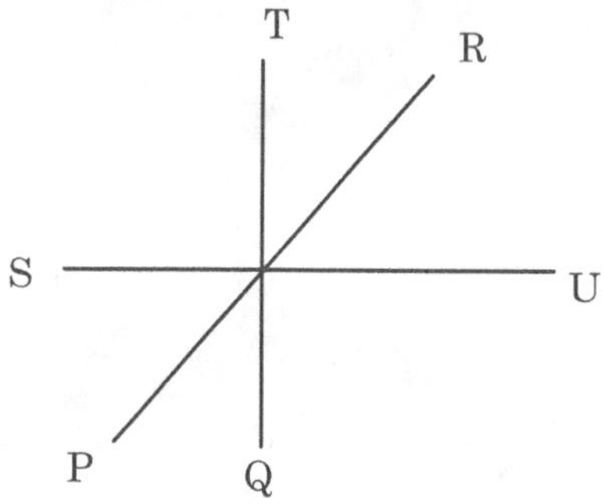

It P and U nor with seats,

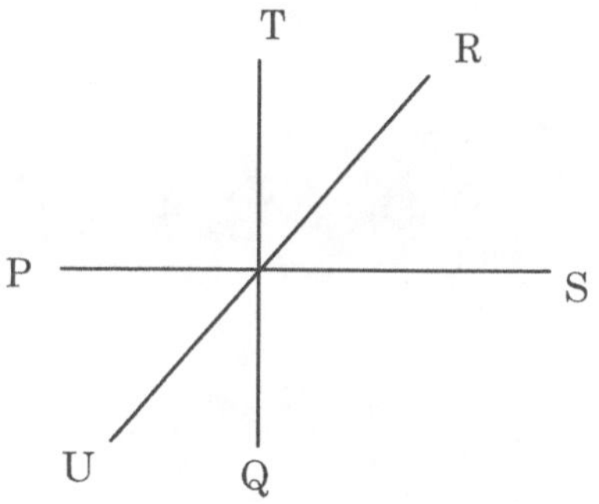

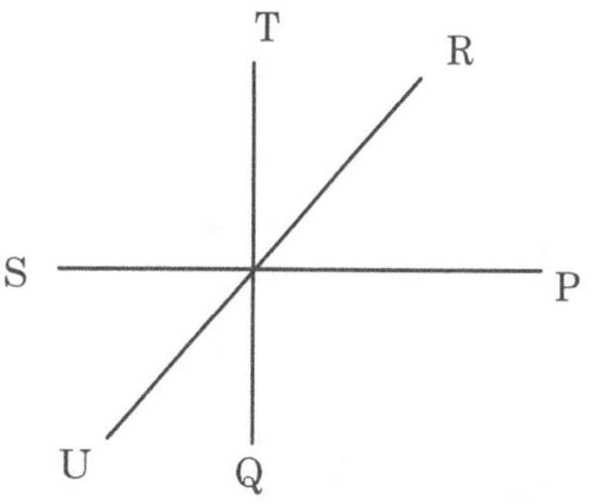

Clearly option C must necessarily be true.

64. Let the speed at which Budhan cycles and walls be s_c and s_w respectively.

$$\therefore\ 19 = s_c \times \frac{1}{2} + \frac{3}{2} \times s_w$$

$$s_c + 3s_w = 38 \qquad \text{...(i)}$$

$$s_c \times 1 + s_w \times 1 = 26$$

$$s_c + s_w = 26 \qquad \text{...(ii)}$$

On solving both eq. (i) and (ii)

s_c = 20 km/hr, s_w = 6 km/hr.

65. Statement (i) is not correct as between 25 and 30 min it moves directly from 2nd floor to 5th floor once.

Statement (ii) is also not correct as it stays on the ground floor for the longest duration.

∴ Neither (i) nor (ii) is correct.

MOCK TEST – 12

AEROSPACE ENGINEERING

Time Allowed: 3:00 Hours | *Maximum Marks: 100*

Read the following instructions carefully before you begin to answer the questions. This Booklet contains questions in English.

INSTRUCTIONS

1. This Booklet contains **65 questions** in total comprising of the following two parts:

   Part -1: Technical Section (85 Marks) *(1-55 Questions)*

   Part -2: General Aptitude Section (15 Marks) *(56-65 Questions)*

2. All questions are compulsory. Q.1-Q.25 & Q.56-Q.60 carry one mark each. Q.26-Q.55 & Q.61-Q.65 carry two marks each.

3. **Types of Questions:**

   (*a*) Multiple Choice Questions(MCQ) carrying 1 or 2 marks each in all papers and sections. These questions are objective in nature, and each will have a choice of four options, out of which the candidate has to mark the correct answer(s).

   (*b*) Numerical Answer Questions of 1 or 2 marks each in all sections. For these questions the answer is a real number, to be entered by the candidate using the virtual keypad. No choices will be shown for these type of questions.

4. For 1-mark multiple-choice questions, 1/3 marks will be deducted for a wrong answer. Likewise, for 2-marks multiple-choice questions, 2/3 marks will be deducted for a wrong answer. There is no negative marking for numerical answer type questions.

MOCK TEST – 12

1. The Shadowgraph flow visualization technique depends on

(a) Variation of density

(b) First derivative of density

(c) Second derivative of density

(d) Fourth derivative of density

2. Mach number of flow is defined by ratio of

(a) Inertia to viscous force

(b) Inertia to gravity force

(c) Inertia to elastic force

(d) Inertia to pressure force

3. The ozone layer lies in

(a) Troposphere (b) Stratosphere

(c) Mesosphere (d) Thermosphere

4. Which option is true regarding cruise flight and take - off flight having same wing

(a) Induced drag during cruise is equal to that of during take - off.

(b) Induced drag during cruise is greater than take - off

(c) Induced drag during cruise is less than that of during take off

(d) None of above

5. The rate of descent and climb of an aircraft is measured by

(a) Airspeed Indicator

(b) Altimeter

(c) Vertical speed Indicator

(d) Turn - bank Indicator

6. Dihedral effect on an aircraft occurs due to

(a) Rolling moment due to rate of roll

(b) Rolling moment due to rate of yaw

(c) Rolling moment due to sideslip

(d) Yawing moment due to rate of roll

7. An aircraft of weight 'w' Lift 'L' and drag 'D' during climbing flight. The rate of climb for the equilibrium climbing is given by

(a) $\frac{\text{Excess thrust}}{\text{weight}}$ (b) $\frac{\text{Excess power}}{\text{weight}}$

(c) $\frac{\text{Excess drag}}{\text{weight}}$ (d) Equivalent to drag

8. An aircraft flies at a speed of 30 m/s in straight and level flight. The slowest speed at which aircraft can execute a level turn at a bank angle of 50° is _______ (m/s).

9. Which of the following PDE is parabalic

(a) $9U_{xx} - 6U_{xy} + U_{yy} + U_x - 2U_y = -30$

(b) $9U_{xx} + U_{yy} + U_x = 0$

(c) $9U_{xx} - 2U_{yy} + 2U_x + 20 = 0$

(d) $2U_{xx} - 6U_{xy} + U_{yy} - 15 = 0$

10. A cantilever beam is shown in fig below load of magnitude 'W' unit distributed uniformly to a length of L/2. The maximum bending moment developed in the beam is

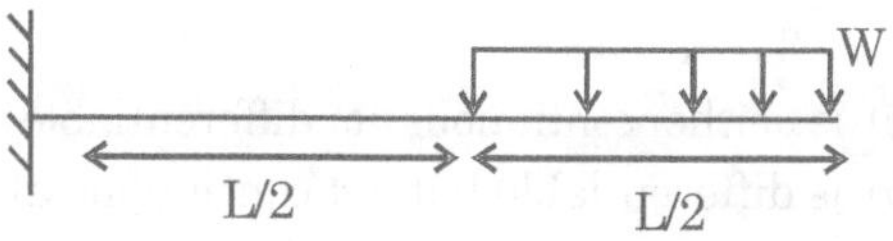

(a) $\frac{WL^2}{8}$ (b) $\frac{3WL^2}{8}$

(c) $\frac{5WL^2}{78}$ (d) $\frac{WL^2}{3}$

11. A pinned ended columns of length 1 m is subjected to a compressive force of magnitude 10 KN at the end. Cross - section of column is 40 mm × 40 mm. The buckling stress of column is (in N/m^2) If E = 200 GPa _______ .

12. For an ideal Ramjet Engine the exit Mach number is

(a) Almost more than double of the inlet

(b) Less than that of inlet

(c) Equal to that of inlet

(d) None of these

13. A turbofan engine operating at an altitude where the ambient pressure is 10 KPa and produces pressure thrust of magnitude 80 KN. If the nozzle exit area is 10m^2 then nozzle exhaust pressure is given by (In N/m^2) _______ .

14. In a phugoid mode of operation of an aircraft, which of the following remains almost constant

(a) Altitude (b) Speed

(c) Pitch angle (d) Angle of attack

15. Range of a jet power aircraft in a steady state and level flight can be maximized by flying at minimum

(a) $\frac{C_L}{C_D}$ (b) $\frac{C_D}{C_L}$

(c) $\frac{C_D}{C_L^{\frac{3}{2}}}$ (d) $\frac{C_L^{\frac{3}{2}}}{C_D}$

16. The drag polar of an aircraft having weight 45000 N and wing span area 40 m² is

$C_D = 0.002 + 0.08C_L^2$

The maximum possible speed of the aircraft at an altitude where density = 1.2 kg/m³ is (m/s) ______ .

17. The function defined by

$f(x) = \sin x, x < 0$

$= 0, x = 0$

$= 3x^3, x > 0$

(a) is neither continuous nor differentiable at x = 0

(b) is differentiable but not continuous at x = 0

(c) is continuous and differentiable at x = 0

(d) is continuous but not differentiable at x = 0

18. The product of the non-zero eigen values of matrix is ______

$$\begin{pmatrix} 1 & 0 & 0 & 0 & 1 \\ 0 & 1 & 1 & 1 & 0 \\ 0 & 1 & 1 & 1 & 0 \\ 0 & 1 & 1 & 1 & 0 \\ 1 & 0 & 0 & 0 & 1 \end{pmatrix}$$

19. The solution of differential equation

$\frac{d^2y}{dx^2} + 2\frac{dy}{dx} + y = 0$ at x = 2 is ______ .

(a) $(C_1 + C_2x)e^{-mx}$ (b) $(C_1 + C_2x)e^{-2mx}$

(c) $(C_1 + C_2x)e^{-3mx}$ (d) $(C_1 + C_2x)e^{-4mx}$

20. The value of K for which the system of equations x + 2y + kz = 1 ; 2x + ky + 8z = 3 has no solution is ______ .

21. If an aircraft takes - off with 10% more fuel in comparison to its standard configuration, its range is

(a) Increase by exactly 10%

(b) Increase by more than 10%

(c) Increase by less than 10%

(d) Remains same

22. Flow past over a wedge shape airfoil of wedge angle 20° with a supersonic speed at Mach 2.5 lift coefficient produced by airfoil is ______ .

23. With increase in flow speed upstream of a normal shock while keeping all parameter same

(a) Total temperature behind the normal shock increases.

(b) Static pressure behind the normal shock decreases

(c) Total pressure behind the normal shock decreases

(d) Total temperature behind the normal shock decreases

24. The ideal static pressure coefficient of a diffuser with an area ratio of 3 is

(a) – 8 (b) 0.89

(c) 1.11 (d) 0.5

25. Stringers and longerons are provided in semimonocoque fuselage, to resist

(a) Shear load (b) Torsional load

(c) Tensile load (d) Bending load

26. Eigenvectors of the matrix $\begin{bmatrix} 2 & 1 \\ -1 & 2 \end{bmatrix}$ are written in the form $\begin{bmatrix} P \\ 1 \end{bmatrix}$ and $\begin{bmatrix} q \\ 1 \end{bmatrix}$. Value of p + q is

(a) 0 (b) $\frac{1}{2}$

(c) 2 (d) Not exist

27. The solution of definite integral $\int_{-5}^{5} |x+2|\,dx$ is ______ .

28. An aircraft has a steady rate of climb of 500m/s at sea level and 250 m/s at 2500 m altitude. The time taken by the aircraft to climb from 500 m to 2000 m is ______ (in seconds).

29. Two overflowing water reservoirs are connected with a circular pipe of length 150 m and Radius R = 2 cm, in such a way that the height difference remains constant as shown in fig.

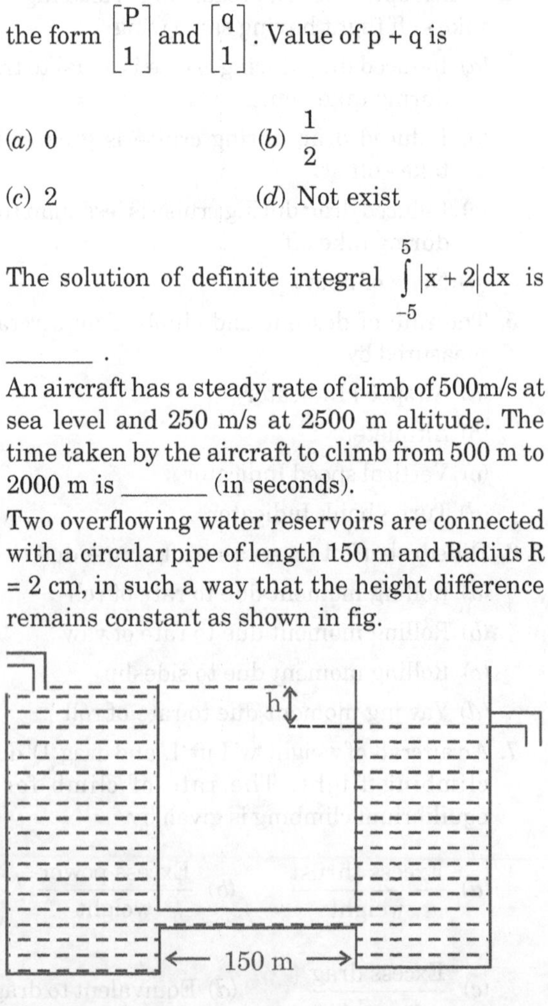

Velocity of the flow through the centre of the pipe is 10 m/s, and the velocity profile inside the pipe is given by $u = \frac{-R^2}{4\mu}\frac{dp}{dx}\left[1-\frac{r^2}{R^2}\right]$

where 'r' is the radial location inside the pipe with reference to the central axis and x is measured from left end of pipe. If ρ_{water} = 1000 kg/m^3, kinematic viscosity = 1×10^{-6} m^2/s, g = 10 m/s^2, then pressure gradient (N/m^3) is given by ______.

30. An airflow passed over the airfoil as shown in fig below:

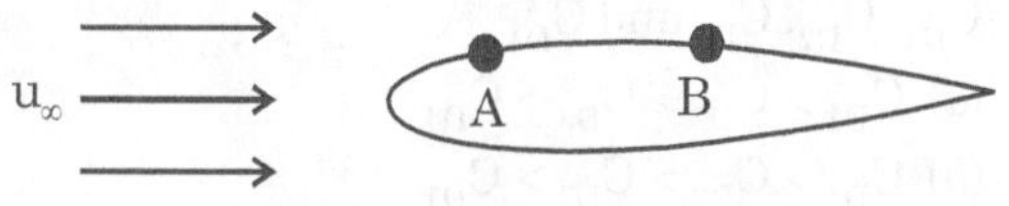

Static pressure at location 'A' is 94350 N/m^2 and co - efficient of pressure at A and B is – 3 and – 1.5 respectively. The value of static pressure at point 'B' is, if freestream pressure is 101325 N/m^2.

(*a*) 246337.5 N/m^2 (*b*) 101325 N/m^2
(*c*) 97838 N/m^2 (*d*) 98373 N/m^2

31. An isentropic supersonic flow passed through a convergent-divergent nozzle with an exit to throat area ratio of 10.25. If the reservoir pressure is 5 atm. Then find exit pressure.

(*a*) 0.57 atm (*b*) 0.057 atm
(*c*) 0.04 atm (*d*) 0.4 atm

32. A satellite in an elliptical orbit has time period equal to a satellite in Geo - Stationary orbit (GEO). What is the Semi - major axis of this orbit?

(*a*) 42176918.51 m (*b*) 82176918.51 m
(*c*) 62176918.51 m (*d*) 72176918.51 m

33. Which of the following quantities are remains constant in an orbit

(P) Angular momentum of satellite about center of planet

(Q) Area swept per unit time by line Joining satellite and center of planet

(R) Linear momentum of satellite

(*a*) P and Q (*b*) Q and R
(*c*) P, Q and R (*d*) Only Q

34. A thin airfoil of chord length 1m is placed in an airstream at an angle of attack of 6°. The value of lift generated per unit span, if equation of strength of vortex r (θ) and circulation (Γ) is given by (Take density = 1.225 kg/m^3)

$$r(\theta) = 200\alpha\frac{(1+\cos\theta)}{\sin\theta}$$

$$\Gamma = \frac{C}{2}\int_0^\pi r(\theta)\sin\theta d\theta$$

(*a*) 20 KN/m (*b*) 10 KN/m
(*c*) 5 KN/m (*d*) 4 KN/m

35. A thin walled box section having dimension 3a × 2a × t is subjected to a twisting moment of T. A solid circular cross section of diameter, d = 3a is also subjected to same twisting moment. Then thickness of the box is, if maximum shear stress developed in both section is same

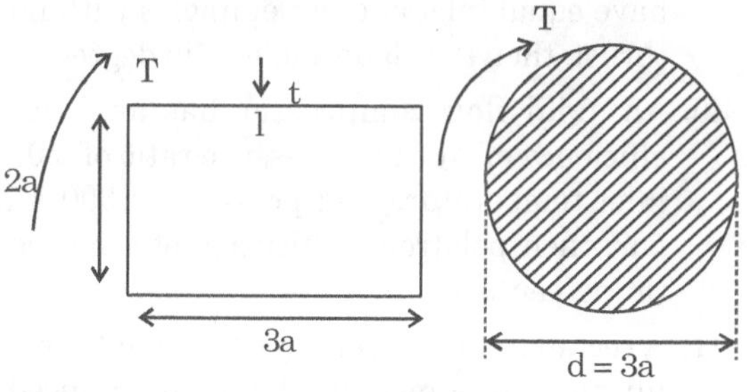

(*a*) 0.4a (*b*) 0.6a
(*c*) 0.04a (*d*) 0.06a

36. A steel tube shown in fig. of length 1 m is subjected to twisting moment. T = 10 KN -m. If modulus of rigidity of steel is given as 76 GPa, then angle of twist is (r = 50 mm) ↓ 8mm

37. Natural frequency of the spring - mass pulley system shown in figure below is given by (mass of pulley is negligible and consider it as frictionless).

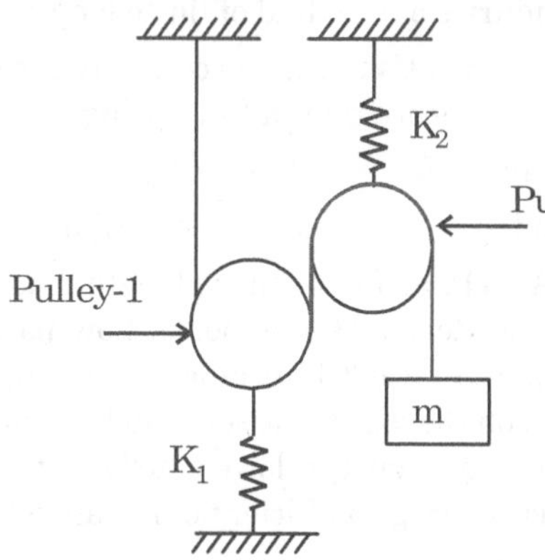

(*a*) $\sqrt{\frac{k_1+k_2}{4m(k_1k_2)}}$ rad/s (*b*) $\sqrt{\frac{4(k_1+k_2)}{k_1k_2}}$ rad/s

(*c*) $\sqrt{\frac{k_1k_2}{4m(k_1+k_2)}}$ rad/s (*d*) $\sqrt{\frac{4mk_1k_2}{(k_1+k_2)}}$ rad/s

38. Stagnation inlet temperature and pressure of nozzle of a turbojet engine is 1046.47 k and 2.2 bar respectively. The engine is operating at a altitude where the ambient pressure and temperature is 1 bar and 293 k respectively. If adiabatic efficiency of nozzle is given as 96.5% then nozzle exhaust velocity is (In m/s) _____.

39. In a single - stage impulse turbine the nozzle discharges the fluid on to the blade at an angle of 70° to the axial direction and the fluid leaves the blade with an absolute velocity of 300 m/s at an angle of 25° to the axial direction. If the blades have equal inlet and outlet angles with no axial thrust, then the blade angle is (in degree) _____.

40. An axial flow compressor has an isentropic efficiency of 85% at a pressure ratio of 6.0. If the inlet temperature and pressure is 300 k and 2 bar, then polytropic efficiency of compressor is (in percentage) _____ .

41. A rectangular wing of aspect ratio '6.0' and zero - lift angle of attack of –2.0° . The induced drag co-efficient at an angle of attack of 3.4° is given 0.01. Then the induced drag co - efficient for the similar wing having same airfoil section but different aspect ratio of 10 is given by

(*a*) 0.0076 (*b*) 0.076
(*c*) 0.096 (*d*) 0.058

42. Consider the statements given below

(i) The strength of a vortex filament is constant along its length.

(ii) A vortex filament can end in a fluid to the boundaries of the fluid of finite length.

which statement/statements correctly define the helmholtz's theorem for a finite wing

(*a*) (i) only (*b*) (ii) only
(*c*) (i) and (ii) (*d*) Neither (i) nor (ii)

43. A NACA 2412 airfoil of chord length is given to 1.5 m. The Reynold's number of flow past over the airfoil is 3.1×10^6. If turbulent boundary layer is formed on the top surface of airfoil and laminar boundary layer on the lower surface. then net skin-friction drag co-efficient of the airfoil is.

(*a*) 0.00744
(*b*) 0.0015
(*c*) 0.00447
(*d*) 0.00754

44. Four aerodynamical shape body of equal vertical height is shown below. The co-efficient of drag based on frontal area of geometry is written below of each geometry

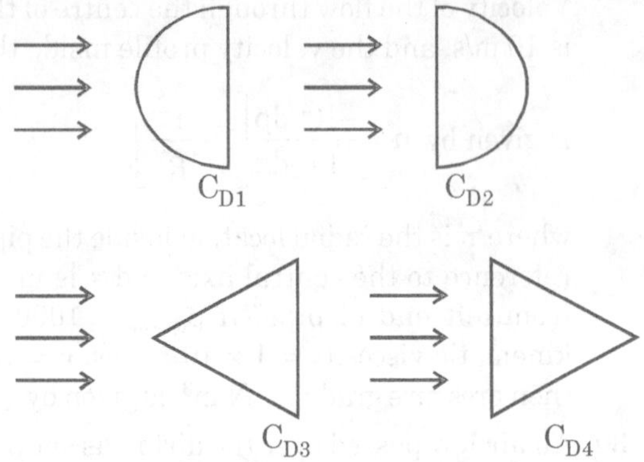

which option correctly states the relation between C_{D1}, C_{D2}, C_{D3} and C_{D4}

(*a*) $C_{D1} > C_{D2} > C_{D3} > C_{D4}$
(*b*) $C_{D4} > C_{D3} > C_{D2} > C_{D1}$
(*c*) $C_{D4} > C_{D2} > C_{D3} > C_{D1}$
(*d*) $C_{D1} > C_{D3} > C_{D2} > C_{D4}$

45. An airplane has a wing area of 19 m^2 and weight of 22700 N. The wind tunnel test results of this airplane is listed below:

lift slope, a = 0.08 per degree

co-efficient of moment at zero lift = 0.06

$\frac{\partial c_{m.cg}}{\partial \alpha_a} = -0.0133, \ \alpha_a = 6.5°$

Tail-volume ratio = 0.34

Elevator control effectiveness = 0.04

Then the elevator deflection angle necessary to trim the airplane at a velocity of 61 m/s at sea level is

(*a*) 1.94° upward (*b*) 1.94 downward
(*c*) 1.49° upward (*d*) 1.49° downward

46. The maximum value of the solution y(0) of the differential equation $\ddot{y}(x) + y(x) = 0$ with initial condition $y(0) = 1$ and $\dot{y}(0) = 1$, for $t \geq 0$ is

(*a*) $\sqrt{2}$ (*b*) 2
(*c*) π (*d*) $\sqrt{\pi}$

47. The solution of the double integration 'I' bounded by surface 'S' is where,

$$I = \iint_S \left(x^3 dydz + x^2 ydzdx + x^2 zdxdy\right)$$

where 'S' is the closed surface consisting of the cylinder $x^2 + y^2 = 25$ $(0 \leq x \leq 3)$ and the circular disk z = 0 and z = 3 $\left(x^2 + y^2 \leq 25\right)$ _____.

48. Two vector 'a' and 'b' separated by an angle 'θ' in such a way that $0 < \theta < 90°$, and $|a| > |b|$. when vector 'a' is projected on vector 'b' the result will be given by

(a) $\left(\frac{a.b}{b.b}\right).a$ (b) $\left(\frac{a.b}{b.b}\right).b$

(c) $\left(\frac{a.b}{a.a}\right).a$ (d) $\left(\frac{a.b}{a.a}\right).b$

49. A solid propellant rocket engine is designed with an end - burnning configuration. Initial grain temperature is 288 k with following characteristics of the propellant; linear burning rate is 1.02×10^{-3} m/s at a combustion pressure of 35 bar and 1.47×10^{-3} m/s at a combustion pressure of 69 bar. The rocket engine is operating at a combustion pressure of 104 bar. After 5 sec the distance of burning surface will be (in mm) ______ .

50. An airfoil with chord length 'C' and the running distance 'x' measured along the chord.
The leading edge is located at x/c = 0 and the trailing edge is located at x/c = 0 and the trailing edge at x/c = 1. The pressure co-efficient variations over the upper and lower surfaces are given as $C_{p.u} = 1 - 300\left(\frac{x}{C}\right)^2, 0 \le \frac{x}{C} \le 1.0$

$C_{p.l} = 1 - \frac{x}{C},\ 0 \le \frac{x}{C} \le 1.0$

The normal force coefficient is given by

(a) 100 (b) 200

(c) 120 (d) 175

51. A cantilevered beam having length 'L' and a point mass 'm' at one end is shown below, the natural frequency of the system would be

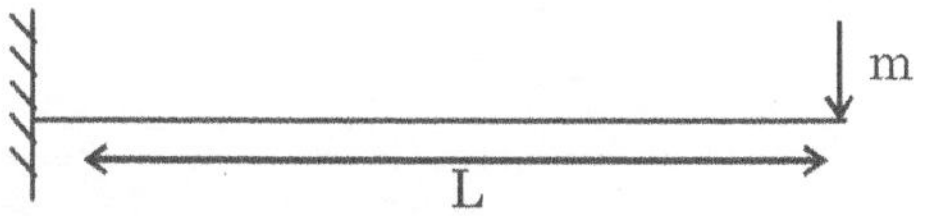

(a) $\sqrt{\frac{2EI}{mL^3}}$ (b) $\sqrt{\frac{EI}{2mL^3}}$

(c) $\sqrt{\frac{3EI}{mL^3}}$ (b) $\sqrt{\frac{mL^3}{3EI}}$

where EI is the flexural rigidity of beam .

52. Combustion between fuel and oxidizer inside a combustion chamber is given by following chemical reaction

$2C_8H_{18} + (25O_2 + 94N_2) \rightarrow 16CO_2 + 18H_2O + 94N_2$

If the combustion takes place with air to fuel ratio of 80 : 1, then value of equivalence ratio is ____.

53. The roots of the characteristic equaction for the longitudinal dynamics of an aircraft are

$\lambda_{1,2} = -0.03 \pm 0.2\hat{i}\ \lambda_{3,4} = -3 \pm 2.5\hat{i}$

The pair of eigen values that represent the phugoid mode is

(a) λ_1 and λ_2 (b) λ_1 and λ_3

(c) λ_2 and λ_4 (d) λ_3 and λ_4

54. The Inverse Laplace transform of

$f(S) = \frac{(S+2)}{(S+3)(S-4)}$ is

(a) $\frac{1}{7}e^{-3t} + \frac{6}{7}e^{4t}$ (b) $\frac{1}{7}e^{3t} + \frac{6}{7}e^{4t}$

(c) $\frac{6}{7}e^{-3t} + \frac{1}{7}e^{4t}$ (d) None of these

55. An airfoil generates a lift of 80 N when operating in a freestream flow of 80 m/s If the ambient pressure and temperature are 80 kpa and 280 k (R = 287 J/kg - k), the circulation on the airfoil (in cm^2/s) ______ .

56. Going by the ______ that many hands make light work, the school ______ involved all the students in the task."

(a) principle, principal (b) principal, principle

(c) principle, principle (d) principal, principal

57. "Her ____ should not be confused with miserliness; she is ever willing to assist those in need."
The word that best fills the blank in the above sentence is

(a) cleanliness (b) punctuality

(c) frugality (d) greatness

58. Seven machines take 7 minutes to make 7 identical toys. At the same rate, how many minutes would it take for 100 machines to make 100 toys?

(a) 1 (b) 7

(c) 100 (d) 700

59. A rectangle becomes a square when its length and breadth are reduced by 10 m and 5 m, respectively, during this process, the rectangle loses 650 m^2 of area. What is the area of the original rectangle in square meters?

(a) 1125 (b) 2250

(c) 2924 (d) 4500

60. A number consists of two digits. The sum of the digits is 9. If 45 is subtracted from the number, its digits are interchanged. What is the number?

(a) 63 (b) 72

(c) 81 (d) 90

61. For integers a, b and c, what would be the minimum and maximum values respectively of a + b + c if log | a | + log | b | + log | c | = 0

(a) –3 and 3 (b) –1 and 1

(c) –1 and 3 (d) 1 and 3

62. Given that a and b are integers and $a + a^2b^3$ is odd, which one of the following statements is correct?

(a) a and b are both odd

(b) a and b are both even

(c) a is even and b is odd

(d) a is odd and b is even

63. From the time the front of train enters a platform, it takes 25 seconds for the back of the train to leave the platform, while travelling at a constant speed of 54 km/h. At the same speed, it takes 14 seconds to pass a man running at 9 km/h in the same direction as the train. What is the length of the train and that of the platform in meters, respectively?

(a) 210 and 140

(b) 162.5 and 187.5

(c) 245 and 130

(d) 175 and 200

64. Which of the following functions describe the graph shown in the below figure.

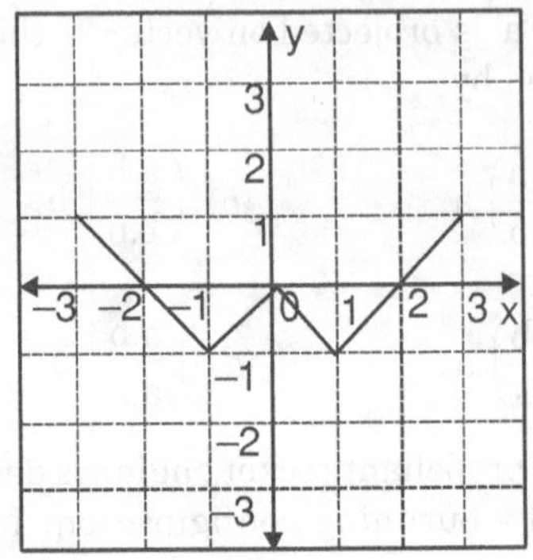

(a) $y = ||x| + 1| - 2$ (b) $y = ||x| - 1| - 1$

(c) $y = ||x| + 1| - 1$ (d) $y = ||x - 1| - 1|$

65. Consider the following three statements:

(i) Some roses are red

(ii) All red flowers fade quickly

(iii) Some roses fade quickly

Which of the following statements can be logically inferred from the above statements?

(a) If (i) is true and (ii) is false, then (iii) is false.

(b) If (i) is true and (ii) is false, then (iii) is true.

(c) If (i) and (ii) are true, then (iii) is true.

(d) If (i) and (ii) are false, then (iii) is false.

OMR SHEET

1. (A) (B) (C) (D)
2. (A) (B) (C) (D)
3. (A) (B) (C) (D)
4. (A) (B) (C) (D)
5. (A) (B) (C) (D)
6. (A) (B) (C) (D)
7. (A) (B) (C) (D)
8.
9. (A) (B) (C) (D)
10. (A) (B) (C) (D)
11.
12. (A) (B) (C) (D)
13.
14. (A) (B) (C) (D)
15. (A) (B) (C) (D)
16.
17. (A) (B) (C) (D)
18. (A) (B) (C) (D)
19. (A) (B) (C) (D)
20.
21. (A) (B) (C) (D)
22.
23. (A) (B) (C) (D)
24. (A) (B) (C) (D)
25. (A) (B) (C) (D)
26. (A) (B) (C) (D)
27. (A) (B) (C) (D)
28.
29.
30. (A) (B) (C) (D)
31. (A) (B) (C) (D)
32. (A) (B) (C) (D)
33. (A) (B) (C) (D)
34. (A) (B) (C) (D)
35. (A) (B) (C) (D)
36. (A) (B) (C) (D)
37. (A) (B) (C) (D)
38.
39.
40.
41. (A) (B) (C) (D)
42. (A) (B) (C) (D)
43. (A) (B) (C) (D)
44. (A) (B) (C) (D)
45. (A) (B) (C) (D)
46. (A) (B) (C) (D)
47.
48. (A) (B) (C) (D)
49.
50. (A) (B) (C) (D)
51. (A) (B) (C) (D)
52.
53. (A) (B) (C) (D)
54. (A) (B) (C) (D)
55.
56. (A) (B) (C) (D)
57. (A) (B) (C) (D)
58. (A) (B) (C) (D)
59. (A) (B) (C) (D)
60. (A) (B) (C) (D)
61. (A) (B) (C) (D)
62. (A) (B) (C) (D)
63. (A) (B) (C) (D)
64. (A) (B) (C) (D)
65. (A) (B) (C) (D)

ANSWERS

1. (*a*)	2. (*c*)	3. (*b*)	4. (*c*)	5. (*c*)	6. (*c*)	7. (*b*)	8. 24.05 m/s	9. (*a*)
10. (*b*)	11. 263189.45 N/mm^2		12. (*c*)	13. 18000 N/m2		14. (*d*)		15. (*)
16. 2165.13 m/s		17. (*d*)	18. 6	19. (*a*)	20. 4	21. (*b*)	22. 0.61	23. (*c*)	24. (*b*)
25. (*d*)	26. (*d*)	27. 25	28. 4.05 sec		29. –100 N/m^3		30. (*a*)	31. (*c*)	32. (*a*)
33. (*a*)	34. (*d*)	35. (*a*)	36. (*a*)	37. (*d*)	38. 591.94 m/s		39. 58.1°	40. 89%	41. (*a*)
42. (*a*)	43. (*c*)	44. (*c*)	45. (*a*)	46. (*a*)	47. 7363.108		48. (*b*)	49. 9.162	50. (*a*)
51. (*c*)	52. 0.188	53. (*a*)	54. (*a*)	55. 1.005 m^2/s		56. (*a*)	57. (*c*)	58. (*b*)	59. (*b*)
60. (*b*)	61. (*a*)	62. (*d*)	63. (*d*)	64. (*b*)	65. (*c*)				

EXPLANATIONS

4. Induced drag is almost 60% of total drag during take off and 25% of total drag during cruise for a typical low-speed aircraft.

7.

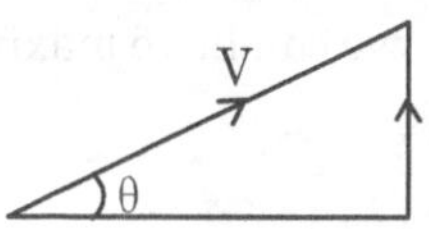

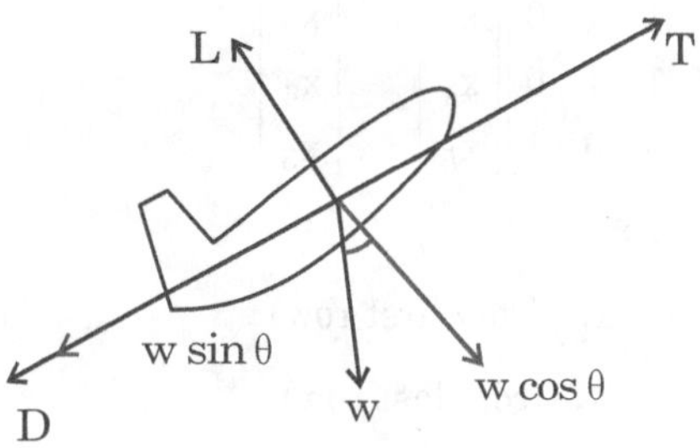

$$T\text{-}D\text{-}w\sin\theta = 0$$

$$\sin\theta = \frac{T - D}{w}$$

$$R/C = v\sin\theta$$

$$= \frac{TV - DV}{w} = \frac{P_{av} - P_{req}}{w}$$

$$R/C = \frac{\text{Excess power}}{w}$$

8.

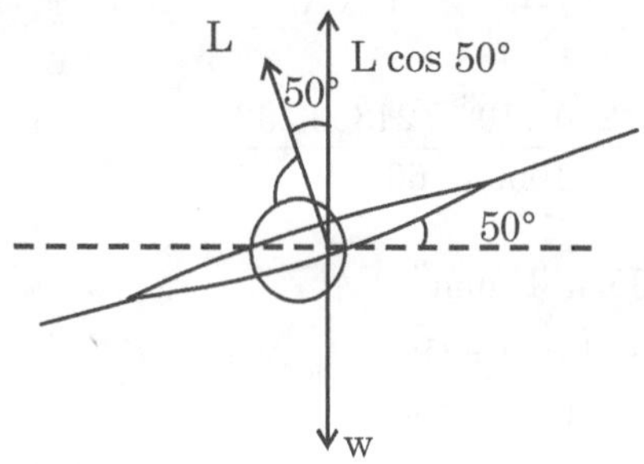

$$(V_{Stall})^2 = \frac{2w}{\rho S\, C_{L\,max}}$$

$$C_{L\,max} = \frac{2w}{\rho S \times (30)^2} = \frac{2L}{\rho S \times 900} \rightarrow \text{for level flight}$$

For level turn

$w = L\cos\phi$

$w = L\cos\phi$

$$V^2 = \frac{2w}{\rho S\, C_{L\,max}}$$

$$= \frac{2L\cos\phi}{\rho S\, C_{L\,max}}$$

$$= \frac{7L\cos\phi}{\rho S \times 2L} \times \rho S \times 900$$

$$V = \sqrt{900 \times \cos 50^\circ} = 24.05 \text{ m/s}$$

9. General form of PDE is

$Au_{xx} + Bu_{xy} + Cu_{yy} + Du_x + Eu_y + f = 0$

so, parabolic if $B^2 - 4AC = 0$

In equation (a), $B = -6, A = 9, C = 1$

$B^2 - 4AC = (-6)^2 - 4 \times 9 = 0$

10. maximum bending moment acts at the cantilever end

$$m_{max} = w \times \frac{1}{2} \times \left(\frac{1}{2} + \frac{1}{4}\right)$$

$$= \frac{3wl^2}{8}$$

11. $$I_{min} = \frac{40 \times 40^3}{12} = 213333.33 \text{ mm}^4$$

$$P = \frac{\pi^2 EI}{l^2} \Rightarrow \frac{P}{A} = \sigma = \frac{\pi^2 EI}{l^2 \times A}$$

$$\sigma = \frac{\pi^2 \times 200 \times 10^3 \times 213333.33}{1000 \times 1600}$$

$= 263189.45 \text{ N/mm}^2$

13. Pressure thrust $= (P_e - P_a) A_e$

$80 \times 10^3 = (P_e - 10000) \times 10$

$10P_e = 80000 + 100000$

$$P_e = \frac{180000}{10} = 18000 \text{ N/m}^2$$

15. $$R = 2\sqrt{\frac{2}{\rho_\infty s}} \times \frac{1}{C_t} \times \frac{C_L^{\frac{1}{2}}}{C_D} \times \left[\sqrt{w_o} - \sqrt{w_f}\right]$$

$$R \propto \frac{\sqrt{C_L}}{C_D}$$

16. max. speed means minimum drag condition

$$C_{L_{md}} = \sqrt{\frac{3C_{Do}}{K}}$$

$$\sqrt{\frac{3 \times 0.02}{0.08}} = 0.866$$

$$V_{max} = \sqrt{\frac{2w}{\rho s C_{L_{md}}}} = \sqrt{\frac{2 \times 45000}{1.2 \times 40 \times 0.866}}$$

$= 2165.13$ m/s

17. $$\underset{x\to 0^+}{Lt} f(x) = \underset{x\to 0}{Lt}\, 3x^3 = 0$$

$$\underset{x\to 0^-}{Lt} f(x) = \underset{x\to 0}{Lt}\, \sin x = 0$$

$f(0) = 0$

since, $\underset{x\to 0^+}{Lt} = f(0) = \underset{x\to 0^-}{Lt}$

So function is continuous at $x = 0$

Left Hand Derivative

$$= \underset{x\to 0^-}{Lt} \frac{f(x) - f(0)}{x - 0}$$

$$= \underset{x\to 0}{Lt} \frac{\sin x - 0}{x - 0} = 1$$

$$\text{Right hand derivative} = \underset{x\to 0^+}{Lt} \frac{f(x) - f(0)}{x - 0}$$

$$= \underset{x\to 0}{Lt} \frac{3x^3 - 0}{x - 0} = 0$$

Since LHD ≠ RHD, so f(x) is not differentiable at $x = 0$

18. Since rank of matrix is 2, we can have maximum 2 non-zero eigen values.

$$\begin{bmatrix} 1 & 0 & 0 & 0 & 1 \\ 0 & 1 & 1 & 1 & 0 \\ 0 & 1 & 1 & 1 & 0 \\ 0 & 1 & 1 & 1 & 0 \\ 1 & 0 & 0 & 0 & 1 \end{bmatrix} \begin{bmatrix} x_1 \\ x_2 \\ x_3 \\ x_4 \\ x_5 \end{bmatrix} = \lambda \begin{bmatrix} x_1 \\ x_2 \\ x_3 \\ x_4 \\ x_5 \end{bmatrix}$$

$\because AX = \lambda X$

$x_1 + x_5 = \lambda x_1$ (from first row)

$x_1 + x_5 = \lambda x_5$ (from last row)

Adding these two,

$2(x_1 + x_5) = \lambda (x_1 + x_5)$

$\Rightarrow \lambda = 2$

Similarly,

$x_2 + x_3 + x_4 = \lambda x_2$

$x_2 + x_3 + x_4 = \lambda x_3$

$x_2 + x_3 + x_4 = \lambda x_4$

Adding these three,

$3(x_2 + x_3 + x_4) = \lambda (x_2 + x_3 + x_4)$

$\Rightarrow \lambda = 3$

So, product of non-zero eigenvalues of the matrix $= 2 \times 3 = 6$.

19. A.E, $m^2 + 2m + 1 = 0$

$(m + 1)^2 = 0$

$m = -1, -1$

Roots are real and equal

solution is, $y = (C_1 + C_2x)\, e^{mx}$.

$y = (C_1 + 2C_2)\, e^{2x}$

20. $x + 2y + Kz = 1$ ...(i)

$2x + Ky + 8Z = 3$

$x + \frac{K}{2} + 4Z = 3$ ...(ii)

for line (i) and (ii) to be parallel

$$\frac{K}{2} = 2 \Rightarrow K = 4$$

21. $$R = 2\sqrt{\frac{2}{\rho_\infty s} \cdot \frac{1}{C_t}} \frac{C_L^{\frac{1}{2}}}{C_D}\left[\sqrt{w_o} - \sqrt{w_f}\right]$$

$$R_1 \propto \left(\sqrt{w_o} - \sqrt{w_f}\right)$$

$$R_2 \propto \left(\sqrt{1.1w_o} - \sqrt{w_f}\right)$$

$$\frac{DR}{R} = \frac{\sqrt{1.1w_o} - \sqrt{w_f}}{\sqrt{w_o} - \sqrt{w_f}}$$

22. From thin airfoil theory

$$C_L = \frac{4\alpha}{\sqrt{M^2 - 1}} = \frac{4 \times 20 \times \pi}{180 \times \sqrt{(2.5)^2 - 1}} = 0.61$$

23.
- Total temperature across normal shock remains same.
- Normal shock is a compression wave which increase the pressure of flow.
- Since static pressure increase across the normal shock so total pressure will reduce.

24. $$C_P = \frac{P - P_\infty}{q_\infty} = 1 - \left(\frac{V}{V_\infty}\right)^2$$

From continuity equation, $AV = A_\infty V_\infty$.

$$\frac{V}{V_\infty} = \frac{A_\infty}{A}$$

$$C_P = 1 - \left(\frac{A_\infty}{A}\right)^2$$

$$= 1 - \left(\frac{1}{3}\right)^2 = 1 - \frac{1}{9}$$

$$= \frac{8}{9} = 0.89$$

25. Stringers and longerous are longitudinal elements of semi monocoque fuselage, which are used to resist bending load due to compressive and tensile forces.

26. Eigen value of matrix is

$$\begin{vmatrix} 2-\lambda & 1 \\ -1 & 2-\lambda \end{vmatrix} = 0$$

$(2 - \lambda)(2 - \lambda) + 1 = 0$

$-4\lambda - \lambda^2 + 4 + 1 = 0$

$\lambda^2 + 4\lambda - 5 = 0$

$$\lambda_{1,2} = \frac{-4 \pm \sqrt{16 + 20}}{2}$$

$$\lambda_{1,2} = \frac{-4 \pm 6}{2}$$

$\lambda = 1, -5$

Let $\hat{X}$ is the eigen vector of matrix

$[A - \lambda I]\, \hat{X} = 0$

$$\begin{bmatrix} 2-\lambda & 1 \\ -1 & 2-\lambda \end{bmatrix}\begin{bmatrix} P \\ 1 \end{bmatrix} = 0$$

For $\lambda = 1$, $\begin{bmatrix} 2 & 1 \\ -1 & 1 \end{bmatrix}\begin{bmatrix} P \\ 1 \end{bmatrix} = 0$

$2P + 1 = 0$

$$\Rightarrow P = -\frac{L}{2}$$

for $\lambda = -5$

$$\begin{bmatrix} 7 & 1 \\ -1 & 7 \end{bmatrix}\begin{bmatrix} q \\ 1 \end{bmatrix} = 0$$

$7q + 1 = 0$

$$q = \frac{-1}{7}$$

$$P + q = -\frac{1}{2} - \frac{1}{7} = \frac{-9}{14}$$

27. For $-5 \le x \le 0$, $|x + 2| = -(x + 2)$

For $0 \le x \le 5$, $|x + 2| = (x + 2)$

$$\int_{-5}^{5} |x+2|\, dx$$

$$\int_{-5}^{0} -(x+2)\, dx + \int_{0}^{5} (x+2)\, dx =$$

$$-\left[\frac{(x+2)^2}{2}\right]_{-5}^{0} + \left[\frac{(x+2)^2}{2}\right]_{0}^{5}$$

$$= -\left[\frac{4}{2} - \frac{9}{2}\right] + \left[\frac{49}{2} - \frac{4}{2}\right]$$

$$= \frac{5}{2} + \frac{45}{2} = 25$$

28. Rate of climb is assuming decreases linearly with altitude

$$R/C = m + nh$$

where, m = 500 m/s

$$n = \frac{250 - 500}{2500} = -0.1$$

$$R/C = \frac{dh}{dt}$$

$$\int_{t_1}^{t_2} dt = \int_{h_1}^{h_2} \frac{dh}{m + nh}$$

$$t_2 - t_1 = \frac{1}{n} \ln \frac{m + nh_2}{m + nh_1}$$

$$= \frac{1}{-0.1} \ln \frac{500 - 0.1 \times 2000}{500 - 0.1 \times 500} = 4.05 \text{ sec}$$

29. $$u = \frac{-R^2}{4\mu}\frac{dP}{dx}\left[1 - \frac{r^2}{R^2}\right]$$

at centerline, $u = u_m$ and $r = 0$

$$u_{mean} = -\frac{R^2}{4\mu}\frac{dP}{dx}$$

$$\frac{dP}{dx} = \frac{u_{mean} \times 4\mu}{-R^2} = \frac{u_{mean} \times 4\nu\rho}{-R^2}$$

$$= \frac{10 \times 4 \times 1 \times 10^{-6} \times 1000}{-(0.02)^2}$$

$$\frac{dP}{dx} = -100 \text{ N/m}^3$$

30. $$C_{PA} = \frac{P_A - P_\infty}{q_\infty}$$

$$P_\infty = P_A - C_{P_A} q_\infty$$

$$C_{PB} = \frac{P_B - P_\infty}{q_\infty}$$

$$C_{PB}\, q_\infty = P_B - (P_A - C_{PA}\, q_\infty)$$

$$q_\infty (C_{PB} - C_{PA}) + P_A = P_B$$

If $q_\infty = 101325$ N/m^2 (Assuming)

$P_B = 101325\,(-1.5 + 3) + 94350$

$= 246337.5$ N/m^2.

31. $$\frac{A_e}{A^*} = 10.25$$

$$\frac{1}{M_e}\left[\frac{5 + M_e^2}{6}\right]^3 = 10.25$$

$M_e = 3.95$

$P_o = 5 \text{ atm} = P_{oe}$

$$\frac{P_{oe}}{P_e} = \left[1 + \frac{r-1}{2} m_e^2\right]^{\frac{r}{(r-1)}}$$

$$P_e = \frac{5}{\left(1 + 0.2 \times 3.95^2\right)^{3.5}} = 0.035 \text{ atm} = 0.04 \text{ atm}$$

32. $$T = 2\pi\sqrt{\frac{a^3}{GM}} \quad \ldots (1)$$

$G = 6.673 \times 10^{-11}$ N-m^2/kg^2.

$m = 5.98 \times 10^{24}$ kg

$Gm = 3.99 \times 10^{14}$ N-m^2/kg

Time period of Geo-stationary satellite

= 1436 min

= 86160 sec

$$T = 2\Pi\sqrt{\frac{a^3}{Gm}}$$

$$a^3 = \left(\frac{T}{2\Pi}\right)^2 \times Gm$$

$$= \left(\frac{86160}{2\Pi}\right)^2 \times 3.99 \times 10^{14}$$

$a = 42176918.51$ m

34. $r(\theta) = 200\ \alpha \dfrac{(1+\cos\theta)}{\sin\theta}$

$$T = \frac{C}{2}\int_0^{\Pi} 200\ \ \alpha \frac{(1+\cos\theta)}{\sin\theta} \sin\theta\ d\theta$$

$$= 100\ C\ \alpha \int_0^{\Pi} (1+\cos\theta)\ d\theta$$

$$= 100\ C\ \alpha \left[\theta + \sin\theta\right]_0^{\Pi}$$

$= 100\ \Pi\ C\ \alpha$

General equation of strength of vortex

$$r(\theta) = 2\ \alpha V_\infty \frac{(1+\cos\theta)}{\sin\theta}$$

when comparing we get

$V_\infty = 100$ m/s

$L' = \rho_\infty V_\infty \Gamma$

$$= 1.225 \times 100 \times 100 \times \pi \times 1 \times 6\ \frac{\Pi}{180}$$

= 4030.09 N/m

= 4.03 KN/m

35. For box section

$q = \tau_{max} t$

$$\tau_{max} = \frac{q}{t}$$

$T = 2Aq = 2A\ \tau_{max} t$

$$\tau_{max} = \frac{T}{2At} = \frac{T}{2 \times 3a \times 2a \times t}$$

$$(\tau_{max})_1 = \frac{T}{12a^2 t}$$

For solid section

$$\frac{T}{I_P} = \frac{\tau}{r} = G\frac{d\theta}{dZ}$$

$$\tau = \frac{T \times r}{I_P} = \frac{T \times d \times 32}{2 \times \Pi d^4}$$

$$(\tau_{max})_2 = \frac{16T}{\pi d^3} = \frac{16T}{\pi \times 27a^3}$$

$$(\tau_{max})_1 = (\tau_{max})_2$$

$$\frac{T}{12a^2 t} = \frac{16T}{\Pi \times 27a^3} \Rightarrow t = \frac{27 \times \Pi a}{12 \times 16}$$

$t = 0.44a$

36. $\dfrac{d\theta}{dZ} = \dfrac{T}{4A^2 G}\oint \dfrac{ds}{t}$

$$A = \frac{\Pi \times 50^2}{2} \times 2 + 100 \times 100$$

$= 17853.98\ mm^2$.

$$\frac{d\theta}{dZ} = \frac{10 \times 10^6}{4 \times (17853.98)^2 \times 76 \times 10^3}$$

$$\left[\frac{2 \times \Pi \times 50}{8} + \frac{2 \times 100}{8}\right]$$

$d\theta = 0.38° \simeq 0.4°$

37.

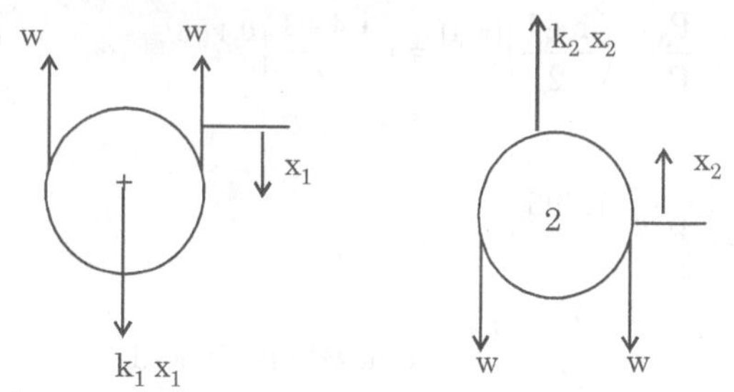

$k_1 x_1 = 2w$

$$x_1 = \frac{2w}{k_1}$$

$k_2 x_2 = 2w$

$$x_2 = \frac{2w}{k_2}$$

$x = 2(x_1 + x_2)$

$$x = 2\left(\frac{2w}{k_1} + \frac{2w}{k_2}\right)$$

$$k_{eq} = \frac{f}{x} = \frac{w}{2 \times 2w\left(\frac{1}{k_1} + \frac{1}{k_2}\right)}$$

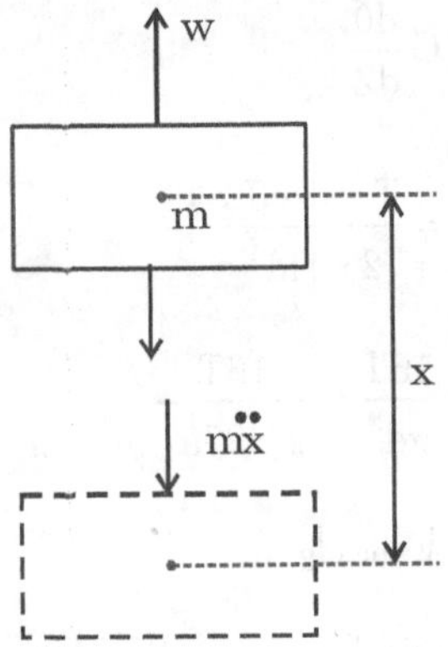

$$k_{eq} = \frac{k_1 k_2}{4(k_1 + k_2)}$$

$$m\ddot{x} + k_{eq}x = 0$$

$$\omega_n = \sqrt{\frac{k_{eq}}{m}}$$

$$= \sqrt{\frac{k_1 k_2}{4m(k_1 + k_2)}} \text{ rad/s}$$

38. $\frac{P_o}{P_e} = 2.2$

Checking choking condition

$$\frac{P_o}{P_c} = \left(\frac{r+1}{2}\right)^{\frac{r}{(r-1)}} = \left(\frac{1.4+1}{2}\right)^{\frac{1.4}{0.4}}$$

$$\frac{P_o}{P_c} = 1.893$$

Since, $\frac{P_o}{P_c} < \frac{P_o}{P_e}$, so nozzle is choked

$$\frac{T_o}{T_c} = \frac{r+1}{2}$$

$$\frac{T_o}{T_c} = \frac{2.4}{2}$$

$T_c = 872.06$ k

$V_e = \sqrt{rRT_c}$

$= \sqrt{1.4 \times 287 \times 872.06}$

$V_e = 591.94$ m/s

39.

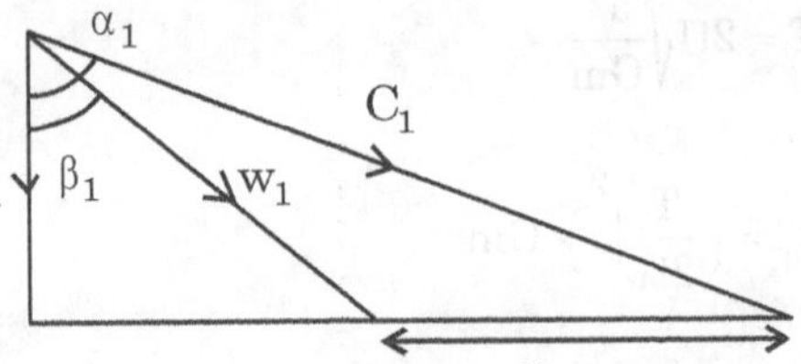

Rotor Inlet

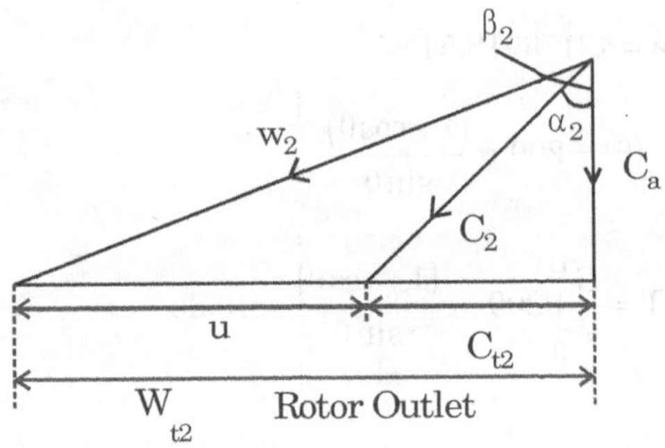

Rotor Outlet

$C_2 = 300$ m/s

$\alpha_2 = 25°$

$\alpha_1 = 70°$

$\beta_1 = 70°$

$C_a = C_2 \cos \alpha_2$.

$= 300 \times \cos 25°$

$Ca = 271.89$ m/s

$W_{t2} - C_{t2} = C_{t1} - W_{t1}$

$C_a(\tan\beta_2 - \tan\alpha_2) = C_a(\tan\alpha_1 - \tan\beta_1)$

$2 \tan \beta_1 = \tan\alpha_1 + \tan \alpha_2$.

$$\tan\beta_1 = \frac{\tan\alpha_1 + \tan\alpha_2}{2}$$

$$\beta_1 = \beta_2 = 58.1°$$

40. $\eta_{iso} = \frac{T'_{02} - T_{01}}{T_{02} - T_{01}}$

$$\eta_{iso} = \frac{r_p^{\frac{r-1}{r}} - 1}{r_p^{\frac{r-1}{r} \times \frac{1}{\eta_P}} - 1}$$

$$r_p^{\frac{r-1}{r} \times \frac{1}{\eta_P}} - 1 = \frac{r_p^{\frac{r-1}{r}} - 1}{\eta_{iso}}$$

$$r_p^{\frac{0.2857}{\eta_P}} = 1 + \frac{(10)^{0.2857} - 1}{0.85}$$

$$r_P^{\frac{0.2857}{\eta_P}} = 2.095$$

$$\frac{0.2857}{\eta_P}\log(r_P) = \log(2.095)$$

$$\eta_P = \frac{0.2857 \times \log 10}{\log(2.095)}$$

$$\eta_P = 0.89 = 89\%$$

41. $C_L^2 = \Pi AR\, C_{Di}$

$= \Pi \times 6 \times 0.01$

$= 0.1885$

$C_L = 0.434$

$$\frac{dC_L}{d\alpha} = \frac{0.434 - 0}{3.4° - (-2°)} = 0.0804/\text{deg}$$

$= 4.605/\text{rad}$

$$\frac{dC_L}{d\alpha} = a = \frac{a_0}{1 + \frac{a_0}{\Pi AR}}$$

$$4.605 = \frac{a_0}{1 + \frac{a_0}{\Pi \times 6}} = \frac{a_0}{1 + 0.053a_0}$$

$a_0 = 6.092/\text{rad}$

a_0 will remain same for both wing

$$a_2 = \frac{a_0}{1 + \frac{a_0}{\Pi AR}} = \frac{6.092}{1 + \frac{6.092}{\Pi \times 10}}$$

$a_2 = 5.102/\text{rad} = 0.089/\text{degree}$

$C_L = a_2\,(\alpha - \alpha_{L=0})$

$= 0.089\,(3.4° - (-2°)) = 0.48$

$$(C_{Di})_2 = \frac{C_L^2}{\Pi AR} = \frac{(0.48)^2}{\Pi \times 10} = 0.0073$$

42. Statement (ii) is incorrect. According to helmholtz's theorem, A vortex filament cannot end in a fluid, it must extends to the boundaries of the fluid of infinite length.

43. For laminar boundary layer

$$C_{f_1} = \frac{1.328}{\sqrt{Re}}$$

$$= \frac{1.328}{\sqrt{3.1 \times 10^6}} = 7.54 \times 10^{-4}$$

For turbulent boundary layer

$$C_{f_2} = \frac{0.074}{(R_e)^{\frac{1}{5}}}$$

$$= \frac{0.074}{\left(3.1 \times 10^6\right)^{\frac{1}{5}}} = 0.00372$$

Net $C_f = C_{f1} + C_{f2}$

$= 7.54 \times 10^{-4} + 0.00372$

$= 0.0045$

45. $$\delta_{trim} = \frac{C_{m,o} + \left(\frac{\partial C_{m,cg}}{\partial \alpha_a}\right)\alpha_a}{V_H\left(\frac{\partial C_{L,t}}{\partial \delta_e}\right)}$$

$$= \frac{0.06 + (-0.0133)(6.5)}{0.34 \times 0.04}$$

$= -1.94°$

$= 1.94°$ upward

46. A.E., $m^2 + 1 = 0$

$m = \pm i$

$y = A\cos x + b\sin x$

$y(0) = 1$

$1 = A \times 1 + B \times 0$

$A = 1$

$$\dot{y} = -A\sin x + B\cos x$$

$$\dot{y}(0) = 1$$

$1 = -A \times 0 + B \times 1$

$B = 1$

$y = \cos x + \sin x$

$y' = -\sin x + \cos x = 0$

$\sin x = \cos x$

$x = 45°$

$y'' = -\cos x - \sin x$

$= -\cos 45 - \sin 45$

$y'' < 0$ at $x = 45°$, maximum value of y

$y_{max} = \cos 45° + \sin 45°$

$$= \frac{1}{\sqrt{2}} + \frac{1}{\sqrt{2}} = \sqrt{2}$$

47. $f_1 = x^3, f_2 - x^2y, f_3 = x^2z$

div f $= 3x^2 + x^2 + x^2 = 5x^2$.

$$I = \iiint (\text{div } f)\, dx\, dy\, dz$$

$$\iiint 5x^2\, dx\, dy\, dz$$

Putting, $x = r\cos\theta$, $y = r\sin\theta$.

$dxdydz = r\, dr\, d\theta\, dz$.

$$= 5\int_0^3\int_0^5\int_0^{2\Pi} r^2\cos^2\theta\, r\, dr\, d\theta\, dz$$

$$= 15\int_0^5\int_0^{2\Pi} r^3\cos^2\theta\, dr\, d\theta$$

$$= 15\times\frac{5^4}{4}\int_0^{2\Pi}\cos^2\theta\, d\theta$$

$$= \frac{15\times 5^4}{4}\times\Pi = 7363.108$$

48.

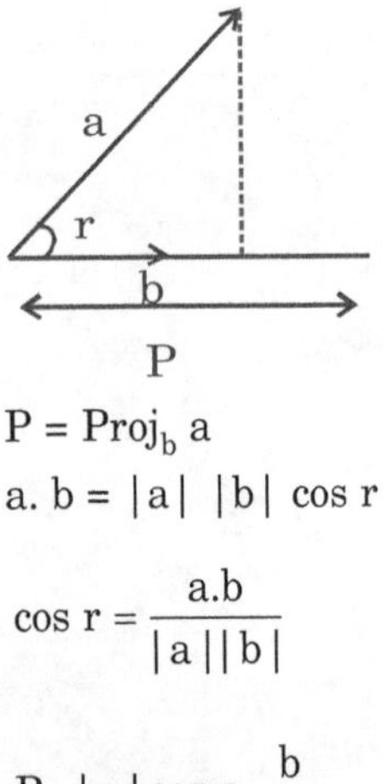

P

P = Proj_b a

a. b = |a| |b| cos r

$$\cos r = \frac{a.b}{|a||b|}$$

$$P = |a|\cos r.\frac{b}{|b|}$$

$$= |a|.\frac{a\cdot b}{|a||b|}.\frac{b}{|b|}$$

$$= \left(\frac{a.b}{b.b}\right)b$$

49. Burning rate $\left(\dot{r}\right) = aP_0^n$

$\Rightarrow 1.02\times 10^{-3} = a\times(35)^n$ ... (a)

$\Rightarrow 1.47\times 10^{-3} = a\times(69)^n$ ... (b)

Dividing (a) and (b)

$$\frac{1.02\times 10^{-3}}{1.47\times 10^{-3}} = \left(\frac{35}{69}\right)^n$$

$$\ln\left(\frac{1.02}{1.47}\right) = n\ln\left(\frac{35}{69}\right)$$

n = 0.538

$$a = \frac{1.02\times 10^{-3}}{\left(35\times 10^5\right)^{0.538}} = 3.075\times 10^{-7}$$

Burning rate for P_o = 104 bar

$$\dot{r} = 3.075\times 10^{-7}\times\left(104\times 10^5\right)^{0.538}$$

$= 1.832\times 10^{-3}$ m/s

In 5 sec, length of burning surface

$L = 1.832\times 10^{-3}\times 5$

$= 9.162\times 10^{-3}$ m

= 9.162 mm

50. $C_n = \int_0^1 (C_{PL} - C_{Pu})\, d\left(\frac{x}{c}\right)\quad \frac{x}{c} = 0\quad \frac{x}{c} = 1$

$$= \int_0^1\left\{1 - \frac{x}{c} - 1 + 300\left(\frac{x}{c}\right)^2\right\}d\left(\frac{x}{c}\right)$$

$$= \left[\frac{x}{c} - \frac{\left(\frac{x}{c}\right)^2}{2} - \left(\frac{x}{c}\right) + 300\frac{\left(\frac{x}{c}\right)^3}{3}\right]_0^1$$

$$= \left[1 - \frac{1}{2} - 1 + \frac{300}{3}\right]$$

$$C_n = \frac{-3+600}{6} = \frac{597}{6} = 99.5$$

51. Deflection at free End $= \dfrac{m\ell^3}{3EI}$

$$K = \frac{f}{\delta} = \frac{m}{m\ell^3/3EI} = \frac{3EI}{\ell^3}$$

$$\omega_n = \sqrt{\frac{K}{m}} = \sqrt{\frac{3EI}{m\ell^3}}$$

52. $\left(\frac{A}{f}\right)_{sti} = \frac{25 \times 2 \times 16 + 94 \times 2 \times 14}{2 \times (12 \times 8 + 1 \times 18)} = 15.053$

$\left(\frac{A}{f}\right)_{actual} = 80$

Equivalence ratio

$= \frac{(fla)_{actual}}{(fla)_{sti}}$

$= \frac{15.053}{80} = 0.188$

53. $\lambda_{1,2} = -p \pm q$

Time period $= \frac{2\Pi\hat{t}}{|q|}$

So, if $|q|$ is small, time period will be large, which will represent phygoid made.

54. $\frac{S+2}{(S+3)(S-4)} = \frac{A}{S+3} + \frac{B}{S-4}$

$(S+2) = A(S-4) + B(S+3)$

if, S = 4, 6 = 7B

$\Rightarrow B = \frac{6}{7}$

if, S = –3, –1 = –7A

$\Rightarrow A = \frac{1}{7}$

$f(S) = \frac{1}{7(S+3)} + \frac{6}{7} \cdot \frac{1}{S-4}$

$= \frac{1}{7}e^{-3t} + \frac{6}{7}e^{4t}$

55. $\rho_\infty = \frac{P_\infty}{RT_\infty} = \frac{80 \times 10^3}{287 \times 280}$

$= 0.995\ kg/m^3$

$L' = \rho_\infty V_\infty \Gamma$

$\Gamma = \frac{L'}{\rho_\infty V_\infty} = \frac{80}{0.995 \times 80} = 1.005\ m^2/s$

56. "Principle" means "ethics."

57. "Frugality" refers to "thriftiness."

58. $\frac{M_1 \times H_1}{\omega_1} = \frac{M_2 \times H_2}{\omega_2}$

$\frac{7 \times 7}{7} = \frac{m \times 100}{100} = 7$

59. Let the length and breadth of the rectangle be l and b respectively.

So, its area = lb

Now, according to the question,

l – 10 = b – 5d which is the side of the square

$\therefore$ lb – (l – 10) (b – 5) = 650

lb = lb + 5l + 106 – 50 = 650

5l + 106 = 700

l + 2b = 140

b + 5 + 2b = 140 $\left(\begin{array}{l} l - 10 = b - 5 \\ \therefore\ l = b + 5 \end{array}\right)$

3b = 135

b = 42 m

l = 50 m

So, area of the original rectangle

$= 50 \times 45 = 2250m^2$

60. Let the number of xy

x + y = 9

10 x + y – 45 = 10y + x

9x = 45 + 9y

x = 5 + y

x – y = 5 ...(i)

and x + y = 9 ...(ii)

on solving eq (i) and (ii)

x = 7

y = 2

$\therefore$ Number is 72.

61. If log |a| + log |b| + log |c| = 0

$\Rightarrow$ log |a| = log |b| = log |c| =0

$\Rightarrow$ a = b = c ± 1

$\therefore$ Minimum value of a + b + c = –1 –1 –1 = –3

Maximum value of a + b + c = 1 + 1 + 1 = 3.

62. $a + a^2 b^3$ = odd.

When a is odd and 6 is even then only at $a^2 b^3$ will give the odd value.

Fore e.g.

If a = 3, b = 2

$a + a^2 b^3 = 3 + 9 \times 8 = 75$ which is odd.

63. Let the lengths of the train and platform be l_t and l_p respectively.

$$\therefore \; l_t + l_p = 54 \times \frac{5}{18} \times 25$$

$$l_t + l_p = 375 \text{ m}$$

$$l_t = (54 - 9) \times \frac{5}{18} \times 14$$

$$45 \times \frac{5}{18} \times 14$$

l_t = 175 m

l_p = 200 m

64. Clearly the function

$||x|-1|-1$ describes the given graph.

At x = 0,	y = 0
x = 1,	y = –1
x = 2,	y = 0
x = –,	y = –1

65. A comes close but if ii is false, we cannot get any information about iii.

MSQ PRACTICE PAPER – 1

1. Which of the relations express that $x_1 = (1,3,4,2)$, $x_2 = (3,–5,2,2)$, and $x_3 = (2,–1,3,2)$ are linearly dependent?

(a) $x_1 + x_2 - 2x_3 = 0$ (b) $\frac{4x_1 + 4x_2 - 8x_3}{\sqrt{3}} = 0$

(c) $2x_1 + x_2 - x_3 = 0$ (d) $x_1 + x_2 + x_3 = 0$

2. The vector which is perpendicular to both $(2\hat{i} + \hat{j} - 3\hat{k})$ and $(-\hat{i} + \hat{j} + 2\hat{k})$ is

(a) If $\vec{A} = xy\hat{i} + x^2\hat{j}$ and c is the path given below

If $\oint_c \vec{A}.d\vec{l}$ = 1 then the values of a and b are respectively

(b) $5\hat{i} + \hat{j} + 3\hat{k}$

(c) $\frac{3\hat{i} - 6\hat{j} - 4\hat{k}}{\sqrt{7}}$

(d) $\frac{20\hat{i} - 4\hat{j} + 12\hat{k}}{\sqrt{21}}$

3. Velocity component of 2-D plane motion of fluid are $u = xy^2 - 2y - \frac{x^3}{3}$; $v = \frac{-y^3}{3} + 2x + x^2y$

(a) Fluid is Solenoidal
(b) Fluid is path independent
(c) Fluid is compressible
(d) Fluid is incompressible

4. If $\vec{A} = xy\hat{i} + x^2\hat{j}$ and c is the path given below.

If $\oint_c \vec{A}.d\vec{l}$ = 1 then the values of a and b are respectively

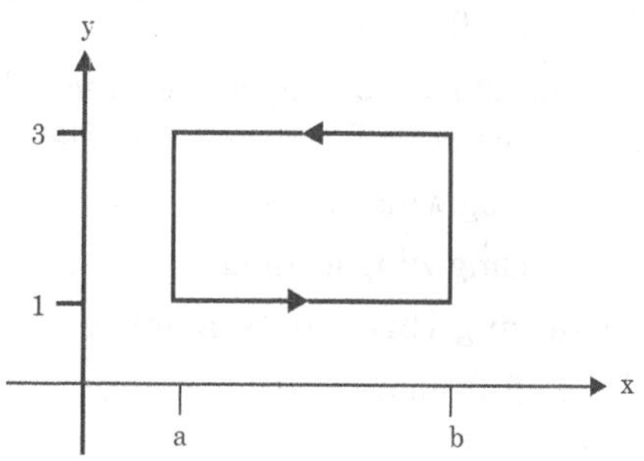

(a) 2,4 (b) $1, \sqrt{2}$

(c) $\frac{1}{\sqrt{3}}, \frac{2}{\sqrt{3}}$ (d) $\sqrt{2}, \sqrt{3}$

5. Line integral, I = $\int \vec{V}.d\vec{r}$ of vector

$\vec{V}(r) = yz^2\hat{i} + xz^2\hat{j} + 2xyz\hat{k}$ along the given curve is

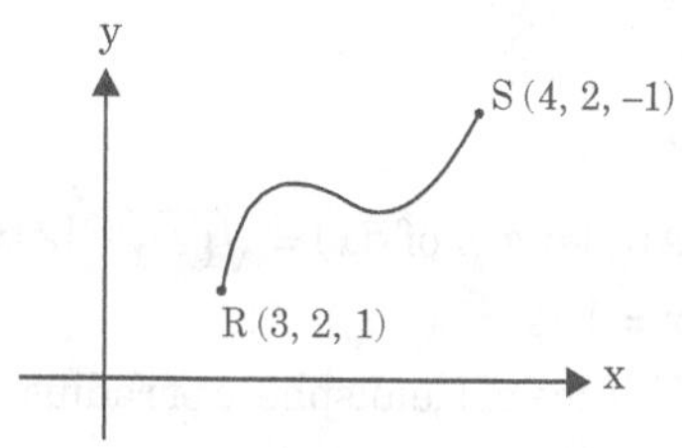

(a) 1 (b) 0
(c) 4 (d) 6

6. The following set of vectors are lieraly dependent when 'a' $(a \in R)$ is equal to

$a\hat{i} + \hat{j} + 2\hat{k}, \hat{i} + a\hat{j} + 2\hat{k}, \hat{i} + a\hat{j} + 5a\hat{k}$

(a) 1 (b) –1
(c) 0.4 (d) 0

7. Which of the following Taylor's series is valid?

(a) $\cos x = \sum_{n=0}^{\infty} (-1)^n \frac{x^{2n}}{(2n)!}$; *about* $x = 0$

(b) $\tan^{-1} x = \sum_{n=0}^{\infty} (-1)^n \frac{x^{2n+1}}{2n+1}$; *about* $x = 0$

(c) $\ln x = \sum_{n=1}^{\infty} (-1)^{n+1} \frac{(x-1)^n}{n}$; *about* $x = 1$

(d) $e^{1/x} = \sum_{n=0}^{\infty} \frac{1}{x^n (n)!}$; *about* $x = 0$

8. Which of the following is the expansion of $f(x) = 3x^2 + 5x - 2$ about $x = 2$ (using Taylor Series)?

(a) $3x^2 + 5x - 2$

(b) $20 + \frac{17(x-2)}{1!} + \frac{6(x-2)^2}{2!}$

(c) $\frac{3(x-2)^2}{2!} + \frac{5(x-2)}{1!} - 2$

(d) $3x^2 + \frac{5x}{1!} - \frac{2}{2!}$

9. Which of the following statement are invalid?

(a) Area bounded by $y^2 = x$ and $x^2 = y$ is $\frac{1}{3}$

(b) Area of $y^2 = x$ bounded by it's latus rectum is $\frac{1}{6}$.

(c) Arc length of $f(x) = \sqrt{1-x^2}$ between $x = 0$ to $x = 1$ is π.

(d) Volume of Hemisphere of radius 3 units is 36π.

10. Let $I_p = \iint_R \frac{1}{\left(\sqrt{1-x^2-y^2}\right)^{P/2}} dxdy$. For what values of "p", I_p will converge?

(a) $P = 2$ (b) $P = \frac{3}{2}$

(c) $P = \frac{5}{2}$ (d) $P = \frac{5}{6}$

11. Which of the following terms is/are appeared in Navier – stokes momentum equation?

(a) Pressure

(b) Body forces

(c) Shear stresses due to viscosity

(d) Internal energy.

12. Bernoulli's equation is valid for

(a) Steady flow (b) Inviscid flow

(c) Compressible flow (d) Incompressible flow

13. Which of the following statement is/are correct for the critical mach number (M_{cr})?

(a) Is the mach number at which flow over the airfoil become sonic for the first time.

(b) M_{cr} for thin airfoil is more than for thick airfoil

(c) M_{cr} for thin airfoil is less than for thick airfoil.

(d) Is the mach number at which there is sudden rise in airfoil drag

14. Which of the following statement are for the flow across a stationary stock wave?

(a) Stagnation temperature remains constant.

(b) Stagnation pressure decreases.

(c) Entropy remains constant.

(d) Stagnation temperature increases.

15. For Prandtl's lifting line theory for a finite wing which are the following statements is true.

(a) The bound vertex is responsible for the lift force.

(b) The trailing edge vortices are responsible for the induced drag.

(c) The bound vortex are responsible for induce drag

(d) The trailing vortices are responsible for lift force.

16. Which of the following statements are true for supersonic flow?

(a) Over a gradual expansion, entropy remains constant.

(b) Over a sharp expansion corner, entropy can increase.

(c) Over a gradual compression, entropy can remain constant.

(d) Over a sharp compression corner, entropy increases.

17. Based on linearised supersonic theory which of the following is/are correct

(a) $C_l = \frac{4\alpha}{\sqrt{M_\infty^2 - 1}}$ (b) $C_d = \frac{4\alpha^2}{\sqrt{M_\infty^2 - 1}}$

(c) $C_l = 2\pi\alpha$ (d) $C_d = \frac{4\alpha}{\sqrt{M_\infty^2 - 1}}$

18. Which of the following statement is/are correct for Maximum Endurance for reciprocating engine propeller driven aircraft?

(a) It should fly at maximum $Cl1.5/Cd$

(b) It Flight at sea level

(c) It should fly at maximum Cl/Cd

(d) It Should fly at 100 ft.

19. Which of the following are correct about for shorten the take-off distance of airplane?

(a) Increasing wing loading

(b) Decreasing wing loading

(c) Increasing Thrust to Weight ratio

(d) Take off from an altitude nearer to mean sea level

20. Lift to drag ratio for glider is 15 . if it is at altitude of 4km and released the which of the following are correct decent angle(θ) and Range(R) of glider.

(a) $\theta = \tan^{-1}(1/15)$ (b) $\theta = \tan^{-1}(1/4)$

(c) R = 60 km (d) R = 30 km

21. Which of the following are correct about short period oscillation mode?

(a) It is longitudinally dynamics mode

(b) It is lateral dynamics mode

(c) Its lightly damped mode

(d) Its heavily damped mode.

22. Which of the following are correct representation about longitudinal dynamics mode?

(a) Phugoid mode (b) Spiral mode

(c) Short period mode (d) Dutch roll mode

23. Which of the following is/ are correct for longitudinal balance and Longitudinal Static Stability?

(a) $C_{M.0}$ must be positive.

(b) $\partial C_{M.cg} / \partial \alpha_a$ must be negative

(c) Thrust line above centre of gravity

(d) Thrust line below centre of gravity

24. Which of the following is/ are correct for Neutral point?

(a) Aerodynamic centre for full aircraft

(b) Centre of pressure of full Aircraft

(c) Where $\partial C_{M.cg} / \partial \alpha_a = 0$

(d) Where $\partial C_{M.cg} / \partial \alpha_a$ is negative

25. Which of the following is /are correct about directional stability?

(a) Wing dihedral increases directional stability.

(b) Wing sweepback increases directional stability.

(c) Short fuselage decrease directional stability.

(d) Short fuselage increase directional stability.

26. An elastic beam with simply supported boundary condition has:

(a) infinite natural frequencies

(b) infinite Mode shapes

(c) 2 Natural frequencies

(d) 2 Mode shapes

27. Maximum Bending Stress in an cantilever beam with rectangular cross-section depends upon:

(a) Cross-section width

(b) Cross-section height

(c) 2nd Moment of Area

(d) Bending Moment

28. For a plain strain problem which of following is not correct :

(a) $\varepsilon_{zz} = 0$ (b) $\varepsilon_{zz} \neq 0$

(c) $\varepsilon_{yy} = 0$ (d) $\varepsilon_{xy} = 0$

29. Number of degree of freedoms (DOFs) and associated natural frequencies for given figure below are:

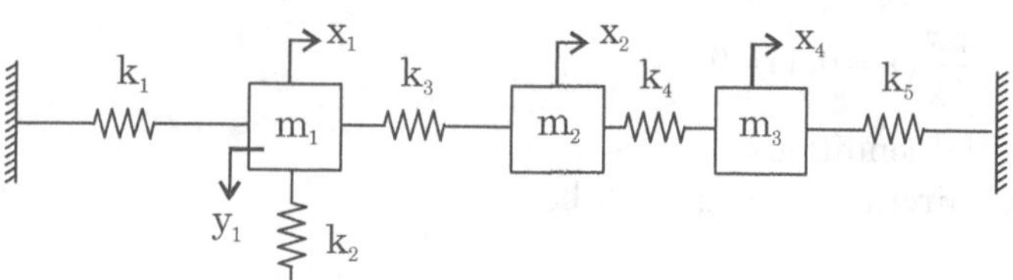

(a) 2 Natural Frequencies, 2 DOF

(b) 3DOF, 3 Natural frequencies

(c) 4 DOF, 4 Natural frequencies

(d) Equal Number of DOF and Natural frequencies.

30. Torsional Constant of a thin walled closed tube entirely depends on:

(a) Tube Material.

(b) Torsional load.

(c) Mean radius.

(d) Thickness of tube.

31. If shear center and centroid of a cross-section lies on same point then which of following is true for a beam under transverse load

(a) Cross-section is symmetric

(b) load at centroid produce twist

(c) Neutral Axis pass through shear center

(d) load at centroid produce No twist

32. Which of the following is true about the spring-mass system given below?

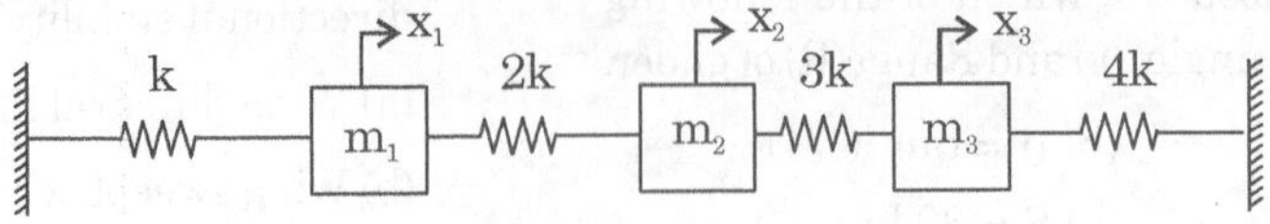

$m_1 = m$
$m_2 = 2m$
$m_3 = 3m$

(a) Stiffness Matrix $k = \begin{bmatrix} 5k & -2k & 0 \\ -2k & 5k & -3k \\ 0 & -3k & 7k \end{bmatrix}$

(b) Mass Matrix $M = \begin{bmatrix} m & 0 & 0 \\ 0 & 2m & 0 \\ 0 & 0 & 3m \end{bmatrix}$

(c) Damping Matrix $C = \begin{bmatrix} 3c & -2c & 0 \\ -2c & 5c & -3c \\ 0 & -3c & 7c \end{bmatrix}$

(d) Stiffness Matrix $k = \begin{bmatrix} 5k & 0 & -7k \\ 0 & 7k & -2k \\ -7k & -2k & 3k \end{bmatrix}$

33. To solve the partial differential equation of a uniform euler-Bernoulli beam which of following is true:

$$EI\frac{\delta^4 w}{\delta x^4} + m\frac{\delta^2 w}{\delta t^2} = 0$$

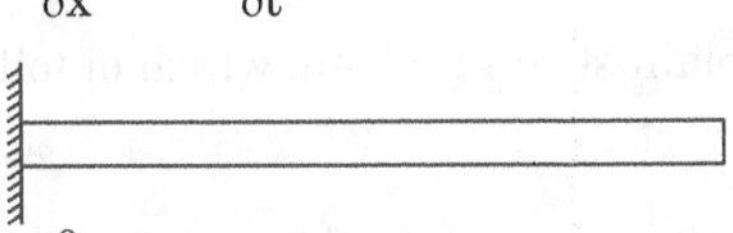

x = 0 x = ℓ

EI = Bending stiffness
m = mass per unit length
w = bending displacement
x = along length coordinate

(a) Required of 6 Boundary conditions
(b) Required of 4 Boundary conditions
(c) $w(x = o, t) = 0, \quad \frac{\delta w}{\delta x}(x = 0, t) = 0$
(d) Required of 2 initial conditions

34. Buckling of skin of aircraft fuselage can be increased by:
(a) Increasing skin thickness
(b) Increasing internal pressure
(c) Placing stiffness or Frame closer
(d) Using material with higher modulus

35. For a two dimensional stress state in an isotropic material which of following is true?

$$[\sigma] = \begin{bmatrix} -80 & 50 \\ 50 & 160 \end{bmatrix} \times b \quad \text{MPa}$$

b is proportional to applied loading
(a) $\sigma_1 = 170b$
(b) $\sigma_2 = -90b$
(c) $\sigma_{1,2} = [40 \pm 130]b$
(d) $\sigma_3 = 0$

36. For a 2 – dimensional stress state

$$\sigma = \begin{bmatrix} -80 & 50 \\ 50 & 160 \end{bmatrix} b \quad \text{MPa}$$

For a isotropic material having yield strength 350 MPa which of following is true:
(a) b = 1.53 (b) $T_{max} = 130c$
(c) b = 15.3 (d) $T_{max} = 13c$

37. A single degree of freedom system with Natural Frequency 15Hz is attached with a viscous clamper with damping ratio $\xi = 0.5$ in 1st case and mass is placed on a Friction-Surface with coefficient of Friction $\mu = 0.02$. Which of the following is correct about these system

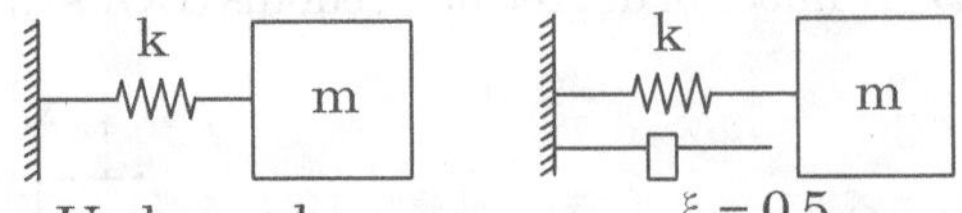

Undamped $\xi = 0.5$

Case - 1

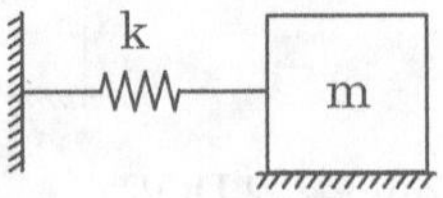

r = 0.02

Case - 2

(a) $(\xi_{eq})_{Case-2} = 1.0398 \times 10^{-3}$
(b) Damped Natural Frequency $(\omega_d)_{Case1} = 25.98u$ rad/sec
(c) Damped Natural Frequency $(\omega_d)_{Case2} = 29.99u$ rad/sec
(d) $(f_d)_{Case1} = 12.99$ Hz

38. In an ideal Ramjet Engine

(a) Inlet velocity is equal to outlet velocity

(b) Inlet mach number is equal to outlet mach number

(c) Inlet velocity is lower than that of outlet velocity

(d) Ratio of total pressure and static pressure at inlet and outlet of engine is equal.

39. When of the following statement is/are correct regarding thrust of air-breathing engine at low speed.

(a) Turbo prop engine has higher thrust than turbojet at take-off

(b) Ramjet has zero thrust at take-off

(c) Ramjet has higher thrust than turboprop at take-off

(d) Turbojet engine has higher thrust than turboprop at take off

40. Ideal Bryton cycle with reheating and intercooling is used to

(a) Reduce the heat supply

(b) Increase the work done

(c) Improve the performance of cycle

(d) Increase the efficiency of cycle

41. A turbofan engine has a by-pass ratio of '6' and total mass flow rate of gas is 20 kg/s. If the mass flow rate of fuel is 1 kg/s, then

(a) Mass flow rate of air through engine is 103 kg/s

(b) Mass flow rate of air through outer nacelle is 114 k/s

(c) Mass flow rate of air through care engine is 19 kg/s

(d) Mass flow rate of air through outer nacelle is 103 kg/s

42. A fighter aircraft equipped with turbojet engine with after burner. If after burner is turned on during operation, then

(a) Thrust of engine increases

(b) SFC decreases

(c) SFC increases

(d) Area of nozzle exhaust increases

43. Air with velocity V_1 enters a turbojet engine and exit the engine with speed v_2. The propulsive efficiency of engine is

(a) maximum when velocity ratio is 1

(b) minimum when $V_2 = 2v_1$

(c) maximum when $V_2 = v_1$

(d) option 'b' and 'c'.

44. The equivalence ratio (ϕ) of a mixture is

(a) Ratio of (o/f) to the (o/f) stoic

(b) Ratio of $(o/f)_{stoic}$ to the (o/f)

(c) Less than 1 for lean mixture

(d) Greater than 1 for rich mixture

45. Combustion between fuel (methane) and oxidizer (air) occurs inside a combustor is given by the following reaction

$$CH_4 + A\,(O_2 + 3.8\,N_2) \rightarrow PCO_2 + QH_2O + RN_2$$

for the stoichiometric reaction:

(a) A = 1, P = 2, Q = 2, R = 7.5

(b) A = 2, P = 1, Q = 2, R = 7.5

(c) $(A/f)_{stoic} = 17.2$

(d) $(A/f)_{stoic} = 8.65$

46. A cruise missile equipped with ideal Ramjet engine is cruising at mach 3.0 at an altitude where the ambient temperature is 200k. Stagnation temperature in the combustion chamber is equal to 1800k. If ratio of specific heat is 1.4 and specific heat at constant pressure is 1 kJ/kg-k, then

(a) Speed os the exhaust gas is 1521 m/s

(b) Speed of the exhaust gas is 1500 m/s

(c) Specific thrust produced by the engine is 651 $\left(\frac{V-S}{kg}\right)$

(d) Specific thrust produced by the engine is 672 $\left(\frac{N-S}{kg}\right)$

47. Air enters to diffuser of an engine with a speed of mach 0.85. The diffuser pressure recovery factor is 0.90. If the ratio of specific heat is 1.4, then

(a) Ram efficiency of diffuser is 73%

(b) Ram efficiency of diffuser is 76%

(c) Adiabatic efficiency of diffuser is 73%

(d) Adiabatic efficiency of diffuser is 76%

48. Diffusion of air in an ideal diffuser is achieved by

(a) Reversible process

(b) Adiabatic process

(c) Non-adiabatic process

(d) Isentropic process

49. Degree of Reaction (R) of an axial flow compressor stage is given by

(a) $R = \frac{1}{2} - \frac{1}{2}\left(\frac{Ca}{u}\right)(\tan\alpha_1 - \tan\beta_2)$

(b) $R = \frac{Ca}{2u}(\tan\beta_1 - \tan\beta_2)$

(c) $R = \frac{1}{2u}(\tan\beta_1 - \tan\beta_1)$

(d) $R = \frac{1}{2} - \frac{Ca}{2u}(\tan\alpha_2 - \tan\beta_1)$

50. Degree of reaction (R) of a turbine stage is given by

(a) $R = \frac{Ca}{2u}(\tan\beta_2 + \tan\beta_3)$

(b) $R = \frac{Ca}{2u}(\tan\beta_3 - \tan\beta_2)$

(c) $R = \frac{Ca}{2u}(\tan\beta_2 - \tan\beta_3)$

(d) $R = \frac{1}{2} + \frac{Ca}{2u}(\tan\beta_3 - \tan\alpha_2)$

51. An axial flow turbine of expansion ratio is '6' and polytropic efficiency of 85%. If the ratio of specific heat is taken as 1.33, then

(a) magnitude of polytorpic gas index is 1.3.

(b) magnitude of polytropic gas index is 3.1.

(c) Isentropic efficiency of turbine is 87%.

(d) Isentropic efficiency of trubine is 82%.

52. Mark the correct statement/statements regarding the pressure thrust produce by the nozzle of engine is

(a) pressure thrust is zero for optimal expansion

(b) pressure thrust is negative for under expansion

(c) pressure thrust is positive for under expansion

(d) pressure thrust is positive for over expansion

53. Velocity of Jet exhaust from the Jet engine is 2500 m/s. The forward flight speed is 1200 m/s and mass flow of air is 70 kg/s. If mass flow rate of fuel is negligible and ratio of specific heat is constant i.e. 1.3, then

(a) Thrust power of engine is 100 mJ/s.

(b) Thrust power of engine is 110 mJ/s.

(c) Propulsive efficiency of engine is 0.6.

(d) Propulsive efficiency of engine is 0.8

54. The aircraft is take-off from an airport, where ambient pressure and temperature is 1 bar and 288 k respectively. Mass flow rate of gas through the nozzle is 25kg/s and its total inlet pressure and temperature is 2.4 bar and 1000 k respectively. Flow through the nozzle is isentropic and Y = 1.3, R = 280 J/kg-k

(a) Exit area of nozzle is 1.1 m^2

(b) Pressure thrust produced by nozzle is 2500N.

(c) momentum thrust produced by nozzle is 1000 N

(d) Exit area of Nozzle is 0.08 m^2

55. A static test of centrifugal commpressor gives the following information

Impeller dia = 70 cm, speed = 11000 rpm

Outlet and inlect total head pressure = 4 bar, 1 bar.

Inlet total head temperature = 25°C

If slip factor of compressor is 0.92, then

(a) work done by compressor is 149.5 kJ/kg

(b) work done of compressor is 162.5 kJ/kg

(c) Isentropic efficiency of compressor is 0.97.

(d) Isentropic efficiency of compressor is 0.89.

56. Which statement/statements correctly match with the impusle machine?

(a) In impulse machine change in static pressure occurs in both rotor as well as stator.

(b) In impulse machine static pressure remains constant across the rotor.

(c) Degree of reaction of impulse machine is always zero.

(d) Degree of reaction of impulse machine is greater than zero.

57. If 'c' denotes absolute velocity and 'u' peripheral speed, then according to Euiler's energy equation of turbomachine.

(a) $\left(\frac{u_1^2 - u_2^2}{2}\right)$ represents impulse effect.

(b) $\left(\frac{u_1^2 - u_2^2}{2}\right)$ represents reaction effect.

(c) $\left(\frac{c_1^2 - c_2^2}{2}\right)$ represents impulse effect.

(d) $\left(\frac{u_1^2 - c_2^2}{2}\right)$ represents centrifugal effect.

58. Which of the following is correct for LEO communication Satellite?

(a) Orbit is elliptical

(b) Orbit altitudes upto 2000 km

(c) Orbit is Geocentric

(d) Orbit is circular

59. What is true for the second law of Kepler?

(a) Conservation of angular momentum

(b) Conservation of linear momentum

(c) Energy is constant

(d) Aerial velocity is constant.

60. A Geostationary Satellite follows

(a) A circular orbit with time period of 90 min.

(b) An altitude approximately 2000 km

(c) An altitude of approximately 35,800 km

(d) A circular orbit with a time period of 24 hours

61. Acceleration due to gravity g in terms of mean density of Earth d (where R is radius of earth and G – universal gravitational constant) is

(a) $g = \dfrac{4\pi G}{3R^2}$ (b) $g = \dfrac{GM}{R^2}$

(c) $g = \dfrac{4}{3}G\pi Rd$ (d) $g = 4\pi R^2 dG$

62. The time period of satellite revolving around Earth in circular orbit is

(a) Independent on the mass of the Earth

(b) Independent on distance from the planet

(c) Independent from the inclination angle

(d) All of the above

63. Which of the following is defines the modified four and five-digit series airfoils with a two-digit code preceded by a hyphen?

(a) First digit describes the roundness of the leading edge.

(b) First digit describes the maximum camber as percentage of chord.

(c) Second digit indicates the distance of maximum thickness from leading edge.

(d) Second digit indicates the maximum thickness in percentage of chord.

64. Airplanes consisting of round nose airfoils consist of: -

(a) Infinite Slope

(b) Zero Slope

(c) Zero 2nd order derivative at leading edge

(d) Infinite 2nd order derivative at leading edge

65. The drag divergence Mach number is usually

(a) Equal to critical Mach number

(b) Less than critical Mach number

(c) Close to critical Mach number

(d) Always greater than critical Mach number

ANSWERS

1. (a), (b)	**2.** (b), (d)	**3.** (a), (d)	**4.** (b), (c), (d)	**5.** (d)
6. (a), (b), (c)	**7.** (a), (b), (c)	**8.** (a), (b)	**9.** (c), (d)	**10.** (b), (d)
11. (a), (b), (c)	**12.** (a), (b), (d)	**13.** (a), (b)	**14.** (a), (b)	**15.** (a), (b)
16. (b), (c)	**17.** (a), (b)	**18.** (a), (b)	**19.** (b), (c), (d)	**20.** (a), (c)
21. (a) & (d)	**22.** (a) & (c)	**23.** (a), (b), (c)	**24.** (a), (b) & (c)	**25.** (a) (b) & (c)
26. (a), (b)	**27.** (a), (b), (c), (d)	**28.** (b), (c), (d)	**29.** (c), (d)	**30.** (c), (d)
31. (a), (c), (d)	**32.** (a), (b)	**33.** (b), (c), (d)	**34.** (a), (b), (c), (d)	**35.** (a), (b), (c), (d)
36. (a), (b)	**37.** (a), (b), (c), (d)	**38.** (b), (c), (d)	**39.** (a), (b)	**40.** (a), (b), (c), (d)
41. (b), (c)	**42.** (a), (c), (d)	**43.** (a), (c)	**44.** (b), (c), (d)	**45.** (b), (c)
46. (a), (d)	**47.** (a), (d)	**48.** (a), (b), (d)	**49.** (a), (d)	**50.** (b), (d)
51. (a), (c)	**52.** (a), (c)	**53.** (b), (c)	**54.** (b), (d)	**55.** (a), (c)
56. (b), (c)	**57.** (c), (d)	**58.** (b), (c), (d)	**59.** (a), (d)	**60.** (c), (d)
61. (b), (c)	**62.** (a), (c)	**63.** (a), (c)	**64.** (a), (d)	**65.** (c), (d)

EXPLANATIONS

1. As given that, $x_1 = (1, 3, 4, 2)$
$x_2 = (3, -5, 2, 2)$
$x_3 = (2, -1, 3, 2)$

The relation is

$$\lambda_1 x_1 + \lambda_2 x_2 + \lambda_3 x_3 = 0$$

$$\Rightarrow \lambda_1(1, 3, 4, 2) + \lambda_2(3, -5, 2, 2) + \lambda_3(2, -1, 3, 2) = 0$$

which is equivalent to

$$\lambda_1 + 3\lambda_2 + 2\lambda_3 = 0$$
$$3\lambda_1 - 5\lambda_2 - \lambda_3 = 0$$
$$4\lambda_1 + 2\lambda_2 + 3\lambda_3 = 0$$
$$2\lambda_1 + 2\lambda_2 + 3\lambda_3 = 0$$

As these are satisfied by the values

$$\lambda_1 = 1, \lambda_2 = 1, \lambda_3 = -2$$

$\therefore\ x_1 + x_2 - 2x_3 = 0$ is correct

Also,

Multiplying above equation with a constant 'k'

Here, $k = \dfrac{4}{\sqrt{3}}$

which gives,

$$\frac{4}{\sqrt{3}} \times (x_1 + x_2 - 2x_3) = \frac{4x_1 + 4x_2 - 8x_3}{\sqrt{3}} = 0$$

Thus option (*a*) and (*b*) are correct.

2. For the vectors to be perpendicular the dot product of two vectors should be zero

Here, we take

Option-(*a*):

$$(2\hat{i} + \hat{j} - 3\hat{k}) \cdot (2\hat{i} + 3\hat{j} - 4\hat{k}) = 4 + 3 + 12 = 19 \neq 0$$

Taking Option-(*b*):

$$(2\hat{i} + \hat{j} - 3\hat{k}) \cdot (5\hat{i} - \hat{j} + 3\hat{k}) = 10 - 1 - 9 = 0$$

Again,

$$(-\hat{i} + \hat{j} + 2\hat{k}) \cdot (5\hat{i} - \hat{j} + 3\hat{k}) = -5 - 1 + 6 = 0$$

Option (*b*) is correct

Taking Option (*c*)

$$(2\hat{i} + \hat{j} - 3\hat{k}) \cdot \left(\frac{3\hat{i} - 6\hat{j} - 4\hat{k}}{\sqrt{7}}\right) = \frac{1}{\sqrt{7}}(6 - 6 + 12) \neq 0$$

Taking Option (*d*)

$$(2\hat{i} + \hat{j} - 3\hat{k}) \cdot \left(\frac{20\hat{i} - 4\hat{j} + 12\hat{k}}{\sqrt{21}}\right) = \frac{1}{\sqrt{21}}(40 - 4 - 36) = 0$$

Again, $(-\hat{i} + \hat{j} - 2\hat{k}) \cdot \left(\dfrac{20\hat{i} - 4\hat{j} + 12\hat{k}}{\sqrt{21}}\right)$

$$= \frac{1}{\sqrt{21}}(-20 - 4 - 24) = 0$$

Hence, Option (*d*) is also correct.

3. As given that,

$$u = xy^2 - 2y - \frac{x^3}{3}$$

$$v = \frac{-y^3}{3} + 2x + x^2 y$$

A 2-D plane motion

$$\vec{F} = u\hat{i} + v\hat{j}$$

A vector to be solenoidal or incompressible the divergence should be zero

$$\therefore\quad \nabla\vec{F} = \frac{du}{dx} + \frac{dv}{dy} = 0$$

Here, $\dfrac{du}{dx} + \dfrac{dv}{dy} = y^2 - x^2 - y^2 + x^2 = 0$

Therefore, the vector is solenoidal or incompressible

Option (*a*) and (*d*) are correct.

Thus, option (*c*) is incorrect.

Now, A vector to be irrotational the curl should be zero

Here $\nabla \times \vec{F} = \dfrac{dv}{dy} - \dfrac{du}{dx} = -y^2 + x^2 - y^2 + x^2 \neq 0$

Therefore, the flow is rotational

Option (*b*) is incorrect.

4. Given,

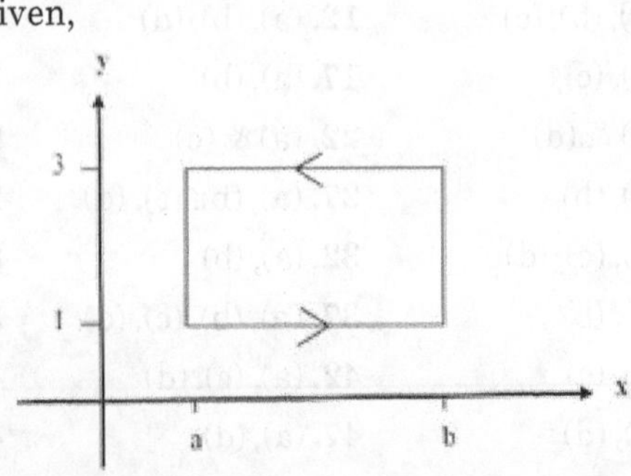

$$\vec{A} = xy\hat{i} + x^2\hat{j}$$

$$\oint_c \vec{A}\,dl = 1$$

$d\vec{l} = dx\hat{i} + dy\hat{j}$

$\therefore \int(xydx + x^2dy)$

As the surface is anticlockwise, closed, 2-D and open.

Now, as per Green's theorem

$$\oint_c (Adx + Bdy) = \iint_s \left(\frac{\partial B}{\partial x} - \frac{\partial A}{\partial y}\right) dxdy$$

Here, $A = xy, B = x^2$

$$\Rightarrow \int_{y=1}^{y=3}\int_{x=a}^{x=b} (2x - x)\, dxdy = \int_{y=1}^{y=3}\int_{x=a}^{x=b} (x)\, dxdy$$

$= 1$

$$\Rightarrow \int_1^3 dy \times \int_a^b xdx = 1$$

$$\Rightarrow (3-1)\times\frac{1}{2}(b^2 - a^2) = 1$$

$\therefore\ b^2 - a^2 = 1$

For, $a = 2, b = 4$

$\therefore\ 16 - 4 = 12 \neq 1$

For, $a = \frac{1}{\sqrt{3}}, b = \frac{2}{\sqrt{3}}$

$\therefore\ \frac{4}{3} - \frac{1}{3} = 1$

For, $a = 1, b = \sqrt{2}$

$\therefore\ 2 - 1 = 1$

For $a = \sqrt{2}, b = \sqrt{3}$

$\therefore\ 3 - 2 = 1$

Therefore, Option (*b*), (*c*) and (*d*) are correct.

5. Given,

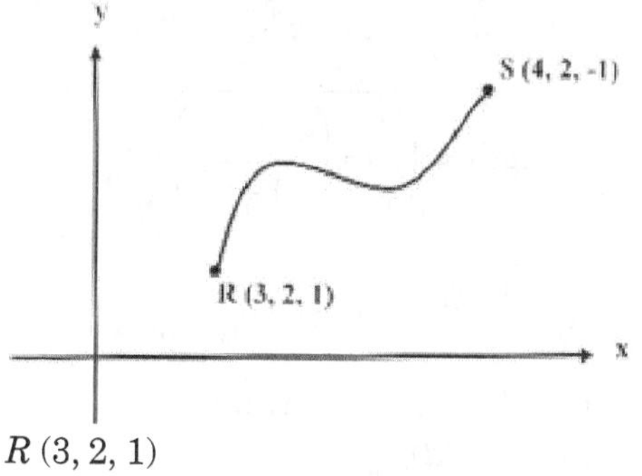

$R\,(3, 2, 1)$

$S\,(4, 3, -1)$

$\vec{V}(r) = yz^2\hat{i} + xz^2\hat{j} + 2xyz\hat{k}$

$\int \vec{V}dr = \int yz^2dx + xz^2dy + 2xyzdz$

$\int \vec{V}dr = \int_R^S \nabla(xyz^2)$

$= (xyz^2)_{(4,3,-1)} - (xyz^2)_{(3,2,1)}$

$= 12 - 6 = 6$

6. For the vectors to be linearly dependent its determinent should be zero

$$\therefore \quad A = \begin{vmatrix} a & 1 & 2 \\ 1 & a & 2 \\ 1 & a & 5a \end{vmatrix}$$

$= a(5a^2 - 2a) - (5a - 2) + 2(a - a)$

$= 5a^3 - 2a^2 - 5a + 2$

For, $a = 1$

$5 - 2 - 5 + 2 = 0$

For, $a = -1$

$-5 - 2 + 5 + 2 = 0$

For, $a = 0.4$

$0.32 - 0.32 - 2 + 2 = 0$

For, $a = 0$

$0 + 2 = 2 \neq 0$

Therefore, Option (*a*), (*b*) and (*c*) are correct.

7. As we know,

Taylor's series is given as

$$f(x) = f(a) + \frac{(x-a)f'(a)}{1!} + \frac{(x-a)^2 f''(a)}{2!}$$

Now,

The taylor series for cos *x* is

$$\cos x = 1 + \frac{x^2}{2!} + \frac{x^4}{4!} - \frac{x^6}{6!} + \ldots$$

$$\cos x = \sum_{n=0}^{\infty}(-1)^n \frac{x^{2n}}{(2n)!};\ \text{about } x = 0$$

Putting $n = 0$, $\sum_{n=0}^{\infty}(-1)^0 \frac{x^0}{(0)!} = 1$

For $n = 1$, $\sum_{n=0}^{\infty}(-1)^2 \frac{x^2}{(2)!} = \frac{x^2}{2}$ and so on

Option (*a*) is valid

The Taylor series for $\tan^{-1} x$ is

$$\tan^{-1} x = x - \frac{x^3}{3} + \frac{x^5}{5}$$

$$\tan^{-1} x = \sum_{n=0}^{\infty}(-1)^n \frac{x^{2n+1}}{2n+1};\ \text{about x} = 0$$

Putting $n = 0$, $\sum_{n=0}^{\infty}(-1)^0 \frac{x^{0+1}}{0+1} = x$

For $n = 1$, $\sum_{n=0}^{\infty}(-1)^1 \frac{x^{2+1}}{2+1} = \frac{-x^3}{3}$ and so on

Option (b) is also valid

Taylor series for $\ln x$ for $x = 1$ is

$$\ln x = (x-1) - \frac{(x-1)^2}{2} + \frac{(x-1)^3}{3} + \ldots +$$

Now,

$$\ln x = \sum_{n=1}^{\infty}(-1)^{n+1}\frac{(x-1)^n}{n}; \text{ about } x = 1$$

For $n = 1$, $\sum_{n=1}^{\infty}(-1)^{1+1}\frac{(x-1)^1}{1}$

For n = 2,

$$\sum_{n=1}^{\infty}(-1)^{2+1}\frac{(x-1)^2}{1} = \frac{-(x-1)^2}{2!} \text{ and so on}$$

$\therefore$ Option (c) is valid

Taylor series for $e^{1/x}$ at $x = 0$ is

$\therefore$ $f(0) = \infty$

Thus, series does not exist.

8. As we know,

Taylor's series is given as

$$f(x) = f(a) + \frac{(x-a)f'(a)}{1!} + \frac{(x-a)^2 f''(a)}{2!}$$

Here, taylor series is used to break a component into x series which already in its expanded form as:

$f(x) = 3x^2 + 5x - 2$

Now, expand option (b) we get,

$3x^2 + 5x - 2$

$\therefore$ It is also true.

9. 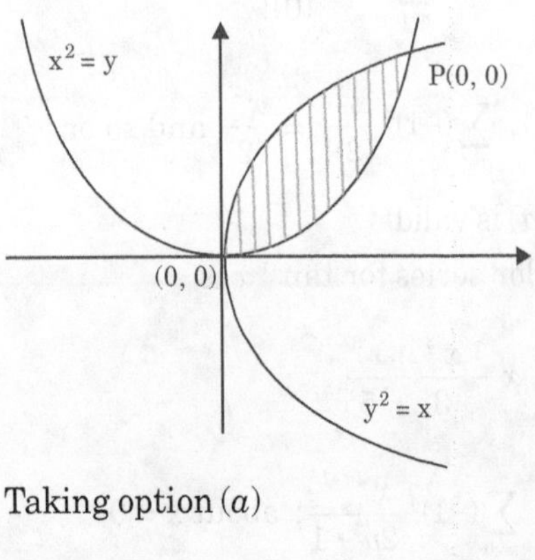

Taking option (a)

$$\text{Area} = \int_{y=0}^{y=1}\int_{x=y^2}^{x=\sqrt{y}} (1)dxdy = \frac{1}{3}, \text{ option } (a) \text{ is valid.}$$

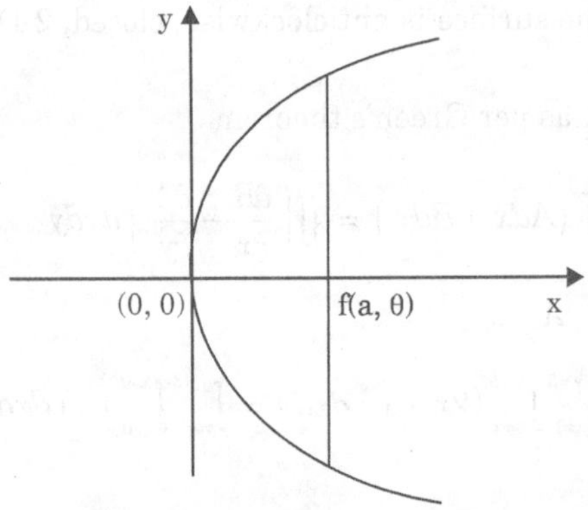

Taking option (b)

Parabola is given as

$y^2 = 4ax$

Putting $x = a$

$y^2 = 4a^2$

$y = 2a$

Given, $y^2 = x$

$\therefore$ $a = \frac{1}{4}x$ and $y = \frac{1}{2}$

$$\text{Area} = \int_{y=0}^{y=1/2}\int_{x=y^2}^{x=1/4} (1)dxdy = \frac{1}{6}$$

$\therefore$ Option (b) is valid.

Taking option (c)

Arc length is given as

$$\int_{x=0}^{x=1}\sqrt{1+\left(\frac{dy}{dx}\right)^2}\,dx$$

$\because$ $f(x) = \sqrt{1-x^2}$

$$\frac{dy}{dx} = \frac{-x}{\sqrt{1-x^2}}$$

$$\therefore \text{ Arc length} = \int_0^1\sqrt{1+\left(\frac{-x}{\sqrt{1-x^2}}\right)^2}\,dx$$

$$= \int_0^1\sqrt{\frac{1}{1-x^2}}dx = \left(\sin^{-1}x\right)_0^1 = \frac{\pi}{2}$$

Therefore, Option (c) is invalid.

Taking Option (d)

Volume of hemisphere

$$= \frac{1}{2}\times\frac{4}{3}\pi r^3 = \frac{2}{3}\pi(3)^3 = 18\pi$$

Therefore, Option (d) is invalid.

10. Given,

$$I_P = \iint_R \frac{1}{\left(\sqrt{1-x^2-y^2}\right)^{P/2}} dxdy$$

In polar coordinates

$$r^2 = x^2 + y^2$$

$$dxdy = rdrd\theta$$

Now,

$$I_P = \int_{\theta=0}^{\theta=2\pi} \int_{r=0}^{r=\infty} \frac{1}{\left(1-r^2\right)^{P/2}} rdrd\theta$$

Let, $1 - r^2 = t$

$\therefore$ $-2rdr = dt$

$$rdr = \frac{-dt}{2}$$

Now,
$$I_P = \int_0^{2\pi} d\theta \times \int_0^{\infty} \frac{r}{\left(1-r^2\right)^{P/2}} dr$$

$$I_P = 2\pi \times \frac{-1}{2} \int_1^{-\infty} \frac{dt}{t^{P/2}}$$

Let, $k = \dfrac{P}{2}$

$\therefore$
$$I_P = -\pi \int_1^{-\infty} \frac{dt}{t^k} = -\pi \left(\frac{t^{-k+1}}{-k+1} \right)_1^{-\infty}$$

For, $k = 1$

Denominator is zero.

Therefore, it diverges

To converge, $k > 1$

Here, $\dfrac{P}{2} > 1$

$\because$ $k = \dfrac{P}{2}$

$P < 2$

Thus, option (*b*) and (*d*) are correct.

11. Navier – stokes momentum equation

$$\underset{\substack{\uparrow \\ \text{Total} \\ \text{derivative}}}{\frac{eD\vec{V}}{Dt}} = \underset{\substack{\uparrow \\ \text{Pressure} \\ \text{gradient}}}{-\bar{\nabla}p} + \underset{\substack{\uparrow \\ \text{Body} \\ \text{force} \\ \text{term}}}{e\bar{g}} + \underset{\substack{\uparrow \\ \text{Diffusion} \\ \text{term} \\ \text{(shere stress} \\ \text{due to} \\ \text{viscosity)}}}{\mu\bar{\nabla}^2\vec{V}}$$

12. Bernoulli's equation is valid for steady inviscid incompressible flow.

14.

M_1	M_2	$M_1 > M_2$
P_1	P_2	$P_2 > P_1$
P_{01}	P_{02}	$P_{01} > P_{02}$
S_1	S_2	$T_{01} = T_{02}$
		$S_2 > S_1$

16. For supersonic flows, over a gradual compression entropy remains constant as no shock wave is formed. Over a sharp compression corner oblique shock wave is formed and entropy increases for expansion, sharp or gradual entropy remains constant.

18.
$$E = \frac{\eta}{C} \frac{C_L^{3/2}}{C_D} (2e_\infty s)^{\frac{1}{2}} \left(W_1^{-\frac{1}{2}} - W_0^{-\frac{1}{2}} \right)$$

$$E \,\alpha\, \frac{C_L^{\frac{3}{2}}}{C_D}$$

$$\alpha\; e_\infty^{\frac{1}{2}}$$

Where e_∞ = density of air at the altitude where aircraft is flying.

19. Take off distance is given by:

$$S_{LO} = \frac{1.44W^2}{ge_\infty SC_{L,max} T}$$

Where, $\dfrac{W}{S}$ = wing loading

$\dfrac{T}{W}$ = Thrust to weight ratio

20.

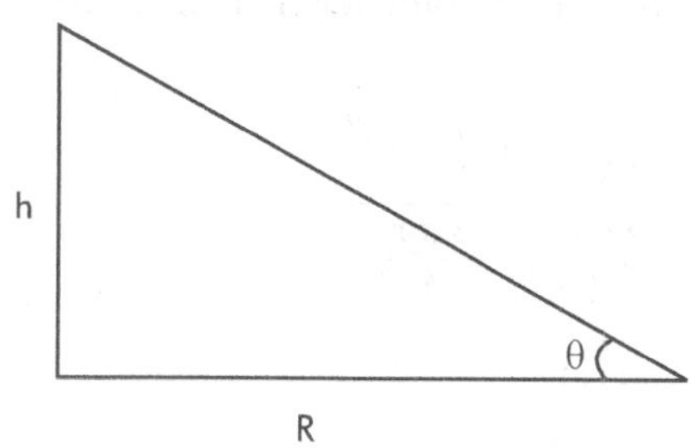

Decent angle: $\tan\theta = \dfrac{1}{\left(\dfrac{L}{D}\right)}$

$$\theta = \tan^{-1}\left(\frac{1}{15}\right)$$

$$\text{Range} = h \times \left(\frac{L}{D}\right)$$

$= 4 \times 15$

$= 60$ km

26. An elastic beam is a continuous system and it has infinite degree of freedom. Therefore it can vibrate with infinite Natural frequencies and mode shapes, regardless of Boundary conditions

27. Bending Stress $\sigma_{max} = \frac{M_y}{I}$

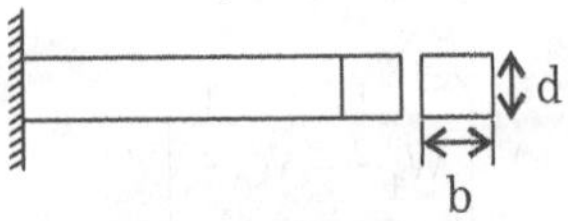

M = Bending Moment

y = distance from centroidal axis to farthest point of cross section.

I = 2nd Moment of Area and it depends on cross-section height and width.

$$= \frac{bd^3}{12}$$

28. For a plain strain problem

$\varepsilon_{xx}, \varepsilon_{yy}, \varepsilon_{xy} \neq 0$

$\varepsilon_{zz}, \varepsilon_{xz}, \varepsilon_{yz} = 0$

29. For a discrete ϕ or a continuous system, number of natural frequencies are always equal to number of degree of freedoms. In the figure, there are 3 masses which on independently move in x_1, x_2 and x_3 coordinates giving 3 DOFs But mass m_1 can also move in y_1 coordinate independently. thus total of 4 DOFs and 4 Natural frequencies.

30. Torsional constant for a then walled closed-section

$$J = \frac{4\bar{A}^2}{\oint \frac{ds}{t}} = \frac{4 \times (\pi r^2)^2}{\frac{2\pi r}{t}}$$

$= 2\pi r^3 . t$

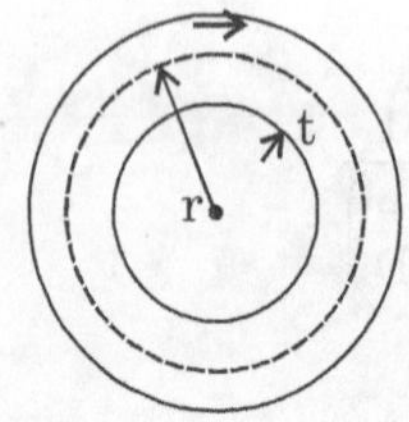

t = thickness

r= mean radius

$\bar{A}$ = mean Area of cross-section

S = Perimeter mean

Thus J depends on mean radius r and thickness t of tube.

31. For symmetric cross-section shear center and centroid lies on same point. Shear center is a point of cross-section about which twist happens. If there is a ecentricity between shear center and centroid then load at centroid produces a twisting moment. If both are at same point A/O twist is generated. We know neutral axis for symmetric cross-section passes through centroid, hence in this case it also passes through shear center.

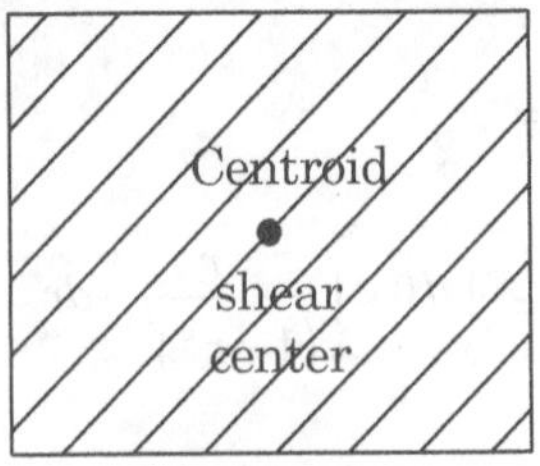

Symmetric Cross-section

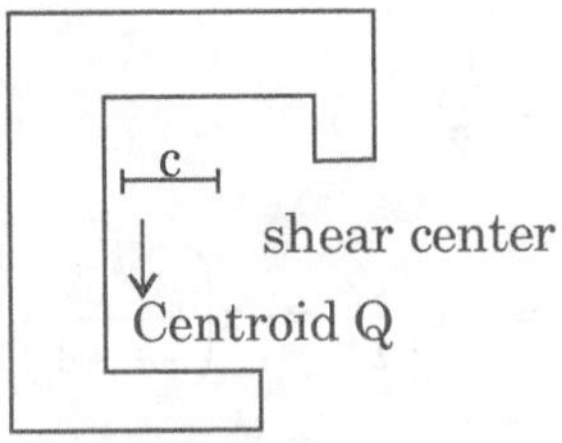

A symmetric Cross-section

32. To get stiffness and Mass Matrices, we need to write governing differential equation of above system:

$$\text{Total K.E., } T = \frac{1}{2} m \dot{x}_1^2 + \frac{1}{2}(2m)\dot{x}_2^2 + \frac{1}{2}(3m)\dot{x}_3^2$$

$$\text{P.E. } U = \frac{1}{2}kx_1^2 + \frac{1}{2}(2k)(x_2 - x_1)^2$$

$$+\frac{1}{2}(3k)(x_3 - x_2)^2 + \frac{1}{2}(4k)x_3^2$$

Using Langrun Equation:

$$\frac{d}{dl}\left(\frac{dT}{d\dot{x}}\right) + \frac{dv}{dx} = f(t).$$

For x_1 : $m\ddot{x}_1 + kx_1 + (2k)(x_2 - x_1)(-1) = 0$

$m\ddot{x}_1 + 3kx_1 - 2kx_2 = 0$ ......(i)

For x_2 :

$$(2m)\ddot{x}_2 + (2k)(x_2 - x_1) + (3k)(x_3 - x_2)(-1) = 0$$

$$\Rightarrow 2m\,\ddot{x}_2 + 5kx_2 - 2kx_1 - 3kx_3 = 0 \quad ......(ii)$$

For x_3 : $3m\,\ddot{x}_3 + 3k(x_3 - x_2) + 4k\,x_3 = 0$

$$3m\,\ddot{x}_3 + 7k\,x_3 - 3k\,x_2 = 0 \quad .....(iii)$$

Using (i), (ii), and (iii), we can get equation in matrix form

$$\underset{\substack{\uparrow \\ \text{Mass Matrix} \\ M}}{\begin{bmatrix} m & 0 & 0 \\ 0 & 2m & 0 \\ 0 & 0 & 3m \end{bmatrix}} \begin{bmatrix} \ddot{x}_1 \\ \ddot{x}_2 \\ \ddot{x}_3 \end{bmatrix} + \underset{\substack{\uparrow \\ \text{Stiffness k Matrix}}}{\begin{bmatrix} 3k & -2k & 0 \\ -2k & 5k & -3k \\ 0 & -3k & 7k \end{bmatrix}} \begin{bmatrix} x_1 \\ x_2 \\ x_3 \end{bmatrix} = \begin{bmatrix} 0 \\ 0 \\ 0 \end{bmatrix}$$

We don't have any damper attached in system

33. As partial differential equation is 4 order in space and 2nd order in time we will get 4 constant in space and 2 constant in time. To calculate constant 4 Buondary conditions and 2 initial condition are required.

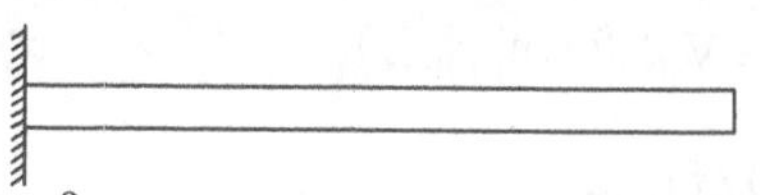

$x = 0$ $\qquad$ $x = \ell$

This is a cantilever beam fixed at x = 0 and free at $x = l$. For a fixed boundary condition deflection and slope in always zero which means that w (x = 0) = 0 and $\frac{\delta w}{\delta x}(x = 0) = 0$. For all t at free end we can get two more boundary conditions as moment and shear force is zero at $x = l$.

Initial condition will be given in terms of time dependent displacement and velocity values at t = 0

i.e. $x\,(t = 0)$

and $\dot{x}\,(t = 0)$

34. As we know that buckling strength depends on the young's modulus, 2nd moment of Area and Buckling length.

$$\Rightarrow P_{cr} \propto E, I \text{ and } P_{cr} \propto \frac{1}{\ell}$$

– Buckling strength increases as young's modulus E and 2nd moment of Area IV increases. Further I depends upon thickness of skin t . $I \propto t$

– Buckling strength increases if we reduce the Buckling length. In fuselage, if we place stiffness or framer closer it reduces the buckling length of stain.

– Internal pressure on fuselage cause it to expand which will resist buckling load which is collapsing. Thus buckling strength increases with internal pressure.

35. From above matrix

$\sigma_{xx} = -80b$ MPa

$\sigma_{yy} = 160b$ MPa

$T_{xy} = 50b$ MPa

Principle stresses are given by:

$$\sigma_{1,2} = \left[\frac{\sigma_{xx} - \sigma_{yy}}{2}\right] \pm \sqrt{\frac{(\sigma_{xx} - \sigma_{yy})^2}{(2)^2} + T_{xy}^2} \text{ MPa}$$

$$\Rightarrow \sigma_{1,2} = \left[\frac{-80 + 160}{2}\right] b$$

$$\pm \sqrt{\frac{[-80 - 160]^2 b^2}{4} + (50)^2 b^2} \text{ MPa}$$

$$\Rightarrow \sigma_{1,2} = [40 \pm 130]\, b \text{ MPa}$$

$$\Rightarrow \sigma_1 = 170b \text{ MPa}, \quad \sigma_2 = -90b \text{ MPa}$$

$\sigma_3 = 0$, as it is a 2–D state of stress.

36. $\sigma_{xx} = -80b$ MPa $\qquad \sigma_{yy} = 160b$ MPa

$T_{xy} = 50b$ MPa

We know that Von-miser failure theory for applied stresses,

$$\sigma_{xx}^2 + \sigma_{yy}^2 - \sigma_{xx}\,\sigma_{yy} + 3T_{xy}^2 \le \sigma_{yield}^2$$

$$(-80b) + (160\,b)^2 - (-80b)\,(160b) + 3\,(50b)^2$$

$$= (350)^2$$

$$52300\,b^2 = 350^2$$

$$\Rightarrow \boxed{b = 1.53}$$

We know that,

$$\text{Maximum shear stress } T_{max} = \frac{\sigma_1 - \sigma_2}{2}$$

$$= \frac{\sqrt{(\sigma_{xx} - \sigma_{yy})^2 + 4T_{xy}^2}}{2}$$

$$= \frac{\sqrt{(-80 - 160)^2 b^2 + 4(50)^2 b^2}}{2}$$

$= 130$ C

37. Natural Frequency of original system

$f_n = 15Hz$

$\omega_n = 2\pi f_n = 30\pi$ rod/sec

Case–1: Viscous damping :

damped Natural Frequency $\omega_{d_1} = \omega_n \sqrt{1-\xi^2}$

$= (30 \times \pi)\sqrt{1-0.5^2}$

$= 25.98\ \pi$ rod/sec

$f_{d_1} = \frac{\omega_d}{2\pi} = 12.99$ Hz

Case 2: Frictional damping :

$\xi_{eq.} = \frac{\mu g}{2\omega_n} = \frac{0.02 \times 9.8}{2 \times 30\pi}$

$= 1.0398 \times 10^{-3}$

damped Natural frequency $\omega_{d_2} = \omega_n \sqrt{1-\xi_{eq}^{\ 2}}$

$= (30\pi)\sqrt{1-\left[1.0398 \times 10^{-3}\right]^2}$

$= 29.99\pi$ rad/sec

$f_{d_z} = 14.99$ Hz

38. In an ideal Ramjet engine

$\frac{P_{oa}}{P_a} = \frac{P_{oe}}{P_e}$ (option(d)

$\left[1+\frac{r-1}{2}m_a^2\right]^{\frac{r}{r-1}} = \left[1+\frac{r-1}{2}m_e^2\right]^{\frac{r}{r-1}}$

For ideal engine, r = constant

So, $m_a = m_e$ (option b)

$\frac{Ci}{\sqrt{rRT_i}} = \frac{C_j}{\sqrt{\pi RT_j}}$

$C_i = \sqrt{\frac{T_i}{T_j}}C_j$

$T_j > T_i$ So, $C_i < C_j$ (option c)

39. → At low speed turbo-prop engine has more thrust during take-off

→ Ramjet engine can not start at static condition, so net thrust is zero during take-off

41. $\dot{m}_g = 20kg/s$, $\dot{m}_f = 1kg/s$, $B = 6$

$\dot{m}_n$ = mass flow rate of air through core engine

$= \dot{m}_g - \dot{m}_a = 19kg/s$

$\dot{m}_c$ = mass flow rate of air through outer nacelle

$= B = \frac{\dot{m}_c}{\dot{m}_n} \Rightarrow \dot{m}_c = 6 \times 19 = 114kg/s$

42. → When afterburner is operated then extra fuel is burn which increases the thrust and Sfc also

→ To pass the more mass of gas nozzle area need to increase, to expand the flow correctly to ambient pressure.

43. Equation of propulsive efficiency for negligible mass of fuel is given by

$\eta_p = \frac{2\alpha}{1+\alpha}$, where $\alpha = \frac{V_1}{V_2}$

If $\alpha = 1$, then $\eta_p = 100\%$ (maximum)

So $\frac{V_1}{V_2} = 1$

$\Rightarrow V_1 = V_2$ (for, $\eta_{p_1 max}$)

44. $\phi = \frac{(O/f)_{stoic}}{(O/f}$

→ $\phi < 1$, represents lean mixture

→ $\phi > 1$, represents rich mixture

→ $\phi = 1$, Stoichimetric

45. $CH_4 + A(O_2 + 3.8\ N_2) \rightarrow PCO_2 + QH_2O + RN_2$

Balancing 'C', we get P = 1

Balancing 'H': 4 = 2Q ⇒ Q = 2

Balancing 'O': 2A = 2P + Q

⇒ A = 2

Balancing 'N': 2 × 3.8A = 2R

R = 7.6

Molecular wt. of oxidizer = ϕ

2 × (32 + 3.8 × 2 × 14) = 276.8

Molecular wt. of fuel = 12 + 4 × 1 = 16

$(A/F)_{stoic} = \frac{\text{molecular wt. of oxidizer}}{\text{molecular wt. of fuel}}$

$= \frac{276.8}{160} = 17.3$

46. $m_i = 3$, $T_i = 200k$, $T_{oc} = T_{oe} = 1800$ k

for ideal Ramjet engine, $m_i = m_j$

$$m_i = \frac{C_i}{\sqrt{\gamma RT_i}}$$

$$C_p = \frac{\gamma R}{\gamma - 1} \Rightarrow R = 285.714 \text{ J/kg-k}$$

$$C_i = M_i \times \sqrt{\gamma RT_i} = 848.528 \text{m/s}$$

$$C_j = C_i \times \sqrt{\frac{T_j}{T_i}}$$

$$\frac{T_{oj}}{T_j} = \left[1 + \frac{r-1}{2} m_j^{\ 2}\right]$$

$T_j = 642.857$ K

$C_j = 1521.277$ m/s

$f = \dot{m}(C_j - C_i)$

$$\frac{f}{\dot{m}} = C_j - C_i = 672.75 \frac{\text{N} - \text{S}}{\text{kg}}$$

47. Ram efficiency $(\eta_{ram}) = \dfrac{P_{o1} - P_a}{P_{oa} - P_a} = \left[\dfrac{\frac{P_{o1}}{P_a} - 1}{\frac{P_{oa}}{P_a} - 1}\right]$

$$= \left[\frac{\frac{P_{o1}}{P_{oa}} \times \frac{P_{oa}}{P_a} - 1}{\frac{P_{oa}}{P_a} - 1}\right]$$

$$\frac{P_{o1}}{P_{oa}} = 0.9, \ \frac{P_{oa}}{Pa} = \left[1 + \frac{r-1}{2} m_a^2\right]^{\frac{r}{r-1}} = 1.604$$

$$\eta_{ram} = \frac{0.9 \times 1.604 - 1}{1.604 - 1} \times 100\% = 73.4\%$$

Adiabatic efficiency $(\eta_{ad}) = \dfrac{T_{o1}' - Ta}{T_{oa} - T_a}$

$$= \frac{\frac{T_{o1}'}{T_a} - 1}{\frac{T_{oa}}{T_a} - 1} = \left[\frac{\frac{T_{o1}'}{T_{oa}} \times \frac{T_{oa}}{T_a} - 1}{\frac{Toa}{T_a} - 1}\right]$$

$$\frac{T_{o1}'}{T_{oa}} = \left(\frac{P_{o1}}{P_{oa}}\right)^{\frac{\gamma-1}{\gamma}} = 0.97$$

$$\frac{T_{oa}}{T_a} = 1 + \frac{\gamma - 1}{2} m_a^2 = 1.1445$$

$$\eta_{ad} = \frac{0.97 \times 1.1445 - 1}{1.1445 - 1} \times 100\% = 76.2\%$$

48. → Air is diffused iscentropically is an ideal diffusser.

→ Isentropic process is adiabatic and reversible.

49. $R = \dfrac{Ca}{2u}(\tan\beta_1 + \tan\beta_2)$

$$= \frac{Ca}{2u}(\tan\alpha_1 + \tan\beta_1 + \tan\beta_2 - \tan\alpha_1)$$

$$= \frac{1}{2u}[C_a(\tan\alpha_1 + \tan\beta_1) - C_a(\tan\alpha_1 - \tan\beta_1)$$

$$= \frac{1}{2u}[u - C_a(\tan\alpha_1 - \tan\beta_2)]$$

Also,

$$= \frac{1}{2} - \frac{Ca}{2u}[\tan\alpha, -\tan\beta_2] \text{ (option (a))}$$

$$R = \frac{C_a}{2u}(\tan\beta_1 + \tan\alpha_2 + \tan\beta_2 - \tan\alpha_2)$$

$$= \frac{1}{2u}[C_a(\tan\alpha_2 + \tan\beta_2) - C_a(\tan\alpha_2 - \tan\beta_2)]$$

$$= \frac{1}{24}[u - C_a(\tan\alpha_2 - \tan\beta_1)]$$

$$= \frac{1}{2} - \frac{C_a}{2u}(\tan\alpha_2 - \tan\beta_1) \text{ Option (d)}$$

50. $R = \dfrac{C_a}{2u}(\tan\beta_3 - \tan\beta_2)$

$C_{t2} = w_{t2} = u$

$C_a \tan\alpha_2 = C_a \tan\beta_2 + u$

$$\tan\beta_2 = \tan\alpha_2 - \frac{u}{C_a}$$

$$R = \frac{1}{2} + \frac{Ca}{2u}(\tan\beta_3 - \tan\alpha_2)$$

51. For expansion process:

Polytropic efficiency $(\eta_{poly}) = \dfrac{\gamma}{\gamma - 1} \times \dfrac{n-1}{n}$

$$0.85 = \frac{1.33}{0.3} \times \frac{n-1}{n}$$

$$\frac{n-1}{n} = 0.211$$

$\text{I}_{\text{sentropic efficiency}}(\eta_{ise}) = \dfrac{\text{work actual}}{\text{work ideal}}$

$$\eta_{ise} = \frac{T_{o1} - T_{o2}}{T_{o1} - T_{o2}} = \left[\frac{1 - \frac{T_{o2}}{T_{o1}}}{1 - \frac{T_{o2}'}{T_{o1}}}\right]$$

$$\frac{T_{o2}'}{T_{o1}} = \left(\frac{P_{o2}}{P_{o1}}\right)^{\frac{n-1}{n}} = \left(\frac{1}{6}\right)^{\frac{1.267-1}{1.267}} = 0.685$$

$$\frac{T_{o2}'}{T_{o1}} = \left(\frac{P_{o2}}{P_{o1}}\right)^{\frac{\gamma-1}{\gamma}} = \left(\frac{1}{6}\right)^{\frac{1.33-1}{1.33}} = 0.641$$

$$\eta_{isen} = \frac{1-0.685}{1-0.641} \times 100\% = 87.7\%$$

52. Pressure thrust = $A_e (P_e - P_a)$

→ for optimal expansion, $P_e = P_a$, f = 0

→ for under expansion, $P_e > P_a$, f = +ve

→ for over expansion, $P_e < P_a$, f = –ve

53. V_e = 2500 m/s, V_o = 1200 m/s, $\dot{m}_a = 70 \text{kg/s}$

$\gamma = 1.3$, $\dot{m}_f << \dot{m}_a$

Thrust power = Thrust × Inlet velocity

$= \dot{m}_a (V_e - V_a) \times V_a$

= 70 × (2500 – 1200) × 1200

= 109.2 mJ/s

$$\text{Propulsive efficiency} = \frac{2\alpha}{1+\alpha}$$

$$\alpha = \frac{V_a}{V_e} = \frac{1200}{2500} = 0.48$$

$$\eta_P = \frac{2 \times 0.48}{1+0.48} = 0.65$$

54. Nozzle choked condition, If $P_c > P_a$

$$\gamma \frac{P_{ou}}{P_c} = \left[1 + \frac{\gamma-1}{2} \times 1\right]^{\frac{\gamma}{\gamma-1}} = \left(\frac{\gamma+1}{2}\right)^{\frac{\gamma}{\gamma-1}}$$

$$\frac{2.4}{P_c} = \left(\frac{1.3+1}{2}\right)^{\frac{1.3}{1.3-1}}$$

P_c = 1.3 bar and P_a = 1 bar

So $P_c < P_a$ → Nozzle is choked

$$\frac{T_{ou}}{T_c} = \frac{\gamma+1}{2}$$

$\Rightarrow T_c$ = 869.56k

$$\rho_c = \frac{P_c}{RT_c} = \frac{1.3 \times 10^5}{280 \times 869.56} = 0.533 \text{ kg/m}^3$$

$$V_c = \sqrt{\gamma R T_c} = 562.6 \text{ m/s}$$

$$\dot{m} = \rho_c A_c V_c$$

$$A_c = \frac{\dot{m}}{\rho_c V_c}$$

$$A_c = \frac{25}{0.533 \times 562.6} = 0.0833 \text{ m}^2$$

Pressure thrust = $(P_c - P_a) A_c$

= (1.3 – 1) × 10^5 × 0.0833

= 2501.12 N

55. $u = \frac{\pi DN}{60} = \frac{\pi \times 0.70 \times 11000}{60} = 403.17 \text{m/s}$

w = μu^2, μ = slip factor

= 0.92 × 403.17^2 = 149.542 kJ/kg

$$\eta_{isen} = \frac{T_{o2}' - T_{o1}}{T_{o2} - T_{o1}}$$

$w = C_p(T_{o2} - T_{o1}) = 149542$

$$T_{o2} = \left[\frac{149542}{1005} + 298\right] = 446.8\text{k}$$

$$\eta_{isen} = \left[\frac{\frac{T_{o2}'}{T_{o1}-1}}{\frac{T_{o2}}{T_{o1}-1}}\right] = \frac{\left(\frac{P_{o2}}{P_{o1}}\right) - 1}{\frac{T_{o2}}{T_{o1}} - 1}$$

= 0.973

56. → In impulse machine there is no pressure change in rotor so degree of reaction is zero.

57. Euler's Energy equation:

$$E = \frac{1}{2}(C_1^2 - C_2^2) + \frac{1}{2}(u_1^2 - u_2^2) + \frac{1}{2}(\omega_2^2 - \omega_1^2)$$

$\frac{C_1^2 - C_2^2}{2}$ represents impurse effect.

$\frac{u_1^2 - u_2^2}{2}$ represents centrifugal effect.

58. A low Earth orbit (LEO) typically is a geocentric circular orbit about 160 to 2,000 kilometers (99 to 1,243 mi) above the earth's surface and, correspondingly, a period (time to revolve around the earth) of about 90 minutes.

59. According to Kepler's second law, Area swept out by radial vector from the primary planet to satellite in equal interval of time is constant,

i.e. $\frac{dA}{dt}$ = Constant

$$h = \vec{v} \times \vec{r}$$

$$h = vr\sin\theta$$

$$\therefore \frac{dA}{dt} = \frac{1}{2} \times v \times r\sin\theta$$

$$= \frac{1}{2}h = \text{Constant}$$

where, h is the angular momentum which is conserved.

60. A geostationary orbit, also referred to as a geosynchronous equatorial orbit (GEO), is a circular geosynchronous orbit 35,786 kilometers (22,236 miles) above Earth's equator and following the direction of Earth's rotation. It has a time period of 24 hours to revolve around the earth.

61. As per given formula,

$$g = \frac{GM}{R^2}$$

and $g = \frac{4G\pi R^3 d}{3 \times R^2}$, where d is the mean density

$$\therefore g = \frac{4}{3}G\pi Rd$$

62. The time period of satellite revolving around Earth in circular orbit is independent of mass and inclination angle.

63. The Four and five-digit series airfoils can be modified with two-digit code preceded by hyphen as first digit indicating the roundness of the leading edge and second digit describes the distance of maximum thickness from the leading edge in tens of percent of the chord.

For example, NACA 4221-03

The above is a NACA-4221 airfoil with sharp leading edge '0' and maximum thickness of 30 % of the chord from the leading edge.

64. Round nose airfoils have infinite slope and infinite 2nd order derivative at leading edge.

65. The drag divergence Mach number is usually close to and always greater than critical Mach number.

MSQ PRACTICE PAPER – 2

1. $\lim_{x\to\infty}\left\{\frac{x^2+1}{x+1}-(ax+b)\right\}=\infty$ if

(a) $a=-3, b=2$
(b) $a=2, b=-2$
(c) $a=0.3, b=7.33$
(d) $a=-3, b=5$

2. If $\lim_{x\to 0}(\cos x+a\sin bx)^{1/x}=e^2$ then the values of a and b are:

(a) $a=1, b=2$

(b) $a=2, b=\frac{1}{2}$

(c) $a=2\sqrt{2}, b=\frac{1}{\sqrt{2}}$

(d) $a=4, b=2$

3. $\lim_{n\to\infty}\left(\frac{1}{\sqrt{n^2a^2-1}}+\frac{1}{\sqrt{n^2a^2-2^2}}+\frac{1}{\sqrt{n^2a^2-3^2}}+\ldots+\frac{1}{\sqrt{n^2a^2-n^2}}\right)$

(a) $sin^{-1}\left(1/a\right)$ (b) $cot^{-1}\left(1/a\right)$

(c) $\frac{\pi}{2}-\cot^{-1}(a)$ (d) $\frac{\pi}{2}-\sec^{-1}(a)$

4. Which of the following limit tends to unity?

(a) $\lim_{t\to 0}\frac{\sin(\tan t)}{\sin t}$

(b) $\lim_{x\to\pi/2}\frac{\sin(\cos x)}{\cos x}$

(c) $\lim_{x\to 0}\frac{\sqrt{1+x}-\sqrt{1-x}}{x}$

(d) $\lim_{x\to 0}\frac{\sqrt{x^2}}{x}$

5. Let a, b, c, d $\in$ N, then the values of a, b, c and d respectively, for which $\lim_{x\to\infty}\frac{a^{\sin x}+bx+c}{\cos x-\sqrt{1+dx^2}}=-2$ are

(a) 3, 4, 2, 4 (b) 4, 2, 7, 1
(c) 3, 6, 12, 9 (d) 13, 14, 33, 49

6. Let α and β are roots of quadratic equation $ax^2+bx+c=0$. Where, a,b,c $\in$ R, then $\lim_{x\to a}\frac{1-\cos(ax^2+bx+c)}{(x-a)^2}=0$ if

(a) $\alpha=\beta, a=0, \beta\neq 0$

(b) $\alpha\neq\beta, a=0, \beta\neq 0$

(c) $\alpha\neq\beta, a\neq 0, \beta=0$

(d) $\alpha=\beta, a=0, \beta=0$

7. Let $f(x)=\frac{e^x}{x}$

(a) $f(x)$ is strictly increasing for all x
(b) $f(x)$ is strictly increasing in $(1,\infty)$
(c) $f(x)$ is strictly decreasing in $(-\infty,1)$
(d) $f(x)$ have only one optimum point

8. $f(x)=\begin{cases}|2x-3|[x]; x\geq 1\\ \sin\left(\frac{\pi x}{2}\right); x<0\end{cases}$ is

(a) continuous at $x=2$
(b) diffentiable at $x=1$
(c) continuous but not differentiable at $x=1$
(d) none of these

9. Let $A=\begin{bmatrix}1&2\\3&4\end{bmatrix}$ and $B=\begin{bmatrix}1&2\\2&4\end{bmatrix}$ and $f(x)=ax^2+bx+c$

(a) $\rho(A)=1$
(b) B is symmetric matrix
(c) The product of zero of f(x) is $\frac{c}{a}$
(d) f(x) is increasing when $x\in\left[-b/a,\infty\right]$

10. Which of the following statements are correct for any square matrix A?

(a) $A + A^T$ is symmetric matrix

(b) $A - A^T$ is skew symmetric matrix

(c) AA^T is orthogonal matrix

(d) AA^H is unitary matrix

11. Which of the following statement is correct about the flow across oblique shock wave?

(a) Total temperature remains constant

(b) Total pressure decreases

(c) Static temperature increases

(d) Total temperature increases

12. Which of the following statement is/are correct?

(a) Potential function exists only for irrotational flow.

(b) Stream function (ψ) applies to both the rotational and irrotational flows. Which are steady and incompressible.

(c) In case of irrotational flow, both the stream function and velocity potential function satisfy Laplace equation.

(d) Potential function exists for both rotational and irrotational flow.

13. Velocity distribution is the boundary layer is given by $\frac{u}{V} = \left(\frac{y}{\delta}\right)^{\frac{1}{7}}$. If δ^* represents displacement thickness, θ represents momentum thickness, then which of the following is/are correct?

(a) $\delta^* = \frac{\delta}{8}$ (b) $\delta^* = \frac{\delta}{10}$

(c) $\theta = \frac{7}{72}\delta$ (d) $\theta = \frac{19}{72}\delta$

14. The flow field of fluid is given by $V = xy\hat{i} + 2yz\hat{j} - \left(yz + z^2\right)\hat{k}$ which of the following is/are correct about the flow.

(a) Represents a possible three dimensional steady incompressible continuous flow.

(b) Represents rotational flow

(c) Represents irrotational flow

(d) Represents a possible three dimensional steady compressible continuous flow.

15. The steam function $\psi = 4xy$, in which ψ is in cm²/sec and x and y are in meter. If flow is incompressible, then if V = velocity and a = convective acceleration then,

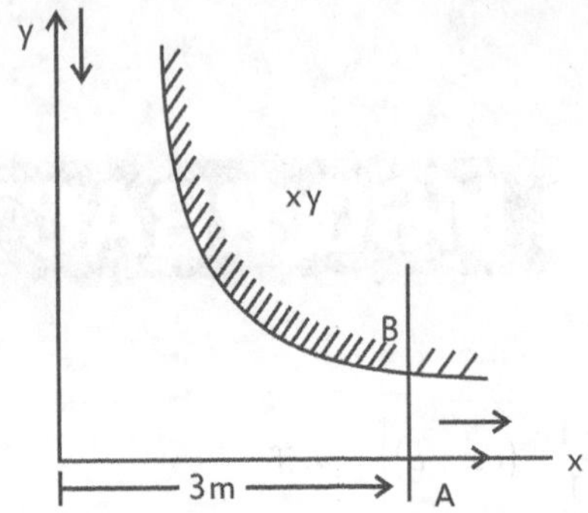

(a) V_B = 13 cm/sec (b) V_B = 20 cm/sec

(c) a_B = 50 cm²/sec (d) a_B = 100 cm²/sec

16. Separation of the boundary can be prevented by

(a) Providing small divergence in a diffuser

(b) Providing a trip - wire ring in the laminar region for the flow over a sphere.

(c) Providing a bypass in the slotted wing

(d) Suction of the slow moving fluid by a suction slot.

17. The critical value of Reynolds number at which boundary layer changes from laminar to turbulent depends upon:

(a) Turbulence in ambient flow

(b) Surface Roughness

(c) Pressure gradient

(d) Plate carvature

18. Chocking of the Nozzle fitted to a pressure tank containing gas implies that

(a) Sonic velocity at throat

(b) Increase of mass flow rate

(c) Obstruction of flow

(d) Supersonic velocity at throat

19. Which of the following is/are correct about design features to support lateral stability?

(a) Wing Dihedral

(b) High Wing

(c) High centre of gravity

(d) Low centre of gravity

20. Which of the following is/are correct for Aircraft motion?

(a) For short period mode speed remains constant

(b) For phugoid mode angle of attack remains constant.

(c) Dutch roll is couple roll-yaw oscillation

(d) Wing rock is roll oscillation.

21. Which of the following statement is/are correct for Maximum Range for reciprocating engine propeller driven aircraft?

(a) Should have largest possible propeller efficiency.

(b) Should have lowest possible specific fuel consumption.

(c) Should fly at maximum L/D

(d) Should have largest fuel weight.

22. Which of the following is /are correct about stalling Velocity of the Aircraft?

(a) Stalling Velocity increases with increasing altitude.

(b) Stalling velocity increases with increasing wing loading.

(c) stalling Velocity increases with increasing maximum lift coefficient

(d) Velocity stall increases with decreasing maximum lift coefficient.

23. Which of the following is/are correct about largest possible turn rate for level turn maneuver of Aircraft?

(a) Should have highest possible load factor

(b) Should have lowest possible load factor

(c) Should fly at highest possible velocity

(d) Should fly at lowest possible velocity

24. For NACA 23012 series which of the following is / are correct?

(a) the design lift coefficient is 0.3.

(b) the location of maximum camber is at 0.15c.

(c) the airfoil has 12 percent maximum thickness in percentage of chord.

(d) the location of maximum camber is at 0.3c.

25. Which of the following is/are Aerodynamic coefficient?

(a) Pressure coefficient.

(b) Absolute temperature.

(c) Lift coefficient.

(d) Drag coefficient.

26. For a critically damped system which of the given option are true:

(a) damping ratio = 1 (b) damping ratio < 1

(c) $C = C_c$ (d) $c = 2\sqrt{km}$

27. which of the Boundary conditions for a shaft fixed at both ends are correct?

(a) $\theta(o) = 0$ (b) $\theta(o) \neq 0$

(c) $\theta(\ell) = 0$ (d) $\theta(\ell) \neq 0$

28. Buckling strength of a simply supported column decreases as we:

(a) increase the applied load

(b) increase the length of column

(c) increase the young's modulus, E

(d) decrease the 2nd moment of Area

29. A metallic rod with both ends free, if applied with Thermal load will not have:

(a) Tensile stress (b) Thermal strain

(c) displacement (d) compressive stress

30. Which of the following represents the stiffness of structural elements?

(a) EA (b) EI

(c) GJ (d) $\frac{I}{y}$

31. Which of the following represent Von-miser failure criteria?

(a) $\sigma_1^2 + \sigma_2^2 + \sigma_1\sigma_2 = \sigma y^2$

(b) $\sigma_x^2 + \sigma_y^2 - \sigma x \sigma y + 3T^2{}_{xy} = \sigma_y^2$

(c) $\sigma_1^2 + \sigma_2^2 + \sigma_3^2 - \sigma_1\sigma_2 - \sigma_2\sigma_3 - \sigma_1\sigma_3 = \sigma_y^2$

(d) $\sigma_1^2 + 3T^2{}_{xy} = \sigma_y^2$

32. Maximum shear stress on a cantilever beam with end point load and circular cross-section depends on

(a) Shear Force

(b) Bending moment

(c) Length of beam

(d) Radius of cross-section

33. Which of following is true about below given system?

Figure (a)

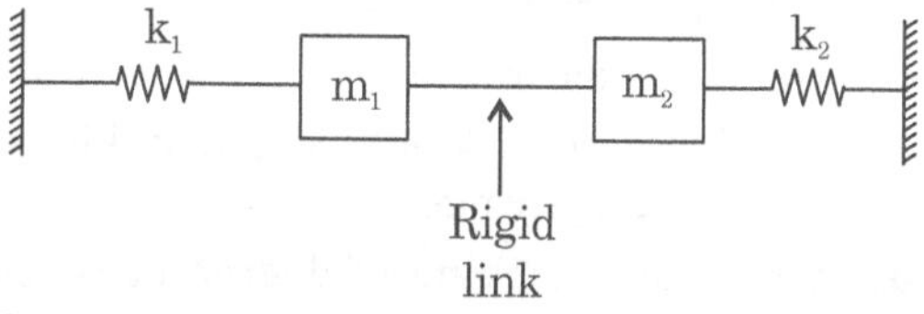

$k_1 = 25$ N/m

$k_2 = 25$ N/m

$m_1 = 5$ kg

$m_2 = 7$ kg

(a) It is a 1 DOF system

(b) It is a 2 DOF system

(c) $\omega_n = 2.04$ rod/sec

(d) $f_n = 2.04$ Hz

34. Fundamental natural frequency of the fixed-fixed rod torsional vibration depends on:

(a) Young's Modulus, (E).

(b) Shear Modulus, (G).

(c) length of shaft.

(d) applied load.

35. For a body under plain strain condition with applied stremer ε_{xx} = 300 MPa, ε_{yy} = 450 MPa Txy = 75 MPa, Txx = 0 , Tyx = 0 which of following is true about stremer and strains [Isotropic material and linearly elastic]
E = 200 GPa Poisson's Ratio V = 0.33

(a) $\varepsilon_{zz} = 0$ (b) $\varepsilon_{zx} = 0$

(c) $\varepsilon_{xx} = 0.34 \times 10^{-3}$ (d) $\varepsilon_{zz} \neq 0$

36. For a 2–D state stress

$\sigma_{xx} = 4x^2y + 5y^2$

$\sigma_{yy} = y^3 + 7xy$

$T_{xy} = -4xy^2 + 2x^2$

This state of stress

(a) Satisfies equilibrium equation.

(b) doesn't satisfies equilibrium equation.

(c) satisfies compatibility equation.

(d) doesn't satisfies compatibility equation

37. A cylindrical pressure venel is subjected to internal pressure of 7 bar. Then which of following is true if vessel material has yield strength of 300 MPa and it is made up of 2mm thick sheet, has an internal dia of 300mm.

(a) Longitudinal stress 6_L = 26.25 MPa

(b) Hoop stress 6_c = 52.5 MPa

(c) $[FOS]_{von\text{-}miser}$ = 6.59

(d) Hoop stress = longitudinal stress

38. Main uses of Ribs in an aircraft wing is to

(a) Maintain Aerodynamic shape

(b) Take bending load

(c) Take shear loads

(d) Increase bucking strength of skin, spar flanges, stringers.

39. A reaction machine of degree of reaction '1' represents.

(a) All energy transfer occurs due to change in static pressure at rotor.

(b) The fluid enters and leaves the compressor with same absolute velocity.

(c) The fluid enters and leaves the compressor with same relative velocity.

(d) static pressure rise across the stator blade is zero.

40. The flight operation represents the off-design condition is

(a) Landing flight (b) Take-off flight

(c) Climbing flight (d) Cruising flight

41. Mark the option/options which needed for off design component matching of engine operation

(a) RPM of rotating components.

(b) mass flow between the components.

(c) Energy between the components.

(d) Geometrical matching of component size.

42. A 50% reaction, axial flow compressor has inlet and exit blade angles of 45° and 15° respectively. The axial flow velocity is 200 m/s and power required is 300 kw. If the inlet total pressure and temperture is 1 bar and 300 k respectively, then

(a) mass flow rate of air through compressor is 10 kg/s.

(b) Mass flow rate of air through compressor is 8 kg/s

(c) Diameter of rotor at its mean radius is 0.9 m if it is rotated at 5000 rpm

(d) Diameter of roter at its mean radius is 0.7 m if it is rotated at 7000 rpm

43. In question (24), it is assumed that ratio of specific heat is constant (1.4) and gas constant is 287 J/kg-k. If rottor rotates at 5000 rmp then.

(a) Height of rotor blade is 0.017 m

(b) Height of rotor blade is 0.117 m

(c) Radius of tip of rotor blade is 0.543 m

(d) Radius of tip of rotor blade is 0.493 m

44. In a turbine stage having efficiency 80%, examine of flow is achieved isentropically. If blade to gas speed is 0.45 and nozzle efflux angle is 65° then

(a) Blade efficiency is 82%.

(b) Blade efficiency is 72%.

(c) Isentropic velocity ratio is 0.45.

(d) Isentropic velocity ratio is 0.40.

45. The blades of turbine is designed as free vortex design, in which $C_t \times r$ = constant. Where 'C_t' is tangential component of absalute velocity and 'r' is radius of blade at any section. The nozzle blade delivered the flow to the rotor blade at angle 62° and velocity 740 m/s. If the diameter of blade at mean section is 50 cm and height of blade is 5 cm, then

(a) Nozzle efflux angle at root is 64°.

(b) Nozzle efflux angle at tip is 59.6°.

(c) Absolute velocity of flow at root is 805 m/s.

(d) Absolute velocity of flow at tip is 600 m/s.

46. Degree of reaction of 50% for a turbine implies that (Suffix '2' and '3' represents rotor inlet and outlet respectively).

(a) $\alpha_2 = \alpha_3$ (b) $\alpha_2 = \beta_3$

(c) $\alpha_2 \neq \alpha_3$ (d) $\alpha_3 = \beta_2$

47. Mark the correct statement related to propulsive efficiency of gas turbine engine at row speed:

(a) Propulsive efficiency of ramjet is greater than turbojet.

(b) propulsive efficiency of ramjet is smaller than turbojet.

(c) propulsive efficiency of turbojet is greater than turbofan.

(d) propulsive efficiency of turbojet is smaller than turbofan.

48. The isentropic efficiency of components of aircraft engine such as diffuser, compressor, turbine and nozzle is represented as η_d, η_c, η_t and η_n respectively. Which options are correct.

(a) $\eta_n > \eta_c$ (b) $\eta_n < \eta_c$

(c) $\eta_c < \eta_d$ (d) $\eta_c > \eta_d$

49. The diameter of the propeller of an aircraft is 2.5 m. If flies at a speed of 500 kmph at an altitude of 7000 m. If flight to jet speed ratio is 0.75, and density is 0.5 kg/m³, then

(a) The flow rate of air through the propeller is 450 kg/s.

(b) The flow rate of air through the propeller is 397.7 kg/s.

(c) Thrust produced by engine is 18.41 kN.

(d) Thrust produced by engine is 20.83 kN.

50. The purpose of turbine blade cooling is

(a) To protect the blade from over heating.

(b) To increase turbine inlet temperature

(c) To reduce thrust produce by engine.

(d) All of the above.

51. Which statements correctly represents for a turboshaft engine?

(a) An auxiliary combustion chamber is used between two turbine.

(b) One turbine is used to derive compressor and other provides power to rotor.

(c) If is mostly used in helicopter

(d) If is used in fighter aircraft.

52. A rocket engine has the following data

Propellant flow rate = 4 kg/s

Nozzle exit area = 70 cm²

Nozzle exit pressure = 1 bar

Ambient pressure = 0.9 bar

Thrust = 7 kN, g = 9.81 m/s²

(a) Effective Jet velocity is 1732 m/s

(b) Effective Jet velocity is 1750 m/s

(c) Actual Jet velocity is 1750 m/s

(d) Actual Jet velocity is 1732 m/s

53. Mark the correct statements regarding the thrust co-efficient of rocket engine.

(a) If depends on the temperature ratio.

(b) If depends on pressure ratio.

(c) for γ = 1.2 and pressure ratio 5, its magnitude is less than 1.

(d) for γ = 1.2 and pressure ratio 5, its magnitude is greater than 1.

54. A missile has maximum flight to jet speed ratio is 0.2 and specific impulse is 200 sec. If burn-out time is 10 sec then, (g = 10 m/s²)

(a) Effective jet velocity is 2200 m/s

(b) mass ratio is 0.78.

(c) propellant mass fraction is 0.22.

(d) All of the above

55. A rocket nozzle has throat area of 20 cm², propellant weight flow rate is 45 N/s, specific impulse is 130 seconds. If the combustion chamber pressure is 25 bar, then

(a) Characteristic velocity is 1111.11 m/s

(b) Characteristic velocity is 2222.22 m/s

(c) Thrust co-efficient is 1.17

(d) (a) and (c)

56. A rocket nozzle is designed to produce maximum thrust at an altitude of 11 km from the sea level. The nozzle oprates in

(a) under expanded condition for H > 11 km

(b) over expanded condition for H > 11 km

(c) unchocked condition for H = 11 km

(d) All of these.

57. A solid rocket motor has an end burning grain of cross sectional area is 0.5 m², and length is 1 m. The propellant grain is stored in a cylindrical motor. Mass of propellant store in motor is 750 kg. Time taken to burn all the propellant is 200 sec. If specific impulse is also 200 sec then

(a) regression rate for linear burning is 5 mm/s.

(b) mass flow rate of propellant is 3.75 kg/sec

(c) Thrust propuced by rocket is 7.5 kN for g = 10 m/s².

(d) All of the above.

58. Consider a single stage rocket of propellant mass m_p, structural mass m_s and payload mass is m_L. If σ and λ reprsents the structural ratio and payload ratio respectively, then equation of burn-out velocity is given by

(a) $\Delta V_b = -V_e \ln\left[\dfrac{m_L + m_S}{m_L + m_S + m_P}\right]$

(b) $\Delta V_b = V_e \ln\left[\dfrac{1}{[\sigma + \lambda(1-\sigma)]}\right]$

(c) $\Delta V_b = V_e \ln[\sigma + \lambda(1-\sigma)]$

(d) $\Delta V_b = V_e \ln\left[\dfrac{1}{\lambda + \sigma(1-\sigma)}\right]$

59. Which of the following is not constant for a satellite in an elliptical orbit around the earth?

(a) Kinetic Energy

(b) Product of speed and radial distance from center of earth

(c) Rate of area swept by the radial vector from the center of the orbit

(d) Rate of area swept by the radial vector from the center of the earth

60. Two identical earth satellite A and B are in circular orbits at altitude h_A and h_B above the Earth's surface respectively, with $h_A > h_B$. If E denotes the total mechanical energy, T the kinetic energy and V the gravitational potential energy of a satellite, then:

(a) $V_A < V_B$ (b) $E_A > E_B$

(c) $E_A < E_B$ (d) $T_A < T_B$

61. The velocity required for a spacecraft to escape earths gravitational field depends on

(a) The mass of the spacecraft

(b) The distance between earth's Centre and the spacecraft

(c) The earth's rotational speed about its own axis

(d) The earth orbital speed

62. Which of the following statements is true as the altitude increases in stratosphere of international standard Atmosphere?

(a) Temperature increases and dynamic viscosity decreases

(b) Temperature remains constant and pressure decreases.

(c) Temperature remains constant and sound speed remains constant.

(d) Temperature decreases and density increases.

63. If eccentricity $e \geq 1$, then the orbit is

(a) Circular (b) Elliptical

(c) Parabolic (d) Hyperbolic

64. Which of the following is true for supercritical airfoils?

(a) NACA 812A218 is a supercritical airfoil

(b) Supercritical airfoil delays the wave drag in transonic range.

(c) The maximum thickness is close to the leading edge.

(d) The maximum thickness is close to the middle of chamber line.

65. Which of the following is used to measure velocity of a moving gas?

(a) Hot Wire Anemometry

(b) Laser Doppler Velocimetry

(c) Particle image Velocimetry

(d) Dot Matrix

ANSWERS

1. (a), (c), (d)	**2.** (a), (c)	**3.** (a), (d)	**4.** (a), (b), (c)	**5.** (a), (b), (c), (d)
6. (a), (b), (d)	**7.** (b), (d)	**8.** (a), (c)	**9.** (b), (c)	**10.** (a), (b)
11. (a), (b), (c)	**12.** (a), (b), (c)	**13.** (a), (c)	**14.** (a), (b)	**15.** (a), (c)
16. (a), (b), (c), (d)	**17.** (a), (b), (c), (d)	**18.** (a), (b), (c)	**19.** (a) (b) & (d)	**20.** (a) (b) (c) & (d)
21.(a) (b)(c) & (d)	**22.** (a) (b) & (d)	**23.** (a) & (d)	**24.** (a), (b) & (c)	**25.** (a), (c), (d)
26. (a), (c), (d)	**27.** (a), (c)	**28.** (b), (d)	**29.** (a), (c)	**30.** (a), (b), (c)
31. (a), (b), (c)	**32.** (a), (d)	**33.** (a), (c)	**34.** (b), (c)	**35.** (a), (b), (c)
36. (b), (d)	**37.** (a), (b), (c)	**38.** (a), (c), (d)	**39.** (a), (b), (d)	**40.** (a), (c), (d)
41. (a), (b), (c), (d)	**42.** (b), (c), (d)	**43.** (a), (d)	**44.** (a), (d)	**45.** (a), (b), (c)
46. (b), (c), (d)	**47.** (b), (d)	**48.** (a), (c)	**49.** (b), (c)	**50.** (a), (b)
51. (b), (c)	**52.** (b), (d)	**53.** (b), (d)	**54.** (b), (c)	**55.** (a), (c), (d)
56. (a) and (c)	**57.** (a), (b), (c), (d)	**58.** (a), (b)	**59.** (a), (b)	**60.** (b), (d)
61. (a), (b)	**62.** (b), (c)	**63.** (c), (d)	**64.** (a), (b), (c)	**65.** (a), (b), (c), (d)

EXPLANATIONS

1. Given,

$$\Rightarrow \lim_{x\to\infty}\left\{\frac{x^2+1}{x+1}-(ax+b)\right\}$$

$$\Rightarrow \lim_{x\to\infty}\left\{\frac{(x^2+1)-(ax+b)(x+1)}{x+1}\right\}$$

$$\Rightarrow \lim_{x\to\infty}\left\{\frac{x^2+1-ax^2-ax-bx-b}{x+1}\right\}$$

$$\Rightarrow \lim_{x\to\infty}\left\{\frac{x^2(1-a)-(a+b)x+(1-b)}{x+1}\right\}$$

To make $x\to\infty$

First term, $x^2(1-a)\neq 0$

$\therefore\ 1-a\neq 1$

Also, if negative then $x\to-\infty$

$\therefore\ a<1; a\in(-\infty, 1)$

Thus, option (a, c, d) are correct.

2. For,

$$\lim_{x\to a}[f(x)]^{g(x)}=e^{\lim_{x\to a} g(x)[f(x)-1]}=1^\infty$$

we get,

$$\lim_{x\to 0}(\cos x+a\sin bx)^{1/x}=e^{\lim_{x\to 0}\frac{1}{x}\left[\frac{\cos x+a\sin bx-1}{x}\right]}$$

Now,

$$e^{\lim_{x\to 0}(-\sin x+ab\cos bx)^{1/x}}=e^{ab}$$

For, L.H.S = R.H.S

$a.b=2$

For, $a=1, b=2$

$a.b=2$

For, $a=2, b=\frac{1}{2}$

$a.b=1$

For, $a=2\sqrt{2}, b=\frac{1}{\sqrt{2}}$

$a.b=2$

For, $a=4, b=2$

$a.b=8$

3. Given,

$$\lim_{n\to\infty}\left(\frac{1}{\sqrt{n^2a^2-1}}+\frac{1}{\sqrt{n^2a^2-2^2}}+\frac{1}{\sqrt{n^2a^2-3^2}}+\ldots+\frac{1}{\sqrt{n^2a^2-n^2}}\right)$$

writing in summation form

$$\lim_{n\to\infty}\sum_{r=1}^{n}\frac{1}{\sqrt{n^2a^2-r^2}}$$

we know,

$$\lim_{n\to\infty}\frac{1}{n}\sum_{r=1}^{n}f\left(\frac{r}{n}\right)=\int_0^\infty f(x)dx$$

Here,

$$\lim_{n\to\infty}\sum_{r=1}^{n}\frac{1}{n\sqrt{a^2-\left(\frac{r}{n}\right)^2}}=\lim_{n\to\infty}\frac{1}{n}\sum_{r=1}^{n}\frac{1}{n\sqrt{a^2-\left(\frac{r}{n}\right)^2}}$$

As per above formula:

$$\int_0^\infty\frac{1}{\sqrt{a^2-x^2}}dx=\sin^{-1}\left(\frac{x}{a}\right)$$

$$\therefore\ \lim_{n\to 0}\sin^{-1}\left(\frac{x}{a}\right)=\sin^{-1}\left(\frac{1}{a}\right)$$

$$\sin x=\frac{\pi}{2}-\cos^{-1}x$$

$$\sin^{-1}\left(\frac{1}{a}\right)=\frac{\pi}{2}-\cos^{-1}\left(\frac{1}{a}\right)=\frac{\pi}{2}-\sec^{-1}a$$

Therefore, Option (a) and (d) are correct.

4. Taking option (a):

$$\lim_{t\to 0}\frac{\sin(\tan t)}{\sin t}$$

$$\Rightarrow \lim_{t\to 0}\frac{d}{dt}\left(\frac{\sin(\tan t)}{\sin t}\right)=\lim_{t\to 0}\frac{\cos(\tan t)\sec^2 t}{\cos t}=1$$

Taking option (b)

$$\lim_{t\to 0}\frac{\sin(\cos x)}{\cos x}=\lim_{x\to\pi/2}\frac{\sin[\sin(\pi/2-x)]}{\sin(\pi/2-x)}$$

Let $\frac{\pi}{2}-x=t$

$$\lim_{t\to 0}\frac{\sin[\sin t]}{\sin t}=1$$

Taking option (c)

$$\lim_{x\to 0}\frac{\sqrt{1-x}-\sqrt{1-x}}{x}=\frac{1}{2\sqrt{1+x}}+\frac{1}{2\sqrt{1-x}}$$

$$=\frac{1}{2}+\frac{1}{2}=1$$

Taking option (d)

$$\lim_{x\to 0}\frac{\sqrt{x^2}}{x}=\frac{|x|}{x}$$

$$\lim_{x\to 0^-}\frac{|x|}{x}=-1$$

$$\lim_{x\to 0^+}\frac{|x|}{x}=1$$

$\therefore$ L.H.S $\neq$ R.H.S

Thus, limit does not exit.

5. Given,

$$\lim_{x\to\infty}\frac{a^{\sin x}+bx+c}{\cos x-\sqrt{1+dx^2}}=-2$$

Dividing Numerator and denominator by x

$$\Rightarrow \lim_{x\to\infty}\frac{\frac{a^{\sin x}}{x}+b+\frac{c}{x}}{\frac{\cos x}{x}-\sqrt{\frac{1}{x^2}+d}}=-2$$

$$\Rightarrow \frac{0+b+0}{0-\sqrt{0+d}}=\frac{-b}{\sqrt{d}}=-2$$

$\Rightarrow b=2\sqrt{d}$

Taking option (a)

$b=4, d=4$

$\therefore\ b=2\sqrt{d}$

Taking option (b)

$b=2, d=1$

$\therefore\ b=2\sqrt{d}$

Taking option (c)

$b=6, d=9$

$\therefore\ b=2\sqrt{d}$

Taking option (d)

$b=14, d=49$

$\therefore\ b=2\sqrt{d}$

6. For,

$$\Rightarrow \lim_{x\to\infty}\frac{1-\cos\left(ax^2+bx+c\right)}{(x-\alpha)^2}=0$$

The roots will be

$a(x-\alpha)(x-\beta)=0$

$$\Rightarrow \lim_{x\to\infty}\frac{1-\cos[a(x-\alpha)(x-\beta)]}{(x-\alpha)^2}$$

we know,

$1-\cos 2\theta = 2\sin^2\theta$

Applying above

$$\Rightarrow \lim_{x\to\alpha}\frac{2\sin^2\left[\frac{a(x-\alpha)(x-\beta)}{2}\right]}{(x-\alpha)^2}$$

$$\Rightarrow \lim_{x\to\alpha}2\left\{\frac{\sin\left[\frac{a(x-\alpha)(x-\beta)}{2}\right]}{\frac{a(x-\alpha)(x-\beta)}{2}}\right\}^2\times\frac{a^2(x-\beta)^2}{4}$$

$\therefore\ 2a^2(\alpha-\beta)^2=0$

For this to be true

$a=0$ or $\alpha=\beta$

Therefore, Option (a, b, d) are correct.

7. The domain of $\frac{e^x}{x}$ is all real number except 0.

Here, in option (c) we get $(-\infty, 1)$ which includes 0. Therefore, option (c) is wrong.

Now,

$$f(x)=\frac{e^x}{x}$$

$$f'(x)=\frac{xe^x-e^x}{x^2}=\frac{(x-1)e^x}{x^2}$$

$f'(x)>0$; it is strictly increasing

$x-1>0$

$x>1$

$\therefore$ Option (a) is wrong. So, option (b) is correct.

To find optimum point, we need to find critical point.

Here, critical point is a point at which derivative is 0.

Here,

$$f'(x)=\frac{(x-1)e^x}{x^2}$$

If $x=1$, we get $f'(x)=0$

Therefore, Option (d) is also correct.

8. For function to be continuous

L.H.S = R.H.S = $f(x)$

Here, L.H.S is

$$\sin\left(\frac{\pi x}{2}\right)=1$$

R.H.S is

$$|2x-3|[x]=1$$

$$f(x)=|2x-3|[x] \text{ at } x=1$$

$=1$

$\therefore$ L.H.S = R.H.S = $f(x)$

Thus, the function is continuous at $x = 1$

Also, for $x = 2$, the function is continuous

Taking derrivative

R.H.S is

$$f(x) = \sin\left(\frac{\pi x}{2}\right)$$

$$f'(x) = \frac{\pi}{2}\cos\left(\frac{\pi x}{2}\right)$$

L.H.S

$$f(x) = |2x-3|[x]$$

$$|2x-3|\times 0+[x]\frac{|2x-3|\times 2}{2x-3} = -2$$

$\therefore$ L.H.S $\neq$ R.H.S

Therefore, it is not differentiable at $x = 1$

Thus, Option (*a*) and (*c*) are correct.

9. Taking Option (*a*)

Determinant of $A = -2$

$\therefore$ The order is 2, also the rank is also 2

Option (*a*) is wrong

Taking option (*b*)

$$B = \begin{bmatrix} 1 & 2 \\ 2 & 4 \end{bmatrix}$$

$$B^T = \begin{bmatrix} 1 & 2 \\ 2 & 4 \end{bmatrix}$$

$\therefore \quad B = B^T$

Thus, B is symmetric matrix

Taking option (*c*)

The products of zeros indicates the roots.

$$f(x) = ax^2 + bx + c$$

Here, the roots are

$$\alpha = \frac{-b+\sqrt{b^2-4ac}}{2a}$$

$$\beta = \frac{-b-\sqrt{b^2-4ac}}{2a}$$

$$\therefore \quad \alpha + \beta = \frac{c}{a}$$

Thus, option (*c*) is correct.

Taking option (*d*)

$$f(x) = ax^2 + bx + c$$

$$f'(x) = 2ax + b$$

For critical point

$$f'(x) = 0$$

$$0 = 2ax + b$$

$$x = \frac{-b}{2a}$$

As we don't know the value of (*a*) and (*b*) we cannot determine the increasing interval.

Option (*d*) is wrong.

10. A matrix is symmetric when

$$M^T = M$$

Taking option (*a*)

$$(A + A^T)^T = A^T + (A^T)^T = A^T + A$$

$\therefore$ $A + A^T$ is symmetric matrix

Thus, option (*a*) is correct.

A matrix to be skew symmetric

$$M^T = -M$$

Here,

$(A - A^T)^T = A^T - A = -(A - A^T)$

$\therefore$ Option (*b*) is correct.

A matrix to be orthogonal when

$$MM^T = I$$

Here, $(AA^T)(AA^T)^T = AA^T(A^T)^T A^T = AA^T AA^T \neq I$

A matrix is said to be unitary when

$$MM^H = I$$

Here, $(AA^H)(AA^H)^H = AA^H(A^H)^H A^H = AA^H AA^H \neq I$

Option (*c*) and (*d*) are not correct.

11. $T_{01} = T_{02}$

$P_{01} > P_{02}$

$T_1 < T_2$

13. $\dfrac{u}{v} = \left(\dfrac{y}{\delta}\right)^{\frac{1}{7}}$

$$\delta^* = \int_0^\delta\left(1-\frac{u}{v}\right)dy$$

$$= \int_0^\delta\left(1-\left(\frac{y}{\delta}\right)^{\frac{1}{7}}\right)dy = \left[y-\frac{7}{8}\frac{y^{\frac{8}{7}}}{\delta^{\frac{1}{7}}}\right]$$

$$= \left[\delta-\frac{7}{8}\times\frac{\delta^{\frac{8}{7}}}{\delta^{\frac{1}{7}}}\right] = \frac{\delta}{8}$$

$$\theta = \int_0^\delta\frac{u}{v}\left(1-\frac{u}{v}\right)dy$$

$$= \int_0^\delta\left(\frac{y}{\delta}\right)^{\frac{1}{7}}\left(1-\left(\frac{y}{\delta}\right)^{\frac{1}{7}}\right)dy$$

$$= \int_0^{\delta}\left[\left(\frac{4}{\delta}\right)^{\frac{1}{7}} - \left(\frac{y}{\delta}\right)^{\frac{2}{7}}\right]dy$$

$$= \left[\frac{7}{8}\frac{y^{\frac{8}{7}}}{\delta^{\frac{1}{7}}} - \frac{7}{9}\cdot\frac{y^{\frac{9}{7}}}{\delta^{\frac{2}{7}}}\right]_0^{\delta} = \frac{7}{72}\delta.$$

14. $u = xy$, $v = 2yz$, $w = -(yz + z^2)$

Flow will be a three – dimensional steady in compressible continuous. If continuity equation is satisfied.

$$\overline{V}.V = \frac{\partial u}{\partial x} + \frac{\partial v}{\partial y} + \frac{\partial w}{\partial z} = 0$$

$$= y + 2z - (y + 2z) = 0$$

$$\overline{V}.V = 0$$

For checking the rationality $(\omega_z \neq 0)$

Where $\omega_z = \frac{1}{2}\left(\frac{\partial u}{\partial y} - \frac{\partial v}{\partial x}\right)$

$$\frac{\partial v}{\partial y} = x, \frac{\partial v}{\partial x} = 0.$$

$$\because \frac{\partial v}{\partial y} \neq \frac{\partial v}{\partial x}$$

Flow is rotational.

15.

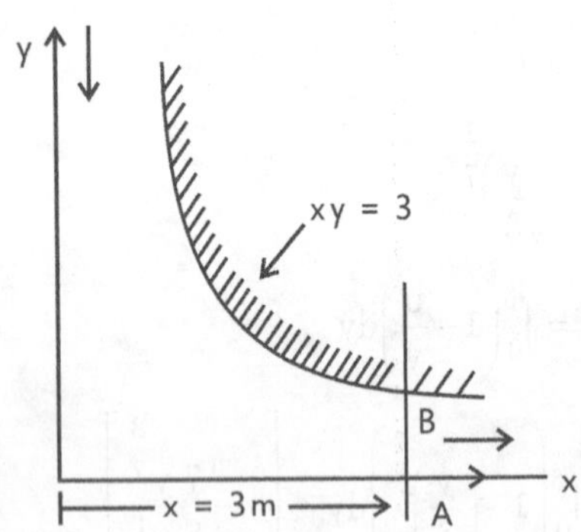

$\psi = 4xy$

ψ is in cm^2/sec

co-ordinate of A are (3, 0)

co-ordinate of B are

$$x = 3m,\ y = \frac{3}{x} = 1m$$

Velocity

$$u = \frac{\partial \psi}{\partial y} = \frac{\partial}{\partial y}(4xy) = 4x$$

$$V = -\frac{\partial \psi}{\partial x} = -\frac{\partial}{\partial x}(4xy) = -4y$$

At point B,

$u = 4 \times 3 = 12$ cm/sec

$V = -4 \times 1 = 4$ cm/sec

$$V = \sqrt{u^2 + v^2} = \sqrt{(12)^2 + (-4)^2} = 12.65 \text{ cm/sec}$$

$$a = \sqrt{{a_x}^2 + {a_y}^2}$$

$$a_x = u\frac{\partial u}{\partial x} + v\frac{\partial u}{\partial y}$$

$$= (4x)(4) + (-4y) \times 0 = 16x$$

$$a_y = u\frac{\partial v}{\partial x} + v\frac{\partial v}{\partial y}$$

$$= (4x)(0) + (-4y)(-4) = 16y$$

$$a = \sqrt{{a_x}^2 + {a_y}^2}$$

$$= \sqrt{(16x)^2 + (16y)^2}$$

$$a_{(3,1)} = \sqrt{(16 \times 3)^2 + (16 \times 1)^2}$$

$$= 16\sqrt{3^2 + 1} = 50.6 \text{ cm/sec}^2.$$

21. For propeller driven A/C range is given by

$$R = \frac{\eta}{C}\frac{C_L}{C_D}\ln\frac{W_0}{W_1}$$

22. Stalling velocity:

$$V_{stall} = \sqrt{\frac{2}{e_\infty}\left(\frac{W}{S}\right)\frac{1}{(C_L)_{max}}}$$

23. Turn rate for level turn maneuver is given by

$$\omega = \frac{9\sqrt{n^2 - 1}}{V_\infty}$$

n = load factor

V_∞ = velocity

For W_{max}, highest possible load factor and lowest possible velocity

24. In NACA five digit series. First digit when multiplied by 0.15 gives the design list coff. 2nd and 3rd divided by 2 gives distance of maximum camber from leading edge in tenths of chord. Last two digit maximum thickness is percentage of chord.

25. Absolute temperature is not an aerodynamic coefficient. However, pressure coefficient, lift coefficient and drag coefficient are all aerodynamic coefficients.

26. A dynamic system is called critically damped when damping ratio ξ is equal to 1.

$\therefore$ $\boxed{\xi = 1}$

also $\xi = \dfrac{C}{C_c} \Rightarrow \dfrac{C}{C_C} = 1 \Rightarrow \boxed{C = C_c}$

also we know that $C_c = 2\sqrt{km}$

C_c = Critical damping coefficient

$\therefore$ $\boxed{C = 2\sqrt{km}}$

27. shaft fixed at both ends

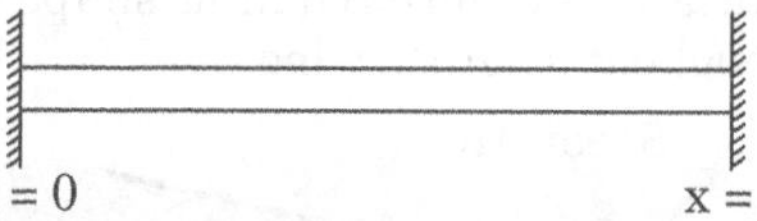

Fixed boundary condition does not allow angular displacement to happen at ends. Thus angular displacement Q at x = 0 and x = l are zero

28. Critical buckling load for column with simply supported case is : $P_{CT} = \dfrac{\pi^2 EI}{\ell^2}$

if we increase length l, P_{cr} decreases thus beam buckler at lower applied load.

similarly if we decrease 2nd moment of Area, I, P_{cr} decrease.

This both these case reduce buckling strength where as increasing E will increase strength as P_{cr} increases applied load does not have any effect of buckling strength.

29. Any metallic structural component cannot have stresses if it is free to expant in all direction. It will have displacement and strain because of expansion. Thermal stresses only develops when there is a resistance to the displacement [expansion of rod].

30. EA = Axial stiffness of rod

EI = Bending stiffness of beam

GJ = Torsional stiffness of shaft

$\dfrac{I}{y}$ is section modulus and doesn't represent stiffness.

31. Option (a) is Von-Miser failure criteria in terms of Principle stresses for 2–D case.

Option (b) is Von-Miser failure criteria in terms of applied stresses for 2–D case

Option (c) is Von-Miser failure criteria in terms of Principle stresses for 3–D case

32. Maximum stress for circular cross-section is given by

$$T_{max} = \frac{4}{3}\left[\frac{P}{\pi r^2}\right]$$

P = Shear Force

r = radius

Thus formula comes from $\boxed{T = \dfrac{\theta A \bar{y}}{I b}}$

33. In figure (A) masses are connected using a Rigid link thus can be considered as single mass

$= m_1 + m_2 = 5 + 7 = 12 \text{ kg} = M$

Governing differential equation of motion (GDEOM):

$$M\ddot{x}_1 + k_1 x_1 + k_2 x_1 = 0$$

$$\ddot{x}_1 + \frac{(k_1 + k_2)}{M} x_1 = 0$$

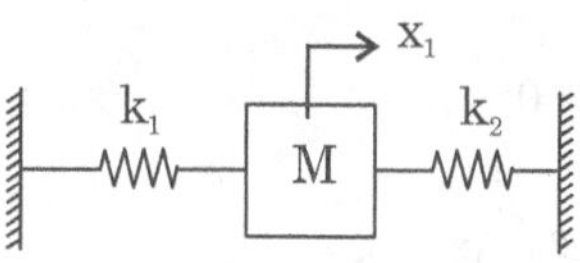

Mass is allowed to move in 1 direction independently It is a 1 DOF system.

$\therefore$ Natural Frequency $\omega_n = \sqrt{\left(\dfrac{k_1 + k_2}{M}\right)}$

= 2.04 rad/sec

34. Natural frequency of rod under torsional vibration is given by

$$\omega_n = \frac{n\pi C}{\ell}; \qquad n = 1, 2, 3$$

here $C = \sqrt{\dfrac{GV}{I_0}}$

For fundamental Natural Frequency n = 1

$$\omega_1 = \frac{\pi}{\ell}\sqrt{\frac{G}{\rho}}$$

I = f V

J = Torsional Constant

ρ = mass density

G = Shear modulus

ℓ = length of shaft

hence, It can be seen that ω_1 depends on length of shaft mass density ρ and shear modulus G

35. For a plain strain problem

$\varepsilon_{zz} = \varepsilon_{zx} = \varepsilon_{zy} = 0$

From stress–strain Relations:

$$\varepsilon_{xx} = \frac{1}{E}\left[\sigma_{xx} - V\left(\sigma_{yy} + \sigma_{zz}\right)\right]$$

Here $\sigma_{zz} = \nu\left(\sigma_{xx} + \sigma_{yy}\right) = 0.33 \times [300 + 450]$

= 247.5 MPa

$$\varepsilon_{xx} = \frac{1}{200 \times 10^3 \text{ MPa}}[300 - 0.33(450 + 247.5)]$$

$= 0.349 \times 10^{-3}$

36. Equilibrium equation

$$\frac{\delta\sigma_{xx}}{\delta x} + \frac{\delta T_{xy}}{\delta y} = 0$$

L.H.S. $\Rightarrow 8xy + [-8xy] = 0 =$ R.H.S

$$\frac{\delta T_{xy}}{\delta x} + \frac{\delta\sigma_{yy}}{\delta y} = 0$$

L.H.S $\Rightarrow -4y^2 + 4x + 3y^2 + 7x$

$= -y^2 + 11x \neq 0$

Thus state of stress doesn't satisfy equilibrium equation.

Compatibilitly equation in terms of stresses:

$$\left(\frac{\delta^2\sigma_{xx}}{\delta x^2} + \frac{\delta^2\sigma_{xx}}{\delta y^2} + \frac{\delta^2\sigma_{yy}}{\delta y^2} + \frac{\delta^2\sigma_{yy}}{\delta x^2}\right) = 0$$

L.H.S. $8y + 10 + 6y + 0 \neq 0$

Thus this state doesn't satisfy compatibility also Once equilibrium eq. is not satisfied, it is straight forward that compatibility will not be satisfied without doing any calculations.

37. Pressure P = 7 bar

$= 7 \times 10^5$ Pa

Hoop stress $\sigma_H = \frac{Pd}{2t} = \frac{7 \times 10^5 \times 0.3}{2 \times 0.002}$

= 52500000 Pa

= 52.5 MPa

Longitudinal stress:

$$\sigma_1 = \frac{Pd}{4t} = \frac{1}{2} \times \sigma_H = 26.25 \text{ MPa}$$

Von–Miser criteria :

equivalent stress = $\sqrt{\sigma_H^2 + \sigma_L^2 - \sigma_H \sigma_L}$

$= \sqrt{52.5^2 + 26.25^2 - 52.5 \times 26.25}$

= 45.466 MPa

$$\therefore [\text{Fo}\bar{\text{s}}]_{\text{Vonmnet}} = \frac{\sigma_{\text{yeild}}}{\text{equivalent stress}}$$

$$= \frac{300}{45.466} = 6.598$$

38. Ribs are airfoil shaped stiffness in the wing which maintain the aerodynamic shape of wing throughout length of wing

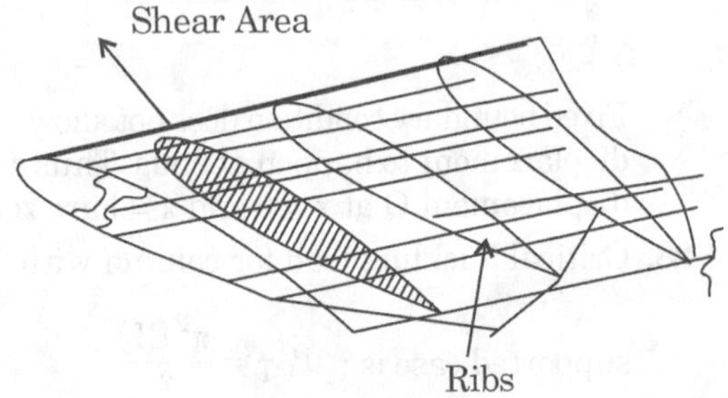

These also contribute in the shearing area for torsional shear, and Normal shear due to drag. Thus there take shear loads.

As these are placed throughout length of wing at some distances apart, These reduce the effective length of skin, stringers, spar flanges which increase the buckling strength

$$\boxed{P_{cr} \propto \frac{1}{\ell}}$$

40. → Design condition represents maximum operating condition. Beyond which operation is not possible.

→ During take-off flight engine is operating at full throttle means design condition and other flight operation occurs at off-design condition.

42. $R = \frac{1}{2} = \frac{C_a}{2.u}(\tan\beta_1 + \tan\beta_2)$

$$1 = \frac{200}{u}(\tan 45 + \tan 15)$$

u = 253.589 m/s

$\omega = uC_a(\tan\alpha_2 - \tan\alpha_1) = uC_a(\tan\beta_1 - \beta_2)$

$P = \dot{m}\omega$

$300 \times 10^3 = \dot{m} \times 253.589 \times 200(\tan 45° - \tan 5°)$

$\dot{m} = 8.08$ kg/s

$$u = \frac{\pi DN}{60}$$

If N = 5000 rpm, D = 0.969m

If N = 7000 rpm, D = 0.692m

43. $P_{o1} = 1$ bar, $T_{o1} = 300k$

$\alpha_1 = \beta_1 = 15°$, u = 253.589 m/s

$$\cos\alpha_1 = \frac{C_a}{C_1} \Rightarrow C_1 = \frac{200}{\cos 15°} = 207.05 \text{m/s}$$

$\beta_1 = \alpha_2 = 45°$

$$\cos\beta_1 = \frac{C_a}{\omega_1} \Rightarrow \omega_1 = 282.843 \text{ m/s}$$

$$T_{01} = T_1 + \frac{C_1^2}{2C_P}$$

$T_1 = 278.67K$

$$m_1 = \frac{\omega_1}{\sqrt{rRT_1}} = \frac{282.843}{\sqrt{1.4 \times 2.87 \times 278.67}} = 0.845$$

$$\frac{P_{01}}{P_1} = \left[1 + \frac{\gamma - 1}{2} m_1^2\right]^{\frac{\gamma}{\gamma-1}}$$

$P_1 = 0.627$ bar

$$\rho_1 = \frac{P_1}{RT_1} = 0.784 \text{ kg/m}^3$$

$$\dot{m} = \rho_1 \pi D_m h C_a$$

h = 0.017 m

$$R_{tip} = R_m + \frac{h}{2} = \frac{0.969}{2} + \frac{0.017}{2} = 0.493\text{m}$$

44. $\eta_b = 4(\sigma \sin\alpha_2 - \sigma^2) \times 100\% = 82.1\%$

$$\sigma = \frac{\sigma_s}{\sqrt{\eta_{nozz}}}$$

$$\sigma_s = \sigma \times \sqrt{\eta_{nozz}} = 0.45 \times \sqrt{0.8} = 0.40$$

45. At min section:

$\alpha_{2,m} = 62°$, $C_{2,m} = 740$ m/s

$\gamma_m = 25$ cm, h = 5 cm

$$\sin\alpha_{2,m} = \frac{C_{t,2m}}{C_{2,m}}$$

$C_{t2,m} = 653.38$ m/s

$$\gamma_{root} = \left[r_m - \frac{h}{2}\right] = 22.5\text{cm}$$

$$\gamma_{tip} = \left[r_m + \frac{h}{2}\right] = 27.5\text{cm}$$

$C_{t2m} \times r_m = C_{t2root} \times r_{root} = C_{t2,tip} \times r_{tip}$

$C_{t2,root} = 725.97$ m/s

$C_{t2,tip} = 593.98$ m/s

$$\cos\alpha_{2,m} = \frac{C_a}{C_2}$$

$C_a = 347.4$ m/s

$$\alpha_{2,root} = \tan^{-1}\left(\frac{C_{t2,root}}{C_a}\right) = 64.43°$$

$$\alpha_{2,tip} = \tan^{-1}\left(\frac{C_{t2,tip}}{C_a}\right) = 59.68°$$

$$\cos_{\alpha_{2,root}} = \frac{Ca}{C_{2,root}}$$

$C_{2,root} = 804.89$ m/s

$$\cos_{\alpha_{2,tip}} = \frac{Ca}{C_{2,tip}}$$

49. $\Rightarrow A = \frac{\pi}{4}d^2 = \frac{\pi}{4} \times (2.5)^2 = 4.909\text{m}^2$

$$\Rightarrow u = 500 \text{ kmph} = \frac{500 \times 1000}{3600} = 138.89\text{m/s}$$

$$\Rightarrow \sigma = 0.75 = \frac{u}{C_j}$$

$C_j = 185.19$ m/s

Velocity of air flow at the propeller (c)

$$\Rightarrow C = \frac{1}{2}(u + C_j) = 162.04\text{m/s}$$

$$\Rightarrow \dot{m}_a = \rho AC = 397.7\text{kg/s}$$

$$f = \dot{m}_a(C_j - u) = 18.41\text{kN}$$

52. Weight flow rrate of propellant $(\dot{w}_P) = \dot{m}_P \times g$

= 4 × 9.81 = 39.24 N/s

Effective jet velocity:

$$f = \dot{m}_P \times C_j$$

$$C_j = \frac{7000}{4} = 1750\text{m/s}$$

$$f = \dot{m}_P C_a + (P_e - P_a)A_e$$

$7000 = 4 \times C_a + (1 - 0.9) \times 10^5 \times 70 \times 10^{-4}$

C_a = actual jet velocity = 1732 m/s

53. Thrust coefficient, $C_f = \dfrac{f}{P_0 A^*}$

$$C_f = \gamma\left(\frac{2}{\gamma-1}\right)^{1/2}\left(\frac{2}{\gamma+1}\right)^{\frac{\gamma+1}{2(\gamma-1)}} \times \sqrt{1-\left(\frac{Pe}{P_0}\right)^{\frac{\gamma-1}{\gamma}}}$$

→ If does not depends on temperature ratio.

→ for $\gamma = 1.2$ and $\dfrac{P_o}{P_e} = 5$

$C_f = 1.09$

54. $I_{SP} = \dfrac{f}{\dot{w}_P} = \left[\dfrac{\dot{m}_p \times C_e}{\dot{m}_p \times g}\right]$

$C_e = g \times J_{SP} = 10 \times 200 = 2000$ m/s

$$V_b = C_e \ln\frac{1}{MR} - gt_p$$

$$\frac{V_b}{C_e} = \ln\frac{1}{MR} - \frac{gt_p}{C_e}$$

$$0.2 = \ln\frac{1}{MR} - \frac{10\times10}{2000}$$

$$\ln\frac{1}{MR} = 0.2 + \frac{100}{2000}$$

$MR = 0.78$

Propellant mass fraction $(\xi) = 1 - MR = 0.22$

55. $C^* = \dfrac{\text{Effective jet velocity}}{\text{thrust co-efficient} + (C_f)}$

$$C_f = \frac{f}{P_o \times A^*}$$

$$I_{SP} = \frac{f}{\dot{w}_P}$$

$f = 130 \times 45 = 5850$ N

$$C_f = \frac{5850}{25\times10^5 \times 20\times10^{-4}} = 1.17$$

$f = \dot{m}_P \times V_c$

$$V_e = \frac{f}{\dot{m}_P}$$

$$= \frac{f\times g}{\dot{w}_P} = \frac{5850\times10}{45}$$

$V_e = 1300$ m/s

$$V^* = \frac{V_e}{C_f} = \frac{1300}{1.17} = 1111.11 \text{m/s}$$

56. at 11 km altitude, $P_e = P_a$

→ with increase in altitude ambient pressure (P_a) reduces. At altitude grater than 11 km $P_e > P_a$ (underexpanded)

→ At altitude lower than 11 km, $P_e < P_a$ (overexpanded)

→ Since At 11 km, P_e = Pa so nozzle is unchoked.

57. Regression rate $(\dot{r}) = \left[\dfrac{\text{Length of motor}}{\text{Time}}\right]$

$$= \frac{1000}{200} = 5 \text{ mm/s}$$

Density of propellant $= \left[\dfrac{\text{mass}}{\text{volume of motor}}\right]$

$$= \frac{\text{mass}}{A\times L} = \left[\frac{750}{0.5\times1}\right] = 1500 \text{kg/m}^3$$

$$\dot{m}_P = \rho_\rho \times A_{CS} \times \dot{r}$$

$$= 1500 \times 0.5 \times \frac{5}{1000} = 3.75 \text{kg/s}$$

$f = \dot{m}_P \times V_e = \dot{m}_P \times g \times J_{SP}$

$= 3.75 \times 10 \times 200$

$= 75000$ N

58. $\Delta V_b = V_e \ln\dfrac{m_i}{m_f}$

$$= Ve \ln\frac{1}{\frac{m_f}{m_i}} = -V_b \ln\frac{m_f}{m_i}$$

$$\frac{m_f}{m_i} = \frac{m_L + m_S}{m_L + m_S + m_P}$$

$$\Delta V_b = -V_b \ln\frac{m_L + m_S}{m_L + m_S + m_P}$$

$$\frac{m_f}{m_i} = \frac{m_L + m_S}{m_i} = \left[\frac{(m_L + m_S)(m_S + m_P)}{m_i \times (m_S + m_P)}\right]$$

$$\frac{m_f}{m_i} = \frac{m_L m_S + m_L m_P + m_S^2 + m_S m_P}{m_i \times (m_S + m_P)}$$

$$= \frac{m_S(m_L + m_S + m_P)}{m_i \times (m_S + m_P)} + \frac{m_L}{m_i} \times \frac{m_P}{(m_S + m_P)}$$

$$= \frac{m_S}{m_S + m_P} + \frac{m_L}{m_i}\left(1 - \frac{m_S}{(m_S + m_P)}\right)$$

$= \sigma + \lambda(1-\sigma)$

$$\Delta V_b = V_e \ln\frac{1}{[\sigma + \lambda(1-\sigma)]}$$

59. The Kinetic Energy changes as the speed of satellite changes. The angular momentum remains same, but it is vector product of radial vector from earth to satellite and its velocity vector multiplied by the mass of satellite. Hence, the scaler product of speed and distance need not be constant. By Kepler's second law the rate of area swept by satellite from the center of earth remains constant.

60. Potential Energy $= \frac{-GMm}{R}$

Kinetic Energy $= \frac{1}{2}mu^2 = \frac{1}{2}m\frac{gR_e^2}{R} = \frac{GMm}{2R}$

Total Energy $= \frac{-GMm}{R} + \frac{GMm}{2R} = \frac{-GMm}{2R}$

Potential energy increases as satellite goes away from Earth and kinetic Energy decreases.

$h_A > h_B$

$E_A > E_B$, $T_A < T_B$

61. Escape velocity of the object is given by

$$ve = \sqrt{\frac{2GM}{r}}$$

where, M is the mass of the Earth r is the distance from Earth's center.

62. According to ISA model temperature remain constant in stratosphere and so does speed of sound and dynamic viscosity, which depends only on temperature. Pressure and density decreases with altitude.

63.

e	*Orbit*
0	Circular
0<e<1	Elliptical
1	Parabolic
>1	Hyperbolic

Since, eccentricity $e \geq 1$ thus according to the above table

The orbit can be parabolic or hyperbolic

64. Supercritical airfoil is a general nomenclature for 8 series airfoils. It has its maximum thickness close to the leading edge. In the transonic range the supercritical airfoil is designed to delay the wave drag.

65. The velocity of a moving gas is measured by: -

Hot wire Anemometry : They are used to perform detailed, high frequency or high resolution flow field velocity measurements. Also, they are used to measure the variable that occurs in the turbulent flows such as mean and fluctuating velocity components, mean and fluctuating temperature and vorticity.

Laser Doppler Velocimetry : The frequency shift of a laser beam that is traversed by a moving object helps in the determination of velocity by using LDVs.

Particle image Velocimetry : It is a photographic technique that measures the velocity by determining the displacement of the seeded particles over time.

Dot Matrix : This is a mechanical method of velocity measurement in which the body is wrapped with a paper over which ink droplets are there and as the flow comes, the ink droplets move with the flow forming a line. This line is measured with reference line which tells us about the velocity and direction of the flow.

SOLVED PAPER - 2020

GENERAL APTITUDE

Q 1 – 5 carry one mark each.

1. The untimely loss of life is a cause of serious global concern as thousands of people get killed _____ accidents every year while many other die ______ diseases like cardio vascular disease, cancer, etc.

(*a*) in,of (*b*) from, of

(*c*) during, from (*d*) from, from

2. He was not only accused of theft______of conspiracy.

(*a*) rather (*b*) but also

(*c*) but even (*d*) rather than

3. Select the word that fits the analogy :

Explicit : Implicit : : Express :_____

(*a*) Impress

(*b*) Repress

(*c*) Compress

(*d*) Suppress

4. The Canadian constitution requires that equal importance be given to English and French. Last year, Air Canada lost a lawsuit, and had to pay a six-figure fine to a French-speaking couple after they filed complaints about formal in-flight announcements in English lasting 15 seconds, as opposed to informal 5 second messages in French.

The French-speaking couple were upset at ______.

(*a*) the in-flight announcements being made in English.

(*b*) the English announcements being clearer than the French ones.

(*c*) the English announcements being longer than the French ones.

(*d*) equal importance being given to English and French.

5. A supperadditive function $f(\cdot)$ satisfies the following property

$$f(x_1 + x_2) \geq f(x_1) + f(x_2)$$

Which of the following functions is a supperadditive function for > 1?

(*a*) e^X

(*b*) $\sqrt{x}$

(*c*) 1/x

(*d*) e^{-X}

Q 6 - Q 10 carry two marks each.

6. The global financial crisis in 2008 is considered to be the most serious world -wide financial crisis, which started with the sub-prime lending crisis in USA in 2007. The sub-prime lending crisis led to banking crisis in 2008 with the collapse of Lehman Brothers in 2008. The sub-prime lending refers to the provision of loans to those borrowers who may have difficulties in repaying loans, and it arises because of excess liquidity following the east Asian crisis.

Which one of the following sequences show the correct precedence as per the given passage?

(*a*) East Asian crisis → subprime lending crisis → banking crisis → global financial crisis.

(*b*) Subprime lending crisis → global financial crisis → banking crisis → East Asian crisis.

(*c*) Banking crisis → subprime lending crisis → global financial crisis → East Asian crisis..

(*d*) Global financial crisis → East Asian crisis → banking crisis → subprime lending crisis.

7. It is a quarter past three in your watch. The angle between the hour hand and the minute hand is

(*a*) 0° (*b*) 7.5°

(*c*) 15° (*d*) 22.5°

8. A circle with centre O is shown in figure. A rectangle PQRS of maximum possible area in inscribed in the circle. If the radius of the circle is *a*, then the area of the shaded portion is_______.

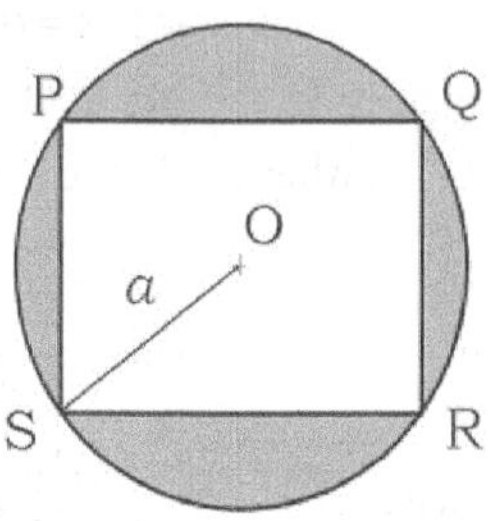

(*a*) $\pi a^2 - a^2$ (*b*) $\pi a^2 - \sqrt{2}a^2$

(*c*) $\pi a^2 - 2a^2$ (*d*) $\pi a^2 - 3a^2$

9. a, b, c are the real numbers. The quadratic equation $ax^2 - bx + c = 0$ has equal roots, which is β, then

(*a*) $\beta = b/a$ (*b*) $\beta^2 = ac$

(*c*) $\beta^3 = bc/(2a^2)$ (*d*) $b^2 \neq 4ac$

10. The following figure shows the data of students enrolled in 5 years (2014 to 2018) for two schools P and Q. During this period, the ratio of the average number of the students enrolled in school P to the average of the difference of the number of students enrolled in schools P and Q is _____.

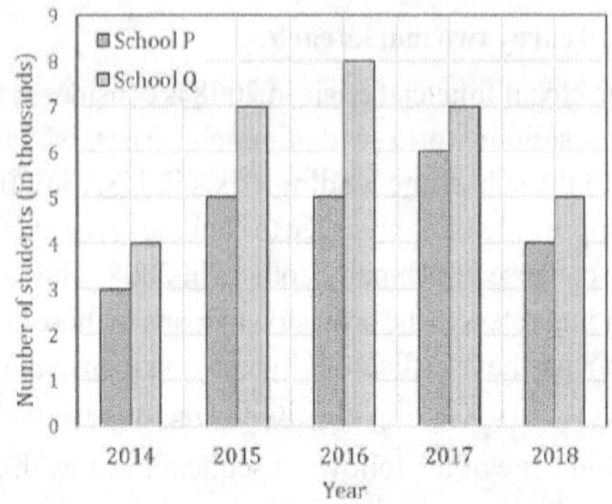

(*a*) 8 : 23 (*b*) 23 : 8

(*c*) 23 : 31 (*d*) 31 : 23

AE : AEROSPACE ENGINEERING

Q 1 – 25 carry one mark each.

1. For f(x) = |x|, with $\frac{df}{dx}$ denoting the derivative, the mean value theorem is not applicable because

(*a*) f(x) is not continuous at x = 0

(*b*) f(x) = 0 at x = 0

(*c*) $\frac{df}{dx}$ is not defined at x = 0

(*d*) $\frac{df}{dx} = 0$ at x = 0

2. For the function $f(x) = \frac{e^{-\lambda}}{\sigma\sqrt{2\pi}}$, where $\lambda = \frac{1}{2\sigma^2}(x-\mu)^2$, and σ and μ are constants, the maximum occurs at

(*a*) $x = \sigma$ (*b*) $x = \sigma\sqrt{2\pi}$

(*c*) $x = 2\sigma^2$ (*d*) $x = \mu$

3. $y = Ae^{mx} + Be^{-mx}$, where A, B and m are constants, is a solution of

(*a*) $\frac{d^2y}{dx^2} - m^2y = 0$ (*b*) $A\frac{d^2y}{dx^2} + m^2y = 0$

(*c*) $B\frac{d^2y}{dx^2} + Ay = 0$ (*d*) $\frac{d^2y}{dx^2} + my = m^2$

4. Which of the following statements is true about the effect of increase in temperature on dynamic viscosity of water and air, at room temperature?

(*a*) It increases for both water and air.

(*b*) It increases for water and decreases for air.

(*c*) It decreases for water and increases for air.

(*d*) It decreases for both water and air.

5. Given access to the complete geometry, surface pressure and shear stress distribution over a body placed in a uniform flow, one can estimate

(*a*) the moment coefficient, and the force on the body.

(*b*) the force coefficient, and the force on the body.

(*c*) the moment coefficient, and the moment of the body.

(*d*) the force and the moment on the body.

6. A pair of infinitely long, counter-rotating line vortices of the same circulation strength Γ are situated a distance h apart in a fluid, as shown in the figure. The vortices will

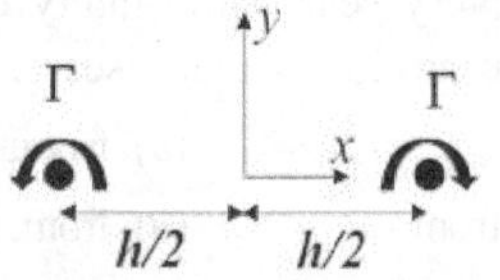

(*a*) rotate counter-clockwise about the midpoint with the tangential velocity at the line vortex equal to $\frac{\Gamma}{2\pi h}$

(*b*) rotate counter-clockwise about the midpoint with the tangential velocity at the line vortex equal to $\frac{\Gamma}{4\pi h}$

(*c*) translate along + y direction with velocity at the line vortex equal to $\frac{\Gamma}{2\pi h}$

(*d*) translate along + y direction with velocity at the line vortex equal to $\frac{\Gamma}{4\pi h}$

7. The streamlines of a steady two dimensional flow through a channel of height 0.2 m are plotted in the figure, where ψ is the stream function in m^2/s. The volumetric flow rate per unit depth is

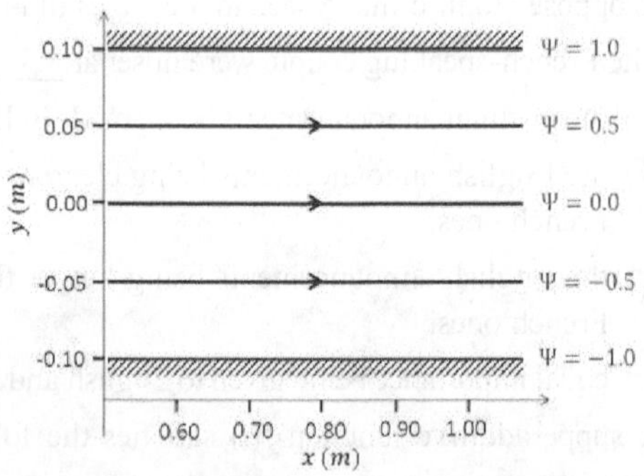

(*a*) $1.0\ m^2/s$ (*b*) $2.0\ m^2/s$

(*c*) $0.5\ m^2/s$ (*d*) $0.1\ m^2/s$

8. Which of the following options can result in an increase in the Mach number of a supersonic flow in a duct?

(*a*) Increasing the length of the duct

(*b*) Adding heat to the flow

(*c*) Removing heat from the flow

(*d*) Inserting a convergent-divergent section with the same cross-sectional area at its inlet and exit planes

9. Which one of the following conditions needs to be satisfied for $\phi = Ax^4 + By^4 + Cxy^3$ to be considered as an Airy's stress function?

(*a*) $A - B = 0$ (*b*) $A + B = 0$

(*c*) $A - C = 0$ (*d*) $A + C = 0$

10. Consider the plane strain field given by $\varepsilon_{xx} = Ay^2 + x$, $\varepsilon_{yy} = Ax^2 + y$, $\gamma_{xy} = Bxy + y$. The relation between A and B needed for this strain field to satisfy the compatibility condition is

(*a*) $B = A$ (*b*) $B = 2A$

(*c*) $B = 3A$ (*d*) $B = 4A$

11. For hyperbolic trajectory of a satellite of mass m having velocity V at a distance r from the center of earth (G: gravitational constant, M: mass of earth), which one of the following relations is true?

(*a*) $\frac{1}{2}mV^2 > \frac{GMm}{r}$ (*b*) $\frac{1}{2}mV^2 < \frac{GMm}{r}$

(*c*) $\frac{1}{2}mV^2 = \frac{GMm}{r}$ (*d*) $\frac{1}{2}mV^2 < \frac{2GMm}{r}$

12. For conventional airplanes, which one of the following is true regarding roll control derivative $\left(C_{l\delta_r} = \frac{\partial C_l}{\partial \delta_r}\right)$ and yaw control derivative $\left(C_{n\delta_r} = \frac{\partial C_n}{\partial \delta_r}\right)$, where δ_r is rudder deflection?

(*a*) $C_{l\delta_r} > 0$ and $C_{n\delta_r} < 0$ (*b*) $C_{l\delta_r} < 0$ and $C_{n\delta_r} > 0$

(*c*) $C_{l\delta_r} < 0$ and $C_{n\delta_r} < 0$ (*d*) $C_{l\delta_r} > 0$ and $C_{n\delta_r} > 0$

13. The ratio of exit stagnation pressure to inlet stagnation pressure across the rotating impeller of a centrifugal compressor, operating with a closed exit, is

(*a*) 0 (*b*) 1

(*c*) >1 (*d*) 0.5

14. Which one of the following is a hypergolic propellant combination used in rocket engines?

(*a*) Liquid hydrogen – liquid oxygen

(*b*) Unsymmetrical dimethyl hydrazine – nitrogen tetroxide

(*c*) Rocket fuel RP-1 – liquid oxygen

(*d*) Liquid hydrogen – liquid fluorine

15. In aircraft engine thermodynamic cycle analysis, *perfectly expanded flow* in the nozzle means that the static pressure in the flow at the nozzle exit is equal to

(*a*) the stagnation pressure at the engine inlet.

(*b*) the stagnation pressure at the nozzle exit.

(*c*) the ambient pressure at the nozzle exit.

(*d*) the static pressure at the nozzle inlet.

16. Three long and slender aluminum bars of identical length are subjected to an axial tensile force. These bars have circular, triangular and rectangular cross sections, with same cross sectional area. If they yield at F_{circle}, $F_{triangle}$ and $F_{rectangle}$, respectively, which one of the following is true?

(*a*) $F_{circle} > F_{triangle} > F_{rectangle}$

(*b*) $F_{circle} < F_{triangle} < F_{rectangle}$

(*c*) $F_{triangle} > F_{circle} > F_{rectangle}$

(*d*) $F_{circle} = F_{triangle} = F_{rectangle}$

17. The positive high angle-of-attack condition is obtained in a steady pull-out maneuver at the largest permissible angle-of-attack of the wing. Under this condition, at which of the following regions of the wing does the maximum tension occur?

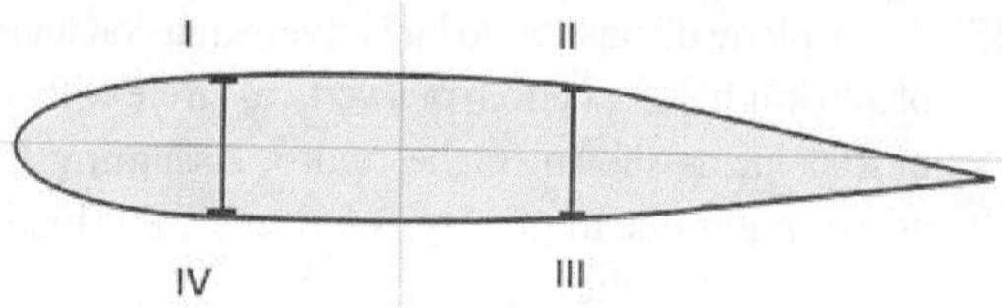

(*a*) I (*b*) II

(*c*) III (*d*) IV

18. The natural frequency of the first mode of a rectangular cross section cantilever aluminum beam is ω rad/s. If the material and cross-section remain the same, but the length of the beam is doubled, the first mode frequency will become

(*a*) $\frac{\omega}{4}$ rad/s (*b*) 4 ω rad/s

(*c*) $\frac{\omega}{16}$ rad/s (*d*) 16 ω rad/s

19. Given $A = \begin{pmatrix} \sin\theta & \tan\theta \\ 0 & \cos\theta \end{pmatrix}$, the sum of squares of eigen values of A is

(*a*) $\tan^2\theta$ (*b*) 1

(*c*) $\sin^2\theta$ (*d*) $\cos^2\theta$

20. Burnout velocity of a space vehicle in a circular orbit at an angle 5 degrees above the local horizon around earth is 13.5 km/s. Tangential velocity of the space vehicle in the orbit is _____ km/s (*round off to two decimal places*).

21. Velocity of an airplane in the body fixed axes is given as [100 – 10 20] m/s. The sideslip angle is ________ degress (*round off to two decimal places*).

22. The similarity solution for the diffusion equation, $\frac{\partial u}{\partial t} = \alpha \frac{\partial^2 u}{\partial x^2}$ is $u(x, t) = u(\eta)$, where similarity variable, $\eta = \frac{x}{\sqrt{\alpha\, t}}$. If $u(x, 0) = e^{-x^2}$ the ratio, $\frac{u(0,1)}{u(0,4)} =$ _______ (*round off to one decimal place*).

23. Air enters the rotor of an axial compressor stage with no pre-whirl ($C_\theta = 0$) and exits the rotor with whirl velocity, $C_\theta = 150$ m/s. The velocity of rotor vanes, U is 200 m/s. Assuming $C_p = 1005$ J/(kg K), the stagnation temperature rise across the rotor is _______ K (*round off to one decimal place*).

24. A thin walled beam of constant thickness shown in the figure is subjected to a torque of 3.2 kNm. If the shear modulus is 25 GPa, the angle of twist per unit length is _____ rad/m (*round off to three decimal places*).

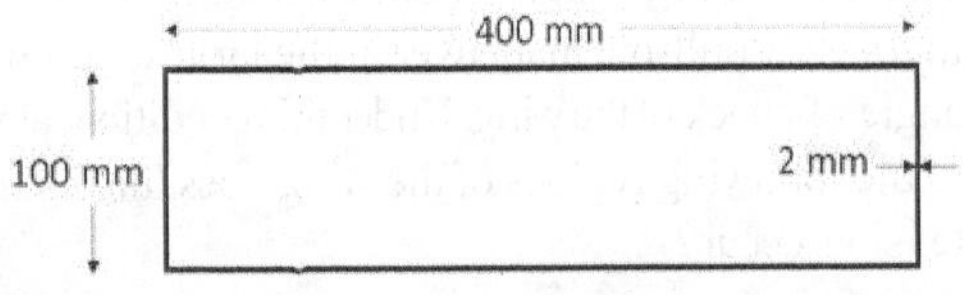

25. An airplane of mass 5000 kg is flying at a constant speed of 360 km/h at the bottom of a vertical circle with a radius of 400 m, as shown in the figure. Assuming that the acceleration due to gravity is 9.8 m/s^2, the load factor experienced at the center of gravity of the airplane is _____ (*round off to two decimal places*).

Q 26 – 55 carry two marks each.

26. The equation $x\frac{dx}{dy} + y = c$, where c is a constant, represents a family of

(*a*) exponential curves (*b*) parabolas

(*c*) circles (*d*) hyperbolas

27. A wedge shaped airfoil is placed in a supersonic flow as shown in the figure (not to scale). The corners of the wedge are at $x = x_A$, $x = x_B$, $x = x_C$, respectively.

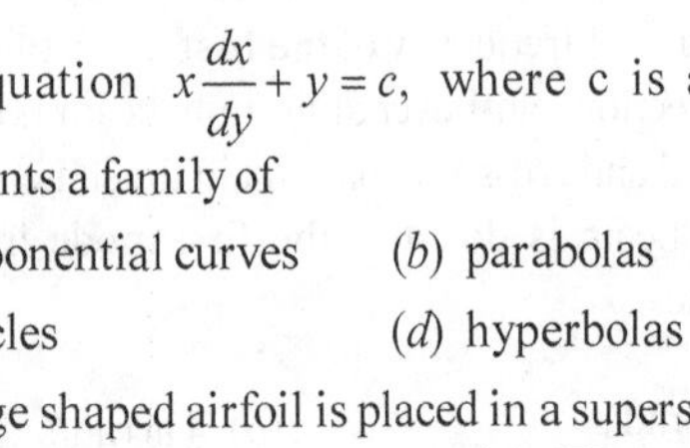

Which one of the following represents the correct static pressure profiles along $y = Y_I$ and $y = Y_{II}$?

(*a*)

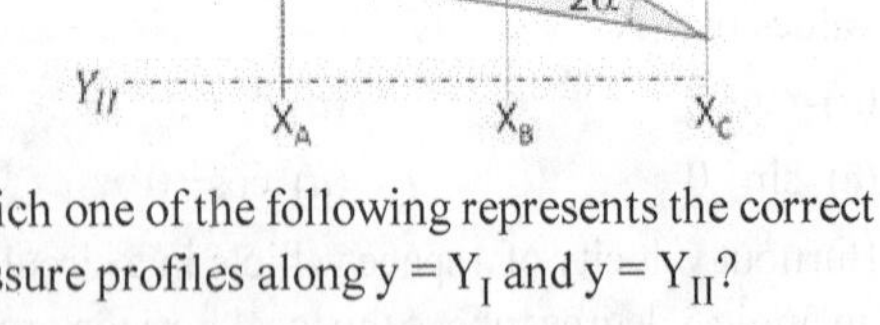

(*b*)

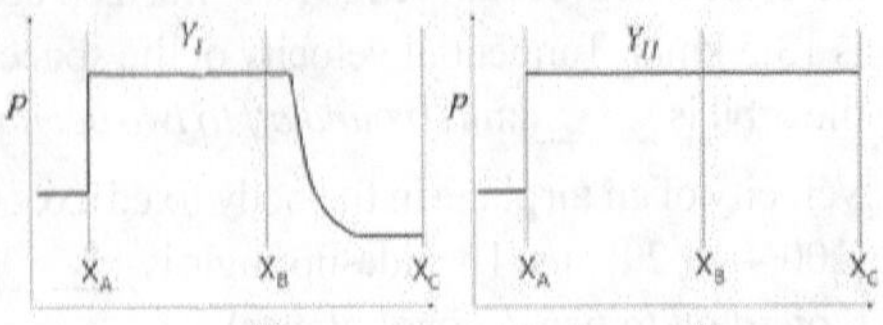

(*c*)

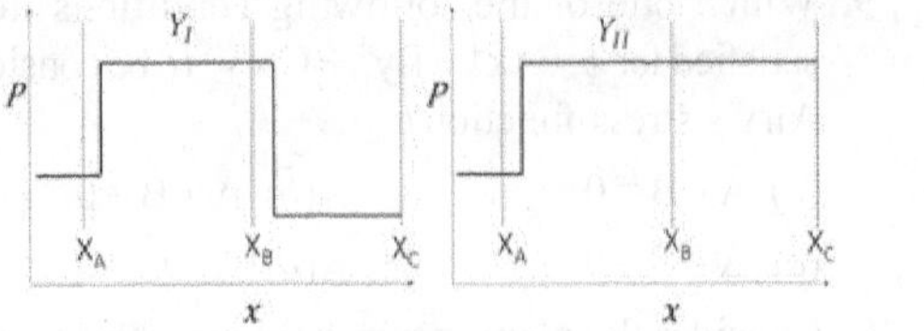

(*d*)

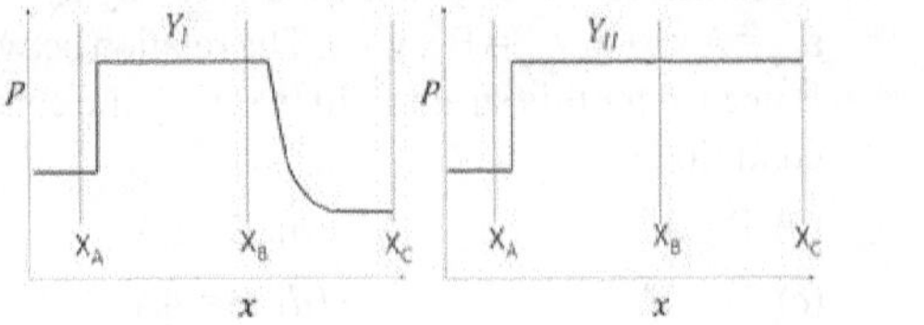

28. The value of Poisson's ratio at which the shear modulus of an isotropic material is equal to the bulk modulus is

(*a*) $\frac{1}{2}$ (*b*) $\frac{1}{4}$

(*c*) $\frac{1}{6}$ (*d*) $\frac{1}{8}$

29. A load P is applied to the free end of a stepped cantilever beam as shown in the figure. The Young's modulus of the material is E, and the moments of inertia of the two sections of length 2 m and 1 m are I and 3I, respectively. Ignoring transverse shear and stress concentration effects, the deflection at the point where the load is applied at the free end of the cantilever is

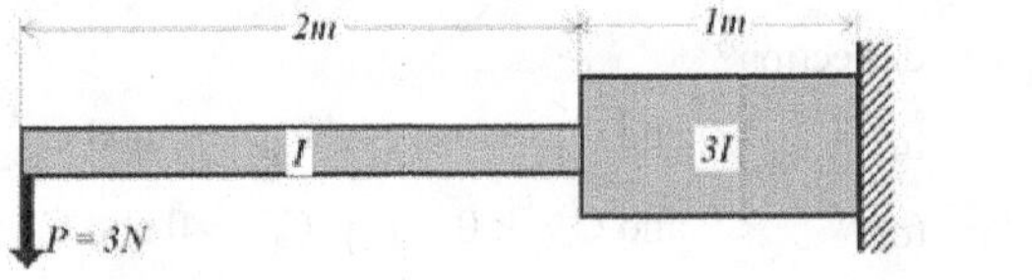

(*a*) $\frac{23}{243EI}$ (*b*) $\frac{1}{3EI}$

(*c*) $\frac{43}{3EI}$ (*d*) $\frac{23}{3EI}$

30. The three dimensional strain-stress relation for an isotropic material, written in a general matrix form, is

$$\begin{Bmatrix} \varepsilon_{xx} \\ \varepsilon_{yy} \\ \varepsilon_{zz} \\ \gamma_{yz} \\ \gamma_{xz} \\ \gamma_{xy} \end{Bmatrix} = \begin{bmatrix} A & C & C & 0 & 0 & 0 \\ C & A & C & 0 & 0 & 0 \\ C & C & A & 0 & 0 & 0 \\ 0 & 0 & 0 & B & 0 & 0 \\ 0 & 0 & 0 & 0 & B & 0 \\ 0 & 0 & 0 & 0 & 0 & B \end{bmatrix} \begin{Bmatrix} \sigma_{xx} \\ \sigma_{yy} \\ \sigma_{zz} \\ \tau_{yz} \\ \tau_{xz} \\ \tau_{xy} \end{Bmatrix}$$

A, B and C are compliances which depend on the elastic properties of the material. Which one of the following is correct?

(*a*) $C = \frac{A}{2} - B$ (*b*) $C = \frac{A}{2} + B$

(*c*) $C = A + \frac{B}{2}$ (*d*) $C = A - \frac{B}{2}$

31. For three different airplanes A, B and C, the yawing moment coefficient (C_n) was measured in a wind-tunnel for three settings of sideslip angle β and tabulated as

	Airplane A	Airplane B	Airplane C
$\beta = -5$ deg	−0.030	−0.025	0.040
$\beta = 0$ deg	0	0	0
$\beta = 5$ deg	0.030	0.025	−0.040

Which one of the following statements is true regarding directional static stability of the airplanes A, B and C?

(*a*) All three airplanes A, B and C are stable.

(*b*) Only airplane C is stable, while both A and B are unstable.

(*c*) Airplane C is unstable, A and B are stable with A being more stable than B.

(*d*) Airplane C is unstable, A and B are both stable with A less stable than B.

32. A closed curve is expressed in parametric form as $x = a \cos\theta$ and $y = b \sin\theta$, where $a = 7$ m and $b = 5$m. Approximating $\pi = \frac{22}{7}$, which of the following is the area enclosed by the curve?

(*a*) 110 m^2 (*b*) 74 m^2

(*c*) 35 m^2 (*d*) 144 m^2

33. An axial compressor is designed to operate at a rotor speed of 15000 rpm and an inlet stagnation temperature of 300 K. During compressor testing, the inlet stagnation temperature of the compressor measured was 280 K. What should be the rotor speed for the compressor to develop the same performance characteristics during this test as in the design condition?

(*a*) 14000 rpm (*b*) 14491 rpm

(*c*) 15526 rpm (*d*) 16071 rpm

34. For the state of stress shown in the figure, which one of the following represents the correct free body diagram showing the maximum shear stress and the associated normal stresses?

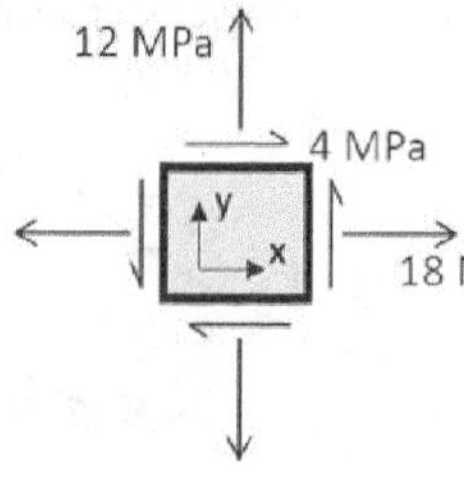

(*a*)

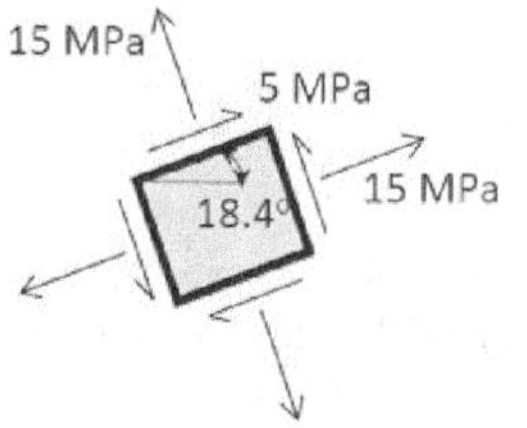

(*b*)

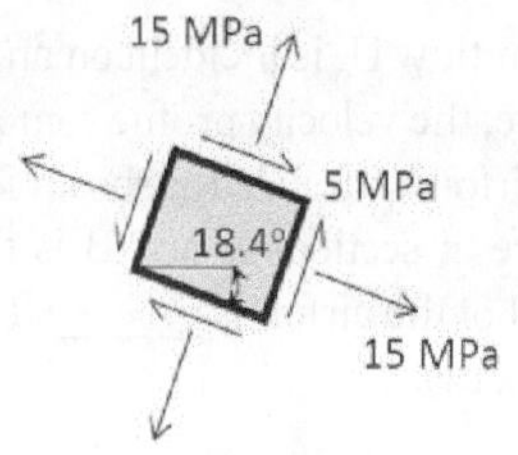

(*c*)

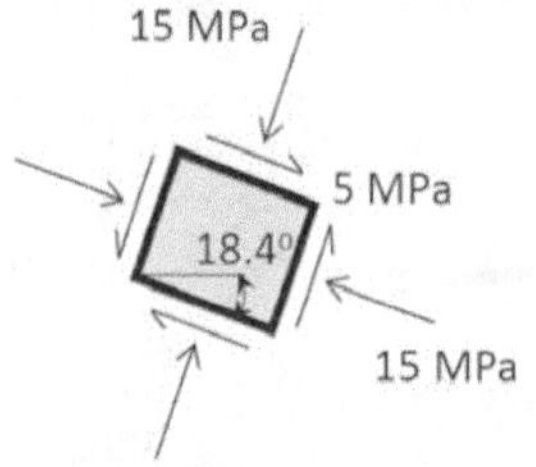

(*d*)

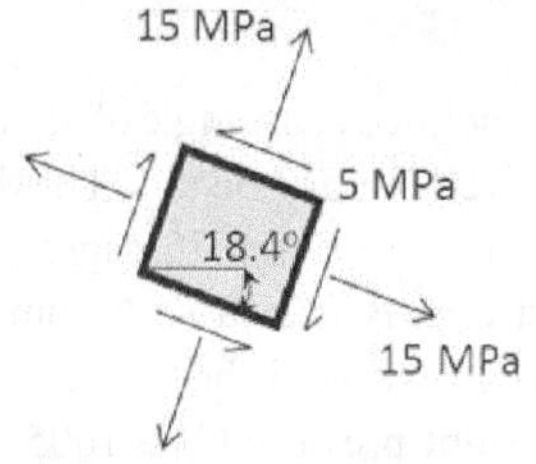

35. In the equation $AX = B$, $A = \begin{bmatrix} \frac{1}{\sqrt{2}} & 0 & \frac{1}{\sqrt{2}} \\ 0 & 1 & 0 \\ \frac{1}{\sqrt{2}} & 0 & \frac{-1}{\sqrt{2}} \end{bmatrix}$, $X = \begin{bmatrix} x \\ y \\ z \end{bmatrix}$

$B = \begin{bmatrix} 0 \\ 1 \\ -\sqrt{2} \end{bmatrix}$, where A is an orthogonal matrix, the sum of the unknowns, $x + y + z =$ _____ (*round off to one decimal place*).

36. If $\int_0^1 (x^2 - 2x + 1)dx$ is evaluated numerically using trapezoidal rule with four intervals, the difference between the numerically evaluated value and the analytical value of the integral is equal to _______ (*round off to three decimal places*).

37. The table shows the lift characteristics of an airfoil at low speeds. The maximum lift coefficient occurs at 16 degrees.

Angle of attack α (in degrees)	Lift coefficient C_l
0	0.10
4	0.53

Using Prandtl-Glauert rule, the lift coefficient for the airfoil at the angle of attack of 6 degrees and free stream Mach number of 0.6 is ______ (*round off to two decimal places*).

38. A low speed uniform flow U_0 is incident on an airfoil of chord c. In the figure, the velocity profile some distance downstream of the airfoil is idealized as shown for section B. The staticpressure at sections A and B is the same. The drag coefficient of the airfoil is _______ *(round off to three decimal places).*

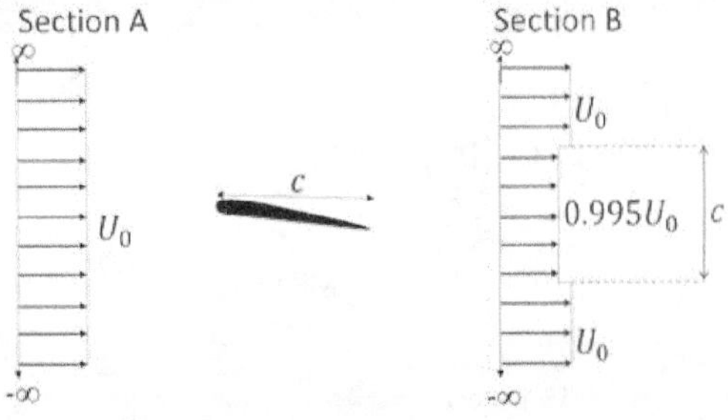

39. An oblique shock is inclined at an angle of 35 degrees to the upstream flow of velocity 517.56 m/s. The deflection of the flow due to this shock is 5.75 degrees and the temperature downstream is 182.46 K. Assume the gas constant R = 287 J/(kg K), specific heat ratio $\gamma = 1.4$, and specific heat at constant pressure C_p = 1005 J/(kg K). Using conservation relations, the Mach number of the upstream flow can be obtained as ______ *(round off to one decimal place).*

40. The thickness of a laminar boundary layer (δ) over a flat plate is, $\frac{\delta}{x} = \frac{5.2}{\sqrt{Re_x}}$, where x is measured from the leading edge along the length of the plate. The velocity profile within the boundary layer is idealized as varying linearly with y. For freestream velocity of 3 m/s and kinematic viscosity of 1.5×10^{-5} m²/s, the displacement thickness at 0.5 m from the leading edge is ________ mm *(round off to two decimal places).*

41. A wing of 15 m span with elliptic lift distribution is generating a lift of 80 kN at a speed of 90 m/s. The density of surrounding air is 1.2 kg/m³. The induced angle of attack at this condition is ________ degrees *(round off to two decimal places).*

42. A solid circular shaft, made of ductile material with yield stress σ_y = 280 MPa, is subjected to a torque of 10 kNm. Using the Tresca failure theory, the smallest radius of the shaft to avoid failure is ______ cm *(round off to two decimal places).*

43. The ratio of tangential velocities of a planet at the perihelion and the aphelion from the sun is 1.0339. Assuming that the planet's orbit around the sun is planar and elliptic, the value of eccentricity of the orbit is _______ *(round off to three decimal places).*

44. The eigenvalues for phugoid mode of a general aviation airplane at a stable cruise flight condition at low angle of attack are $\lambda_{1,2} = -0.02 \pm i\,0.25$. If the acceleration due to gravity is 9.8 m/s², the equilibrium speed of the airplane is _______ m/s *(round off to two decimal places).*

45. For a general aviation airplane with tail efficiency $\eta = 0.95$, horizontal tail volume ratio $V_H = 0.453$, downwash angle slope $\frac{d\varepsilon}{d\alpha} = 0.35$, wing lift curve slope $C^w_{L\alpha} = 4.8$ rad^{-1}, horizontal tail lift curve slope $C^t_{L\alpha} = 4.4$ rad^{-1}, shift in neutral point location as a percentage of mean aerodynamic chord is ______ *(round off to two decimal places).*

46. A single engine, propeller driven, general aviation airplane is flying in cruise at sea-level condition (density of air at sea-level is 1.225 kg/m³) with speed to cover maximum range. For drag coefficient $C_D = 0.025 + 0.049\ C_L^2$ and wing loading W/S = 9844 N/m², the speed of the airplane is _______ m/s *(round off to one decimal place).*

47. The design flight Mach number of an ideal ramjet engine is 2.8. The stagnation temperature of air at the exit of the combustor is 2400 K. Assuming the specific heat ratio of 1.4 and gas constant of 287 J/(kg K), the velocity of air at the exit of the engine is ______ m/s *(round off to one decimal place).*

48. The oprating conditions of an aircraft engine combustor are as follows.

The rate of total enthalpy of air entering the combustor = 28.94 MJ/s.

The rate of total enthalpy of air leaving the combustor = 115.42 MJ/s.

Mass flow rate of air = 32 kg/s.

Air to fuel mass ratio = 15.6

Lower heating value of the fuel = 46 MJ/kg.

The efficiency of the combustor is ________ % *(round off to two decimal places).*

49. The figure shows the T-S diagram for an axial turbine stage.

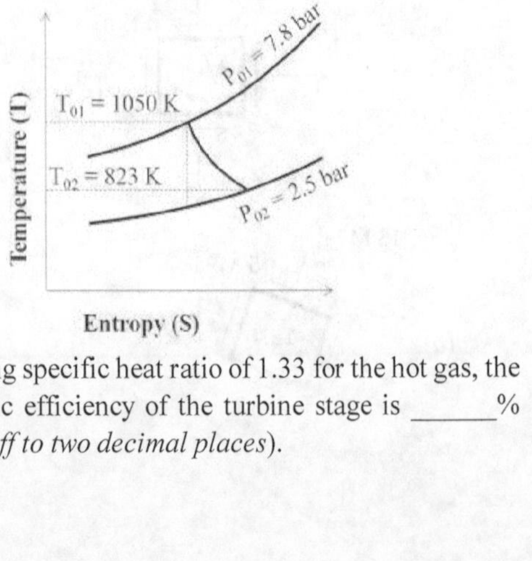

Assuming specific heat ratio of 1.33 for the hot gas, the isentropic efficiency of the turbine stage is ______% *(round off to two decimal places).*

50. A rocket engine has a sea level specific impulse of 210 s and a nozzle throat area of 0.005 m^2. While testing at sea level conditions, the characteristic velocity and pressure for the thrust chamber are 1900 m/s and 50 bar, respectively. Assume the acceleration due to gravity to be 9.8 m/s^2. The thrust produced by the rocket engine is ______ kN (*round off to one decimal place*).

51. A critically damped single degree of freedom spring-mass-damper system used in a door closing mechanism becomes overdamped due to softening of the spring with extended use. If the new damping ratio (ξ_{new}) for overdamped condition is 1.2, the ratio of the original spring stiffness to the new spring stiffness (k_{org}/k_{new}), assuming that the other parameters remain unchanged, is ______ (*round off to two decimal places*).

52. The two masses of the two degree of freedom system shown in the figure are given initial displacements of 2 cm (x_1) and 1.24 cm (x_2). The system starts to vibrate in the first mode. The first mode shape of this system is $\phi_1 = [1 \;\; a]^T$, where a = ____ (*round off to two decimal places*).

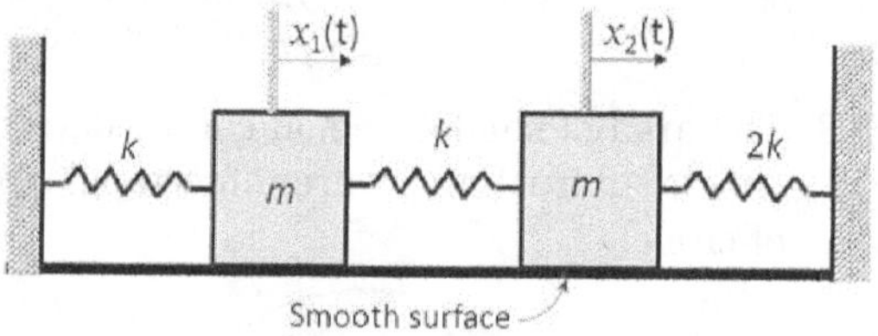

53. As shown in the figure, a beam of length 1 m is rigidly supported at one end and simply supported at the other. Under the action of a uniformly distributed load of 10 N/m, the magnitude of the normal reaction force at the simply supported end is ______ N (*round off to two decimal places*).

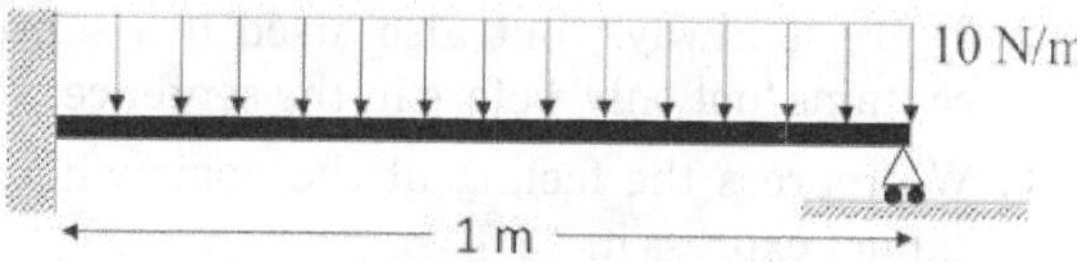

54. An airplane of mass 4000 kg and wing reference area 25 m^2 flying at sea level has a maximum lift coefficient of 1.65. Assume density of air as 1.225 kg/m^3 and acceleration due to gravity as 9.8 m/s^2. Using a factor of safety of 1.25 to account for additional unsteady lift during a sudden pull-up, the speed at which the airplane reaches a load factor of 3.2 is ______ m/s (*rounde off to two decimal places*).

55. A Pitot tube mounted on the wing tip of an airplane flying at an altitude of 3 km measures a pressure of 0.72 bar, and the outside air temperature is 268.66 K. Take the sea level conditions as, pressure = 1.01 bar, temperature = 288.16 K, and density = 1.225 kg/m^3. The acceleration due to gravity is 9.8 m/s^2 and the gas constant is 287 J/(kg K). Assuming standard atmosphere, the equivalent airspeed for this airplane is ___ m/s (round off to two decimal place).

ANSWERS

General Aptitude

1. (*a*)	**2.** (*b*)	**3.** (*b*)	**4.** (*c*)	**5.** (*a*)	**6.** (*a*)	**7.** (*b*)
8. (*c*)	**9.** (*c*)	**10.** (*b*)				

Aerospace Engineering

1. (*c*)	**2.** (*d*)	**3.** (*a*)	**4.** (*c*)	**5.** (*d*)	**6.** (*c*)	**7.** (*b*)
8. (*c*)	**9.** (*b*)	**10.** (d)	**11.** (a)	**12.** (*a*)	**13.** (c)	**14.** (b)
15. (*c*)	**16.** (d)	**17.** (c)	**18.** (a)	**19.** (b)	**20.** (13.42 to 13.47)	
21. (–5.62 to –5.57)		**22.** (1.9 to 2.1)	**23.** (29.8 to 30.0)	**24.** (0.009 to 0.011)		**25.** (3.50 to 3.60)
26. (*c*)	**27.** (d)	**28.** (*d*)	**29.** (*c*)	**30.** (*d*)	**31.** (c)	**32.** (a)
33. (b)	**34.** (b)	**35.** (0.9 to 1.1)	**36.** (0.010 to 0.012)		**37.** (0.92 to 0.94)	**38.** (0.009 to 0.011)
39. (1.9 to 2.1)	**40.** (4.00 to 4.20)		**41.** (1.29 to 1.38)	**42.** (3.55 to 3.58)	**43.** (0.016 to 0.018)	
44. (55.20 to 55.33)		**45.** (25.60 to 25.70)		**46.** (149.0 to 151.0)		**47.** (1712.0 to 1719.0)
48. (91 to 93)	**49.** (87 to 89)	**50.** (27.0 to 27.2)	**51.** (1.43 to 1.45)		**52.** (0.61 to 0.63)	
53. (3.74 to 3.76)	**54.** (62.95 to 63.08)		**55.** (57.10 to 60.00)			

EXPLANATIONS

GENERAL APTITUDE

1. It is a punctuality checking, there is always 'in' used with accidentand of used with diseases.
2. There is always 'but also' used if a sentence contains 'not only' before in the sentence.
3. We repress the feeling at the point where we cannot express it.
4. In the given section there are 15 minutes used for English announcement and only 5 minutes used for French announcement.
5. Since x > 1 then let $x_1 = 2$ and $x_2 = 3$

Let $f(x) = \frac{1}{x}$, then $f(x_1) = \frac{1}{x_1}$

$$f(2) = \frac{1}{2}$$

Similarly, $$f(3) = \frac{1}{3}$$

$$f(x_1 + x_2) = f(2+3) = f(5) = \frac{1}{5} = 0.2$$

$$f(x_1) + f(x_2) = \left(\frac{1}{2}\right) + \left(\frac{1}{3}\right) = \frac{5}{6} = 0.833$$

But for super additive function

$$f(x_1 + x_2) \geq f(x_1) + f(x_2).$$

Let $$f(x) = e^x$$

Then $$f(x_1) = e^2 \text{ and } f(x_2) = e^3$$

$$f(x_1 + x_2) = f(2+3) = f(5) = e^5 = 148.413$$

$$f(x_1) + f(x_2) = f(2) + f(3) = e^2 + e^3 = 27.475$$

Because $f(x_1 + x_2) > f(x_1) + f(x_2)$

Therefore $f(x) = e^x$ is a super additive function.

6. As we can see in the given section global crisis happened in 2008 while subprime landing crisis happened in USA in 2007. The subprime lending crisis led to the banking crisis in 2008 and subprime lending crisis happened due to excess liquidity followingthe East Asian crisis. So making correct arrangement of crisis we shall get.

East Asian crisis → subprime lending crisis → banking crisis → global financial crisis

7. Hour needle rotates 3600 in 12 hours.

Hour needle rotates $3600 \times \frac{1}{(12 \times 60)}$ in 1 minute

Hour needle rotates $3600 \times \frac{15}{(12 \times 60)}$ in 15 minutes.

The angle between the hour needle and minute needle

$$= \frac{(360 \times 15)}{(12 \times 60)} = 7.50$$

8. The area of shaded region can be determine by simple subtracting the area of rectangle from area of circle.

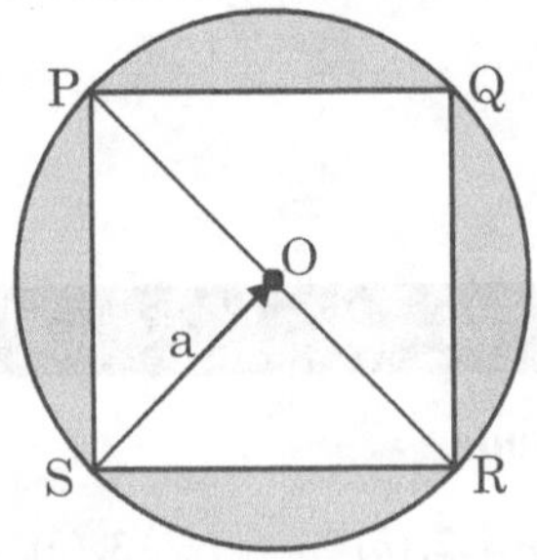

Area of the circle is πa^2 and the area of rectangle will be PQ*PS. Now we have to find out the area of rectangle so that we can determine the area of shaded region. If we draw a line or hypotaneous PR in the rectangle then we see two triangles POS and SOR are similar and their sides are equals to each other respectively.

Using Pythagoras theorem

Now we have PR = 2a and PS = SR

$$PS^2 + SR^2 = PR^2$$

$$PS^2 + PS^2 = (2a)^2$$

$$2PS^2 = 4a^2$$

$PS^2 = 2a^2$ then $PS = \sqrt{2}a$ and $PS = SR = \sqrt{2}a$

The area of shaded region

$= \text{area of circle} - \text{area of rectangle}$

$= \pi a^2 - PS \times SR$

$= \pi a^2 \times \sqrt{2}a \times \sqrt{2}a$

$= \pi a^2 - 2a^2$

9. The quadratic equation is $ax^2 - bx + c = 0$.

Let assume the equal roots are β then

Sum of roots $\beta + \beta$ = coefficient of x/coefficient of x^2

$$2\beta = \frac{b}{a}$$

Product of roots β^2 = constant value/coefficient of x^2

$$= \frac{c}{a}$$

Now $$2\beta \times \beta^2 = \frac{b}{a} \times \frac{c}{a}$$

$$\beta^3 = \frac{bc}{2a^2}$$

10. Total number of students enrolled in school P during 2014-2018 are 23

Average students in School P

$$= \frac{(3+5+5+6+4)}{5}$$

$$= \frac{23}{5}$$

Average of Difference of students enrolled in school P and school Q are 8

$$\frac{(\text{School P} - \text{School Q})}{5} = \frac{(1+2+3+1+1)}{5}$$

$$= \frac{8}{5}$$

Now the ratio will be

$$= \left(\frac{\text{Average students in school P}}{\text{Average of difference of students Enrolled in school P and school Q}} \right)$$

$$= \frac{\left(\frac{23}{5}\right)}{\left(\frac{8}{5}\right)}$$

$$= \frac{23}{8} = 23:8$$

AE : AEROSPACE ENGINEERING

1.

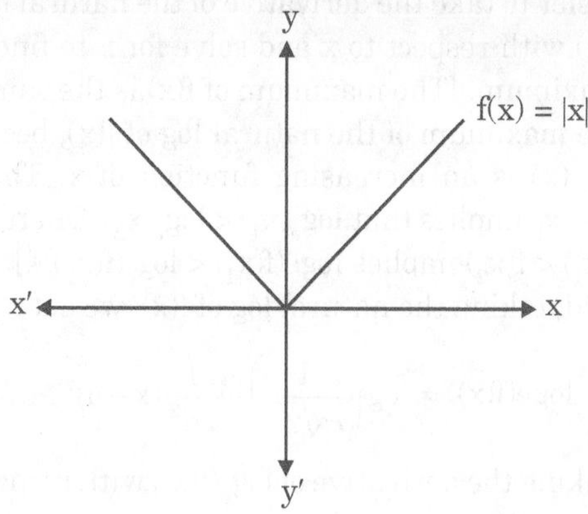

$$f(x) = |x|$$

$$|x| = \begin{cases} x, & x \ge 0 \\ -x, & x < 0 \end{cases}$$

$$f'(0) = \lim_{x\to 0} \frac{f(x) - f(0)}{x - 0}$$

$$= \lim_{x\to 0} \frac{|x| - |0|}{x - 0}$$

$$= \lim_{x\to 0} \frac{|x|}{0}$$

Check : $f(x) = |x|$

$$\lim_{x\to 0^+} \frac{|x|}{x} = \lim_{x\to 0^+} \frac{x}{x}$$

$$= \lim_{x\to 0^+}(1) = 1 \quad ...(1)$$

$$\lim_{x\to 0^-} \frac{|x|}{x} = \lim_{x\to 0^-} \frac{-x}{x}$$

$$= \lim_{x\to 0^-}(-1) = -1 \quad ...(2)$$

$$(1) \neq (2)$$

$$f'(x) = \begin{cases} 1 & \text{if } x > 0 \\ -1 & \text{if } x < 0 \end{cases}$$

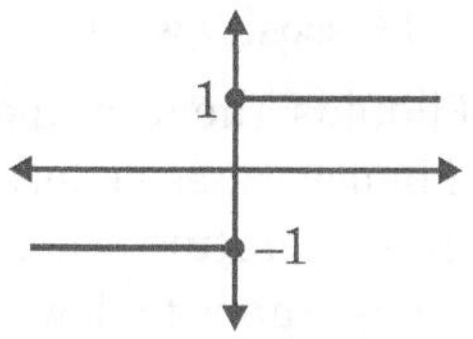

$\frac{\partial f}{\partial x}$ is not defined at x = 0 because derivative does not exist at the point

2. Differentiate f(x) with respect to x and solve for x to find the maximum. Because our f(x) contains the natural exponential function, however, it is easier to take the derivative of the natural log of f(x) with respect to x and solve for x to find the maximum. [The maximum of f(x) is the same as the maximum of the natural log of f(x), because $\log_e(x)$ is an increasing function of x. That is $x_1 < x_2$ implies that $\log_e(x_1) < \log_e(x_2)$. Therefore, $f(x_1) < f(x_2)$ implies $loge(f(x_1) < \log_e(f(x_2)))$.] That said, taking the natural log of f(x), we get :

$$loge(f(x)) = \log\left(\frac{1}{\sigma\sqrt{2\pi}}\right) - \frac{1}{2\sigma^2}(x-\mu)^2$$

Taking the derivative of $\log_e(f(x))$ with respect to x, we get :

$$\frac{d \log f(x)}{dx} = -\frac{1}{2\sigma^2}.2(x-\mu)$$

Now, setting the derivative of loge(f(x)) to 0 :

$$\frac{d \log f(x)}{dx} = -\frac{1}{2\sigma^2}.2(x-\mu) \overset{\equiv}{skr}\ 0$$

and solving for x, we get that x = μ. Taking the second derivative of $\log_e(f(x))$ with respect to x, we get :

$$\frac{d^2 \log f(x)}{dx^2} = \frac{1}{\sigma^2}$$

Because the second derivative of $\log_e(f(x))$ is negative (for all x, in fact), the point x = μ is deemed a local maximum.

3.
$$y = Ae^{mx} + Be^{-mx}$$
$$y' = mAe^{mx} + (-m)Be^{-mx}$$
$$y'' = m^2Ae^{mx} + (-m \times -m)Be^{-mx}$$
$$y'' = m^2(Ae^{mx} + Be^{-mx})$$
$$y'' = m^2y$$
$$y'' - m^2y = 0$$
$$\frac{d^2y}{dx^2} - m^2y = 0$$

4. Variation of viscosity with temperature :

In case of liquids, the inter mdecular distance is small and hence cohesive forces are large. With increase in temperature cohesive forces decrease and hence desistance to flow also decreases.

So, viscosity of a liquid decreases with increase in temperature.

In case of Gases, as the inter molecular distance is large and hence cohesive forces are negligible. With increase in temperature the molecular agitation (or) molecular disturbance increase and the resistance to flow also increase.

Viscosity of a gas increases with increase in temperature.

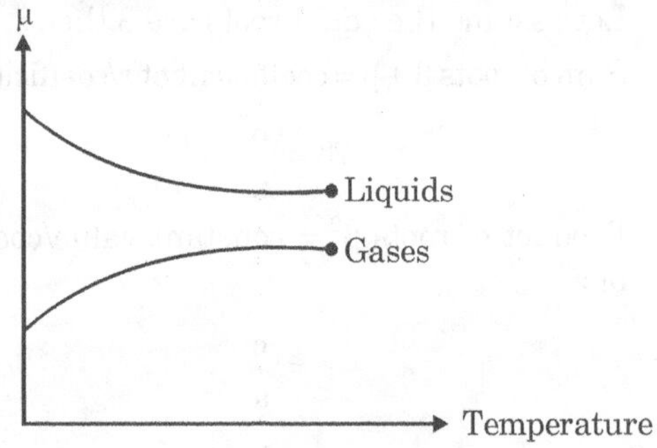

5. Given, access to complete geometry, surface pressure and shear stress distribution over a body in uniform flow, we can estimate only force and moment on body. If we want to get coefficients, we need parameter of flow like density, velocity for normalizing.

6. $\text{Velocity} = \frac{\Gamma}{4\pi h} + \frac{\Gamma}{4\pi h} = \frac{\Gamma}{2\pi h}$

7.

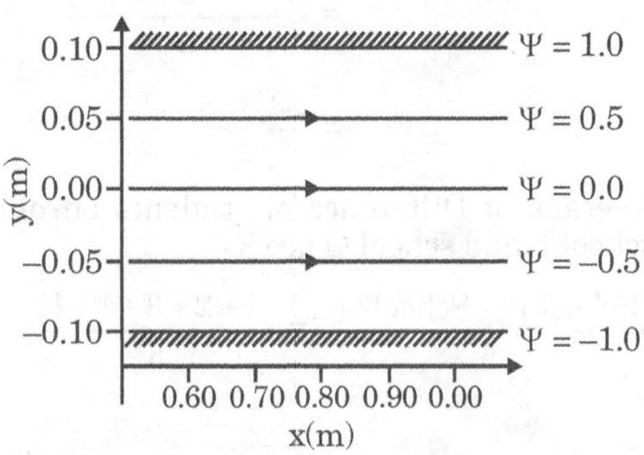

Change in stream function between two streamlines, dΨ

= Volumetric flow rate between two streamlines

$= 1 - (-1) = 2\ m^3/s$

8. From Rayleigh flow the mach number in supersonic flow can be increased by removing heat from the flow.

9.
$$\phi = Ax^4 + By^4 + Cxy^3$$

For Airy's stress function to be valid

$$\nabla^4\phi = 0$$

$$\therefore \quad \nabla^4\phi = \frac{\partial^4\phi}{\partial x^4} + \frac{\partial^4\phi}{\partial y^4} + \frac{2\partial^4\phi}{\partial x^2\partial y^2} = 0 \quad ...(1)$$

$$\therefore \quad \frac{\partial^2 \phi}{\partial x^4} = 24\text{ A}$$

$$\frac{\partial^2 \phi}{\partial y^4} = 24\text{ B}$$

$$\frac{\partial^2 \phi}{\partial x^2 \partial y^2} = 0$$

Put above values in equation (1), we get

$$24A + 24B + 0 = 0$$

$$24(A + B) = 0$$

$$A + B = 0$$

10. $\varepsilon_{xx} = Ay^2 + x;$

$$\varepsilon_{yy} = Ax^2 + y;$$

$$\gamma_{xy} = Bxy + y$$

Compatibility equation for plain strain condition is

$$\frac{\partial^2 \gamma_{xy}}{\partial x \partial y} = \frac{\partial^2 \varepsilon_{yy}}{\partial x^2} + \frac{\partial^2 \varepsilon_{xx}}{\partial y^2}$$

$$B = 2A + 2A$$

$$B = 4A$$

11. If e > 1

$\frac{1}{2}mv^2 > \frac{GMm}{r}$, Trajectory is a hyperbolic

If e = 1

$\frac{mv^2}{2} = \frac{GMm}{r}$, Trajectory is a parabola

If e < 1

$\frac{mv^2}{2} < \frac{GMm}{r}$, Trajectory is an ellipse

12. For lateral and directional stability of aircraft,

Roll control derivative, $C_{l\delta r} > 0$

Yaw control derivative, $C_{n\delta r} < 0$

13. For centrifugal compressor,

$$\frac{P_{oe}}{P_{oi}} > 1$$

14. A hypergolic propellant combination used in a rocket engine is one whose components spontaneously ignite when they come into contact with each other.

The two propellant components usually consist of a fuel and an oxidizer. Although commonly used hypergolic propellants are difficult to handle because of their extreme toxicity and/or corrosiveness, they can be stored as liquids at room temperature and hypergolic engines are easy to ignite reliably and repeatedly.

In contemporary usage, the terms "hypergol" or "hypergolic propellant" usually mean the most common such propellant combination, dinitrogen tetroxide plus hydrazine and/or its relatives monomethylhydrazine and unsymmetrical demethylhydrazine.

15. When perfectly expanded condition exit in the nozzle, means that static pressure in the flow at the exit of nozzle is equal to ambient pressure at the nozzle exit.

$P_a = P_e$ (optimum expended)

P_a = Ambient pressure

P_e = static pressure in the flow at nozzle exit.

If $P_a > P_e$ (over – expanded condition)

$P_a < P_e$ (under – expanded condition)

16. Since tensile yield strength is a material property, therefore it remains same for all the cases.

$$F_{circle} = (\sigma_{yt})_{circle} \cdot A_{circle}$$

$$F_{triangle} = (\sigma_{yt})_{triangle} \cdot A_{triangle}$$

$$F_{rectangle} = (\sigma_{yt})_{rectangle} \cdot A_{rectangle}$$

$$A_{circle} = A_{triangle} = A_{rectangle} \text{ and}$$

$$(\sigma_{yt})_{circle} = (\sigma_{yt})_{triangle} = (\sigma_{yt})_{rectangle}$$

$$\therefore \quad F_{circle} = F_{triangle} = F_{rectangle}$$

17. At positive high angle of attack the beading moment from the normal force produce compressive stress on the upper side of the wings flanges of the front spar and adjacent stringers and tensile stress at low after region of the wing crossections, were is case of positive low angle of attack wing bending moment produce the maximum compressive stress on the upper rear spar flanges and stringers and maximum tension stress on.

The lower front spar f largos and stagers.

18. For a cantilever beam subjected to free vibration, and the system is considered as continuous system in which the beam mass in considered as distributed along with the stiffness of the shaft, the equation of motion can be written as (Meirovitch, 1967)

$$\frac{d^2}{dx^2}\left\{EI(x)\frac{d^2Y(x)}{dx^2}\right\} = \omega^2 m(x)Y(x) \quad ...(1)$$

Where, E is the modulus of rigidity of beam material, I is the moment of inertia of the beam cross-section, Y(x) is displacement in y direction at distance x from fixed end, ω is the circular natural frequency, m is the mass per unit length, $m = \rho A(x)$, ρ is the material density, x is the distance measured from the fixed end.

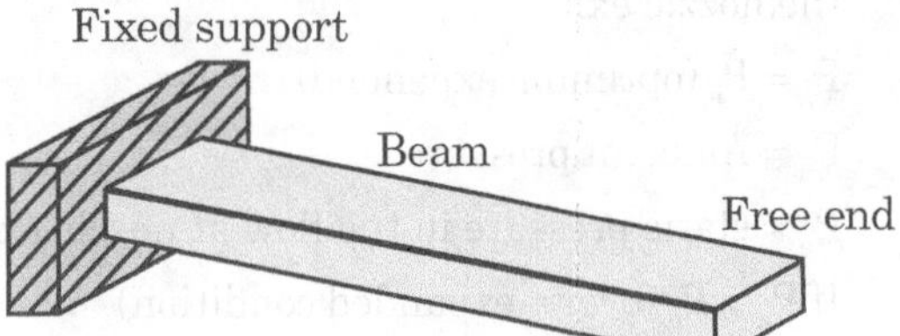

Fig. 1(a) : A cantilever beam

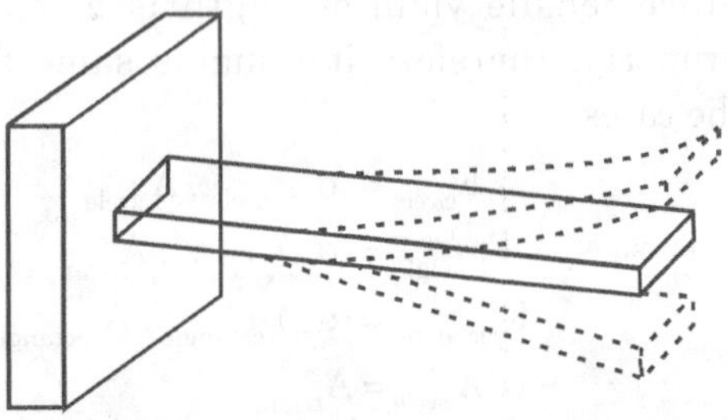

Fig. 1(b) : The beam under free vibration

Fig. 1(a) shows of a cantilever beam with rectangular cross section, which can be subjected to bending vibration by giving a small initial displacement at the free end; and Fig. 1(b) depicts of cantilever beam under the free vibration.

We have following boundary conditions for a cantilever beam (Fig. 1)

at $x = 0$, $Y(x) = 0$,

$$\frac{dY(x)}{dx} = 0 \quad ...(2)$$

at x = 1,

$$\frac{d^2Y(x)}{dx^2} = 0,$$

$$\frac{d^3Y(x)}{dx^3} = 0 \quad ...(3)$$

For a uniform beam under free vibration from equation (1), we get

$$\frac{d^4Y(x)}{dx^4} - \beta^4 Y(x) = 0 \quad ...(4)$$

with $\beta^4 = \dfrac{\omega^4 m}{El}$

The mode shapes for a continuous cantilever beam is given as

$$f_n(x) = A_n\{(\sin\beta_n L - \sin h\,\beta_n L)(\sin\beta_n x - \sin h\,\beta_n x) + (\cos\beta_n L - \cos h\,\beta_n L)(\cos\beta_n x - \cosh\beta_n x)\} \quad ...(5)$$

Where n = 1, 2, 3....∞ and $\beta_n L = n\pi$

A closed form of the circular natural frequency ω_{nf}, from above equation of motion and boundary conditions can be written as,

$$\omega_{nf} = \alpha_n^2\sqrt{\frac{El}{mL^4}} \quad ...(6)$$

Where $\alpha_n = 1.875, 4.694, 7.885$

So, First natural frequency

$$\omega_{nf} = 1.875^2\sqrt{\frac{El}{\rho AL^4}} \quad ...(7)$$

Second natural frequency

$$\omega_{nf} = 4.694^2\sqrt{\frac{El}{\rho AL^4}} \quad ...(8)$$

Third natural frequency

$$\omega_{nf} = 7.855^2\sqrt{\frac{El}{\rho AL^4}} \quad ...(9)$$

First natural frequency

$$\omega_{n1} = 1.875^2\sqrt{\frac{El}{\delta AL^4}}$$

If L becomes 2L, then

$$\omega'_{n1} = 1.875^2\sqrt{\frac{EI}{16\delta AL^4}}$$

$$\frac{\omega_{n1}}{\omega'_{n1}} = \frac{1}{\sqrt{16}}$$

$$\frac{\omega_{n1}}{\omega'_{n1}} = \frac{1}{4}$$

$$\omega_{n1} = \omega$$

$\therefore$ $\omega'_{n1} = 4\omega$ rad/sec

19. $A = \begin{pmatrix} \sin\theta & \tan\theta \\ 0 & \cos\theta \end{pmatrix}$

In case of diagonal matrix, diagonal elements are eigen values (roots)

$$\lambda_1 = \sin\theta,\ \lambda_2 = \cos\theta$$

$\therefore$ $\lambda_1^2 + \lambda_2^2 = \sin^2\theta + \cos^2\theta = 1$

20.

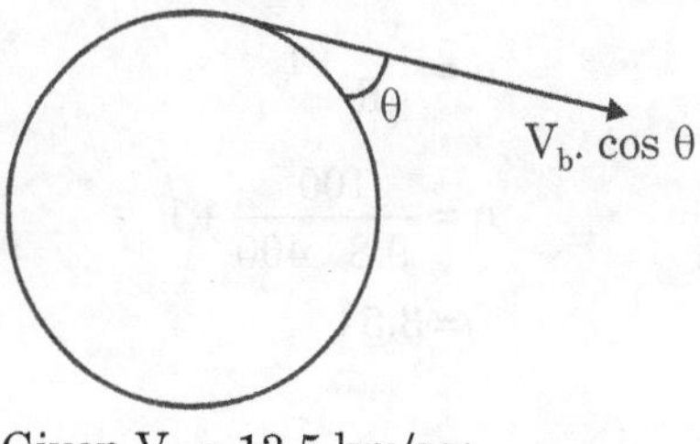

Given $V_b = 13.5$ km/sec

Tangential velocity of the space vehicle in orbit

$$= V_b .\cos\theta$$
$$= 13.5 \times \cos$$
$$= 13.448 \text{ km/sec}$$

21.

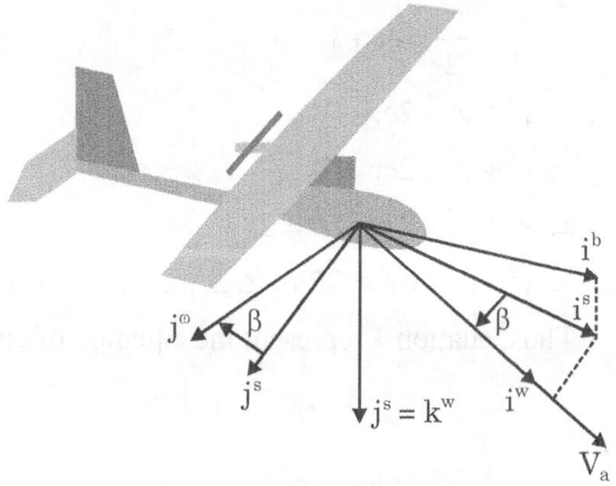

Airspeed, Angle of Attack, Sideslip Angle

$$V_a^b = \begin{pmatrix} u_r \\ v_r \\ w_r \end{pmatrix}$$

$$= R_w^b \begin{pmatrix} V_a \\ 0 \\ 0 \end{pmatrix}$$

$$= \begin{pmatrix} \cos\beta\cos\alpha & -\sin\beta\cos\alpha & -\sin\alpha \\ \sin\beta & \cos\beta & -\sin\beta\sin\alpha \\ \cos\beta\sin\alpha & 0 & \cos\alpha \end{pmatrix} \begin{pmatrix} V_a \\ 0 \\ 0 \end{pmatrix}$$

$$\begin{pmatrix} u_r \\ v_r \\ w_r \end{pmatrix} = V_a \begin{pmatrix} \cos\alpha\cos\beta \\ \sin\beta \\ \sin\alpha\cos\beta \end{pmatrix}$$

$$V_a = \sqrt{u_r^2 + v_r^2 + w_r^2}$$

$$\alpha = \tan^{-1}\left(\frac{w_r}{u_r}\right)$$

$$\beta = \sin^{-1}\left(\frac{u_r}{\sqrt{u_r^2 + v_r^2 + w_r^2}}\right)$$

$$\beta = \sin^{-1}\left(\frac{-10}{\sqrt{(100)^2 + (-10)^2 + (20)^2}}\right)$$

$$\beta = 5.6°$$

22. Given,

$$u(x, 0) = e^{-x^2}$$

$$\eta = \frac{x}{\sqrt{\alpha t}}$$

$$u(x, t) = u(\eta)$$

Now, $u(0, t) = e^0 = 1$

Therefore, $\dfrac{u(0,1)}{u(0,4)} = \dfrac{1}{\sqrt{\alpha}} \times \dfrac{\sqrt{\alpha 4}}{1} = 2$

23.

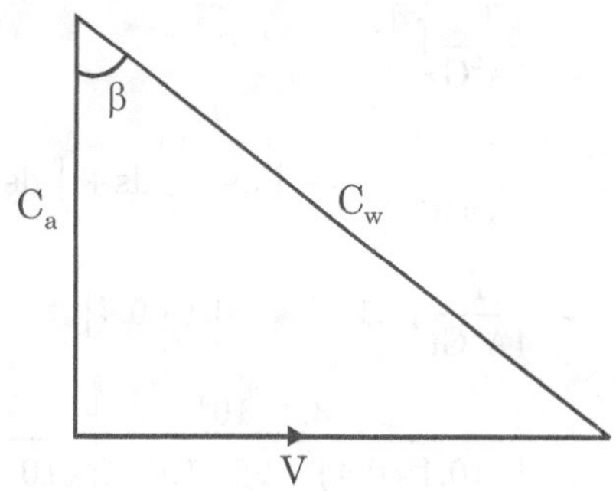

$$C_\theta = 0$$
$$C_{w1} = 0$$
$$C_{w2} = 150 \text{ m/sec}$$
$$V = 200 \text{ m/sec}$$
$$C_p = 1005 \frac{J}{kg\ K}$$

Work done by the compresor (W)

$$= \dot{m}V(C_{w2} - C_{w1})$$

$$\Rightarrow \quad \dot{m}C_p.\Delta T = \dot{m}V(C_{w2} - C_{w1})$$

$$\Delta T = \frac{V(C_{w2} - C_{w1})}{C_p}$$

$$= \frac{200 \times (150 - 0)}{1005}$$

$$\Delta T = 30°C \text{ or } 30 \text{ K}$$

24. Given,

$$G = 25 \text{ Gpa} = 25 \times 10^{-9} \text{ Pa}$$
$$T = 3.2 \text{ kNm} = 3.2 \times 10^3 \text{ Nm}$$
$$A = 0.1 \times 0.4 \text{ m}^2$$
$$t = 2 \text{ mm} = 2 \times 10^{-3} \text{ m}$$

Angle of twist per unit length,

$$\alpha = \frac{\theta}{L}$$

$$= \frac{q}{2AG}\int_0^s \frac{ds}{t} \quad ...(i)$$

From Bredth-Batho theory

$$T = 2\,qA$$

$$q = \frac{T}{2A} \quad ...(ii)$$

Shear flow $q = T \times t$ ...(iii)

Putting the value in equation (i), we get

$$\alpha = \frac{\theta}{L}$$

$$= \frac{T}{4A^2G}\int_0^s \frac{ds}{t}$$

$$= \frac{T}{4A^2Gt}\left[\int_0^{0.1} ds + \int_0^{0.4} ds + \int_0^{0.1} ds + \int_0^{0.4} ds\right]$$

$$= \frac{T}{4A^2Gt}[0.1 + 0.4 + 0.1 + 0.4]$$

$$= \frac{3.2\times 10^3}{4\times(0.1\times 0.4)^2\times 25\times 10^9\times 2\times 10^{-3}}\times 1$$

$$= 0.010 \text{ rad/m}$$

25.

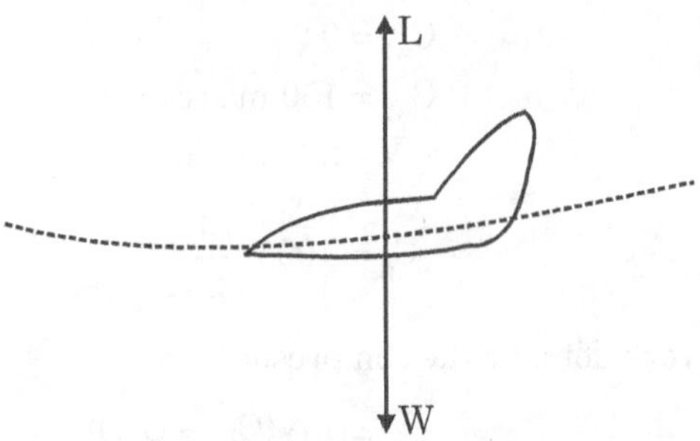

$$V = 360 \text{ km/h}$$

$$= 360 \times \frac{5}{18}$$

$$= 100 \text{ m/sec}$$

$$g = 9.8 \text{ m/sec}^2$$

$$R = 400 \text{ m}$$

Writing equation of motion of airplane

$$L - W = \frac{mv^2}{R}$$

$$\frac{L-W}{W} = \frac{v^2}{gR} \qquad [W = mg]$$

$$n - 1 = \frac{v^2}{gR}$$

$$n = \frac{v^2}{gR} + 1$$

$$n = \frac{100^2}{9.8\times 400} + 1$$

$$= 3.54$$

26. $x\dfrac{dx}{dy} + y = c$

$\Rightarrow xdx + ydy = cy$

Integrating we get,

$\Rightarrow \dfrac{x^2}{2} + \dfrac{y^2}{2} = cy + k$

$\Rightarrow x^2 + y^2 - 2cy + 2k$

$\Rightarrow x^2 + y^2 - 2cy = 2k$

$\Rightarrow x^2 + y^2 - 2cy + c^2 = 2k + c^2$

$\Rightarrow x^2 + (y-c)^2 = 2k + c^2$ ....(i)

Thus, euation 1 represent the equation of circle.

27.

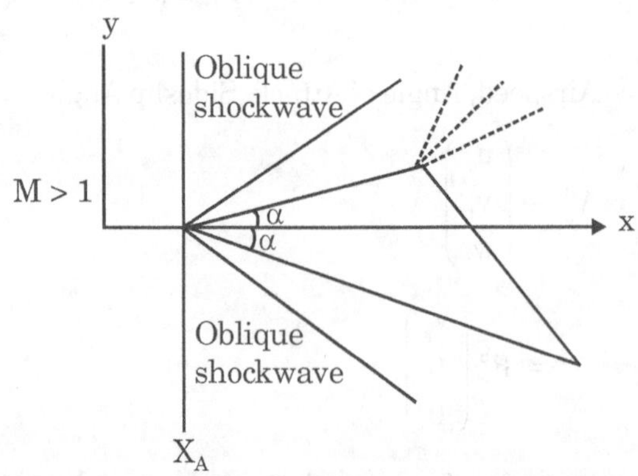

When the formation of shock waves takes place, the static pressure increases discontinuously across the shock wave and across the expansion wave the static pressure decreases continuously. Thus, option D gives the clear picture of the variation of the static pressure.

28. We have,

$$E = 2G(1 + \nu) \quad ...(i)$$

$$E = 3K(1 - 2\nu) \quad ...(ii)$$

From (i) and (ii), we get

$$2G(1 + \nu) = 3K(1 - 2\nu)$$

$$\Rightarrow \quad \frac{G}{K} = \frac{3}{2}\frac{(1-2\nu)}{(1+\nu)}$$

As per question

$$G = K$$

$$\therefore \quad 1 = \frac{3}{2}\frac{(1-2\nu)}{(1+\nu)}$$

$$\Rightarrow \quad 2 + 2\nu = 3 - 6\nu$$

$$\Rightarrow \quad 8\nu - 1 = 0$$

$$\Rightarrow \quad \nu = \frac{1}{8}$$

29. Given,

W = 3 N

Moment of inertia of CB = I

Moment of inertia of AC = 3I

Length of AC = 1 m

Length of CB = 2 m

Now as per formula

Strain Energy stored,

$$E = \int \frac{(\text{Bending moment})^2\,dx}{2(\text{Flexural rigidity})}$$

$$= \int_0^2 \frac{(Wx)^2\,dx}{2EI} + \int_2^3 \frac{(Wx)^2\,dx}{2E(3I)}$$

$$= \frac{W^2}{2EI}\int_0^2 x^2dx + \frac{W^2}{6EI}\int_2^3 x^2dx$$

$$= \frac{W^2}{2EI}\left[\frac{x^3}{3}\right]_0^2 + \frac{W^2}{6EI}\left[\frac{x^3}{3}\right]_2^3$$

$$= \frac{W^2}{2EI}\left[\frac{8}{3}\right] + \frac{W^2}{6EI}\left[\frac{27}{3} - \frac{8}{3}\right]$$

$$= \frac{8W^2}{6EI} + \frac{19W^2}{18EI} = \frac{43W^2}{18EI}$$

Also, $E = \frac{1W\delta}{2}$, where δ is the deflection

Equating both sides

$$\frac{1W\delta}{2} = \frac{43W^2}{18EI}$$

$$\Rightarrow \quad \delta = \frac{43}{3EI}$$

30.
$$\begin{bmatrix} \varepsilon_{xx} \\ \varepsilon_{yy} \\ \varepsilon_{zz} \\ \gamma_{yz} \\ \gamma_{xz} \\ \gamma_{xy} \end{bmatrix} = \begin{bmatrix} A & C & C & 0 & 0 & 0 \\ C & A & C & 0 & 0 & 0 \\ C & C & A & 0 & 0 & 0 \\ 0 & 0 & 0 & B & 0 & 0 \\ 0 & 0 & 0 & 0 & B & 0 \\ 0 & 0 & 0 & 0 & 0 & b \end{bmatrix} \begin{bmatrix} \alpha_{xx} \\ \alpha_{yy} \\ \alpha_{zz} \\ \tau_{yz} \\ \tau_{xz} \\ \tau_{xy} \end{bmatrix} \quad \text{...(i)}$$

Strain – stress relation in three dim ension

$$\begin{bmatrix} \varepsilon_{xx} \\ \varepsilon_{yy} \\ \varepsilon_{zz} \\ \gamma_{yz} \\ \gamma_{xz} \\ \gamma_{xy} \end{bmatrix} = \begin{bmatrix} \frac{1}{E} & \frac{-\gamma}{E} & \frac{-\gamma}{E} & 0 & 0 & 0 \\ \frac{-\gamma}{E} & \frac{1}{E} & \frac{-\gamma}{E} & 0 & 0 & 0 \\ \frac{-\gamma}{E} & \frac{-\gamma}{E} & \frac{1}{E} & 0 & 0 & 0 \\ 0 & 0 & 0 & \frac{1}{G} & 0 & 0 \\ 0 & 0 & 0 & 0 & \frac{1}{G} & 0 \\ 0 & 0 & 0 & 0 & 0 & \frac{1}{G} \end{bmatrix} \begin{bmatrix} \sigma_{xx} \\ \sigma_{yy} \\ \sigma_{zz} \\ \tau_{yz} \\ \tau_{xz} \\ \tau_{xy} \end{bmatrix} \quad \text{...(ii)}$$

From equation (i) and (ii), we get

$$A = \frac{1}{E}$$

$$B = \frac{1}{G}$$

$$C = \frac{-\nu}{E}$$

We have

$$E = 2G(1 + v)$$

$$\frac{1}{2G} = \frac{(1+\nu)}{E}$$

$$\Rightarrow \quad \frac{1}{2G} = \frac{1}{E} + \frac{\nu}{E} = A + (-C)$$

$$\Rightarrow \quad \frac{B}{2} = A - C$$

$$\Rightarrow \quad C = A - \frac{B}{2}$$

31. For direction stability of Aircraft $C_{\eta\beta} > 0$ plotting the given data,

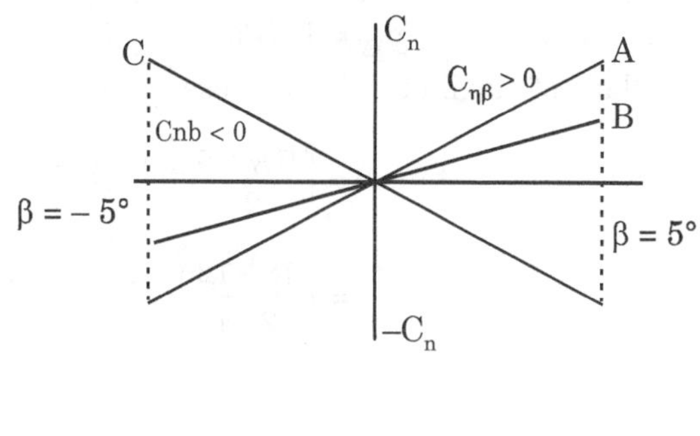

From the above discussion, we can say that C is unstable. A and B are stable with A being more stable than B.

32. Given,

$x = a\cos\theta$

$y = b\sin\theta$

$a = 7, b = 4$

$$\frac{x}{a} = \cos\theta$$

$$\frac{y}{b} = \sin\theta$$

$$\frac{x^2}{a^2}+\frac{y^2}{b^2} = \cos^2\theta + \sin^2\theta = 1$$

$$\frac{x^2}{a^2}+\frac{y^2}{b^2} = 1$$

It is equation of ellipse, hence the area enclosed by ellipse = $\pi ab = \pi \times 7 \times 5 = 110\ m^2$

33. Given,

$N_1 = 15000$ rpm

$T_1 = 300$ K

$T_2 = 280$ K

As per formula

$$\frac{N_1}{\sqrt{\gamma RT_1}} = \frac{N_2}{\sqrt{\gamma RT_2}}$$

$$\Rightarrow \quad N_2 = N_1\sqrt{\frac{T_2}{T_1}}$$

$$= 15000\times\sqrt{\frac{280}{300}}$$

$$= 14491 \text{ rpm}$$

34. All normal stresses are tensile stresses, at any angle compressive normal stresses can't occur and looking direction of shear stress τ_{xy} we can directly neglect (C) and (D).

Given,

$\sigma_{XX} = 18$ MPa

$\sigma_{YY} = 12$ MPa

$\sigma_{XY} = 4$ MPa

Plane of maximum shear stress

$$\tan 2\theta = \left(\frac{\sigma_{YY}-\sigma_{XX}}{2\tau_{XY}}\right)$$

$$= \left(\frac{18-12}{2\times 4}\right)$$

$$= \frac{3}{4}$$

$$2\theta = \tan^{-1}\left(3/4\right)$$

$$\theta = \frac{\tan^{-1}\left(3/4\right)}{2}$$

$$= 18.4°$$

Now, we know Normal stress at any angle θ for given loading is

$$\sigma_n = \tau_{XY}\sin 2\theta + \frac{\sigma_{XX}+\sigma_{YY}}{2} + \frac{\sigma_{XX}-\sigma_{YY}}{2}\cos 2\theta$$

Here, $\theta = 90° - 18.4°$

$$\Rightarrow \sigma_n = 4\times\sin(2(90°-18.4°)) + \frac{18+12}{2} + \frac{18-12}{2}\cos(2(90°-18.4°))$$

$$= 14.99 \approx 15 \text{ MPa}$$

$$\tau_{max} = \left(\frac{\sigma_{XX}-\sigma_{YY}}{2}\right)\sin 2\theta - \tau_{XY}\cos 2\theta$$

$$= \left(\frac{18-12}{2}\right)\sin(2(90°-18.4°)) - 4\times\cos(2(90°-18.4°))$$

$$= 4.99 \approx 5 \text{ MPa}$$

35.

$$A = \begin{bmatrix} \frac{1}{\sqrt{2}} & 0 & \frac{1}{\sqrt{2}} \\ 0 & 1 & 0 \\ \frac{1}{\sqrt{2}} & 0 & \frac{-1}{\sqrt{2}} \end{bmatrix}, X = \begin{bmatrix} x \\ y \\ z \end{bmatrix}$$

$$B = \begin{bmatrix} 0 \\ 1 \\ -\sqrt{2} \end{bmatrix}$$

Now, $AX = B$

$$\begin{bmatrix} \frac{1}{\sqrt{2}} & 0 & \frac{1}{\sqrt{2}} \\ 0 & 1 & 0 \\ \frac{1}{\sqrt{2}} & 0 & \frac{-1}{\sqrt{2}} \end{bmatrix}\begin{bmatrix} x \\ y \\ z \end{bmatrix} = \begin{bmatrix} 0 \\ 1 \\ -\sqrt{2} \end{bmatrix}$$

$$\begin{bmatrix} \frac{x}{\sqrt{2}}+\frac{z}{\sqrt{2}} \\ y \\ \frac{x}{\sqrt{2}}-\frac{z}{\sqrt{2}} \end{bmatrix} = \begin{bmatrix} 0 \\ 1 \\ -\sqrt{2} \end{bmatrix}$$

$$\frac{x}{\sqrt{2}}+\frac{z}{\sqrt{2}} = 0 \quad ...(i)$$

$$y = 1 \quad ...(ii)$$

$$\frac{x}{\sqrt{2}}-\frac{z}{\sqrt{2}} = -\sqrt{2} \quad ...(iii)$$

From equation (i) and (ii), we get

$x = -1$ and by then putting $x = -1$ in equation (i), we get

$$z = 1$$

$$\therefore \quad x + y + z = -1 + 1 + 1 = 1$$

36. $\int_0^1 (x^2 - 2x + 1)dx$

As given to evaluate with four intervals

$$f(x) = x^2 - 2x + 1$$

x	0	0.25	0.5	0.75	1
f(x)	1	0.562	0.25	0.062	0

Trapezoidal rule in given by

$$\int_a^b f(x)dx = \frac{b-a}{2n}\left[f(x_o)+f(x_n)+2[f(x_1)+f(x_2)]+......+f(x_{n-1})\right]$$

For given Interval

$$\int_a^b f(x)dx = \frac{0.25}{2}\left[f(0)+f(1)+2\left[f(0.25)+f(0.5)+f(0.75)\right]\right]$$

$$= 0.3435$$

For analytical value

$$\int_0^1 (x^2-2x+1)dx = \int_0^1 x^2dx - 2\int_0^1 xdx + \int_0^1 dx$$

$$= \left|\frac{x^3}{3}\right|_0^1 - 2\left|\frac{x^2}{2}\right|_0^1 + |x|_0^1$$

$$= \frac{1}{3} - 1 + 1$$

$$= \frac{1}{3}$$

$$\text{Difference} = \left|0.3435 - \frac{1}{3}\right| = 0.0102$$

37. Given,

$C_L = 0.10$ at $\alpha = 0$

$\therefore$ $C_L = C_{Lo} = 0.10$

$M = 0.6$

Now, $\quad C_{Lo(\alpha=6°)} = C_{Lo} + 2\pi\alpha$

$$= 0.10 + 2\pi \times \frac{6\pi}{180} = 0.75$$

As per Prandtl – Glauert rule

$$C_L = \frac{C_{Lo}}{\sqrt{1-M^2}}$$

$$= \frac{0.75}{\sqrt{1-0.6^2}}$$

$$= 0.937$$

38. Given,

Static pressure are same

$$P_A = P_B = P$$

Total pressure $\quad P_o = P + \frac{\rho U_o^2}{2}$

As per formula

pressure force = drag force

$$(P_{oA} - P_{oB})A = \frac{1}{2}\rho U_o^2 C_d A$$

$$P_A + \frac{\rho U_o^2}{2} - \left(P_B + \frac{\rho(0.99U_o)^2}{2}\right) = \frac{1}{2}\rho U_o^2 C_d A$$

$$U_0^2(1-0.9801) = C_d U_0^2$$

$$\Rightarrow \quad C_d = 0.01$$

39. Given,

$\beta = 35°$

$V = 517.56$ m/s

$\theta = 5.75°$

$T_2 = 182.46$ K

$R = 287$ J/kg K

$\gamma = 1.4$

Using θ-β relation for oblique shock wave, we get

$$\tan\theta = 2\cot\beta\left[\frac{M_1^2\sin^2\beta - 1}{M_1^2(\gamma+\cos 2\beta)+2}\right]$$

$$\Rightarrow \tan 5.75° = 2\cot(35°)\times\left[\frac{M_1^2\sin^2 35° - 1}{M_1^2(1.4+\cos 70°)+2}\right]$$

$$\Rightarrow \quad 0.035 = \left(\frac{0.35M_1^2 - 1}{1.74M_1^2 + 1}\right)$$

$$\Rightarrow \quad M_1 = 1.99 \approx 2$$

40. Given,

$v = 1.5 \times 10^{-5}$ m²/s

$x = 0.5$ m

$V = 3$ m/s

$$\frac{\delta}{x} = \frac{5.2}{\sqrt{Re}}$$

$$Re = \frac{V \times x}{\nu}$$

$$= \frac{3 \times 0.5}{1.5 \times 10^{-5}} = 1 \times 10^{-5}$$

$$\delta = \frac{5.2 \times x}{\sqrt{Re}}$$

$$= \frac{5.2 \times 0.5}{\sqrt{1 \times 10^{-5}}} = 8.22 \times 10^{-3} \text{ m}$$

As per formula

Displacement thickness

$$\delta^* = \frac{\delta}{2}$$

$$= \frac{8.22 \times 10^{-3}}{2}$$

$$= 4.11 \times 10^{-3} \text{ m} = 4.11 \text{ mm}$$

41. Given,

$b = 15$ m

$L = 80$ kN

$V_\infty = 90$ m/s

$\rho_\infty = 1.2$ kg/m^3

For elliptical lift distribution, induce angle of attack is

$$\alpha_i = \frac{\Gamma}{2bV_\infty}$$

$$= \frac{\frac{L}{\rho_\infty V}}{2bV_\infty}$$

$$= \frac{L}{\rho_\infty V_\infty^2 b}$$

$$= \frac{80}{1.2 \times 90^2 \times 15} = 1.334$$

42. Given,

Yield stress, $\sigma_y = 280$ MPa

Torque, $T = 10$ kNm

Using tresca failure theory or maximum shear stress theory

$$\tau_{max} = \frac{\sigma_y}{2}$$

$$= \frac{280 \times 10^6}{2} = 140 \times 10^6 \text{ Pa}$$

Also, $$\tau_{max} = \frac{16T}{\pi d^3}$$

$$\Rightarrow \quad d = \sqrt[3]{\frac{16T}{\pi \times \tau_{max}}}$$

$$= \sqrt[3]{\frac{16 \times 10 \times 10^3}{\pi \times 140 \times 10^6}} = 0.071 \text{ m}$$

$$\Rightarrow \quad r = \frac{d}{2}$$

$$= \frac{0.071}{2} = 0.035 \text{ m} = 3.5 \text{ cm}$$

43. As per formula

Velocity at perihelion $(V_p) = \sqrt{\frac{\mu(1+e)}{a(1-e)}}$

Velocity at aphelion $(V_a) = \sqrt{\frac{\mu(1-e)}{a(1+e)}}$

Now, According to problem

$$\frac{V_p}{V_a} = 1.0339$$

$$\Rightarrow \quad \frac{\sqrt{\frac{\mu(1+e)}{a(1-e)}}}{\sqrt{\frac{\mu(1-e)}{a(1+e)}}} = 1.0339$$

$$\Rightarrow \quad \sqrt{\frac{(1+e)^2}{(1-e)^2}} = 1.0339$$

$$\Rightarrow \quad \frac{1+e}{1-e} = 1.0339$$

$$\Rightarrow \quad 1 + e = 1.0339(1-e)$$

$$\Rightarrow \quad e + 1.0339\,e = 1.0339 - 1$$

$$\Rightarrow \quad e = \frac{0.0339}{2.0339} = 0.017$$

44. Eigen value for plugoid mode $\lambda_{1,2} = -0.02 \pm i0.25$

$$g = 9.8 \text{ m/s}^2$$

$$\lambda_{1,2} = \sqrt{0.02^2 + 0.25^2} = 0.251$$

Frequency of phugoid mode is given as

$$\omega = \frac{\sqrt{2}g}{V}$$

$$V = \frac{\sqrt{2}g}{\omega}$$

$$= \frac{\sqrt{2} \times 9.8}{0.251} = 55.27 \text{ m/s}$$

45. Given

$\eta = 0.95$

$V_H = 0.453$

$\frac{\partial \varepsilon}{\partial \alpha} = 0.35$

$C_{L\alpha}^{Wb} = 4.8 \text{ rad}^{-1}$

$C_{lx}^{t} = 4.4 \text{ rad}^{-1}$

Expression for neutral point

$$\Rightarrow \quad x_n = x_{ac} + \eta V_H \frac{C_{L\alpha}^{Wb}}{C_{lx}^{t}}\left(1 - \frac{\partial \varepsilon}{\partial \alpha}\right)$$

$$\Rightarrow x_n - x_{ac} = \eta V_H \frac{C_{L\alpha}^{Wb}}{C_{lx}^{t}}\left(1 - \frac{\partial \varepsilon}{\partial \alpha}\right)$$

$$= 0.95 \times 0.453 \times \frac{4.8}{4.4}(1 - 0.35)$$

$$= 0.2565$$

In precentage

$(x_n - x_{ac}) = 0.2565 \times 100$

$= 25.65\%$ of mean Aerodynamic chord.

46. Given,

$C_D = 0.025 + 0.049\, C_L^2$

$\rho_\infty = 1.225 \text{ kg/m}^3$

$\frac{W}{S} = 9844 \text{ N/m}^2$

For maximum range

$$C_{Do} = C_{Di} = KC_L^2$$

$$C_L = \sqrt{\frac{C_{Do}}{K}} = \sqrt{\frac{0.025}{0.049}} = 0.7143$$

and $\quad L = W$

$$\Rightarrow \quad V_\infty = \sqrt{\frac{2W}{C_L \rho S}}$$

$$= \sqrt{\frac{2W}{S} \times \frac{1}{C_L} \times \frac{1}{\rho}}$$

$$= \sqrt{2 \times 9844 \times \frac{1}{1.225} \times \frac{1}{0.7145}}$$

$$= 150 \text{ m/s}$$

47. For ramjet engine

$M_i = M_e$ [Mach number at inlet = Mach number at exit]

$\therefore \quad T_{oe} = T_{oi}$

$$\Rightarrow \quad M_i = \frac{V_e}{\sqrt{\gamma R T_e}}$$

$$\Rightarrow \quad V_e = M_i \sqrt{\gamma RT}$$

$$= M_i \sqrt{\gamma R \left(\frac{T_{oe}}{1 + \frac{\gamma - 1}{2} M_e^2}\right)}$$

$$= 2.8 \times \sqrt{1.4 \times 287 \times \left(\frac{2400}{1 + 0.2 \times 2.8^2}\right)}$$

$$= 1715.82 \text{ m/s}$$

48. Given,

$h_{03} = 115.42$ MJ/sec

$h_{02} = 28.94$ MJ/sec

$m_a = 32$ kg/s

$\frac{\dot{m}_f}{\dot{m}_a} = 15.6$ kg/s

$m_f = \frac{32}{15.6} = 2.05$ kg/s

CV = 46 MJ/kg

Efficiency of combustor is given by

$$\eta_c = \frac{(\dot{m}_a + \dot{m}_f)h_{03} - \dot{m}_a h_{02}}{\dot{m}_f \, CV}$$

$$= \frac{(32 + 2.05) \times 115.42 - 32 \times 28.94}{2.05 \times 46 \times 10^6}$$

$$= 91.65\% \approx 92\%$$

49. Efficiency, $\eta_T = \frac{T_{01} - T_{02}}{T_{01} - T_{02}'}$

$$= \frac{1050 - 825}{1050 - T_{02}'}$$

Now, isentropic relation

$$\left(\frac{P_{01}}{P_{02}}\right)^{\frac{\gamma}{\gamma-1}} = \frac{T_{01}}{T_{02}'}$$

$$\Rightarrow \quad T_{02}' = \frac{T_{01}}{\left(\frac{P_{01}}{P_{02}}\right)^{\frac{\gamma-1}{\gamma}}}$$

$$= \frac{1050}{\left(\frac{7.8}{2.5}\right)^{\frac{1.33-1}{1.33}}} = 789.5 \text{ K}$$

$$\therefore \quad \eta_T = \frac{1050 - 823}{1050 - 789.5}$$

$$= 0.8789 = 87.89 \approx 88\%$$

50. Given,

$I_{Sp} = 210$ sec

$A_t = 0.005\ m^2$

$C^* = 1900\ m/s$

$P_C = 50\ bar = 50 \times 10^5\ bar$

$g = 9.8\ m/s^1$

We have

$$I_{Sp} = \frac{T}{\dot{m}g}$$

$$T = \dot{m}\, gI_{sp} \qquad ...(i)$$

and expression for m is given by

$$\dot{m} = \frac{P_o A_t}{C^*} \qquad ...(ii)$$

Putting value of m in equation (i) we get,

$$T = \frac{P_o A_t I_{sp}\, g}{C^*}$$

$$= \frac{50 \times 10^5 \times 0.005 \times 210 \times 9.8}{1900}$$

$$= 27.1\ kN$$

51. Given $\xi_{new} = 1.2$

$$\xi = \frac{C}{C_m}$$

$$= \frac{C}{2\sqrt{km}}$$

$$= \frac{C}{2\sqrt{m}} \times \frac{1}{\sqrt{k}}$$

$$= \frac{C_1}{\sqrt{k}}$$

$$\frac{\xi_{original}}{\xi_{new}} = \frac{C_1}{\sqrt{k_{org}}} \times \frac{\sqrt{k_{new}}}{C_1}$$

$$= \frac{\sqrt{k_{new}}}{\sqrt{k_{org}}}$$

$$\frac{\xi_{new}}{\xi_{original}} = \sqrt{\frac{k_{org}}{k_{new}}}$$

$$(1.2)^2 = \frac{k_{org}}{k_{new}}$$

$$\frac{k_{org}}{k_{new}} = 1.44$$

52. Equation of motion

$$K.E = \frac{1}{2}m\dot{x}_1^2 + \frac{1}{2}m\dot{x}_2^2$$

$$P.E = \frac{1}{2}kx_1^2 + \frac{1}{2}k(x_1 - x_2)^2 + \frac{1}{2}(2k)(x_2)^2$$

Using lagrangian equation

$$\frac{d}{dt}\left[\frac{d(K.E)}{d\dot{x}}\right] + \frac{d(P.E)}{dx} = 0$$

For, x_1

$$m\ddot{x}_1 + kx_1 + k(x_1 - x_2) = 0$$

$$\Rightarrow m\ddot{x}_1 + 2kx_1 - kx_2 = 0 \qquad ...(i)$$

For, x_2

$$m\ddot{x}_2 + k(x_1 - x_2)(-1) + 2kx_2 = 0$$

$$m\ddot{x}_2 + 3kx_2 + kx_1 = 0 \qquad ...(ii)$$

$$\begin{bmatrix} m & 0 \\ 0 & m \end{bmatrix}\begin{bmatrix} \ddot{x}_1 \\ \ddot{x}_2 \end{bmatrix} + \begin{bmatrix} 2k & -k \\ -k & 3k \end{bmatrix}\begin{bmatrix} x_1 \\ x_2 \end{bmatrix} = 0$$

For getting mode shape and natural frequency

$$A = M^{-1}k$$

$$A = \begin{bmatrix} \frac{1}{m} & 0 \\ 0 & \frac{1}{m} \end{bmatrix}\begin{bmatrix} 2k & -k \\ -k & 3k \end{bmatrix} = \begin{bmatrix} \frac{2k}{m} & \frac{-k}{m} \\ \frac{-k}{m} & \frac{3k}{m} \end{bmatrix}$$

$$|A - \lambda I| = 0$$

$$\begin{vmatrix} \frac{2k}{m} - \lambda & \frac{-k}{m} \\ \frac{-k}{m} & \frac{3k}{m} - \lambda \end{vmatrix} = 0$$

$$\left(\frac{2k}{m} - \lambda\right)\left(\frac{3k}{m} - \lambda\right) - \left(\frac{k}{m}\right)^2 = 0$$

$$\lambda^2 - \frac{5k}{m}\lambda + 5\left(\frac{\lambda}{m}\right)^2 = 0$$

$$\lambda_{1,2} = \frac{-\frac{5k}{m} \pm \sqrt{\left(\frac{-5k}{m}\right)^2 \pm 4 \times 1 \times 5\left(\frac{k}{m}\right)^2}}{2}$$

$$= \left(1.38\frac{k}{m}, 3.61\frac{k}{m}\right)$$

$$\Rightarrow \omega_{1,2} = \sqrt{\lambda_{1,2}} = 1.174\sqrt{\frac{k}{m}}, 1.9\sqrt{\frac{k}{m}}$$

To get 1st mode shape

Putting λ in $[A - \lambda I][\phi] = 0$

$$\begin{bmatrix} \frac{2k}{m} - \frac{1.38k}{m} & \frac{-k}{m} \\ \frac{-k}{m} & \frac{3k}{m} - \frac{1.38k}{m} \end{bmatrix}\begin{bmatrix} X_{m1} \\ X_{m2} \end{bmatrix} = \begin{bmatrix} 0 \\ 0 \end{bmatrix}$$

Where, X_{m1} and X_{m2} are amplitude of mass 1 and mass 2 respectively,

$$\begin{bmatrix} 2-1.38 & -1 \\ -1 & 3-1.38 \end{bmatrix}\begin{bmatrix} X_{m1} \\ X_{m2} \end{bmatrix} = \begin{bmatrix} 0 \\ 0 \end{bmatrix}$$

$$\begin{bmatrix} 0.62 & -1 \\ -1 & 1.62 \end{bmatrix}\begin{bmatrix} X_{m1} \\ X_{m2} \end{bmatrix} = \begin{bmatrix} 0 \\ 0 \end{bmatrix}$$

$\therefore$ $X_{m1} = 1$ and $X_{m2} = a = 0.62$

53. As we know,

$$\gamma_{udl} = \frac{wl^4}{8EI}$$

$$\gamma_{cantiliver} = \frac{Rl^3}{3EI}$$

$$\therefore \quad \gamma_{udl} = \gamma_{cantiliver}$$

$$\Rightarrow \quad \frac{wl^4}{8EI} = \frac{Rl^3}{3EI}$$

$$\Rightarrow \quad R = \frac{wl^4}{8EI} \times \frac{3EI}{Rl^3}$$

$$= \frac{3}{8}Wl$$

$$= \frac{3}{8} \times 10 \times 1 = 3.7 \text{ N}$$

54. Given,

$m = 4000$ kg

$S = 25$ m^2

$C_L = 1.65$

$\rho = 1.225$ kg/m^3

$g = 9.8$ m/s^2

$fs = 1.25$

$\eta = 3.2$

$$\text{Load factor, } \eta = \frac{L}{W}$$

$$\text{Safe load factor, } \eta_{safe} = \frac{\eta}{fs}$$

$$L = \eta_{safe} W$$

$$C_L \times \frac{1}{2}\rho_\infty S V_\infty^2 = \eta_{safe} W$$

$$\Rightarrow \quad V_\infty = \sqrt{\frac{\eta_{safe} W}{C_L \rho_\infty S}}$$

$$= \sqrt{\frac{\eta W}{C_L \times \rho_\infty \times S \times fs}}$$

$$= \sqrt{\frac{3.2 \times 4000 \times 9.8}{1.65 \times 1.225 \times 25 \times 1.25}}$$

$$= 70.46 \text{ m/s}$$

55. Given,

$h = 3$ km $= 3 \times 10^3$ m

$P_{oa} = 0.72$ bar

$T_o = 268.66$ K

P infinity = 1.01 bar

T infinity = 288.16 K

$g = 9.8$ m/s^2

$\rho_\infty = 1.225$ kg/m^3

$R = 287$ J/(kgK)

As per formula

$$P_a = P_{infinity}\, e^{\frac{-g\Delta h}{RT_{infinity}}}$$

$$= 1.01 \times e^{\frac{-9.8 \times (3000-0)}{287 \times 288.16}} = 0.70 \text{ bar}$$

Equivalent velocity is given as

$$V_{eq} = \sqrt{\frac{2(P_{oa} - P_a)}{\rho_\infty}}$$

$$= \sqrt{\frac{2 \times (0.72 - 0.70) \times 10^5}{1.225}}$$

$$= 57.14 \text{ m/s}$$

SOLVED PAPER - 2021

GENERAL APTITUDE (GA)

Q.1-5: Multiple Choice Question (MCQ), carry ONE mark each (for each wrong answer: –1/3).

1. (i) Arun and Aparna are here.

(ii) Arun and Aparna is here.

(iii) Arun's families is here.

(iv) Arun's family is here.

Which of the above sentences are grammatically CORRECT?

(A) (i) and (ii) (B) (i) and (iv)

(C) (ii) and (iv) (D) (iii) and (iv)

2.

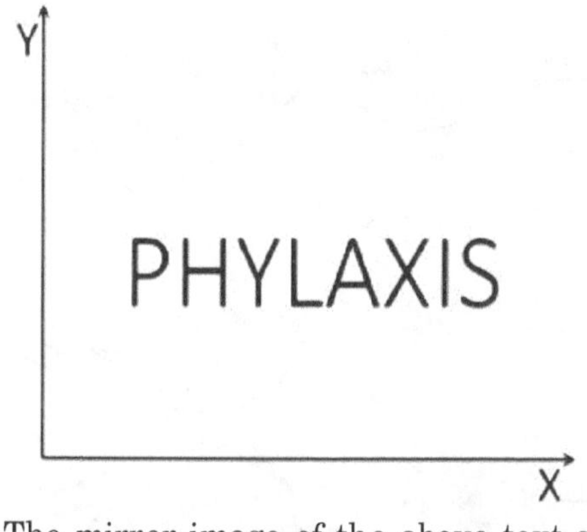

The mirror image of the above text about the x-axis is

(A)  (B) 

(C)  (D) 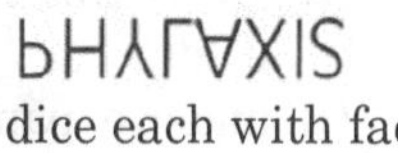

3. Two identical cube shaped dice each with faces numbered 1 to 6 are rolled simultaneously. The probability that an even number is rolled out on each dice is:

(A) $\frac{1}{36}$ (B) $\frac{1}{12}$

(C) $\frac{1}{8}$ (D) $\frac{1}{4}$

4. $\oplus$ and $\odot$ are two operators on numbers p and q such that $p \odot q = p - q$, and $p \oplus q = p \times q$

Then, $(9 \odot (6 \oplus 7)) \odot (7 \oplus (6 \odot 5)) =$

(A) 40 (B) –26

(C) –33 (D) –40

5. Four persons P, Q, R and S are to be seated in a row. R should not be seated at the second position from the left end of the row. The number of distinct seating arrangements possible is:

(A) 6 (B) 9

(C) 18 (D) 24

Q. 6–10: Multiple Choice Question (MCQ), carry TWO marks each (for each wrong answer: –2/3).

6. On a planar field, you travelled 3 units East from a point O. Next you travelled 4 units South to arrive at point P. Then you travelled from P in the North-East direction such that you arrive at a point that is 6 units East of point O. Next, you travelled in the North-West direction, so that you arrive at point Q that is 8 units North of point P.

The distance of point Q to point O, in the same units, should be ______

(A) 3 (B) 4

(C) 5 (D) 6

7. The author said, "Musicians rehearse before their concerts. Actors rehearse their roles before the opening of a new play. On the other hand, I find it strange that many public speakers think they can just walk on to the stage and start speaking. In my opinion, it is no less important for public speakers to rehearse their talks."

Based on the above passage, which one of the following is TRUE?

(A) The author is of the opinion that rehearsing is important for musicians, actors and public speakers.

(B) The author is of the opinion that rehearsing is less important for public speakers than for musicians and actors.

(C) The author is of the opinion that rehearsing is more important only for musicians than public speakers.

(D) The author is of the opinion that rehearsal is more important for actors than musicians.

8. 1. Some football players play cricket.
2. All cricket players play hockey.

Among the options given below, the statement that logically follows from the two statements 1 and 2 above, is:

(A) No football player plays hockey.
(B) Some football players play hockey.
(C) All football players play hockey.
(D) All hockey players play football.

9.

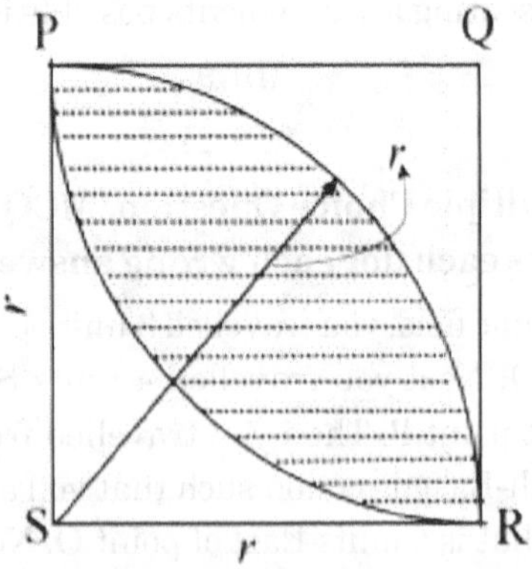

In the figure shown above, PQRS is a square. The shaded portion is formed by the intersection of sectors of circles with radius equal to the side of the square and centers at S and Q.
The probability that any point picked randomly within the square falls in the shaded area is ______

(A) $4-\frac{\pi}{2}$ (B) $\frac{1}{2}$
(C) $\frac{\pi}{2}-1$ (D) $\frac{\pi}{4}$

10. In an equilateral triangle PQR, side PQ is divided into four equal parts, side QR is divided into six equal parts and side PR is divided into eight equal parts. The length of each subdivided part in cm is an integer. The minimum area of the triangle PQR possible, in cm^2, is

(A) 18 (B) 24
(C) $48\sqrt{3}$ (D) $144\sqrt{3}$

AEROSPACE ENGINEERING (AE)

Q.1-13: Multiple Choice Question (MCQ), carry ONE mark each (for each wrong answer: – 1/3).

1. Consider the differential equation $\frac{d^2y}{dx^2}+8\frac{dy}{dx}+16y=0$ and the boundary conditions $y(0)=1$ and $\frac{dy}{dx}(0)=0$. The solution to this equation is:

(A) $y=(1+2x)e^{-4x}$ (B) $y=(1-4x)e^{-4x}$
(C) $y=(1+8x)e^{-4x}$ (D) $y=(1+4x)e^{-4x}$

2. $u(x, y)$ is governed by the following equation

$$\frac{\partial^2 u}{\partial x^2}-4\frac{\partial^2 u}{\partial x\partial y}+6\frac{\partial^2 u}{\partial y^2}=x+2y.$$

The nature of this equation is:

(A) linear (B) elliptic
(C) hyperbolic (D) parabolic

3. Consider the velocity field

$\vec{V}=(2x+3y)\hat{i}+(3x+2y)\hat{j}$. The field $\vec{V}$ is

(A) divergence-free and curl-free
(B) curl-free but not divergence-free
(C) divergence-free but not curl-free
(D) neither divergence-free nor curl-free

4. The figure shows schematics of wave patterns at the exit of nozzles A and B operating at different pressure ratios.

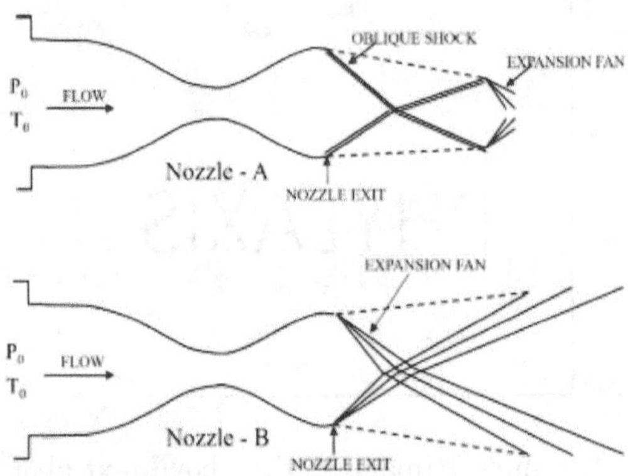

Nozzles A and B, respectively, are said to be operating in:

(A) over-expanded mode and under-expanded mode
(B) under-expanded mode and perfectly expanded mode
(C) perfectly expanded mode and under-expanded mode
(D) under-expanded mode and over-expanded mode

5. The combustion process in a turbo-shaft engine during ideal operation is:

(A) isentropic (B) isobaric
(C) isochoric (D) isothermal

6. How does the specific thrust of a turbojet engine change for a given flight speed with increase in flight altitude?

(A) Increases monotonically
(B) Decreases monotonically
(C) Remains constant
(D) First increases and then decreases

7. How does the propulsion efficiency of a turbofan engine, operating at a given Mach number and a given altitude, change with increase in compressor pressure ratio?

(A) Remains constant

(B) Increases monotonically

(C) Decreases monotonically

(D) First decreases and then increases

8. A solid propellant rocket producing 25 MN thrust is fired for 150 seconds. The specific impulse of the rocket is 2980 Ns/kg. How much propellant is burned during the rocket operation?

(A) 8390 kg (B) 82300 kg

(C) 1.26×10^6 kg (D) 11.2×10^6 kg

9. The shape of a supersonic diffuser that slows down a supersonic flow to subsonic flow is

(A) converging

(B) diverging

(C) diverging-converging

(D) converging-diverging

10. Uniaxial tension test (see the figure) is conducted on two different samples prepared with homogeneous, Isotropic materials. One of the materials is brittle, whereas the other is ductile.

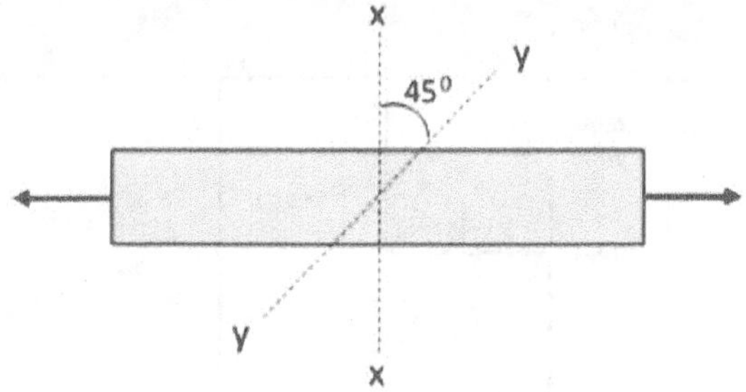

Assuming that there is no stress concentration at loading points, the failure would initiate:

(A) along x-x in both materials

(B) along x-x in brittle material and along y-y in ductile material

(C) along y-y in brittle material and along x-x in ductile material

(D) along y-y in both materials

11. For the state of stress as shown in the figure, what is the orientation of the plane with maximum shear stress with respect to the x-axis?

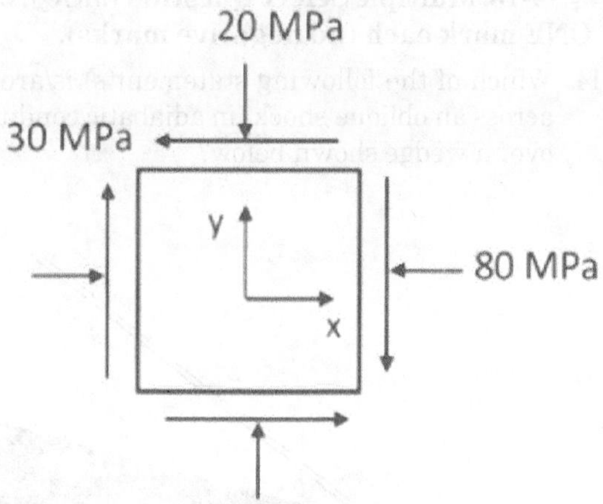

(A) 45° (B) –45°

(C) 22.5° (D) –22.5°

12. Let V_{TAS} be the true airspeed of an aircraft flying at a certain altitude where the density of air is ρ, and V_{EAS} be the equivalent airspeed. If ρ_0 is the density of air at sea-level, what is the ratio $\frac{V_{TAS}}{V_{EAS}}$ equal to?

(A) $\frac{\rho}{\rho_0}$ (B) $\frac{\rho_0}{\rho}$

(C) $\sqrt{\frac{\rho_0}{\rho}}$ (D) $\sqrt{\frac{\rho}{\rho_0}}$

13. $C_m - \alpha$ variation for a certain aircraft is shown in the figure. Which one of the following statements is true for this aircraft?

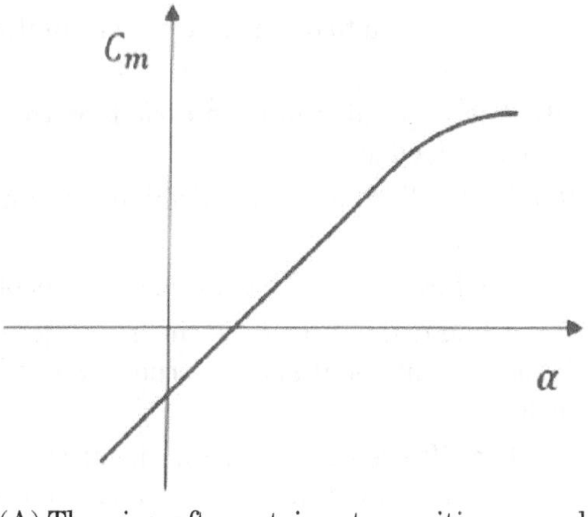

(A) The aircraft can trim at a positive α and it is stable.

(B) The aircraft can trim at a positive α, but it is unstable

(C) The aircraft can trim at a negative α and it is stable.

(D) The aircraft can trim at a negative α, but it is unstable.

Q.14-16: Multiple Select Question (MSQ), carry ONE mark each (no negative marks).

14. Which of the following statement(s) is/are true across an oblique shock (in adiabatic conditions) over a wedge shown below?

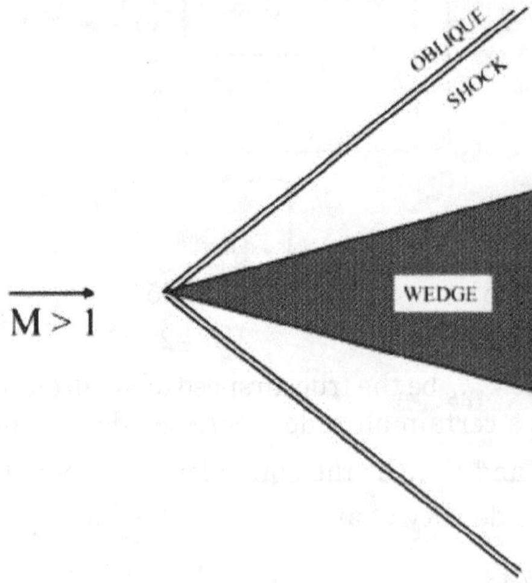

(A) Total pressure decreases

(B) Mach number based on velocity tangential to the shock decreases

(C) Total temperature remains constant

(D) Mach number based on velocity tangential to the shock remains the same and that based on velocity normal to the shock decreases

15. Which of the following statement(s) is/are true with regards to Kutta condition for flow past airfoils?

(A) It is utilized to determine the circulation on an airfoil.

(B) It is applicable only to airfoils with sharp trailing edge.

(C) The trailing edge of an airfoil is a stagnation point.

(D) The flow leaves the trailing edge smoothly.

16. According to the thin airfoil theory, which of the following statement(s) is/are true for a cambered airfoil?

(A) The lift coefficient for an airfoil is directly proportional to the absolute angle of attack.

(B) The aerodynamic center lies at quarter chord point.

(C) The center of pressure lies at quarter chord point.

(D) Drag coefficient is proportional to the square of lift coefficient.

Q.17-25: Numerical Answer Type (NAT), carry ONE mark each (no negative marks).

17. $\lim_{x\to 0}\left(\frac{1}{\sin x}-\frac{1}{x}\right)=$ ______ *(round off to nearest integer).*

18. Given that ς is the unit circle in the counter-clockwise direction with its center at origin, the integral $\oint_{\varsigma}\frac{z^3}{4z-i}\,dz=$ ______ *(round off to three decimal place).*

19. A single degree of freedom spring-mass-damper system is designed to ensure that the system returns to its original undisturbed position in minimum possible time without overshooting. If the mass of the system is 10 kg, spring stiffness is 17400 N/m and the natural frequency is 13.2 rad/s, the coefficient of damping of the system in Ns/m is ______ *(round off to nearest integer).*

20. Two cantilever beams (Beam 1 and Beam 2) are made of same homogenous material and have identical cross sections. Beam 1 has length l and Beam 2 has length $2l$. Ratio of the first natural frequency of Beam 1 to that of Beam 2 is ______ *(round off to nearest integer).*

21. A free vortex filament (oriented along Z-axis) of strength $K = 5\ m^2/s$ is placed at the origin as shown in the figure. The circulation around the closed loop ABCDEFA for this flow is ______.

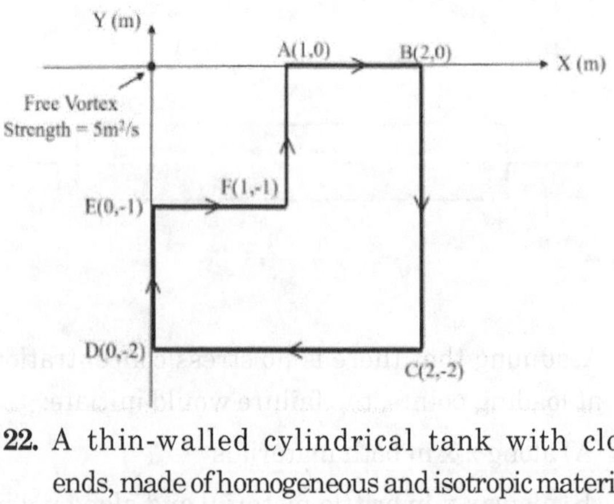

22. A thin-walled cylindrical tank with closed ends, made of homogeneous and isotropic material, is pressurized internally. If the hoop (circumferential) strain developed in the material is thrice the value of the axial strain then the Poisson's ratio of the material is ______ *(correct up to one decimal place).*

23. A jet aircraft has the following specifications: wing loading = 1800 N/m², wing area = 30 m², drag polar $C_D = 0.02 + 0.04\,C_L^2$, and $C_{L,max} = 1.6$.

Take density of air at sea level as 1.225 kg/m^3. The speed at which the aircraft achieves maximum endurance in a steady and level flight at sea level is ______ m/s (*round off to two decimal places*).

24. An aircraft with twin jet engines has the following specifications:

Thrust produced (per engine) = 8000 N

Spanwise distance between the two engines = 10 m

Wing area = 50 m^2, Wing span = 10 m

Rudder effectiveness, $C_{n_{\delta r}} = -0.002$ /deg

Density of air at sea level = 1.225 kg/m^3

The rudder deflection, in degrees, required to maintain zero sideslip at 100 m/s in steady and level flight at sea level with a non-functional right engine is______ (*round off to two decimal places*).

25. The velocity required to launch a space shuttle from the surface of the earth to achieve a circular orbit of 250 km altitude is _____ (*round off to two decimal places*).

For earth, Gm_e = 398,600.4 km^3/s^2 and surface radius R_0 = 6378.14 km.

Q.26-30: Multiple Choice Question (MCQ), carry TWO marks each (for each wrong answer: –2/3).

26. A rigid massless rod pinned at one end has a mass *m* attached to its other end. The rod is supported by a linear spring of stiffness *k* as shown in the figure.

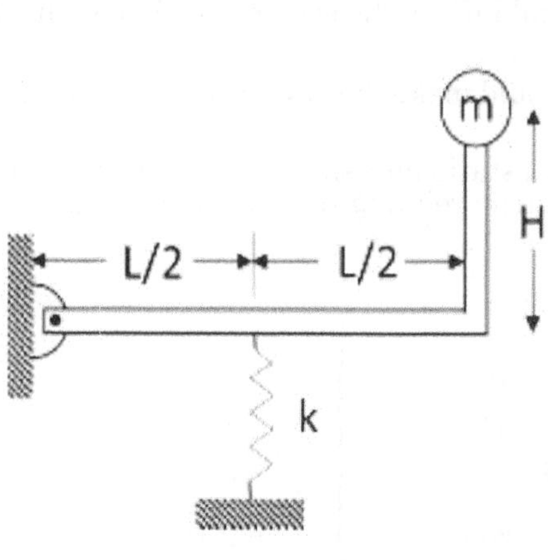

The natural frequency of this system is:

(A) $\frac{1}{2\pi}\sqrt{\frac{kL^2}{4m\left(L^2+H^2\right)}}$

(B) $\frac{1}{2\pi}\sqrt{\frac{kL^2}{m\left(L^2+H^2\right)}}$

(C) $\frac{1}{2\pi}\sqrt{\frac{4kL^2}{m\left(L^2+H^2\right)}}$

(D) $\frac{1}{2\pi}\sqrt{\frac{k\left(L^2+H^2\right)}{4mL^2}}$

27. The figure shows three glasses P, Q and R with water and floating ice cube. Glass P has a solid ice cube, glass Q has an ice cube with a small solid steel ball embedded in it and glass R has an ice cube with an air bubble. After the ice cube melts, the level of water in glasses P, Q and R, respectively:

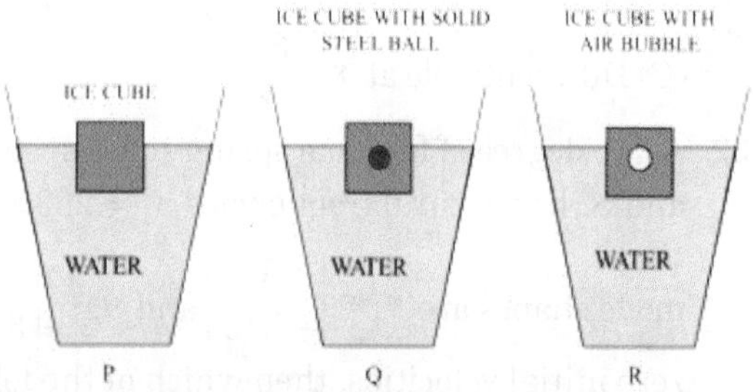

(A) remains same, increases, and decreases

(B) increases, decreases, and increases

(C) remains same, decreases, and decreases

(D) remains same, decreases, and increases

28. To estimate aerodynamic loads on an aircraft flying at 100 km/h at standard sea-level conditions, a one-fifth scale model is tested in a variable-density wind tunnel ensuring similarity of inertial and viscous forces. The pressure used in the wind tunnel is 10 times the atmospheric pressure. Assuming ideal gas law to hold and the same temperature conditions in model and prototype, the velocity needed in the wind tunnel test-section is ______.

(A) 25 km/h (B) 50 km/h

(C) 100 km/h (D) 20 km/h

29. The figure shows schematic of a set-up for visualization of non-uniform density field in the test section of a supersonic wind tunnel. This technique of visualization of high speed flows is known as:

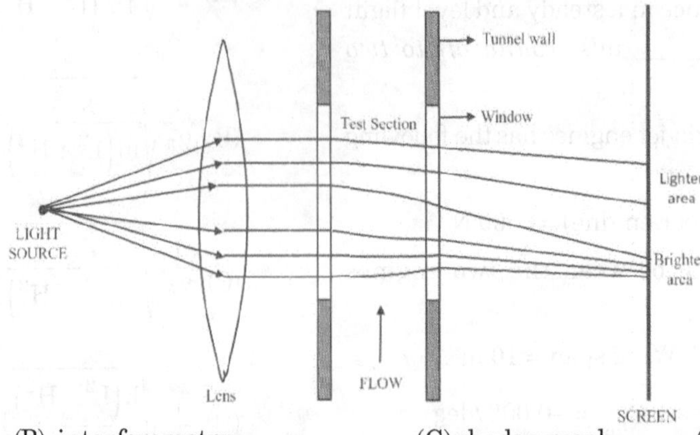

(A) schlieren (B) interferometry (C) shadowgraph (D) holography

30. For a conventional fixed-wing aircraft in a 360° inverted vertical loop maneuver, what is the load factor (n) at the topmost point of the loop?
Assume the flight to be steady at the topmost point.

(A) n =1 (B) n < 1 (C) n = – 1 (D) n > – 1

Q.31-36: Multiple Select Question (MSQ), carry TWO marks each (no negative marks).

31. Which of the following statement(s) is/are true about the function defined as $f(x) = e^{-x}|\cos x|$ for $x > 0$?

(A) Differentiable at $x = \frac{\pi}{2}$ (B) Differentiable at $x = \pi$

(C) Differentiable at $x = \frac{3\pi}{2}$ (D) Continuous at $x = 2\pi$

32. A two degree of freedom spring-mass system undergoing free vibration with generalized coordinates x_1 and x_2 has natural frequencies $\omega_1 = 233.9$ rad/s and $\omega_2 = 324.5$ rad/s, respectively. The corresponding mode shapes are $\phi_1 = \begin{bmatrix} 1 \\ -3.16 \end{bmatrix}$ and $\phi_2 = \begin{bmatrix} 1 \\ 3.16 \end{bmatrix}$. If the system is disturbed with certain deflections and zero initial velocities, then which of the following statement(s) is/are true?

(A) An initial deflection of $x_1(0) = 6.32$ cm and $x_2(0) = -3.16$ cm would make the system oscillate with only the second natural frequency.

(B) An initial deflection of $x_1(0) = 2$ cm and $x_2(0) = -6.32$ cm would make the system oscillate with only the first natural frequency.

(C) An initial deflection of $x_1(0) = 2$ cm and $x_2(0) = -2$ cm would make the system oscillate with a linear combination of first and second natural frequencies.

(D) An initial deflection of $x_1(0) = 1$ cm and $x_2(0) = -6.32$ cm would make the system oscillate with only the first natural frequency.

33. A shock moving into a stationary gas can be transformed to a stationary shock by a change in reference frame, as shown in the figure. Which of the following is/are true relating the flow properties in the two reference frames?

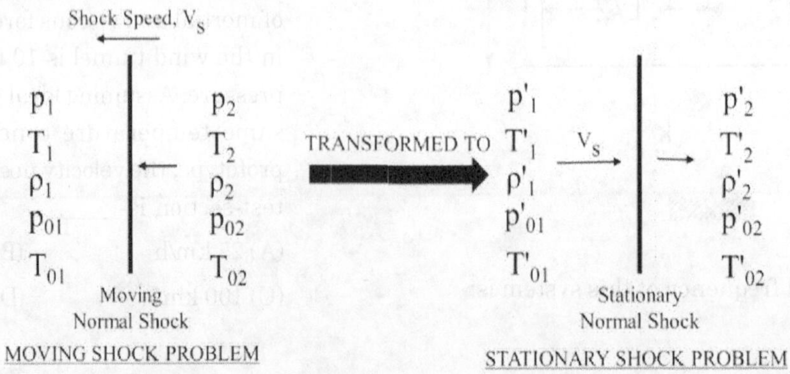

(A) $T_1' > T_1, T_{01}' > T_{01}, p_{01}' > p_{01}, \rho_2' > \rho_1'$

(B) $T_1' = T_1, T_2' < T_{01}, p_{01}' > p_{01}, \rho_2' = \rho_2$

(C) $T_1' < T_1, p_1' > p_1, p_{01}' > p_{01}, \rho_2' > \rho_1$

(D) $T_1' = T_1, p_2 > p_{01}, T_{01}' > T_{01}, p_{01}' > p_{01}$

34. For a conventional fixed-wing aircraft, which of the following statements are true?

(A) Making C_{m_α} more negative leads to an increase in the frequency of its short-period mode.

(B) Making C_{m_q} more negative leads to a decreased damping of the short-period mode.

(C) The primary contribution towards C_{lp} is from the aircraft wing.

(D) Increasing the size of the vertical fin leads to a higher yaw damping,

35. Which of the following statement(s) is/are true?

(A) Service ceiling is higher than absolute ceiling for a piston-propeller aircraft.

(B) For a given aircraft, the stall speed increases with increase in altitude.

(C) Everything else remaining the same, a tailwind increases the range of an aircraft.

(D) For a jet aircraft constrained to fly at constant altitude, there exists an altitude where its range is maximum.

36. A conventional fixed-wing aircraft, with a horizontal tail and vertical fin, in steady and level flight is subjected to small perturbations. Which of the following statement(s) is/are true?

(A) Vertical fin has a stabilizing effect on the lateral stability of the aircraft.

(B) Vertical fin has a destabilizing effect on the directional stability of the aircraft.

(C) Presence of wing anhedral increases the lateral stability of the aircraft.

(D) Horizontal tail has a stabilizing effect on the longitudinal static stability of the aircraft.

Q.37-55: Numerical Answer Type (NAT), carry TWO mark each (no negative marks).

37. The ratio of the product of eigenvalues to the sum of the eigenvalues of the given matrix

$\begin{bmatrix} 3 & 1 & 2 \\ 2 & -3 & -1 \\ 1 & 2 & 1 \end{bmatrix}$ is ______ *(round off to nearest integer).*

38. The definite Integral $\int_1^5 x^2 dx$ is evaluated using four equal intervals by two methods-first by the trapezoidal rule and then by the Simpson's one - third rule. The absolute value of the difference between the two calculations is ______ *(round off to the decimal places).*

39. The deflection y of a certain beam of length l and uniform weight per unit length w, is given as

$y = \frac{\omega}{48EI}\left(2x^4 - 3lx^3 + l^3x\right)$, where x is the distance from the point of support and EI is a constant. The non-dimensional location $\frac{x}{l}$, where the deflection of the beam is maximum, is ______ *(round off to two decimal places).*

40. A large water tank is fixed on a cart with wheels and a vane (see figure). The wheels of the cart provide negligible resistance for rolling on the fixed support. The cart is tied to the fixed support with a thin horizontal rope. There is a hole of diameter 5 cm on the side of the tank through which a jet of constant velocity of 10 m/s emerges. The jet of water is deflected by the attached vane by 60° (see figure). Assume that the jet velocity remains constant at 10 m/s after emerging from the vane. Take density of water to be 1000 kg/m^3. The tension in the connecting rope is ______ N *(round off to one decimal place).*

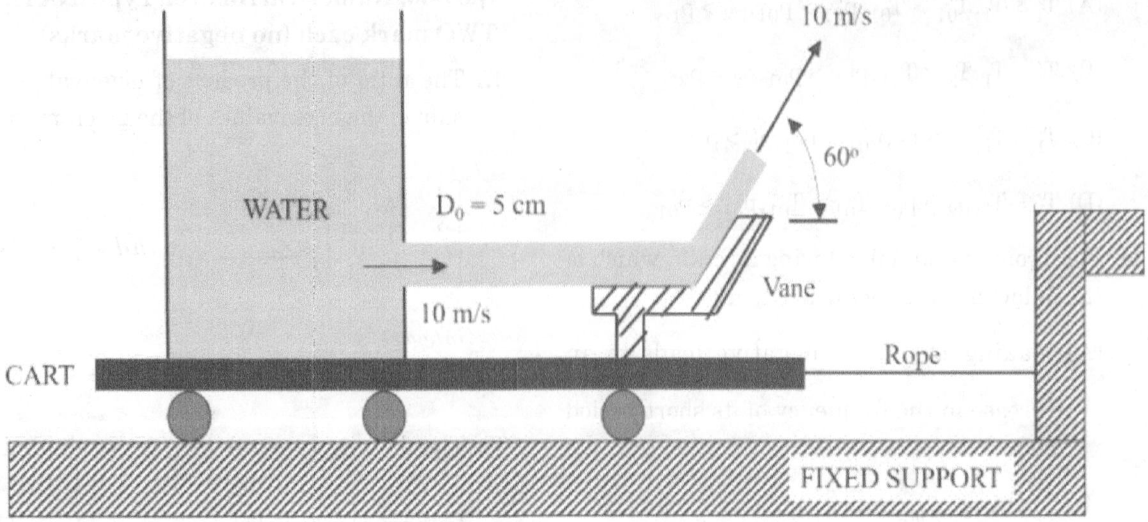

41. A finite wing of elliptic planform with aspect ratio 10, whose section is a symmetric airfoil, is placed in a uniform flow at 5 degrees angle of attack. The induced drag coefficient for the wing is _____ (*round off to three decimal places*).

42. Consider a model of a boundary layer with the following velocity profile:

$$\frac{u}{U}=\begin{cases}\left(\frac{y}{\delta}\right)^2 & y \le \delta \\ 1 & y > \delta\end{cases}$$

The shape factor, defined as the ratio of the displacement thickness to momentum thickness, for this profile is _____ (*round off to 2 decimal places*).

43. An aircraft with a turbojet engine is flying at 270 m/s. The enthalpy of the Incoming air at the intake is 260 kJ/kg and the enthalpy of the exhaust gases at the nozzle exit is 912 kJ/kg. The ratio of mass flow rates of fuel and air is equal to 0.019. The chemical energy (heating value) of fuel is 44.5 MJ/kg and the combustion process is ideal. The total loss of heat from the engine to the ambient is 25 kJ per kg of air. The velocity of the exhaust jet is _____ m/s (*round off to two decimal places*).

44. Hot gases are generated at a temperature of 2100 K and a pressure of 14 MPa in a rocket chamber. The hot gases are expanded ideally to the ambient pressure of 0.1 MPa in a convergent-divergent nozzle having a throat area of 0.1 m^2. The molecular mass of the gas is 22 kg/kmol. The ratio of specific heats (γ) of the gas is 1.32. The value of the universal gas constant (R_0) is 8314 J/kmol-K. The acceleration due to gravity, g, is 9.8 m/s^2. The specific impulse of the rocket is _____ seconds (*round off to two decimal places*).

45. A twin-spool turbofan engine is operated at sea level (P_a = 1 bar, T_a = 288 K). The engine has separate cold and hot nozzles. During static thrust test at sea level, the overall mass flow rate of air through the engine and the cold exhaust temperature are measured to be 100 kg/s and 288 K, respectively. The parameters for the engine are:

Fan pressure ratio = 1.6

Overall pressure ratio = 20

Bypass ratio = 3.0

Turbine entry temperature=1800 K.

The specific heat at constant pressure (C_p) is 1.005 kJ/kg-K and the ratio of specific heats (γ) is 1.4 for air.

Assuming ideal fan and ideal expansion in the nozzle, the sea-level static thrust from the cold nozzle is _____ kN (*round off to two decimal places*).

46. At the design conditions, the velocity triangle at the mean radius of a single stage axial compressor is such that the blade angle at the rotor exit is equal to 30°. The absolute velocities at the rotor inlet and exit are equal to 140 m/s and 240 m/s, respectively. The flow velocities relative to the rotor at inlet and exit of the rotor are equal to 240 m/s and 140 m/s, respectively.

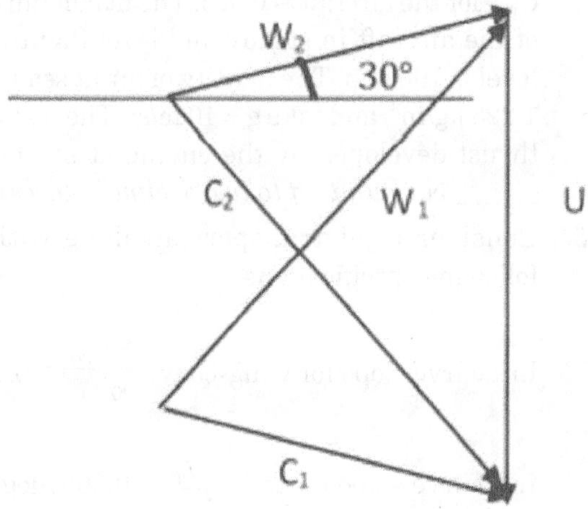

The blade speed (U) at the mean radius of the rotor is _____ m/s (*round off to two decimal places*).

47. A single stage axial turbine has a mean blade speed of 340 m/s at design condition with blade angles at inlet and exit of the rotor being 21° and 55° respectively. The degree of reaction at the mean radius of the rotor is equal to 0.4. The annulus area at the rotor inlet is 0.08 m^2 and the density of gas at the rotor inlet is 0.9 kg/m^3. The flow rate through the turbine at these conditions is _______ kg/s (*round off to two decimal places*).

48. The air flow rate through the gas generator of a turboprop engine is 100 kg/s. The stagnation temperatures at inlet and exit of the combustor are 600 K and 1200 K, respectively. The burner efficiency is 90% and the heating value of the fuel is 40 MJ/kg. The specific heats at constant pressure (Cp) for air and burned gases are 1000 J/kg-K and 1200 J/kg-K, respectively. The flow rate of the fuel being used is ______ kg/s (*round off to two decimal places*).

(Note: Do not neglect the fuel flow rate with respect to the air flow rate)

49. A rigid horizontal bar ABC, with roller support at A, is pinned to the columns BD and CE at points B and C, respectively as shown in figure. The other end of the column BD is fixed at D, whereas the column CE is pinned at E. A vertical load P is applied on the bar at a distance 'a' from point B. The two columns are made of steel with elastic modulus 200 GPa and have a cross section of 1.5 cm × 1.5 cm. The value of 'a' for which both columns buckle simultaneously, is ______ cm (*round off to one decimal place*).

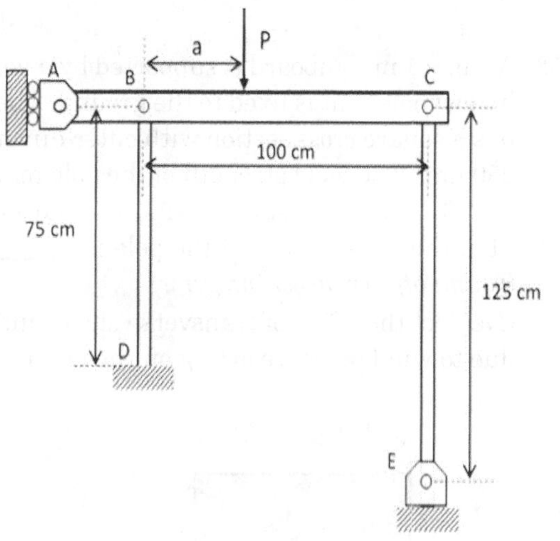

50. A two-cell wing box is shown in the figure. The cell walls are 1.5 mm thick and the shear modulus G = 27 GPa. If the structure is subjected to a torque of 12 kNm, then the wall AD will experience a shear stress of magnitude ______ MPa (*round off to one decimal place*).

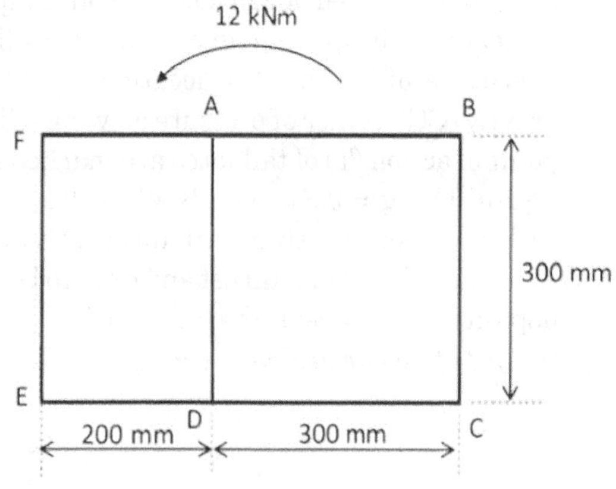

51. Two cantilever beams AB and DC are in contact with each other at their free ends through a roller as shown in the figure. Both beams have a square cross section of 50 mm × 50 mm, and the elastic modulus E = 70 GPa. If beam AB is subjected to a uniformly distributed load of 20 kN/m, then the compressive force experienced by the roller is ______ kN (*round off to one decimal place*).

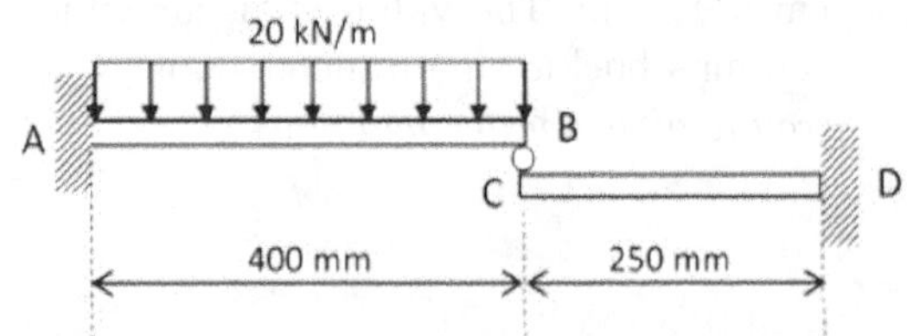

52. A 3 m × 1 m signboard is supported by a vertical hollow pole that is fixed to the ground. The pole has a square cross section with outer dimension 250 mm. The yield strength of the pole material is 240 MPa. To sustain a wind pressure of 7.5 kPa, the dimension d of the pole is ______ mm (*round off to nearest integer*).

(Neglect the effect of transverse shear and load due to wind pressure acting on the pole)

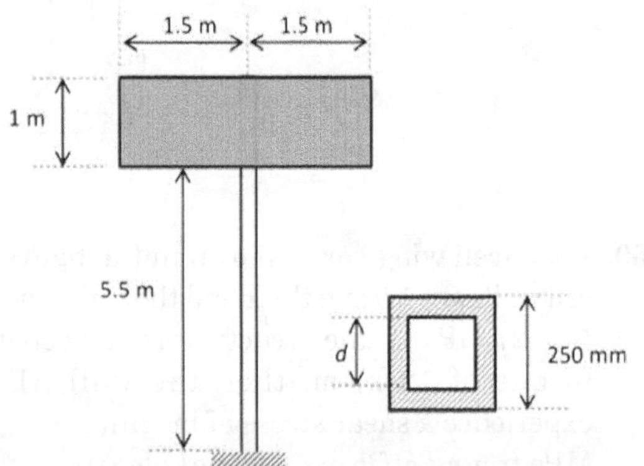

53. An airplane weighing 5500 kg is in a steady level flight with a speed of 225 m/s. The pilot initiates a steady pull-up maneuver with a radius of curvature of 775 m. The location of center of gravity (CG), center of pressure on wing (CP) and point of action (T) of tail force are marked in the figure. Use g = 9.81 m/s^2. Neglect drag on the tail and assume that tail force is vertical. Assuming the engine thrust and drag to be equal, opposite and collinear, the tail force is ______ kN (*round off to one decimal place*).

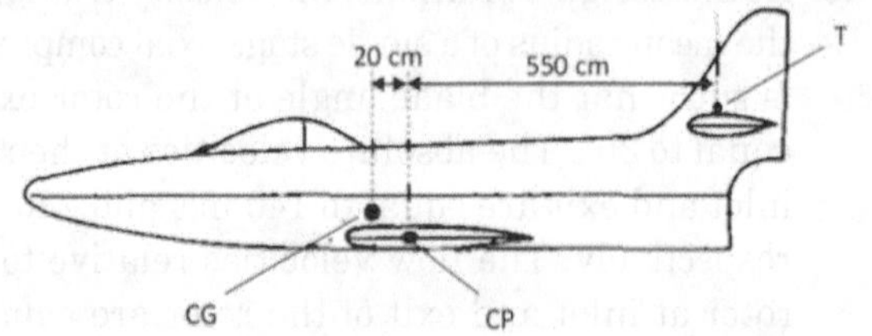

54. A jet aircraft weighing 10,000 kg has an elliptic wing with a span of 10 m and area 30 m^2. The C_{D0}, for the aircraft is 0.025. The maximum speed of the aircraft in steady and level flight at sea level is 100 m/s. The density of air at sea level is 1.225 kg/m^3, and take g = 10 m/s^2. The maximum thrust developed by the engine at sea level is ______ N (*round off to two decimal places*).

55. Consider a jet transport airplane with the following specifications:

Lift curve slope for wing-body $\dfrac{\partial C_{L_{wb}}}{\partial \alpha_{wb}} = 0.1$ /deg

Lift curve slope for tail $\dfrac{\partial C_{L_t}}{\partial \alpha_t} = 0.068$ /deg

Tail area S_t = 80 m^2

Wing area S = 350 m^2

Distance between mean aerodynamic centers of tail and wing-body

l_t = 28 m

Mean aerodynamic chord $\bar{c} = 9m$

Downwash $\varepsilon = 0.4\alpha$

Axial location of the wing-body mean aerodynamic center $x_{cg}/\bar{c} = 0.25$

Axial location of the center of gravity $x_{cg}/\bar{c} = 0.3$

All axial locations are with respect to the leading edge of the root chord and along the body x-axis. Ignore propulsive effects.

The pitching-moment-coefficient curve slope $(C_{m\alpha})$ is ______ /deg (*round off to three decimal places*).

ANSWERS

General Aptitude (GA)

1. (B)	**2.** (B)	**3.** (D)	**4.** (D)	**5.** (C)	**6.** (C)
7. (A)	**8.** (B)	**9.** (C)	**10.** (D)		

Aerospace Engineering (AE)

1. (D)	**2.** (B)	**3.** (B)	**4.** (A)	**5.** (B)	**6.** (A)
7. (D)	**8.** (C)	**9.** (D)	**10.** (B)	**11.** (D)	**12.** (C)
13. (B)	**14.** (A; B; C)	**15.** (A; B; D)	**16.** (A; B)	**17.** (0 to 0)	
18. (0.02 to 0.03)	**19.** (264 to 264)	**20.** (4 to 4)	**21.** (0 to 0)		
22. (0.2 to 0.2)	**23.** (64.30 to 64.60)	**24.** (6.50 to 6.60)	**25.** (7.75 to 7.77)	**26.** (A)	**27.** (C)
28. (B)	**29.** (C)	**30.** (D)	**31.** (B; D)	**32.** (B; C)	**33.** (D)
34. (A; C; D)	**35.** (B; C; D)	**36.** (A; D)	**37.** (8 to 8)	**38.** (0.66 to 0.68)	
39. (0.41 to 0.43)	**40.** (98.0 to 98.5)	**41.** (0.005 to 0.009)	**42.** (4.8 to 5.2)	**43.** (605 to 607)	
44. (216 to 220)	**45.** (20 to 22)	**46.** (274 to 282)	**47.** (18.0 to 19.5)	**48.** (2.35 to 2.50)	
49. (14.0 to 16.0)	**50.** (2.1 to 2.5)	**51.** (2.2 to 2.6)	**52.** (232 to 238)		
53. (14.8 to 15.2)	**54.** (9787 to 9795)	**55.** (–0.025 to –0.023)			

EXPLANATIONS

General Aptitude (GA)

1. Option (i) is correct.

Arun and Aparna are two persons, i.e., they are plural. Therefore, are is the appropriate linking verb used here.

Option (iv) is correct

Here arun's denotes a singular person. Therefore, family is the appropriate word used here.

2. As the mirror is in x-axis. The correct option is Option B.

3. Total number of cases = 6 × 6 = 36

For even number of choices we have,

2, 4, 6 i.e., 3 choices

As they are 2 dices

Thus, even choices are = 3 × 3 = 9

$\therefore$ P (even number is rolled) $= \frac{9}{36} = \frac{1}{4}$

4. Given, $p \odot q = p - q$

$p \oplus q = p \times q$

$\therefore$ we get

$\odot = -$

$\oplus = \times$

then, $(9 \odot (6 \oplus 7)) \odot (7 \oplus (6 \odot 5))$

$= (9-(6\times 7))-(7\times(6-5)) = -40$

5. Total number of arrangement in which four persons can sit

= 4 × 3 × 2 × 1

= 24

Number of arrangement for which R is seated at second position is

3 × 2 × 1 = 6

$\therefore$ Number of arrangement for which R is not seated at second position is 24 – 6 = 18

6.

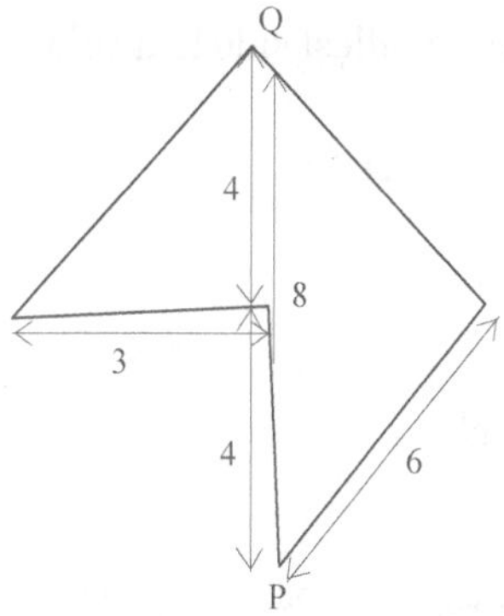

Therefore, distance between OQ is

$OQ = \sqrt{3^2 + 4^2} = 5$ units

7. As per the passage option A is correct.

The author is of the opinion that rehearsing is important for musicians, actors and public speakers.

8. Given,

1. Some football players play cricket
2. All cricket players play hockey.

As per above statement it is clear that option B gives the correct statement as

Some football players play hockey.

9.

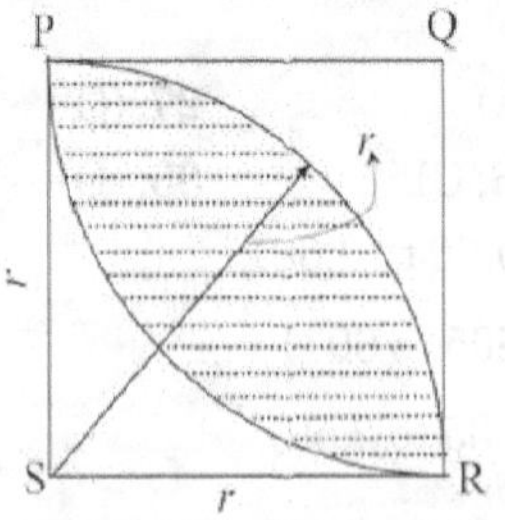

P (points which falls in the shaded area) = $\frac{\text{Shaded area}}{\text{Total area}}$

Total area = r^2

Shaded area = 2 × (area of sector – area of remaing triangle of square)

$$= 2\times\left(\frac{1}{2}\times r^2\times\frac{\pi}{2}-\frac{1}{2}\times r^2\right) = r^2\left(\frac{\pi}{2}-1\right)$$

P (points which falls in the shaded area)

$$= \frac{r^2\left(\frac{\pi}{2}-1\right)}{r^2} = \left(\frac{\pi}{2}-1\right)$$

10. Given,

As equalateral triangle

PQ = QR = PR

$\frac{PQ}{4}, \frac{QR}{6}, \frac{PR}{8}$ and each are integers

Firstly, we find the smallest side length which is computed as

LCM of 4, 6, 8 which is 24

Now,

$$\text{Area} = \frac{1}{2}ab\sin\theta$$

As equalateral triangle

$\therefore\ \theta = 60°$

Thus, Minimum area = $\frac{1}{2}\times 24\times 24\times \sin 60 = 144\sqrt{3}$

Aerospace Engineering (AE)

1. Given,

$$\frac{d^2y}{dx^2}+8\frac{dy}{dx}+16y=0$$

The auxillary equation is

$(D^2 + 8D + 16)y = 0$

$(D + 4)(D + 4) = 0$

$D = -4, -4$

Now, The C.F. is

$y = (c_1 + c_2x)e^{-4x}$

Differentiating we get,

$$\frac{dy}{dx} = (c_1 + c_2x) - 4e^{-4x} + c_2e^{-4x}$$

Now for, $x = 0$, $y = 1$

$\therefore\ 1 = (c_1 + 0)e^0$

$\Rightarrow c_1 = 1$

For $\frac{dy}{dx} = 0, x = 0$

$\Rightarrow 0 = (c_1 + 0) \times -4e^0 + c_2e^0$

$\Rightarrow c_2 = 4c_1$

$\therefore\ c_2 = 4$

$y = (1 + 4x)e^{-4x}$

2. Given,

$$\frac{\partial^2 u}{\partial x^2}-4\frac{\partial^2 u}{\partial x\partial y}+\frac{\partial^2 u}{\partial y^2}=x+2y$$

The standard form is given as

$$A\frac{\partial^2 u}{\partial x^2}+B\frac{\partial^2 u}{\partial x\partial y}+C\frac{\partial^2 u}{\partial y^2}+D\frac{\partial y}{\partial x}+E\frac{\partial x}{\partial y}+F=0$$

Comparing with the standard form

we get, A = 1, B = –4, C = 6

Now, the determinant

$B^2 - 4AC = 16 - 4 \times 1 \times 6 = -8$

As the determinant is negative.

Thus, the function elliptic in nature.

3. Given,

$\vec{V} = (2x+3y)\hat{i} + (3x+2y)\hat{j}$

Case-I

Divergence

$$\nabla.\vec{V} = \frac{\partial}{\partial x}(2x+3y)+\frac{\partial}{\partial y}(3x+2y) = 2+2 = 4 \neq 0$$

Case-II

$$\nabla\times\vec{V} = \begin{vmatrix} \hat{i} & \hat{j} & \hat{k} \\ \frac{\partial}{\partial x} & \frac{\partial}{\partial y} & \frac{\partial}{\partial z} \\ (2x+3y) & (3x+2y) & 0 \end{vmatrix} = 0\hat{i}+0\hat{j}+\hat{k}(3-3)=0$$

Thus, the equation is curl free but not divergence free.

4. A

If $\rho_e = \rho_a$ = Optimum expanded

$\rho_e < \rho_a$ = Under expanded

$\rho_e > \rho_a$ = Over expanded

6. Specific thrust $= \dfrac{T}{\dot{m}_a} = (V_e - V_i)$

For constant V_i, with increase in altitude ambient pressure decreases. In other words magnitude of exhaust velocity from the nozzle increases due to increase in expansion rate with altitude and hence specific thrust increases.

7. d

8. c

Given, T = 25 MN, t = 150 Sec, I_{sp} = 2980 Ns/kg

I_{sp} = Thrust/mass flow rate of propellant

$$\frac{T}{\dot{m}_p}$$

$$\dot{m}_p = \frac{25 \times 10^6}{2980} = 8389.2617 \text{ kg/s}$$

Total mass of propellant burned in 150 seconds

$$= \dot{m}_p \times t = 8389.2617 \times 150 = 1258389.255 \text{ kg}$$

9. D

We have area mach number relation

$$\frac{dA}{A} = -\frac{dv}{v}(1 - M^2)$$

$$\Rightarrow \frac{dv}{v} = -\frac{dA}{A} \times \frac{1}{(1 - M^2)}$$

$$dv = -\frac{dA}{A} \times \frac{1}{(1 - M^2)} \times \frac{1}{v}$$

If M > 1

dv = positive (Area should decrease)

If M < 1

dv = Negative (Area should increase)

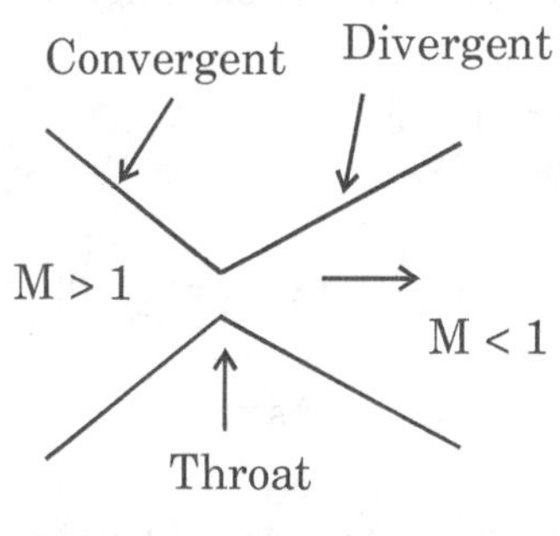

The shape of supersonic diffuser that slows down a supersonic flow to subsonic floe is converging – diverging diffuser.

10.

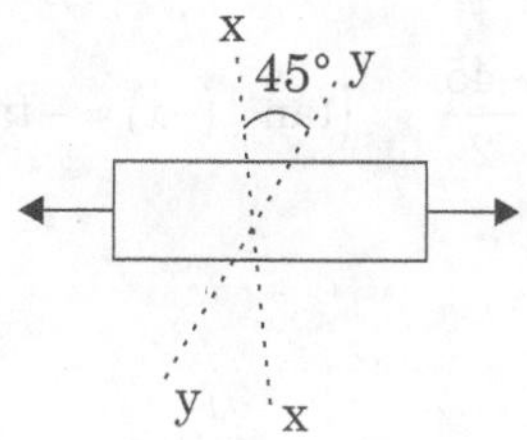

Brittle failure:

Brittle failure occures without any plastic deformation or necking

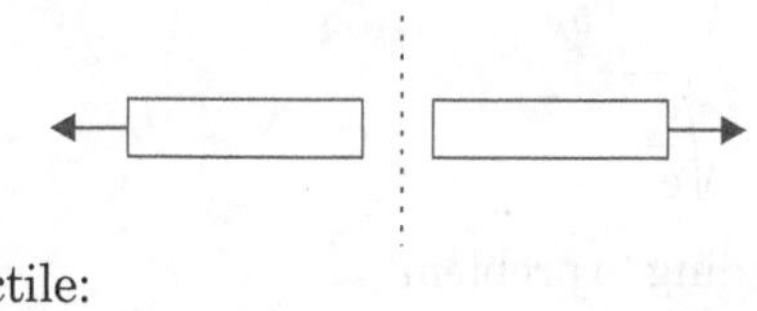

Ductile:

Ductile failure involves plastic deformation and necking mostly it will occure at 45°

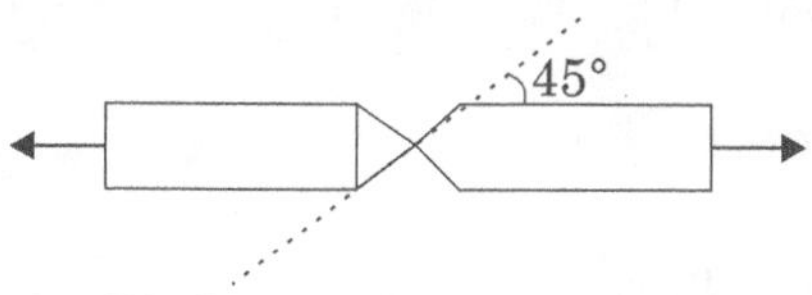

(Option B) along x-x brittle and along yy ductile material

11. State of stress

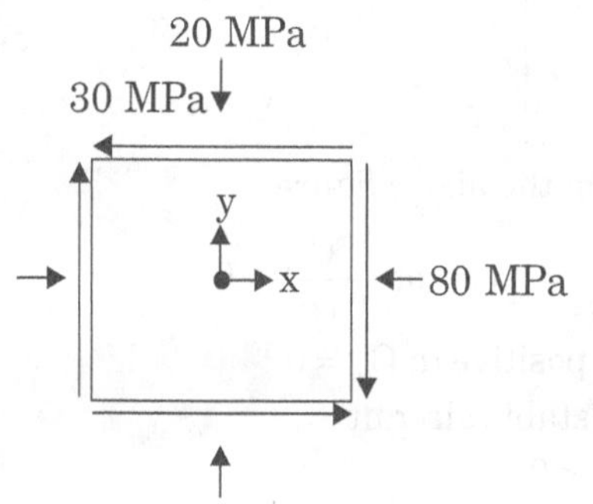

Here, stress

6_x = –80 MPa

6_y = –20 MPa

τ_{xy} = –30 MPa

Angle of maximum shear stress can be calculated using

$$\tan(2\theta_s) = \frac{\dfrac{-(6_x - 6_y)}{2}}{\tau_{xy}}$$

$$\theta_s = \frac{1}{2}\tan^{-1}\left[\frac{\dfrac{-(6_x - 6_y)}{2}}{\tau_{xy}}\right]$$

$$= \frac{1}{2}\tan^{-1}\left[\frac{\dfrac{-[(-80) - (-20)]}{2}}{(-30)}\right]$$

$= -\frac{1}{2}\tan^{-1}(1) = \frac{-45}{2} \qquad \left(\tan^{-1}(-x) = -\tan^{-1}(x)\right)$

$= -22.5°$

12. C

$$V_T = \sqrt{\frac{2(P_T - P)}{e_a}}$$

$$V_{eq} = \sqrt{\frac{2(P_T - P)}{e_5}}$$

$$\frac{V_T}{V_{eq}} = \sqrt{\frac{e_s}{e_a}}$$

According to problem

$$\frac{V_{TAS}}{V_{EAI}} = \sqrt{\frac{e_0}{e}}$$

13. B

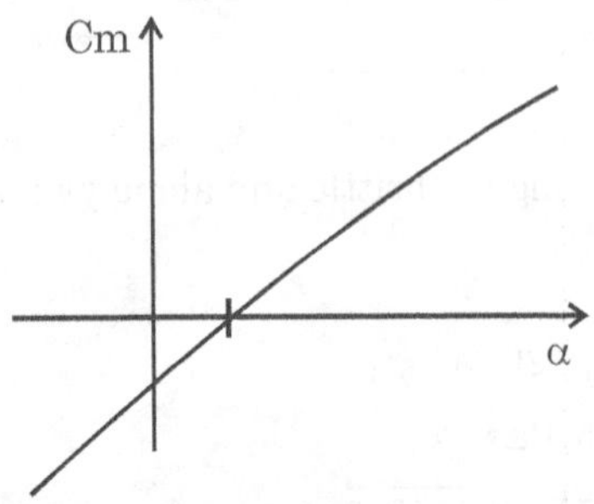

From the above figure

$\frac{\partial C_m}{\partial \alpha} = +ve$ or $\frac{\partial C_m}{\partial \alpha} > 0$

For positive α, $C_m = 0$

For stable aircraft

$C_{m\alpha} < 0$

$\frac{\partial C_m}{\partial \alpha} < 0$

at trimed condition $C_m = 0$

from above discussion we can say that the aircraft can trim at positive α, and Aircraft is unstable.

14. A, B, C

In adiabatic process, total temperature will remains constant.

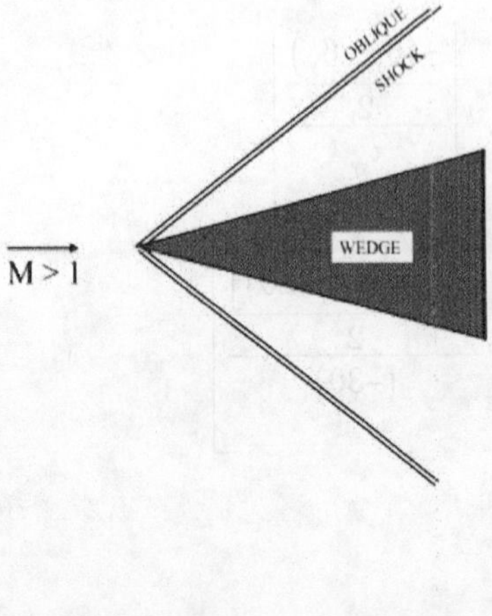

the shock wave

$T_{O_1} = T_{O_2}$

$\rho_{O_1} > \rho_{O_2}$

M> M2 $\qquad \mu_1 < \mu_2$

$\omega_1 = \omega_2$ = tangential velocity remains constant.

Tangential mach number

$$M_{T_1} = \frac{\omega_1}{\sqrt{rRT_1}}$$

$$M_{T_2} = \frac{\omega_2}{\sqrt{rRT_2}} \qquad [T_2 > T_1]$$

$\omega_1 = \omega_2, \; M_{T_2} < M_{T_1}$

Mach number based on tangential velocity of the shock decreases.

15. A,C,D

Kutta condition says the flow will leaves smoothly trailing edge of aerofoil. Kutta condition is applicable for sharp as well as cusp aerofoil, where as trailing edge is stagnation point in case of sharp aerofoil.

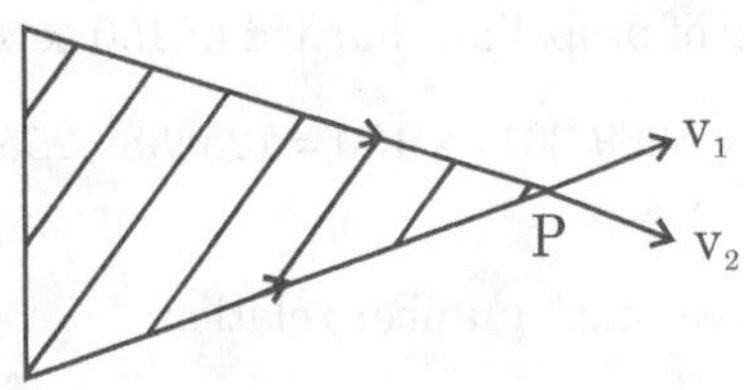

At point P

$v_1 = v_2 = 0$ [stagnation point]

In case of Cusp aerofoil

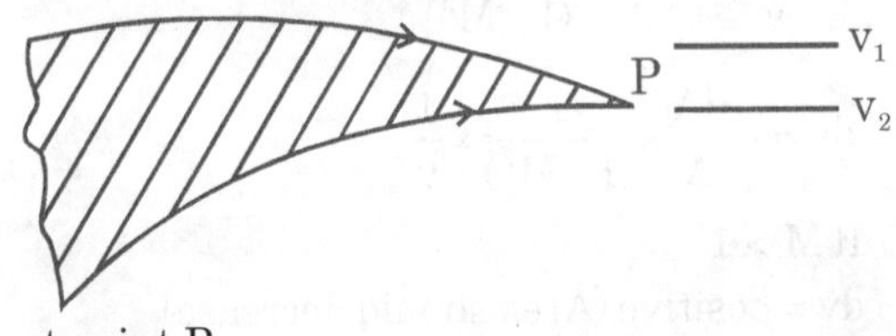

at point P

$v_1 = v_2 \neq 0$

at trailing edge it should have one single velocity $v_1 = v_2$.

Circulation (δ) $= v_T - v_B$

$= v_2 - v_1$

Sharp aerofoil

δ(T.E.) = 0 $\qquad$ [As $v_1 = v_2 = 0$]

Cusp aerofoil

δ(T.E.) = 0 $\qquad$ [As $v_1 = v_2$]

16. A, B

Thin aerofoil theory for cambered aerofoil does not discussed about drag. Prandtt lifting line theory discussed about lift induced drag. Option (D) is incorrect.

According to thin aerofoil theory

$C_l \propto \alpha$

$C_l = 2\pi\alpha$

Aerodynamic centre is a point where moment is indepent of angle of attack.

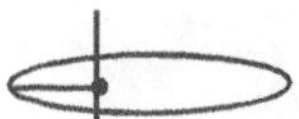

$\frac{c}{4}$

For combersed aerofoil, moment coefficient is given by

$$C_{m_{le}} = -\left[\frac{C_l}{4} + \frac{\pi}{4}(A_1 + A_2)\right]$$

$$C_{m_{c/4}} = C_{m_{le}} + \frac{C_l}{4}$$

$$= -\frac{C_l}{4} - \frac{\pi}{4}(A_1 - A_2) + \frac{C_l}{4}$$

$$= -\frac{\pi}{4}(A_1 - A_2)$$

A_1, A_2 will not change as α changes. So option (B) is correct.

Centre of pressure is a point about which moment is zero, but in the above question.

Moment about quarter chord

$C_{m_{c/4}} = -\frac{\pi}{4}(A_1 - A_2)$ which is constant not a zero.

So the centre of pressure will not lies on quarter chord point. Hence, option (C) is incorrect.

17. $\lim_{x \to 0}\left(\frac{1}{\sin x} - \frac{1}{x}\right)$

$$\lim_{x \to 0}\left(\frac{x - \sin x}{x \sin x}\right)$$

Substituting $x = 0$, we get $\frac{0}{0}$

Thus, applying L. Hospital's rule

$$\lim_{x \to 0}\left(\frac{1 - \cos x}{x\cos x + \sin x}\right), \text{again } \frac{0}{0} \text{ from}$$

Thus, $\lim_{x \to 0}\left(\frac{\sin x}{-x\sin x + \cos x + \cos x}\right) = \frac{0}{0+1+1} = \frac{0}{2} = 0$

18. Given,

$$\oint \frac{z^3}{4z - i} dz$$

$\xi = 1$

Now, $f(z) = \frac{z^3}{4z - i}$

For Poles

$4z - i = 0,$

$\Rightarrow z = \frac{i}{4}$

Using residue theorem

$I = 2\pi i \times$ sum of residues

$$= 2\pi i \times \lim_{z \to \frac{i}{4}}\left(z - \frac{i}{4}\right) \times \frac{z^3}{4z - i}$$

$$= 2\pi i \times \frac{\left(\frac{i}{4}\right)^3}{4} = \frac{2\pi \times 1}{256} = 0.0245 \approx 0.025$$

19. We know that

System returns to its original position in minimum possible time when it is critically damped

$\therefore$ damping ratio $\zeta = 1$

$$\Rightarrow \frac{C}{C_c} = 1$$

$\Rightarrow C = 2m\omega_n$ $\quad C_c = 2\,m\,\omega_n$

$= 2 \times 10 \times 13.2$ $\quad m = 10$ kg

$= 264$ N s/m $\quad \omega_n = 13.2$ rad/sec.

Natural frequency

20. Natural frequency for contiliver beam is given by:

$f = \frac{k}{2\pi l^2}\sqrt{\frac{EI}{\rho A}}$ $\quad$ k = constant

given: both beam have same cross-section and homoganeous material

$\therefore$ E, I, & A are same for both beams

$\therefore$ Ratio of first natural frequencies

$$\frac{(f_l)}{(f_{2l})} = \frac{\left(\frac{k}{2\pi l^2}\right)\sqrt{\frac{EI}{\rho A}}}{\left(\frac{k}{2\pi(2l)^2}\right)\sqrt{\frac{EI}{\rho A}}} = \frac{4l^2}{l^2}$$

K is same for both beams for any natural frequencies

$= 4$

21. 0

Definition of circulation

$$\tau = \int_c \vec{v}.dl$$

In the above question the position of origin is outside of closed loop ABCDEFA.

$\therefore \tau = 0$

22. Cylindrical tank with closed ends:

given: $\varepsilon_c = 3\,\varepsilon_{axial}$ ....(i)

we know that

$$\text{hoop stress } 6_c = \frac{Pd}{2t}$$

$$\text{axial or longitudinal stress } 6_l = \frac{Pd}{4t}$$

writing eq. (i) in terms of stresses

$$\frac{1}{E}(6_c - \upsilon 6_l) = 3 \times \frac{1}{E}[6_l - \upsilon 6_c]$$

$$\Rightarrow \frac{1}{E}\left[\frac{Pd}{2t} - \upsilon\frac{Pd}{4t}\right] = 3 \times \frac{1}{E}\left[\frac{Pd}{4t} - \frac{Pd}{2t}\upsilon\right]$$

$$\Rightarrow \left[1 - \upsilon\frac{1}{2}\right]\frac{Pd}{2t} = 3 \times \left[\frac{1}{2} - \upsilon\right]\frac{Pd}{2t}$$

$$\Rightarrow 1 - \frac{\upsilon}{2} = \frac{3}{2} - 3\upsilon$$

$$\Rightarrow 3\upsilon - \frac{\upsilon}{2} = \frac{3}{2} - 1$$

$$\Rightarrow \frac{5\upsilon}{2} = \frac{1}{2}$$

$$\Rightarrow \boxed{\upsilon = \frac{1}{5} = 0.2}$$

23. 65.5

Given

$\frac{W}{S} = 1800 \text{ N/m}^2$, $S = 30 \text{ m}^2$, $C_D = 0.02 + 0.04\, C_L^2$

$C_{Lmax} = 1.6$, $e_\infty = 1.225 \text{ kg/m}^3$, $C_{DO} = 0.02$, $K = 0.4$

Endurance of Jet Aircraft

$$E_g = \frac{C_L}{C_D}\log_e\frac{W_0}{W_1}$$

As E_g to be maximum Aircraft should fly at $\left(\frac{C_L}{C_D}\right)$ max

$$\text{For } \left(\frac{C_L}{C_D}\right) \text{max} \quad C_L = \sqrt{\frac{C_{DO}}{K}} = \sqrt{\frac{0.02}{0.04}} = \sqrt{\frac{1}{2}}$$

$$\therefore V_\infty = \sqrt{\frac{2W}{e_\infty S C_L}} = \frac{\sqrt{2 \times 1800}}{1.225 \times \sqrt{\frac{1}{2}}} = 64.475 \text{ m/sec}$$

$V\infty = 65.5$ m/sec

24. 6.53°

Calculating

$$\text{Yawing moment} = T \times \frac{b}{2}$$

$= 8000 \times 5$

$= 40000$ Nm

$= 40$ kNm

Coefficient of Yawing moment (C_n)

$$= \frac{\text{Yawing moment}}{9_\infty \text{Sb}}$$

$$= \frac{40 \times 10^3}{\frac{1}{2} \times e_\infty \times V_\infty{}^2 \text{Sb}}$$

$$= \frac{40 \times 10^3}{\frac{1}{2} \times 1.225 \times 100^2 \times 50 \times 10} = 0.01308$$

Now writing C_n is terms of α, θr, β

$$C_n = C_{n\alpha''}\alpha'' + C_{n8_r} 8r + C_{n\beta}\beta \quad \text{...(1)}$$

For straight and level flight

$\alpha'' = 0$

For zero side slip $\beta = 0$

From equation (1)

$$\therefore C_n = C_{n8_r}{}^{8r}$$

$$\Rightarrow 0.01308 = -0.0028r$$

$$8r = -\frac{0.01308}{0.002} = -6.54°$$

25. Given, $h = 250$ km

$Gm_e = 398600.4 \text{ km}^3/\text{s}^2$

$R = 6378.14$ km

As per formula

$$V = \sqrt{\frac{GM}{(R+h)}} = \sqrt{\frac{398600.4}{(250 + 6378.14)}} = 7.75 \text{ km/s}$$

26. Kinetic energy $= \frac{1}{2}I\theta^2$

Potential energy $= \frac{1}{2}kx^2$

$$= \frac{1}{2}k\left[\theta^2\right]\left[\frac{l^2}{4}\right]$$

Using logragian equation:

$$\frac{d}{dt}\left(\frac{\partial k \cdot E}{\partial \dot{\theta}}\right) + \frac{\partial(P \cdot \varepsilon)}{\partial \theta} = f(t)$$

$$I\ddot{\theta} + \frac{kl^2}{4}\theta = 0$$

Moment of inertia $I = mz^2 = m[l^2 + h^2]$

$\therefore$ Natural frequency in Hz

$$= \frac{1}{2\pi}\sqrt{\frac{\left(\frac{kl^2}{4}\right)}{I}} = \frac{1}{2\pi}\sqrt{\frac{kl^2}{4m\left[l^2 + h^2\right]}}$$

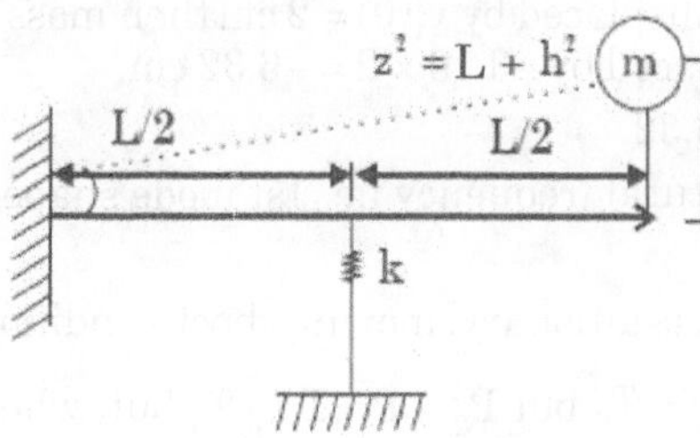

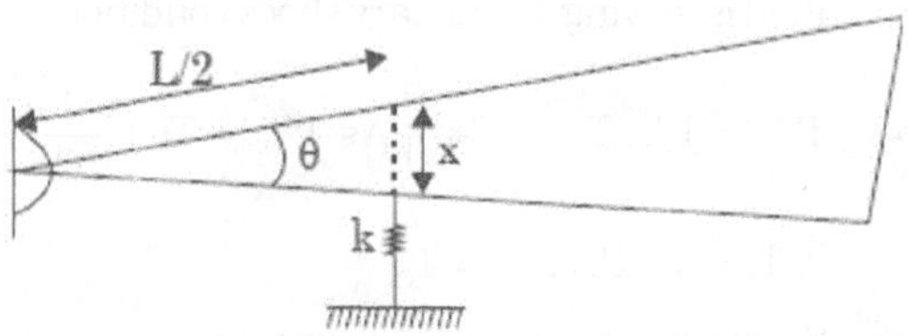

27. C

In the case of P if the ice cube melts, the amount of water that gets displaced in P is amount of the volume of this ice cube. As the ice cube will melt and convert into water.

There is a increase in water level but it will never increase more than that, it will remains the same because the whatever the volume that get displaced that is going to come as the water again after it gets melt. So in P its remains the same.

In Q, volume of ice cube is volume of cube minus volume of solid ball. So the water that will go down because of melting of ice is volume of ice cube, so water will not come out it will decreased. Same stories in R, Q so it will also decreased.

28. B

Given, subscript a used for aircraft, m used for model.

Va = 100 km/h

$$= 1000\ 100 \times \frac{5}{18} = 27.78 = 28 \text{ m/sec}$$

$e_a = 1.225$ kg/m^3

$P_a = 1.01$ bar $\quad P_m = 10$ Pa

$T_a = 288$ k $\quad T_a = T_m$

According to problem

$Re_a = Re_m$

$$\frac{e_a v_a l_a}{\mu_a} = \frac{e_m v_m l_m}{\mu_m}$$

$$\because \mu_a = \mu_m,\ l_m = \frac{l_a}{5}$$

$$\Rightarrow \frac{e_a v_a l_a}{\mu_a} = \frac{e_m v_m \frac{l_a}{5}}{\mu_a}$$

$$\Rightarrow e_a v_a = \frac{e_m v_m}{5}$$

$$\Rightarrow v_m = \frac{e_a v_a \times 5}{e_m}$$

$$= \frac{e_a \times 28 \times 5}{\frac{P_m}{RT_m}}$$

$$= \frac{28 \times 5 \times 287 \times 288 \times 1.225}{10 \times 1.01 \times 10^5}$$

$$v_m = 14 \text{ m/sec} = \frac{14 \times 18}{5} = 50 \text{ km/h.}$$

29. C

Shadow graph – Non-uniform density

Shlieren – Uniform density

Interferometry – Optical beans.

30. D

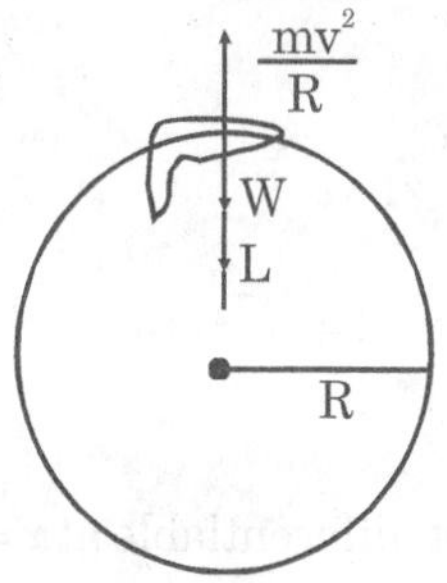

$$L + W = \frac{mv^2}{R}$$

$$\left(\frac{L}{W} + 1\right) = \frac{mv^2}{RW}$$

$$n + 1 = \frac{V^2}{Rg}$$

$$n = \frac{V^2}{Rg} - 1$$

$\frac{V^2}{Rg}$ is always positive

$\therefore n > -1$

$n > -1$

31. Given,

$f(x) = e^{-x}|\cos x|$ for $x > 0$

we can write the equation in two expressions

$$f(x) = \begin{Bmatrix} -e^{-x}\cos x & \cos x < 0 \\ e^{-x}\cos x & \cos x > 0 \end{Bmatrix}$$

Differentiating we get,

$$f'(x) = \begin{Bmatrix} e^{-x}\cos x + e^{-x}\sin x & \cos x < 0 \\ -e^{-x}\cos x - e^{-x}\sin x & \cos x > 0 \end{Bmatrix}$$

Taking option (A)

At $x = \frac{\pi}{2}, \cos\frac{\pi}{2} = 0$

$\therefore$ L.H.S = $e^{-\frac{\pi}{2}} \times 0 + e^{-\frac{\pi}{2}} \times 1 = e^{-\frac{\pi}{2}}$

R.H.S = $-e^{-\frac{\pi}{2}} \times 0 - e^{-\frac{\pi}{2}} \times 1 = -e^{-\frac{\pi}{2}}$

$\therefore$ L.H.S $\neq$ R.H.S

Thus, the function is not differentiable at $x = \frac{\pi}{2}$

Taking option (B)

For $x = \pi$, $\cos \pi = -1 < 0$

The above criteria is valid for only one expression

Thus, the function is differentiable at $x = \pi$

Taking option (C)

$$x = \frac{3\pi}{2}, \cos\frac{3\pi}{2} = 0$$

$\therefore$ L.H.S = $e^{-\frac{3\pi}{2}} \times 0 + e^{-\frac{3\pi}{2}} \times 1 = e^{-\frac{3\pi}{2}}$

R.H.S = $-e^{-\frac{3\pi}{2}} \times 0 - e^{-\frac{3\pi}{2}} \times 1 = -e^{-\frac{3\pi}{2}}$

$\therefore$ L.H.S $\neq$ R.H.S

Thus, the function is not differentiable at $x = \frac{3\pi}{2}$

Taking option (D)

For $x = 2\pi$

$\cos 2\pi = 1 > 0$

The above criteria is valid for only one expression

Thus, the function is continuous at $x = 2\pi$

32. Natural frequency 1, $\omega_{n_1} = 233.9$ rad/s

Natural frequency 2, $\omega_{n_2} = 324.5$ rad/s

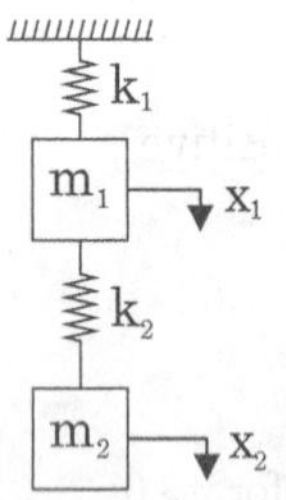

1st mode shape $\phi_1 = \begin{bmatrix} 1 \\ -3.16 \end{bmatrix}$

2nd mode shape $\phi_2 = \begin{bmatrix} 1 \\ 3.16 \end{bmatrix}$

(i) 1st mode shape tells us that if mass 1 in displaced by 1 unit deflection mass 2 will move by 3.16 times unit deflection in opposite side.

(ii) 2nd mode shape tells us if mass 1 is displaced by b unit deflection mass 2 will move with 3.16 times unit deflection in same direction.

Option (B) is correct

If mass 1 is displaced by $x_1(0) = 2$ cm then mass 2 will be displaced by $-3.16 \times 2 = -6.32$ cm

$\Rightarrow x_2(0) = -6.32$

For first natural frequency i.e. 1st mode shape.

33. D

When fluid is stationary, moving shock condition

$P_{O_1} = P_1$, $T_{O_1} = T_1$ but $P_2 > P_1$, $T_2 > T_1$ but, when fluid is moving. Stationary shock condition

$$P'_{O_1} = P'_1,\ T'_{O_1} = T_1' \left(\text{As } T'_{O_1} = T_1' + \frac{1}{2Cp}Vs^2\right)$$

$\therefore T_1 = T'_1 P$, $T'_{O_1} = T_{O_1}$

So option (A) and (C) are wrong.

$P'_{O_1} > P_{O_1}$ which is correct and given in option (B) and (D. In the case of stationary shock P'_{O_1} (Total pressure) which will be always greater than that of moving shock.

$P_2 > P_{O_1}$ $\left(\text{As } P_{O_1} = P_1\right)$

After the shock goes in the case of moving normal shock goes in the case of moving normal shock pressure P_2 will increase, so it will be greater than P_{O_1}.

So all the condition is satisfied by option (D).

34. **A, C, D**

Natural frequency for short period mode is given as

$$W_n = \sqrt{-M_\alpha + Z_\alpha \frac{M_q}{u_0}} \quad ...(1)$$

(i) (ii)

where $m_\alpha = \frac{\partial M}{\partial \alpha}$

$m_q = \frac{\partial M}{\partial q}$

↓

damping parameter

$Z_\alpha = \frac{\partial Z}{\partial \alpha}$

$C_{M\alpha} \alpha M_\alpha$

$\frac{\partial C_M}{\partial \alpha} \alpha \frac{\partial M}{\partial \alpha}$ (C_M = Pitching moment coefficient)

When $C_{M\alpha}$ will be more negative $\frac{\partial M}{\partial \alpha}$ will be more negative, then writing W_n for this case.

$$W_n = \sqrt{-(-M\alpha) + Z\alpha \frac{M_q}{u_0}}$$

$$= \sqrt{M_\alpha + Z_\alpha \frac{M_q}{u_0}}$$

So natural frquency increases.

Looking for 2nd term keeping (i) constant usually

$Z_\alpha = -$ ve

when $m_q \to -$ve

then product of $Z_\alpha M_q = +$ve

Hence W_n increase

$\therefore$ Hence natural frequency increases

Hence option (B) is incorrect.

C_{lp} = rolling moment due to rolling velocity or damping in roll parameter.

C_I = rolling moment coefficient

P = rolling velocity

$$C_{Ip} = \frac{\partial C_L}{\partial\left(\frac{Pb}{2V_\infty}\right)}$$

Usually wing contributes more in roll stability and roll damping for roll stability the primary factor is dihedral wing and swept wing. So option (C) is correct.

Increasing the size of vertical fin, it increases more yawing stability that means more Yaw damping. So option (D) is also correct.

35. B, C, D

Piston powered

Service ceiling $\to (R/C)_{max}$ = 100 ft/min. 0.5 m/sec

Absolute ceiling $\to (R/C)_{max} = 0$

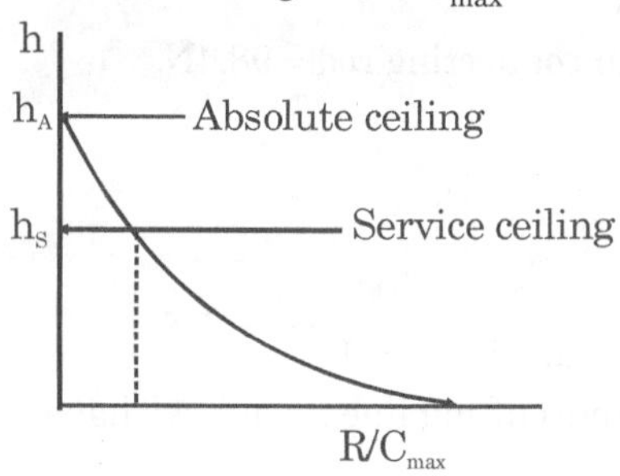

$h_A > h_S$

For

Jet powered

Service ceiling $\to (R/C)_{max}$ = 500 ft/min

Absolute ceiling $\to (R/C)_{max} = 0$

$$\text{Stall speed } V_S = \sqrt{\frac{2\omega}{\rho S C_{L_{max}}}}$$

AS $h \uparrow \rho \downarrow$ $\quad h \to$ Increases

$e \to$ decreases

$\therefore$ AS $e \to$ decreases

$V_S \to$ Increases

In case of tail wind

$V_g = V_\infty + V_{T\omega}$

V_g = ground velocity

V_∞ + free stream velocity

$V_{T\omega}$ = tail wind velocity

General Range formula,

$$R = \frac{V_g}{C_t} = \frac{C_L}{C_D} \ln \frac{\omega_0}{\omega_1}$$

$R \propto V_g$

$V_g \to$ Increases

$R \to$ Increases

For Jet Aircraft

$$R_{Jet} = \frac{1}{C_t}\left(\frac{{C_L}^{1/2}}{C_D}\right) 2\sqrt{\frac{2}{e_\infty S}}\left({\omega_0}^{1/2} - {\omega_1}^{1/2}\right)$$

$R_{Jet} \uparrow$ as $e \downarrow$

$e \to$ decrease

$R_{Jet} \to$ Increases

36. A, D

Primary function of vertical fin is to provide more directional stability but also provides lateral stability. So option (A) is correct and (B) is wrong. Presence of dihedral wing provides lateral stability and presence of anhedral wing decreases the lateral stability of Aircraft. So option (C) is incorrect. Option (D) is correct because Horizontal tail have stablizing effect on longitudinal stability.

37. Given,

$$\text{Let } A = \begin{bmatrix} 3 & 1 & 2 \\ 2 & -3 & -1 \\ 1 & 2 & 1 \end{bmatrix}$$

As we know

Product of eigen values = Determinant of matrix and Sum of eigen values = Trace of matrix

$\therefore$ Determinant of matrix A is

$$\begin{vmatrix} 3 & 1 & 2 \\ 2 & -3 & -1 \\ 1 & 2 & 1 \end{vmatrix} = 3(-3+2) - 1(2+1) + 2(4+3)$$

$$= -3 - 3 + 14 = 8$$

Trace of matrix A is

$$3 - 3 + 1 = 1$$

$$\therefore \frac{\text{Product of eigenvalues}}{\text{Sum of eigen values}} = \frac{8}{1} = 8$$

38. Given,

$$\int_1^5 x^2 dx$$

evaluated at four equal intervals

$$h = \frac{b-a}{n} = \frac{5-1}{4} = 1$$

$y = x^2$

x	1	2	3	4	5
y	1	4	9	16	25

Using trapezoidal method

$$I_t = \frac{h}{2}[(1+25)+2(4+9+16)] = \frac{1}{2}(26+58) = 42$$

Using Simphon's one-third rule

$$I_s = \frac{h}{3}[(1+25)+4(4+16)+2(9)]$$

$$= \frac{1}{3}(26+80+18) = 41.33$$

Difference

$$I_t - I_s = 42 - 41.33 = 0.67$$

39. Deflection function

$$y = \frac{\omega}{48EI}(2x^4 - 3lx^3 + l^3x) \quad ....(A)$$

at maximum deflection of beam slope is zero

Slope $\Rightarrow \frac{dy}{dx}$

$$= \frac{\omega}{48EI}[8x^3 - 9lx^2 + l^3]$$

$$\Rightarrow \left(\frac{dy}{dx}\right)_x = 0$$

$$\frac{\omega}{48EI}[8x^3 - 9lx^2 + l^3] = 0$$

$$8x^3 - 9lx^2 + l^3 = 0$$

$$8\left(\frac{x}{l}\right)^3 - 9\left(\frac{x}{l}\right)^2 + 1 = 0 \quad ....(i)$$

Solving eq. (i)

$$\frac{x}{l} = -0.296, 1, 0.4215$$

$\frac{x}{l}$ cannot be –ve

$\therefore$ –0.296 cannot be ans.

If we put $\frac{x}{l} = 1 \Rightarrow x = l$ in (A)

$$y = \frac{\omega}{48EI}[2l^4 - 3l^4 + l^4] = 0$$

$\therefore$ deflection is zero at $\frac{x}{l} = 1$

$\therefore \frac{x}{l} = 1$ cannot be ans

$$\therefore \boxed{\frac{x}{l} = 0.4215}$$

40. 98.4

In the above problem tension in connecting rope will be due to change in momentum.

Let F_x be the force in x-direction.

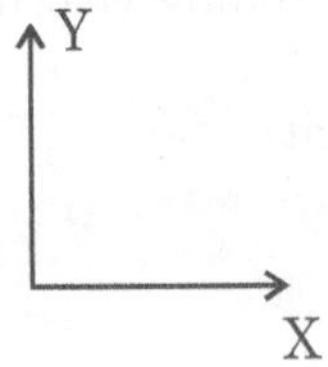

$$F_x = -\{\dot{m} v_{out} - \dot{m} v_{in}\}$$

[–ve sign because tension in negative direction]

$$= -\{eA_{out}v_{out}^2 - eA_{in}v_{in}^2\}$$

$$A_{out} = A_{in} = A$$

$$= -[eAv_{out}^2 - eAv_{in}^2]$$

$$= -eA[(10\cos 60)^2 - 10^2]$$

$$= -1000 \times 3.14 \times \left(\frac{0.05}{2}\right)^2 \times 100[\cos^2 60 - 1]$$

$$= -98.4 \text{ N}$$

Tension in connecting rod = 98.4N

41. 0.0066

Given

AR = 10

$\alpha = 5°$

For ellipilical wing, e = 1

From prandtl lifting line theory, we have

$$C_{Dj} = \frac{C_I^2}{\pi e AR} = \frac{C_I^2}{\pi \times 1 \times 10}$$

$$C_{Dj} = \frac{C_I^2}{10\pi} \quad ...1$$

From thin aerofoil theor

CI = $2\pi\alpha$ (for 2D wing) for 3D wing (finite wing)

$$a = \frac{a_0}{1 + \frac{a_0}{\pi e AR}} = \frac{2\pi}{1 + \frac{2\pi}{\pi \times 1 \times 10}}$$

$$= \frac{2\pi}{1 + \frac{2\pi}{10\pi}} = \frac{2\pi}{1.2} = 5.23 \text{ rod}^{-1}$$

$$C_l = a\alpha = 5.23 \times \frac{5\pi}{180} = 0.4564$$

$\therefore$ From equation (1)

$$C_{Dj} = \frac{(0.4564)^2}{10\pi} = 0.0066.$$

42. 5

$$\text{Shape factor} = \frac{\text{Displacement thickness}}{\text{Momentum thickness}}$$

$$= \frac{\delta *}{Q}$$

$$\delta * = \int_0^\delta \left(1 - \left(\frac{u}{v}\right)\right) dy = \int_0^\delta \left(1 - \frac{y^2}{\delta^2}\right) dy = \left[y - \frac{y^3}{3\delta^2}\right]_0^\delta$$

$$\delta * = \delta - \frac{\delta^3}{3\delta^2} = \frac{2\delta}{3}$$

$$\theta = \int_0^\delta \frac{u}{v}\left(1 - \left(\frac{u}{v}\right)\right) dy = \int_0^\delta \frac{y^2}{\delta^2}\left(1 - \frac{y^2}{\delta^2}\right) dy$$

$$= \int_0^\delta \left(y^2 - \frac{y^2}{\delta^2}\right) dy = \frac{1}{\delta^2}\left[\frac{\delta^3}{3} - \frac{\delta^5}{5}\right]$$

$$= \left(\frac{\delta}{3} - \frac{\delta}{5}\right) = \frac{2\delta}{15}$$

$$\text{Shope factor} = \frac{\frac{2\delta}{3}}{\frac{2\delta}{15}} = 5$$

43. 606.83

Given, $f = \frac{\dot{m}_f}{\dot{m}_a} = 0.019$, $h_1 = 260$ KJ/kg, $h_2 = 912$ KJ/kg, HV = 44.5 MJ/kg

Q_L = heat loss to the surrounding = 25 KJ/kg

From law of conservation of energy,

$(\text{Energy})_{in} = (\text{Energy})_{out}$

$$\dot{m}_a h_1 + \frac{\dot{m}_a v_1^2}{2} + \dot{m}_f \times HV$$

$$= \left(\dot{m}_a + \dot{m}_f\right) h_2 + \left(\dot{m}_a + \dot{m}_f\right)\frac{v_2^2}{2} + \dot{m}_a Q_L$$

Dividing by $\dot{m}_a$ on both sides

$$h_1 + \frac{v_1^2}{2} + f \times HV (1+f) h_2 + (1+f)\frac{v_2^2}{2} + Q_L$$

$$260 \times 1000 + \frac{270^2}{2} + 0.019 \times 44.5 \times 10^6$$

$$= (1+0.019)\,(912 \times 1000 + \frac{v_2^2}{2}) + 25 \times 1000$$

V_2 = 606.83 m/s

44. 218.16

$$R = \frac{\bar{R}}{M} = \frac{8314}{22} = 377.909 \text{ KJ/kg}$$

$$\frac{T_{OC}}{T^*} = \frac{\gamma + 1}{2} = \frac{1.32 + 1}{2}$$

T^* = 1810.345 K

$$\frac{P_{OC}}{P^*} = \frac{\gamma + 1^{\frac{\gamma}{\gamma - 1}}}{2}$$

P^* = 1.367 MPa

$$\rho^* = \frac{P^*}{RT^*} = 1.9981 \text{ kg/m}^3$$

$$a^* = \sqrt{\gamma RT^*} = 950.301 \text{ m/s}$$

$$\dot{m}_p = \rho^* A^* a^* = 1.9981 \times 0.1 \times 950.301$$

= 189.879 kg/s

$$\frac{P_{OC}}{P_e}\left[1 + \frac{(\gamma - 1)M_e^2}{2}\right]^{\frac{\gamma}{\gamma - 1}}$$

$$\frac{14}{0.1}\left[1 + \frac{(1.32 - 1)M_e^2}{2}\right]^{\frac{1.32}{1.32 - 1}}$$

M_e = 3.802

$$\frac{T_{OC}}{T_e}\left[1 + \frac{(\gamma - 1)M_e^2}{2}\right]$$

T_e = 633.899 K

$$V_e = M_e\sqrt{\gamma RT_e} = 2137.975 \text{ m/s}$$

$$T = \dot{m}_p V_e = 405956.555 \text{ N}$$

$$I_{sp} = \frac{T}{\dot{m}_p g} = 218.16 \text{ sec}$$

45. 21.64

$$\dot{m}_a = \dot{m}_c + \dot{m}_h$$

$$B = \text{by-pass ratio} = \frac{\dot{m}_c}{\dot{m}_h}$$

$$3 = \frac{\dot{m}_c}{\dot{m}_a - \dot{m}_c} = \frac{\dot{m}_c}{100 - \dot{m}_c}$$

$\dot{m}_c$ = 75 kg/s

P_{oa} = Total ambient pressure = 1 bar

$P_{o,aftfan}$ = Total pressure after fan = Total pressure at exit of cold nozzle

$$r_F = \text{Fan pressure ratio} = \frac{P_{o,aftfan}}{P_{oa}} = 1.6 = 1.6$$

$P_{o,aftfan}$ = 1.6 bar

$$\frac{P_{oe}}{P_e} = \left[1 + \frac{(\gamma-1)M_e^2}{2}\right]^{\frac{\gamma}{\gamma-1}}$$

$$\frac{1.6}{1} = \left[1 + \frac{(1.4-1)M_e^2}{2}\right]^{\frac{1.4}{1.4-1}}$$

$M_e = 0.848$

$$C_P = \frac{\gamma R}{\gamma - 1}$$

$$1005 = \frac{1.4R}{1.4-1}$$

$R = 287.143$ J/kg-K

$V_{e,\text{cold nozzle}} = M_e\sqrt{\gamma R T_e} = 288.539$ m/s

Thrust from cold nozzle,

$T_{cold} = \dot{m}_c (V_{e,\text{cold nozzle}} - V_{intake})$

Since test is static so, $V_{intake} = 0$

$T_{cold} = \dot{m}_c V_{e,\text{cold nozzle}} = 21640.539$ N

46. 277.125

$C_1 = W_2 = 140$ m/s

$C_2 = W_1 = 240$ m/s

Hence, Degree of reaction, R = ½

And, $\beta_2 = \alpha_1 = 30°$

$C_a = W_2 \times \cos\beta_2 = 121.243$ m/s

$$\beta_1 = \alpha_2 = \cos^{-1}\left(\frac{C_a}{C_2}\right) = 59.657°$$

$$R = \frac{1}{2} = \frac{C_a}{2U}(\tan\beta_1 + \tan\beta_2)$$

$U = 277.125$ m/s

47. 19.12

$\beta_2 = 22°$, $\beta_3 = 55°$

$U = 340$ m/s, $R = 0.4$

$$R = \frac{C_a}{2U}(\tan\beta_3 - \tan\beta_2)$$

$C_a = 265.593$ m/s

$\dot{m} = \rho_i A_i C_a = 0.9 \times 0.08 \times 265.593$

$\dot{m} = 19.123$ kg/s

48. 2.43

Applying Law of conservation of energy across combustion chamber,

$(\text{Energy})_{in} = (\text{Energy})_{out}$

$H_{03} + \eta_{cc} \times Q = H_{04}$

$\dot{m}_a C_{Pa} T_{03} + \eta_{cc} \dot{m}_f HV = (\dot{m}_a + \dot{m}_f)C_{Pg} T_{04}$

$$\dot{m}_f = \frac{\dot{m}_a C_{pg} T_{04} - \dot{m}_a C_{Pg} T_{03}}{HV \times \eta_{cc} - C_{Pg} T_{04}}$$

$$\dot{m}_f = \frac{100 \times 1200 \times 1200 - 100 \times 1000 \times 600}{40 \times 10^6 \times 0.9 - 1200 \times 1200}$$

$\dot{m}_f = 2.43$ kg/s

49.

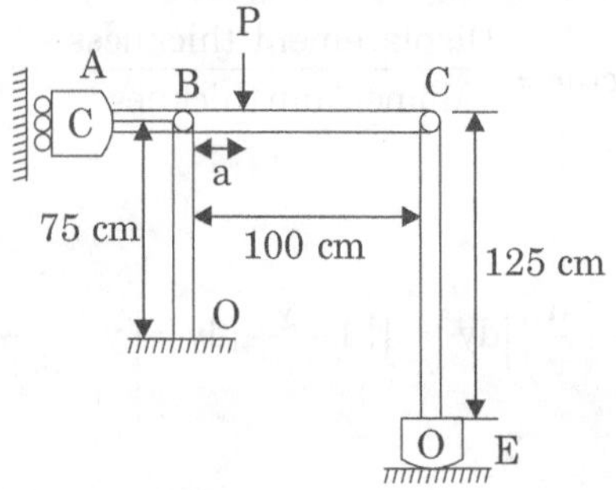

E = 200 GPa

Cross-section = 1.5 × 1.5 cm²

Free body diagram

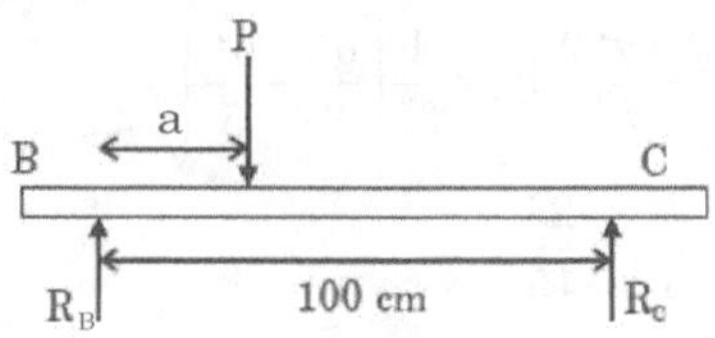

Simply supported beam

Force equilibrum $R_B + R_C = P$ ....(i)

Moment equilibrium about A

$R_C \times 100 = P \times a$

$$R_C = \frac{P \times a}{100}$$

$$\Rightarrow R_B = \frac{P(100-a)}{100}$$

⇒ R_B will act as a buckling load for BD column

R_C will act as a buckling load for CE column

FOr column BD which is fixed-hinged

$$(P_{cr})_{BD} = \frac{\pi^2 EI}{\left(\frac{l}{\sqrt{2}}\right)^2}$$

For column C.E which is Hinged-Hinged

$$(P_{cr})_{CE} = \frac{\pi^2 EI}{l^2}$$

Taking ratio

As EI are same for both materials

$$\frac{(P_{cr})_{BD}}{(P_{cr})_{CE}} = \frac{\frac{2\pi^2 EI}{(75)^2}}{\frac{\pi^2 EI}{(125)^2}} = 5.56$$

$\Rightarrow (P_{Cr})_{BD} = 5.56\,(P_{Cr'})_{CE}$

Now using R_B and R_C as bukling load

$R_B = 5.56\,R_C$

$$\frac{P(100-a)}{100} = 5.56\frac{Pa}{100}$$

$P(100 - a) = 5.56\,Pa$

$\Rightarrow \boxed{a = 15.24}$ in 1 decimal

$\boxed{a = 15.2}$

50. Two cell wing box

– Cell wall thickness = 1.5 mm

– Shear modulus = 127 GPa

– Applied Torque, T = 12 KN m

= 12×10^6 Nm

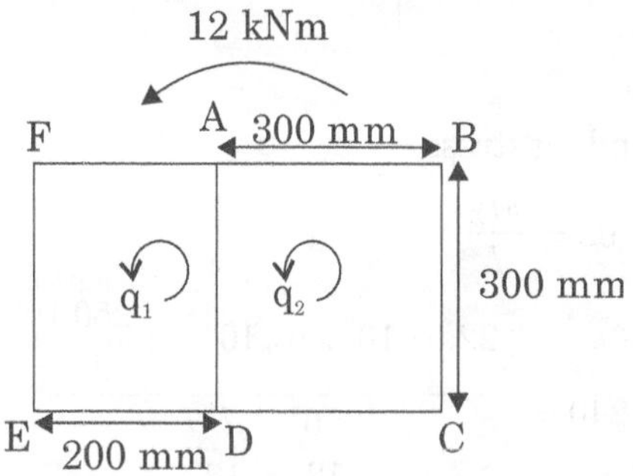

Say q_1 and q_2 are shear flows in cell 1 and cell 2

We know $T = \Sigma 2Aq$

$\therefore T = 2A_1q_1 + 2A_2q_2$

$= 2[200 \times 300]q_1 + 2[300 \times 300]q_2$

$$\Rightarrow 12\times10^6 = 2\left[6\times10^4\right]q_1 + 2\left[9\times10^4\right]q_2$$

$$\Rightarrow 600 = 6q_1 + 9q_2$$

$$\Rightarrow 200 = 2q_1 + 3q_2 \quad ....(i)$$

We know that

$$\beta = \frac{\theta}{l} = \frac{1}{2AG}\oint\frac{qdS}{t}$$

$$\text{Cell 1}: \beta_1 = \left(\frac{\theta}{l}\right)$$

$$= \frac{1}{2A_1Gt}\left[300q_1 + 200q_1 + 300\left(q_1 - q_2\right) + 200q_1\right]$$

$$= \frac{1}{2A_1Gt}\left[1000q_1 - 300q_2\right]$$

Cell 2:

$$\beta_2 = \left(\frac{\theta}{l}\right) = \frac{1}{2A_2Gt}$$

$$\left[300q_2 + 300q_2 + 300\left(q_2 - q_1\right) + 300q_2\right]$$

$$= \frac{1}{2A_2Gt}\left[1200q_2 - 300q_1\right]$$

We know: $\beta_1 = \beta_2$

$$\Rightarrow \frac{1}{2A_1Gt}\left[1000q_1 - 300q_2\right]$$

$$= \frac{1}{2A_2Gt}\left[1200q_2 - 300q_1\right]$$

$$\Rightarrow \frac{1}{\left(6\times10^4\right)}\left[1000q_1 - 300q_2\right]$$

$$= \frac{1}{9\times10^4}\left[1200q_2 - 300q_1\right]$$

$$\Rightarrow \frac{1}{2}\left[10q_1 - 3q_2\right]\times100$$

$$= \frac{1}{3}\left[12q_2 - 3q_1\right]\times100$$

$$\Rightarrow 5q_1 - \frac{3}{2}q_2 = 4q_2 - q_1$$

$$\Rightarrow \frac{11}{2}q_2 = 6q_1$$

$\Rightarrow q_2 = 1.09q_1$ ....(ii)

Putting eq. (ii) in (i)

$200 = 2q_1 + 3q_2$

$\Rightarrow 200 = 2q_1 + (3 \times 1.09q_1)$

$\Rightarrow q_1 = 37.95$ N/mm

$\therefore q_2 = 1.09 \times 37.95 = 41.366$ N/mm

Shear stress along wall AD

$$\tau_{AD} = \frac{q_2 - q_2}{t}$$

$$\frac{3.416}{1.5} = 2.277 \text{ N/mm}^2 \text{ or MPa}$$

$= 2.3$ MPa

51.

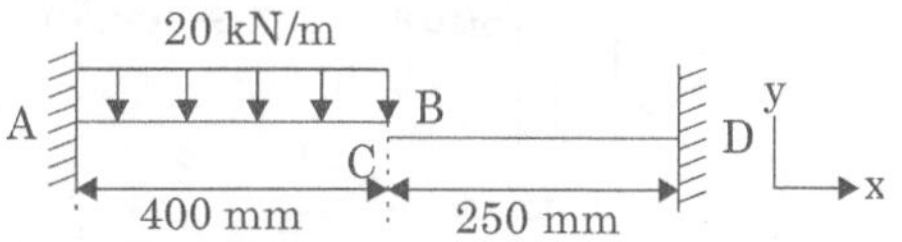

Contilever Beams AB and DC

Square cross-section 50 mm × 50 mm

Elastic modulus E = 70 GPa

$= 70 \times 10^3$ N/mm^2

UDL on beam AB = 20 KN/m

= 20 N/mm

Maximum deflection for beam AB which is at end B is

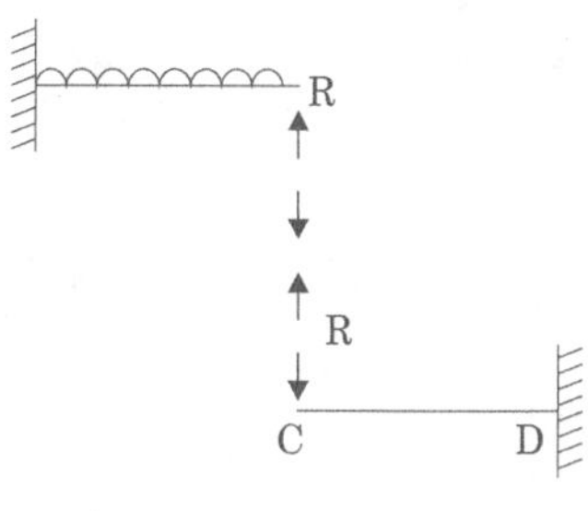

$y_B = \frac{\omega l^4{}_{AB}}{8EI}$

deflection at B should be equal to deflection at C

$[\text{deflection}]_{\text{UDL on AB}} - [\text{deflection}]_{\text{R on AB}} = [\text{deflection}]_{\text{R on DC}}$

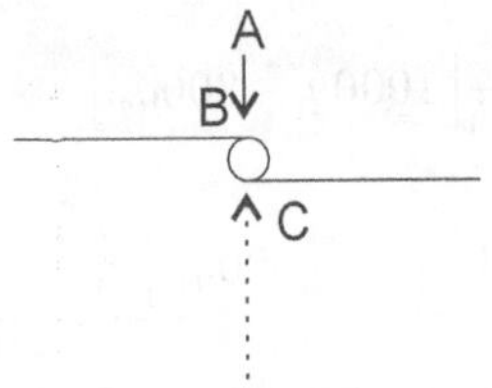

$$\frac{\omega l^4{}_{AB}}{8EI} - \frac{R \times l^3{}_{AB}}{3EI} = \frac{R \times l^3_{DC}}{3EI}$$

$$\Rightarrow \frac{20 \times (400)^4}{8} - \frac{R(400)^3}{3} = \frac{R \times [250]^3}{3}$$

$\Rightarrow$ R = 2411.3 N

= 2.4 kN

Thus compressive force roller = 2.4 kN

52. Length of Pole = 5.5 m

Yield strength = 240 MPa

= 240 N/mm2

Wind pressure, P = 7.5 kPa

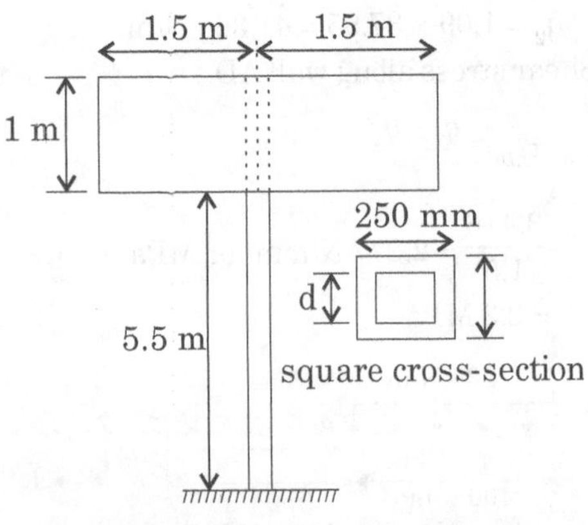

Force acting of Board

F = P × A

$= 7.5 \times 10^3 \times 1 \times [1.5 + 1.5]$

$= 22.5 \times 10^3$ N

This force will act at centroid of plate

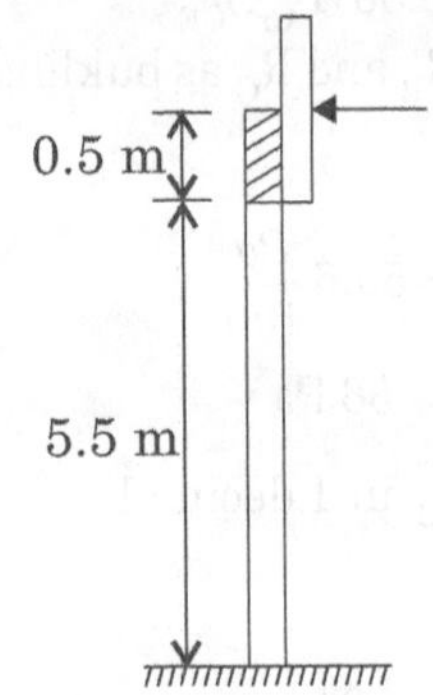

this is a contilever case with end load

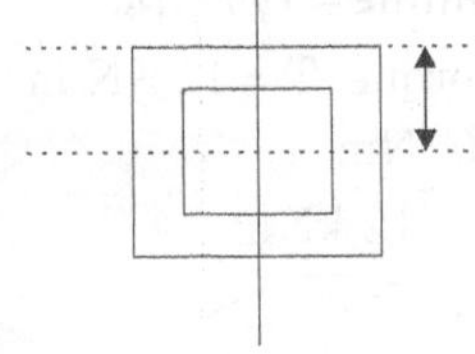

Bending stress:

$$6g = \frac{My}{I}$$

$$\Rightarrow 240 = \frac{\left[22.5 \times 10^3 \times 6 \times 10^3\right] \times \left[\frac{250}{2}\right]}{\frac{250^4}{12} - \frac{d^4}{12}}$$

$\Rightarrow$ d = 235.24 ~ 235 mn

53. 15KN

Given

m = 5500 kg = 5500 × 10 = 55KN

V_∞ = 225 m/sec, R = 775 m, L_t = ?

Pull up maneuver

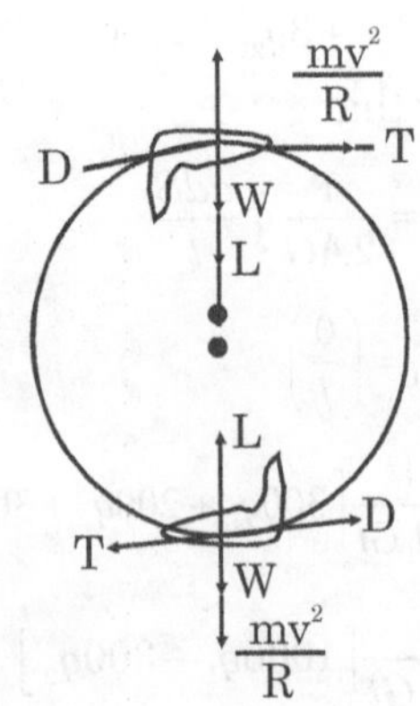

From figure given is problem moment about CP

$$-\left(W + \frac{mv^2}{R}\right) \times 0.02 + L_T \times 5.5 = 0$$

$$-\left(55 \times 10^3 + \frac{5500 \times 225^2}{775}\right) \times 0.02 + L_t \times 5.5 = 0$$

L_t = 15 KN

54. 9790

Given

m = 10000 kg

W = 10000 × 10 = 10^5N

e = 1

b = 10 m

S = $30m^2$

$C_{DO} = 0.025$

V_∞ = 100 m/sec

e = 1.225 kg/m^3

At cruise

L = W, T = D

$C_D = C_{DO} + KC_L^2$

$= 0.025 + \frac{C_L^2}{\pi eAR}$

$= 0.025 + \frac{C_L^2}{\pi \times 1 \times \frac{100}{30}}$ $\left[\because AR = \frac{b^2}{5}\right]$

$= 0.025 + \frac{30C_L^2}{100\pi}$ ...(1)

Now

$$C_L = \frac{2W}{eV_\infty^2 S} = \frac{2 \times 10^5}{1.225 \times 100^2 \times 30} = 0.544$$

Putting the value of C_L in equation (1), we get

$$C_D = \frac{0.025 + 30 \times (0.544)^2}{100 \times 3.14}$$

= 0.053

$$D = C_D \times \frac{1}{2} eV_\infty^2 S$$

$= 0.053 \times \frac{1}{2} \times 1.225 \times 100^2 \times 30 = 97.90$N

55. 0.024

$$\frac{\partial C_m}{\partial_\alpha} = a_w\left[h - h_{ae} - n_t V_H + \frac{a_t}{a_w}\left(1 - \frac{\partial \varepsilon}{\partial_\alpha}\right)\right] \quad ...(1)$$

Where

$$V_H = \frac{S_t l_t}{S\bar{C}} = \frac{80 \times l_t}{350 \times 9}$$

$l_t = (0.3 - 0.25)C$

$$l_t = \bar{l}_t - \left(\frac{x_{cg}}{\bar{c}} - \frac{x_{ac}}{\bar{c}}\right)C$$

= 28 – (0.3 – 0.25) × 9

= 28 – 0.45

= 27.55 m

$$\therefore V_H = \frac{80 \times 27.55}{350 \times 9} = 0.699$$

$n_t = 1$

$\varepsilon = 0.4\alpha$

$$\frac{\partial_\alpha}{\partial_\alpha} = 0.4$$

from equation (1), we get

$$\frac{\partial C_m}{\partial_\alpha} = 0.1\left[0.3 - 0.25 - 1 \times 0.699 \times \frac{0.068}{0.1}(1 - 0.4)\right]$$

= – 0.024 per degree